Introduction to Protein Structure

Second Edition

An aerial view of the European Synchrotron Radiation Facility at Grenoble, France, an advanced source of synchrotron x-ray radiation for use in the study of protein structure, as well as for use in the physical and material sciences. The synchrotron radiation is produced in the circular building in the lower left of the photograph. (Courtesy of ESRF.)

Introduction to Protein Structure

Second Edition

Carl Branden

Microbiology and Tumor Biology Center
Karolinska Institute
Stockholm
Sweden

John Tooze

Imperial Cancer Research Fund Laboratories
Lincolns Inn Fields
London
UK

GARLAND PUBLISHING
ALERE FLAMMAM
Taylor & Francis Group

THE COVER

Front: The structure of the potassium channel from *Streptomyces lividans*, determined by Roderick MacKinnon at the Rockefeller University, New York. As discussed in Chapter 12, this structure—the first of such an ion channel—shows how the channel allows the passage of potassium ions through cell membranes with high efficiency and selectivity. The view is looking down the protein as it sits in the cell membrane, as seen from outside the cell, with a potassium ion shown in gold. This image was produced using the GRASP program (A. Nicholls and B. Honig, Columbia University) from atomic coordinates kindly provided by Roderick MacKinnon.

Back: A hand-drawn image of the potassium channel, in the same view as on the front cover, with each subunit of the tetrameric protein shown in a different color.

Cover design by Christopher Thorpe and Nigel Orme.

Visit the *Introduction to Protein Structure* Web site:
http://www.proteinstructure.com/

For information on other textbooks available from Garland Publishing, visit:
http://www.garlandpub.com/

Library of Congress Cataloging-in-Publication Data
Branden, Carl–Ivar
    Introduction to protein structure / Carl–Ivar Branden, John Tooze.
--2nd ed.
        p. cm.
    Includes bibliographical references and index.
    ISBN 0-8153-2304-2 (hardcover). --ISBN 0-8153-2305-0 (pbk.)
    1. Proteins--Structure. I. Tooze, John. II. Title.
QP551.B7635 1998
572'.633--dc21   98-34487
                                                                      CIP

Published by Garland Publishing, Inc.
19 Union Square West, New York, NY 10003-3382

Printed in the United States of America

15 14 13 12 11 10 9 8 7 6 5 4 3

Preface

The determination of the atomic structures of proteins has seen an enormous increase in impetus since the first edition of this book was published in 1991. The number of new structures reported is close to doubling each year. Technical advances—for example the increased availability of synchrotron x-ray beams and methods for freezing crystals so as to reduce radiation damage to them, the development of multidimensional NMR and NMR machines with ever more powerful magnetic fields, and the exploitation of gene cloning, sequencing and expression systems have all contributed to the growth of protein structure determination. On the one hand, it is becoming increasingly easy to obtain relatively large amounts of naturally rare proteins, on the other hand the crystallographers can work with ever smaller crystals.

The fundamental tenet of molecular biology, namely that one cannot really understand biological reactions without understanding the structure of the participating molecules, is at last being vindicated. As the database of known protein structures rapidly expands, so does the range of biological pathways about which we can ask meaningful questions at close to atomic levels of resolution. An understanding of the principles of protein structure is becoming of ever widening significance to molecular biology.

The pharmaceutical industry has over the past decade become a major user of the protein structure databases, and a major contributor of newly determined structures. Knowledge of an enzyme's or a receptor's atomic structure is invaluable in the search for specific and strongly binding inhibitors. For example the quest for effective inhibitors of HIV protease, to be used in combination therapy for AIDS, led many pharmaceutical companies to determine the structure of that protease with bound inhibitors. Over 120 of these structure determinations have been done so far and at least two inhibitors of HIV protease are now being regularly used to treat AIDS. It seems certain that the determination of the atomic structure of target molecules will play an increasingly important role in drug design.

The commercial exploitation of our increased understanding of protein structure will not, of course, be restricted to the pharmaceutical industry. The industrial use of enzymes in the chemical industry, the development of new and more specific pesticides and herbicides, the modification of enzymes in order to change the composition of plant oils and plant carbohydrates are all examples of other commercial developments that depend, in part, on understanding the structure of particular proteins at high resolution.

As the complete genomes of more and more species are sequenced, the determination of the function of previously unidentified open reading frames is becoming an increasing and challenging problem. The possibility of setting up centers for automated high through-put structure determinations is being seriously discussed. In the absence of any recognizable sequence homology to proteins of known function, this approach, surprising though it may seem, could become an effective way of determining function via structural homology.

The growth in the interest in high-resolution protein structure over the past decade and the reception of the first edition have encouraged us to prepare a new edition of this book. Universities are devoting more time to courses specifically on protein structure, or increasing the amount of time given to protein structure in more generally based biology and biochemistry courses. We hope that this new edition of *Introduction to Protein Structure* will prove useful both to teachers and students.

In 1988 when we began writing the first edition, about 250 protein structures had been determined to medium to high resolution and in those days a professional protein crystallographer was familiar with most of them. We were not therefore faced with a severe problem of what to leave out as we wrote. Today, the coordinates of over 6500 proteins have been deposited in the Protein Data Bank at Brookhaven, New York. Both the number of structures and the variety of biological systems to which they relate are so high that the field of protein structure is becoming more fragmented and specialized. It is becoming increasingly difficult to keep sight of the wood amongst so many trees. The question of what to include and what to omit is, for today's authors, crucially important. We have tried to resist the temptation to describe more and more proteins, adding detail but not increasing understanding of the basic concepts. This edition is inescapably a little larger than its predecessor, but to contain the increase in size we have deleted two chapters while adding four. We run the risk of disappointing not a few structural biologists whose favourite proteins are not mentioned. To them we apologize and ask for their understanding.

Acknowledgements

In preparing the second edition of this book we have again relied heavily on and benefited greatly from the advice and constructive criticism of numerous colleagues. We are particularly grateful to Ken Holmes (Max-Planck Institute, Heidelberg), Lawrence Stern (MIT), Michelle Arkin (Sunesis Pharmaceuticals), and Watson Fuller (Keele University, UK) for their contributions to, respectively, Chapters 14 and 18, Chapter 15, Chapter 17 and Chapter 18. Stephen Harrison (Harvard University) and Paul Sigler (Yale University) provided extensive help and advice on Chapters 8–10, 13 and 16, and Chapters 8–10 and 13, respectively, for which we are especially grateful.

The following, in alphabetical order, have reviewed one or more chapters, correcting our errors of fact or interpretation and helping to ensure they have the appropriate balance and emphasis: Tom Alber (University of California, Berkeley), Tom Blundell (Cambridge University, UK), Stephen Burley (Rockefeller University), Charles Craik (University of California, San Francisco), Ken Dill (University of California, San Francisco), Chris Dobson (Oxford University, UK), Anthony Fink (Unversity of California, Santa Cruz), Robert Fletterick (University of California, San Francisco), Richard Henderson (LMB, Cambridge, UK), Werner Kühlbrandt (MPI, Frankfurt), David Parry (Massey University, New Zealand), Greg Petsko (Brandeis University), and David Trentham (NIMR, London, UK).

The book depends for its accessibility upon its illustrations and we are hugely indebted to Nigel Orme, who, as with the first edition, has converted sketches into lucid figures. Keith Roberts has again advised us on how best graphically to represent chemical and structural phenomena. Jane Richardson (Duke University) has generously produced the Kinemage supplement to this edition and the book relies upon Richardson-type diagrams throughout to render the structures discussed comprehensible. We thank our publishers Garland Publishing, now part of the Taylor and Francis Group, for their support, and in particular Matthew Day for his enthusiastic editing of the complete manuscript. Miranda Robertson, in her inimitable style, has again managed the entire project.

Contents

xiii

xiv

Basic Structural Principles

Part 1

The Building Blocks

1

Recombinant DNA techniques have provided tools for the rapid determination of DNA sequences and, by inference, the amino acid sequences of proteins from structural genes. The number of such sequences is now increasing almost exponentially, but by themselves these sequences tell little more about the biology of the system than a New York City telephone directory tells about the function and marvels of that city.

The proteins we observe in nature have evolved, through selective pressure, to perform specific functions. The functional properties of proteins depend upon their three-dimensional structures. The three-dimensional structure arises because particular sequences of amino acids in polypeptide chains fold to generate, from linear chains, compact domains with specific three-dimensional structures (Figure 1.1). The folded domains can serve as modules for building up large assemblies such as virus particles or muscle fibers, or they can provide specific catalytic or binding sites, as found in enzymes or proteins that carry oxygen or that regulate the function of DNA.

To understand the biological function of proteins we would therefore like to be able to deduce or predict the three-dimensional structure from the amino acid sequence. This we cannot do. In spite of considerable efforts over the past 25 years, this folding problem is still unsolved and remains one of the most basic intellectual challenges in molecular biology.

Figure 1.1 The amino acid sequence of a protein's polypeptide chain is called its **primary** structure. Different regions of the sequence form local regular **secondary** structures, such as alpha (α) helices or beta (β) strands. The **tertiary** structure is formed by packing such structural elements into one or several compact globular units called domains. The final protein may contain several polypeptide chains arranged in a **quaternary** structure. By formation of such tertiary and quaternary structure amino acids far apart in the sequence are brought close together in three dimensions to form a functional region, an **active site**.

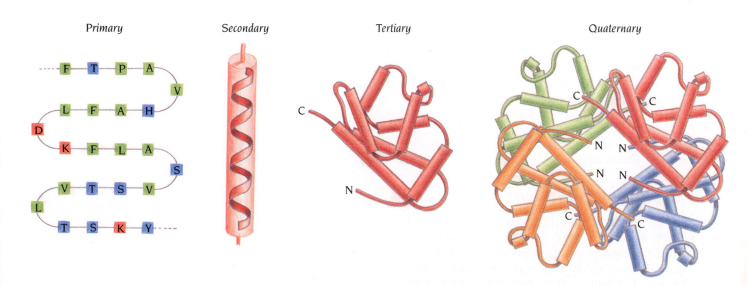

3

Protein folding remains a problem because there are 20 different amino acids that can be combined into many more different proteins than there are atoms in the known universe. In addition there is a vast number of ways in which similar structural domains can be generated in proteins by different amino acid sequences. By contrast, the structure of DNA, made up of only four different nucleotide building blocks that occur in two pairs, is relatively simple, regular, and predictable.

Since the three-dimensional structures of individual proteins cannot be predicted, they must instead be determined experimentally by x-ray crystallography, electron crystallography or nuclear magnetic resonance (NMR) techniques. Over the past 30 years the structures of more than 6000 proteins have been solved, and the sequences of more than 500,000 have been determined. This has generated a body of information from which a set of basic principles of protein structure has emerged. These principles make it easier for us to understand how protein structure is generated, to identify common structural themes, to relate structure to function, and to see fundamental relationships between different proteins. The science of protein structure is at the stage of taxonomy where we can begin to discern patterns and motifs among the relatively small number of proteins whose three-dimensional structure is known.

The first six chapters of this book deal with the basic principles of protein structure as we understand them today, and examples of the different major classes of protein structures are presented. Chapter 7 contains a brief discussion on DNA structures with emphasis on recognition by proteins of specific nucleotide sequences. The remaining chapters illustrate how during evolution different structural solutions have been selected to fulfill particular functions.

Proteins are polypeptide chains

All of the 20 amino acids have in common a central carbon atom (C_α) to which are attached a hydrogen atom, an amino group (NH_2), and a carboxyl group (COOH) (Figure 1.2a). What distinguishes one amino acid from another is the side chain attached to the C_α through its fourth valence. There are 20 different side chains specified by the genetic code; others occur, in rare cases, as the products of enzymatic modifications after translation.

Amino acids are joined end-to-end during protein synthesis by the formation of **peptide bonds** when the carboxyl group of one amino acid condenses with the amino group of the next to eliminate water (Figure 1.2b). This process is repeated as the chain elongates. One consequence is that the amino group of the first amino acid of a polypeptide chain and the carboxyl group of the last amino acid remain intact, and the chain is said to extend from its amino terminus to its carboxy terminus. The formation of a succession of peptide bonds generates a "main chain," or "backbone," from which project the various side chains.

The main-chain atoms are a carbon atom C_α to which the side chain is attached, an NH group bound to C_α, and a carbonyl group C'=O, where the carbon atom C' is attached to C_α. These units, or residues, are linked into a polypeptide by a peptide bond between the C' atom of one residue and the nitrogen atom of the next (see Figure 1.2b). The basic repeating unit along the main chain from a biochemical or genetic viewpoint is thus (NH–C_αH–C'=O), which is the residue of the common parts of amino acids after peptide bonds have been formed (see Figure 1.2b).

The genetic code specifies 20 different amino acid side chains

The 20 different side chains that occur in proteins are shown in Panel 1.1 (pp. 6–7). Their names are abbreviated with both a three-letter and a one-letter code, which are also given in the panel. The one-letter codes are worth memorizing, as they are widely used in the literature. A mnemonic device for linking the one-letter code to the names of the amino acids is given in Panel 1.1.

Figure 1.2 Proteins are built up by amino acids that are linked by peptide bonds to form a polypeptide chain. (a) Schematic diagram of an amino acid, illustrating the nomenclature used in this book. A central carbon atom (C_α) is attached to an amino group (NH_2), a carboxyl group (COOH), a hydrogen atom (H), and a side chain (R). (b) In a polypeptide chain the carboxyl group of amino acid n has formed a peptide bond, C–N, to the amino group of amino acid $n + 1$. One water molecule is eliminated in this process. The repeating units, which are called residues, are divided into main-chain atoms and side chains. The main-chain part, which is identical in all residues, contains a central C_α atom attached to an NH group, a C'=O group, and an H atom. The side chain R, which is different for different residues, is bound to the C_α atom.

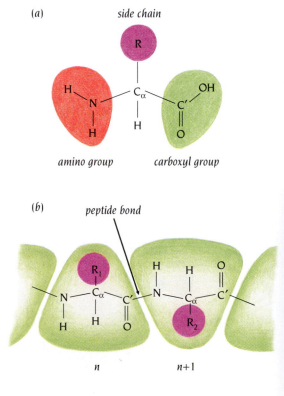

The amino acids are usually divided into three different classes defined by the chemical nature of the side chain. The first class comprises those with strictly hydrophobic side chains: Ala (A), Val (V), Leu (L), Ile (I), Phe (F), Pro (P), and Met (M). The four charged residues, Asp (D), Glu (E), Lys (K), and Arg (R), form the second class. The third class comprises those with polar side chains: Ser (S), Thr (T), Cys (C), Asn (N), Gln (Q), His (H), Tyr (Y), and Trp (W). The amino acid glycine (G), which has only a hydrogen atom as a side chain and so is the simplest of the 20 amino acids, has special properties and is usually considered either to form a fourth class or to belong to the first class.

The four groups attached to the C_α atom are chemically different for all the amino acids except glycine, where two H atoms bind to C_α. All amino acids except glycine are thus chiral molecules that can exist in two different forms with different "hands," L- or D-form (Figure 1.3).

Biological systems depend on specific detailed recognition of molecules that distinguish between chiral forms. The translation machinery for protein synthesis has evolved to utilize only one of the chiral forms of amino acids, the L-form. All amino acids that occur in proteins therefore have the L-form. There is, however, no obvious reason why the L-form was chosen during evolution and not the D-form.

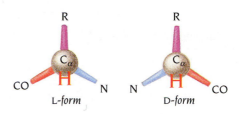

Figure 1.3 The "handedness" of amino acids. Looking down the H–C_α bond from the hydrogen atom, the L-form has CO, R, and N substituents from C_α going in a clockwise direction. There is a mnemonic to remember this; for the L-form the groups read CORN in clockwise direction.

Cysteines can form disulfide bridges

Two cysteine residues in different parts of the polypeptide chain but adjacent in the three-dimensional structure of a protein can be oxidized to form a **disulfide bridge** (Figure 1.4). The disulfide is usually the end product of air oxidation according to the following reaction scheme:

$$2 -CH_2SH + \frac{1}{2} O_2 \rightleftharpoons -CH_2-S-S-CH_2 + H_2O$$

This reaction requires an oxidative environment, and such disulfide bridges are usually not found in intracellular proteins, which spend their lifetime in an essentially reductive environment. Disulfide bridges do, however, occur quite frequently among extracellular proteins that are secreted from cells, and in eucaryotes, formation of these bridges occurs within the lumen of the endoplasmic reticulum, the first compartment of the secretory pathway.

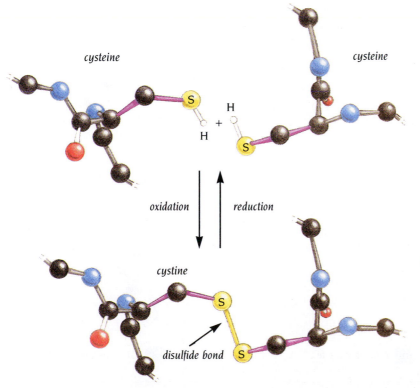

Figure 1.4 The disulfide is usually the end product of air oxidation according to the following schematic reaction scheme:

$$2 -CH_2SH + \frac{1}{2} O_2 \rightleftharpoons -CH_2-S-S-CH_2 + H_2O$$

Disulfide bonds form between the side chains of two cysteine residues. Two SH groups from cysteine residues, which may be in different parts of the amino acid sequence but adjacent in the three-dimensional structure, are oxidized to form one S–S (disulfide) group.

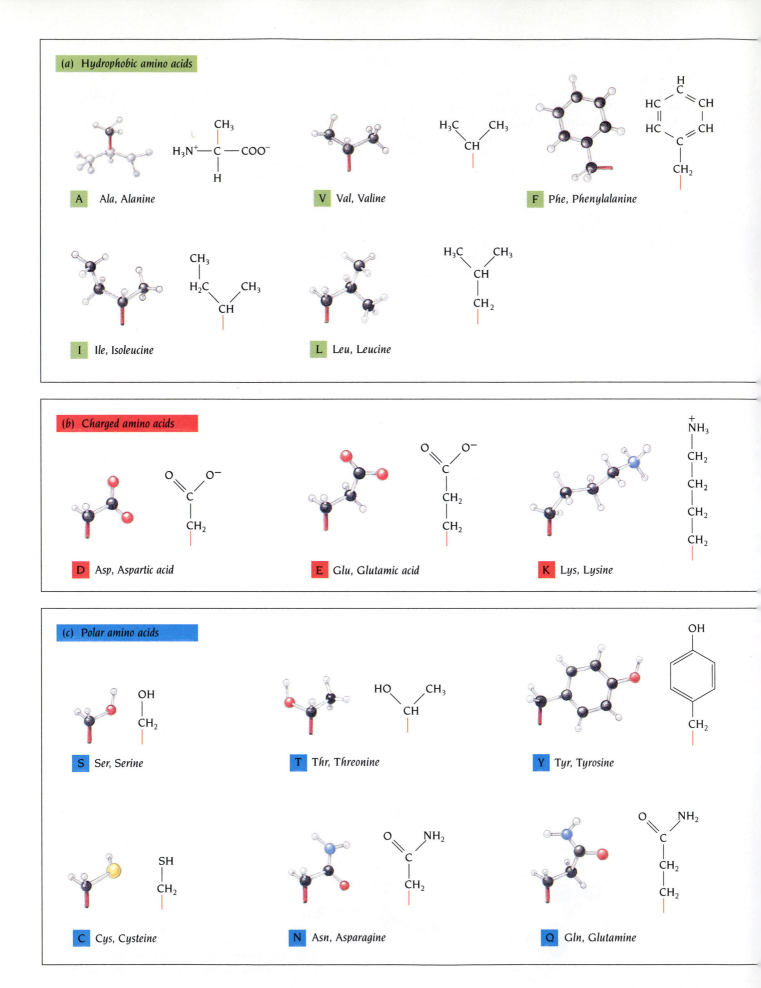

(a) Hydrophobic amino acids

A — Ala, Alanine

V — Val, Valine

F — Phe, Phenylalanine

I — Ile, Isoleucine

L — Leu, Leucine

(b) Charged amino acids

D — Asp, Aspartic acid

E — Glu, Glutamic acid

K — Lys, Lysine

(c) Polar amino acids

S — Ser, Serine

T — Thr, Threonine

Y — Tyr, Tyrosine

C — Cys, Cysteine

N — Asn, Asparagine

Q — Gln, Glutamine

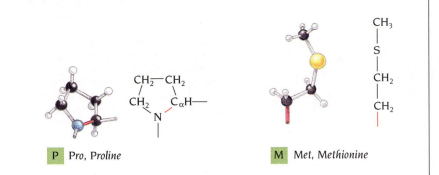

P Pro, Proline

CH₂—CH₂

CH₂   CₐH—
    \   /
     N
     |

M Met, Methionine

CH₃
|
S
|
CH₂
|
CH₂
|

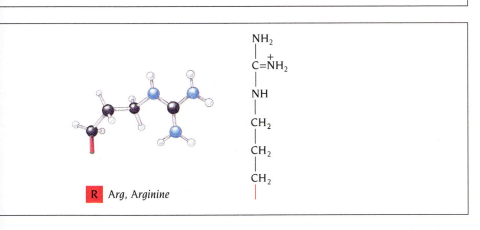

R Arg, Arginine

NH₂
|
C=⁺NH₂
|
NH
|
CH₂
|
CH₂
|
CH₂
|

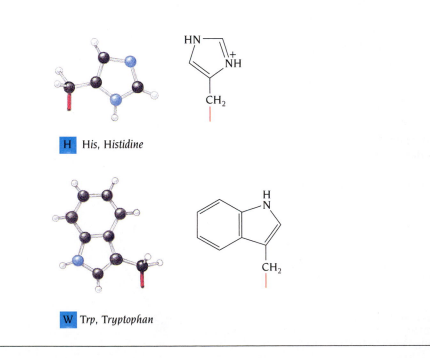

H His, Histidine

HN
    \
     \⁺
     NH
     |
     CH₂
     |

W Trp, Tryptophan

H
N

CH₂
|

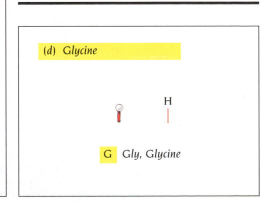

(d) Glycine

H

G Gly, Glycine

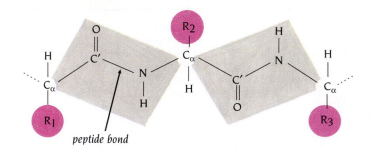

peptide bond

Figure 1.5 Part of a polypeptide chain that is divided into peptide units, represented as blocks in the diagram. Each peptide unit contains the C_α atom and the C'=O group of residue *n* as well as the NH group and the C_α atom of residue *n* + 1. Each such unit is a planar, rigid group with known bond distances and bond angles. R_1, R_2, and R_3 are the side chains attached to the C_α atoms that link the peptide units in the polypeptide chain. The peptide group is planar because the additional electron pair of the C=O bond is delocalized over the peptide group such that rotation around the C–N bond is prevented by an energy barrier.

Disulfide bridges stabilize three-dimensional structure. In some proteins these bridges hold together different polypeptide chains; for example, the A and B chains of insulin are linked by two disulfide bridges between the chains. More frequently intramolecular disulfide bridges stabilize the folding of a single polypeptide chain, making the protein less susceptible to degradation. There are many examples of this, including proteins with short polypeptide chains, such as snake venom toxins and protease inhibitors, that need additional stabilizing factors to produce a stable fold. Much effort is currently spent on introducing extra intramolecular disulfide bridges into enzymes by site-directed mutagenesis in order to make them more thermostable and hence more useful for industrial applications as catalysts, as described in Chapter 17.

Peptide units are building blocks of protein structures

Figure 1.2 shows one way of dividing a polypeptide chain, the biochemist's way. There is, however, a different way to divide the main chain into repeating units that is preferable when we want to describe the structural properties of proteins. For this purpose it is more useful to divide the polypeptide chain into peptide units that go from one C_α atom to the next C_α atom (see Figure 1.5). Each C_α atom, except the first and the last, thus belongs to two such units. The reason for dividing the chain in this way is that all the atoms in such a unit are fixed in a plane with the bond lengths and bond angles very nearly the same in all units in all proteins. Note that the peptide units of the main chain do not involve the different side chains (Figure 1.5). We will use both of these alternative descriptions of polypeptide chains—the biochemical and the structural—and discuss proteins in terms of the sequence of different amino acids and the sequence of planar peptide units.

Since the peptide units are effectively rigid groups that are linked into a chain by covalent bonds at the C_α atoms, the only degrees of freedom they have are rotations around these bonds. Each unit can rotate around two such bonds: the C_α–C' and the N–C_α bonds (Figure 1.6). By convention the angle of rotation around the N–C_α bond is called **phi (ϕ)** and the angle around the C_α–C' bond from the same C_α atom is called **psi (ψ).**

In this way each amino acid residue is associated with two conformational angles ϕ and ψ. Since these are the only degrees of freedom, the conformation of the whole main chain of the polypeptide is completely determined when the ϕ and ψ angles for each amino acid are defined with high accuracy.

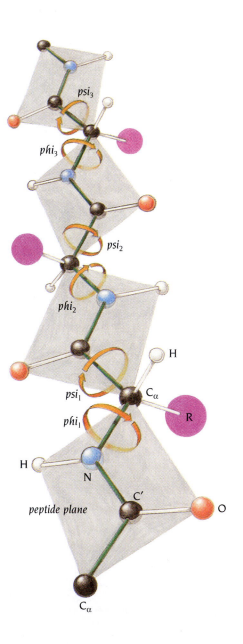

peptide plane

C_α

Figure 1.6 Diagram showing a polypeptide chain where the main-chain atoms are represented as rigid peptide units, linked through the C_α atoms. Each unit has two degrees of freedom; it can rotate around two bonds, its C_α–C' bond and its N–C_α bond. The angle of rotation around the N–C_α bond is called phi (ϕ) and that around the C_α–C' bond is called psi (ψ). The conformation of the main-chain atoms is therefore determined by the values of these two angles for each amino acid.

Glycine residues can adopt many different conformations

Most combinations of φ and ψ angles for an amino acid are not allowed because of steric collisions between the side chains and main chain. It is reasonably straightforward to calculate those combinations that are allowed. Since the D- and L-forms of the amino acids have their side chain oriented differently with respect to the CO group, they have different allowed φ and ψ angles. Proteins built from D-amino acids would thus be expected to have different conformations from those found in nature that are exclusively made of L-amino acids. Since the L- and D-forms of each amino acid are mirror images of one another, would a protein made exclusively of D-form residues produce a structure that is the mirror image of the natural protein? Stephen Kent and his colleagues at the Scripps Institute chemically synthesized both the L- and the D-forms of HIV-1 protease. The D-enzyme proved indeed to be the mirror image of the L-enzyme. Furthermore the D-enzyme and L-enzyme had reciprocal chiral specificity on peptide substrates, the D-enzyme only recognizing and cutting peptides made of D-amino acids. Perhaps the choice of the L-form at the outset of the evolution of life on earth was random and irrevocable.

The angle pairs φ and ψ are usually plotted against each other in a diagram called a **Ramachandran plot** after the Indian biophysicist G.N. Ramachandran who first made calculations of sterically allowed regions. Figure 1.7 shows the results of such calculations and also a plot for all amino

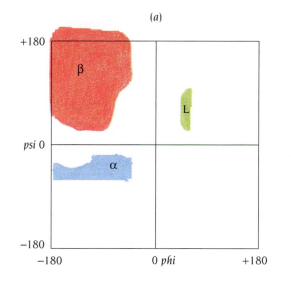

(a)

Figure 1.7 Ramachandran plots showing allowed combinations of the conformational angles phi and psi defined in Figure 1.6. Since phi (φ) and psi (ψ) refer to rotations of two rigid peptide units around the same C_α atom, most combinations produce steric collisions either between atoms in different peptide groups or between a peptide unit and the side chain attached to C_α. These combinations are therefore not allowed. (a) Colored areas show sterically allowed regions. The areas labeled α, β, and L correspond approximately to conformational angles found for the usual right-handed α helices, β strands, and left-handed α helices, respectively. (b) Observed values for all residue types except glycine. Each point represents φ and ψ values for an amino acid residue in a well-refined x-ray structure to high resolution. (c) Observed values for glycine. Notice that the values include combinations of φ and ψ that are not allowed for other amino acids. (From J. Richardson, *Adv. Prot. Chem.* 34: 174–175, 1981.)

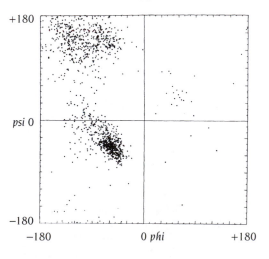

(b)

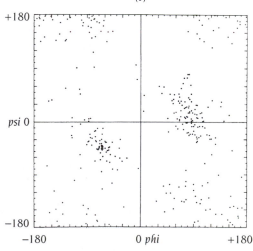

(c)

acids except glycine from a number of accurately determined protein structures. It is apparent that the observed values are clustered in the sterically allowed regions. There is one important exception. Glycine, with only a hydrogen atom as a side chain, can adopt a much wider range of conformations than the other residues, as seen in Figure 1.7c. Glycine thus plays a structurally very important role; it allows unusual main-chain conformations in proteins. This is one of the main reasons why a high proportion of glycine residues are conserved among homologous protein sequences.

Regions in the Ramachandran plot are named after the conformation that results in a peptide if the corresponding φ and ψ angles are repeated in successive amino acids along the chain. The major allowed regions in Figure 1.7a are the right-handed α-helical cluster in the lower left quadrant (see Chapter 3); the broad region of extended β strands of both parallel and antiparallel β structures (see Chapter 4) in the upper left quadrant; and the small, sparsely populated left-handed α-helical region in the upper right quadrant. Left-handed α helices are usually found in loop regions or in small single-turn α helices.

Certain side-chain conformations are energetically favorable

Any side chain longer than that of alanine can in principle have several different conformations because of rotations around the bonds between the side-chain carbon atoms. It is a general rule in chemistry that the most energetically favored arrangements for two tetrahedrally coordinated carbon atoms are the "staggered" conformations, in which the substituents of one carbon atom are between those of the other when viewed along the axis of rotation, as illustrated in Figure 1.8. For each such carbon atom in a side chain, there are three possible staggered conformations, which are related by a three step, 120° rotation around the carbon–carbon bond. These three conformations are indistinguishable for alanine, since all three substituents on the side-chain carbon atom C_β are the same.

For valine, however, the three staggered conformations are energetically different because the substituents on C_β are different; two of them are methyl groups and the other is a hydrogen atom (see Figure 1.8b–d). It is energetically most favorable to have the two methyl groups close to the small hydrogen atom bound to C_α, as shown in Figure 1.8b, and therefore this is the conformation most frequently found in proteins. An analysis of accurately determined protein structures has shown that most side chains have one or a few conformations that occur more frequently than the other possible staggered

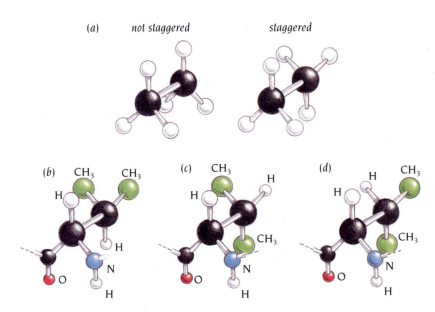

Figure 1.8 The staggered conformations are the most energetically favored conformations of two tetrahedrally coordinated carbon atoms. (a) A view along the C–C bond in ethane (CH_3CH_3) showing how the two carbon atoms can rotate so that their hydrogen atoms are either not staggered (aligned) or staggered. Three indistinguishable staggered conformations are obtained by a rotation of 120° around the C–C bond. (b–d) Similar views as in (a) of valine. The three staggered conformations are different for valine because the three groups attached to C_β are different. The first staggered conformation (b) is less crowded and energetically most favored because the two methyl groups bound to C_β are both close to the small H atom bound to C_α.

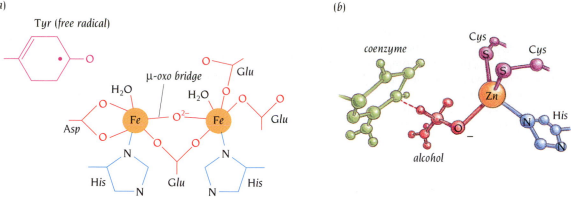

Tyr (free radical)

μ-oxo bridge

Glu

H_2O H_2O

Asp   Fe O^{2-} Fe   Glu

N   N

His   Glu   His

(b)

coenzyme

Cys   Cys

S   S

Zn   His

alcohol

N   N

conformations. These are called **rotamers**. Today, collections of these favored conformations, or rotamer libraries, are a standard tool in computer programs used for modeling protein structures.

Many proteins contain intrinsic metal atoms

The side chains of the 20 different amino acids listed in Panel 1.1 (pp. 6–7) have very different chemical properties and are utilized for a wide variety of biological functions. However, their chemical versatility is not unlimited, and for some functions metal atoms are more suitable and more efficient. Electron-transfer reactions are an important example. Fortunately the side chains of histidine, cysteine, aspartic acid, and glutamic acid are excellent metal ligands, and a fairly large number of proteins have recruited metal atoms as intrinsic parts of their structures; among the frequently used metals are iron, zinc, magnesium, and calcium. Several metallo proteins are discussed in detail in later chapters and it suffices here to mention briefly a few examples of iron and zinc proteins.

The most conspicuous use of iron in biological systems is in our blood, where the erythrocytes are filled with the oxygen-binding protein hemoglobin. The red color of blood is due to the iron atom bound to the heme group in hemoglobin. Similar heme-bound iron atoms are present in a number of proteins involved in electron-transfer reactions, notably cytochromes. A chemically more sophisticated use of iron is found in an enzyme, ribonucleotide reductase, that catalyzes the conversion of ribonucleotides to deoxyribonucleotides, an important step in the synthesis of the building blocks of DNA.

Ribonucleotide reductase from *Escherichia coli* and mammals contains a di-iron center (Figure 1.9a) that reacts with oxygen and oxidizes a nearby tyrosine side chain, producing a tyrosyl free radical that is essential for the catalysis. The two iron atoms in this iron center are close to each other and bridged by the oxygen atoms of a glutamic acid side chain as well as by an oxygen ion called a μ-oxo bridge. Glutamic acid, aspartic acid, and histidine side chains from the protein, as well as water molecules, complete the co-ordination sphere of the octahedrally coordinated iron atom.

Zinc is used to stabilize the DNA-binding regions of a class of transcription factors called zinc fingers, which are discussed in Chapter 10. Zinc ions also participate directly in catalytic reactions in many different enzymes by binding substrate molecules and providing a positive charge that influences the electronic arrangement of the substrate and thereby facilitates the catalytic reaction. One such example is found in the enzyme alcohol dehydrogenase, which in yeast produces alcohol during fermentation and in our livers detoxifies the alcohol we have consumed by oxidizing it. The enzyme provides a scaffold containing three zinc ligands—one histidine and two cysteine side chains—which sequester a zinc atom so that it binds alcohol as a fourth ligand in a tetrahedral coordination (Figure 1.9b).

Figure 1.9 Examples of functionally important intrinsic metal atoms in proteins. (a) The di-iron center of the enzyme ribonucleotide reductase. Two iron atoms form a redox center that produces a free radical in a nearby tyrosine side chain. The iron atoms are bridged by a glutamic acid residue and a negatively charged oxygen atom called a μ-oxo bridge. The coordination of the iron atoms is completed by histidine, aspartic acid, and glutamic acid side chains as well as water molecules. (b) The catalytically active zinc atom in the enzyme alcohol dehydrogenase. The zinc atom is coordinated to the protein by one histidine and two cysteine side chains. During catalysis zinc binds an alcohol molecule in a suitable position for hydride transfer to the coenzyme moiety, a nicotinamide. [(a) Adapted from P. Nordlund et al., *Nature* 345: 593–598, 1990.]

Conclusion

All protein molecules are polymers built up from 20 different amino acids linked end-to-end by peptide bonds. The function of every protein molecule depends on its three-dimensional structure, which in turn is determined by its amino acid sequence, which in turn is determined by the nucleotide sequence of the structural gene.

Each amino acid has atoms in common, and these form the main chain of the protein. The remaining atoms form side chains that can be hydrophobic, polar, or charged.

The conformation of the whole main chain of a protein is determined by two conformational angles, phi (ϕ) and psi (ψ), for each amino acid. Only certain combinations of these angles are allowed because of steric hindrance between main-chain atoms and side-chain atoms, except for glycine.

Certain side-chain conformations are energetically more favorable than others. Computer programs used to model protein structures contain rotamer libraries of such favored conformations.

Many proteins contain intrinsic metal atoms that are functionally important. The most frequently used metals are iron, zinc, magnesium, and calcium. These metal atoms are mainly bound to the protein through the side chains of cysteine, histidine, aspartic acid, and glutamic acid residues.

Selected readings

Alberts, B., et al. *Molecular Biology of the Cell*, 3rd ed. New York: Garland, 1994.

Creighton, T.E. *Proteins: Structures and Molecular Properties*, 2nd ed. New York: Freeman, 1993.

Fletterick, R.J., Schroer, T., Matela, R.J. *Molecular Structure: Macromolecules in Three Dimensions*. Oxford, UK: Blackwell Scientific, 1985.

Judson, H.F. *The Eighth Day of Creation: Makers of the Revolution in Biology*. New York: Simon & Schuster, 1979.

Karlin, K.D. Metalloenzymes, structural motifs, and inorganic models. *Science* 261: 701–708, 1993.

Lesk, A. *Protein Architecture: A Practical Approach*. Oxford, UK: Oxford University Press, 1991.

Mathews, C.K., van Holde, K.E. *Biochemistry*. Menlo Park, CA: Benjamin/Cummings, 1990.

Perutz, M. *Protein Structure: New Approaches to Disease and Therapy*. New York: Freeman, 1992.

Petsko, G.A. On the other hand... *Science* 256: 1403–1404, 1992.

Ramachandran, G.N., Sasisekharan, V. Conformation of polypeptides and proteins. *Adv. Prot. Chem.* 28: 283–437, 1968.

Richardson, J.S., Richardson, D.C. Principles and patterns of protein conformation. In *Prediction of Protein Structure and the Principles of Protein Conformation* (ed. Fasman, G.D.), pp. 1–98. New York: Plenum, 1989.

Schulz, G.E., Schirmer, R.H. *Principles of Protein Structure*. New York: Springer, 1979.

Spiro, T.S. *Zinc Enzymes*. New York: Wiley, 1983.

Stryer, L. *Biochemistry*, 4th ed. New York: Freeman, 1995.

Motifs of Protein Structure

2

X-ray structural studies have played a major role in transforming chemistry from a descriptive science at the beginning of the twentieth century to one in which the properties of novel compounds can be predicted on theoretical grounds. When W.L. Bragg solved the very first crystal structure, that of rock salt, NaCl, the results completely changed prevalent concepts of bonding forces in ionic compounds.

The first x-ray crystallographic structural results on a globular protein molecule were reported for myoglobin (Figure 2.1) in 1958, and came as a shock to those who had hoped for simple, general principles of protein structure and function analogous to the simple and beautiful double-stranded DNA structure that had been determined five years before by James Watson and Francis Crick. John Kendrew at the Medical Research Council Laboratory of Molecular Biology, in Cambridge, UK, who determined the myoglobin structure to low resolution in 1958, expressed his disappointment about the complexity of the structure in the following words: "Perhaps the most remarkable features of the molecule are its complexity and its lack of symmetry. The arrangement seems to be almost totally lacking in the kind of regularities which one instinctively anticipates, and it is more complicated than has been predicted by any theory of protein structure."

In retrospect it is easy to see that such structural irregularity is actually required for proteins to fulfill their diverse functions. Information storage and transfer from DNA is essentially linear, and DNA molecules of very different information content can therefore have essentially the same gross structure. In contrast, proteins must recognize many thousands of different molecules in the cell by detailed three-dimensional interactions, which

Figure 2.1 Kendrew's model of the low-resolution structure of myoglobin shown in three different views. The sausage-shaped regions represent α helices, which are arranged in a seemingly irregular manner to form a compact globular molecule. (Courtesy of J.C. Kendrew.)

10 Å

require diverse and irregular structures of the protein molecules. In spite of these requirements, there are regular features in protein structures, the most important of which is their **secondary structure.**

The interior of proteins is hydrophobic

When high-resolution studies of myoglobin became available, Kendrew noticed that the amino acids in the interior of the protein had almost exclusively hydrophobic side chains. This was one of the first important general principles to emerge from studies of protein structure. The main driving force for folding water-soluble globular protein molecules is to pack hydrophobic side chains into the interior of the molecule, thus creating a **hydrophobic core** and a hydrophilic surface.

The hydrophobic core is surprisingly densely packed with the side chains in the interior of the protein. Given the constraints of the different shapes of the hydrophobic side chains and considering that their positions must be compatible with the regular secondary structure in the interior of the protein, fitting these shapes into a densely packed core is like solving a three-dimensional jigsaw puzzle. In the few cases where there is a hole in the interior, the space is usually occupied by one or more water molecules that hydrogen-bond to internal polar groups. Such firmly bound internal water molecules can be regarded as integral parts of the protein structure.

There is a major problem, however, with creating such a hydrophobic core from a protein chain. To bring the side chains into the core, the main chain must also fold into the interior. The main chain is highly polar and therefore hydrophilic, with one hydrogen bond donor, NH, and one hydrogen bond acceptor, C'=O, for each peptide unit. In a hydrophobic environment, these main-chain polar groups must be neutralized by the formation of hydrogen bonds. This problem is solved in a very elegant way by the formation of regular secondary structure within the interior of the protein molecule. Such secondary structure is usually one of two types: **alpha helices** or **beta sheets.** Both types are characterized by hydrogen-bonding between the main-chain NH and C'=O groups, and they are formed when a number of consecutive residues have the same phi (ϕ), psi (ψ) angles.

The secondary structure elements, formed in this way and held together by the hydrophobic core, provide a rigid and stable framework. They exhibit relatively little flexibility with respect to each other, and they are the best-defined parts of protein structures determined by both x-ray and NMR techniques. Functional groups of the protein are attached to this framework, either directly by their side chains or, more frequently, in loop regions that connect sequentially adjacent secondary structure elements. We will now have a closer look at these structural elements.

The alpha (α) helix is an important element of secondary structure

The α helix is the classic element of protein structure. It was first described in 1951 by Linus Pauling working at the California Institute of Technology. He predicted that it was a structure which would be stable and energetically favorable in proteins. He made this remarkable prediction on the basis of accurate geometrical parameters that he had derived for the peptide unit from the results of crystallographic analyses of the structures of a range of small molecules. This prediction almost immediately received strong experimental support from diffraction patterns obtained by Max Perutz in Cambridge, UK, from hemoglobin crystals and keratin fibers. It was completely verified from John Kendrew's high-resolution structure of myoglobin, where all secondary structure is helical.

Alpha helices in proteins are found when a stretch of consecutive residues all have the ϕ, ψ angle pair approximately −60° and −50°, corresponding to the allowed region in the bottom left quadrant of the

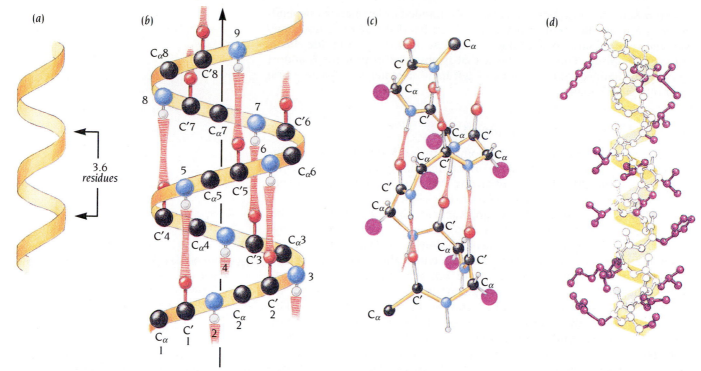

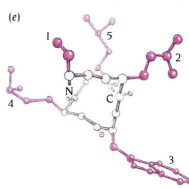

Figure 2.2 The α helix is one of the major elements of secondary structure in proteins. Main-chain N and O atoms are hydrogen-bonded to each other within α helices. (a) Idealized diagram of the path of the main chain in an α helix. Alpha helices are frequently illustrated in this way. There are 3.6 residues per turn in an α helix, which corresponds to 5.4 Å (1.5 Å per residue). (b) The same as (a) but with approximate positions for main-chain atoms and hydrogen bonds included. The arrow denotes the direction from the N-terminus to the C-terminus. (c) Schematic diagram of an α helix. Oxygen atoms are red, and N atoms are blue. Hydrogen bonds between O and N are red and striated. The side chains are represented as purple circles. (d) A ball-and-stick model of one α helix in myoglobin. The path of the main chain is outlined in yellow; side chains are purple. Main-chain atoms are not colored. (e) One turn of an α helix viewed down the helical axis. The purple side chains project out from the α helix.

Ramachandran plot (see Figure 1.7a). The α helix has 3.6 residues per turn with hydrogen bonds between C'=O of residue n and NH of residue $n + 4$ (Figure 2.2). Thus all NH and C'O groups are joined with hydrogen bonds except the first NH groups and the last C'O groups at the ends of the α helix. As a consequence, the ends of α helices are polar and are almost always at the surface of protein molecules.

Variations on the α helix in which the chain is either more loosely or more tightly coiled, with hydrogen bonds to residues $n + 5$ or $n + 3$ instead of $n + 4$ are called the π helix and 3_{10} helix, respectively. The 3_{10} helix has 3 residues per turn and contains 10 atoms between the hydrogen bond donor and acceptor, hence its name. Both the π helix and the 3_{10} helix occur rarely and usually only at the ends of α helices or as single-turn helices. They are not energetically favorable, since the backbone atoms are too tightly packed in the 3_{10} helix and so loosely packed in the π helix that there is a hole through the middle. Only in the α helix are the backbone atoms properly packed to provide a stable structure.

In globular proteins α helices vary considerably in length, ranging from four or five amino acids to over forty residues. The average length is around ten residues, corresponding to three turns. The rise per residue of an α helix is 1.5 Å along the helical axis, which corresponds to about 15 Å from one end to the other of an average α helix.

An α helix can in theory be either right-handed or left-handed depending on the screw direction of the chain. A left-handed α helix is not, however, allowed for L-amino acids due to the close approach of the side chains and the C'O group. Thus the α helix that is observed in proteins is almost always right-handed. Short regions of left-handed α helices (3–5 residues) occur only occasionally.

The α helix has a dipole moment

All the hydrogen bonds in an α helix point in the same direction because the peptide units are aligned in the same orientation along the helical axis. Since a peptide unit has a dipole moment arising from the different polarity of NH and C'O groups, these dipole moments are also aligned along the helical axis (Figure 2.3). The overall effect is a significant net dipole for the α helix that gives a partial positive charge at the amino end and a partial negative charge at the carboxy end of the α helix. The magnitude of this dipole moment corresponds to about 0.5–0.7 unit charge at each end of the helix. These charges would be expected to attract ligands of opposite charge and negatively charged ligands, especially when they contain phosphate groups and frequently bind at the N-termini of α helices. In contrast, positively charged ligands rarely bind at the C-terminus. This may be because, in addition to the dipole effect, the N-terminus of an α helix has free NH groups with favorable geometry to position phosphate groups by specific hydrogen bonds (see Figure 2.3). Such ligand-binding occurs frequently in proteins; it provides examples of specific binding through main-chain conformation in which side chains are not involved.

Some amino acids are preferred in α helices

The amino acid side chains project out from the α helix (see Figure 2.2e) and do not interfere with it, except for proline. The last atom of the proline side

Figure 2.3 Negatively charged groups such as phosphate ions frequently bind to the amino ends of α helices. The dipole moment of an α helix as well as the possibility of hydrogen-bonding to free NH groups at the end of the helix favors such binding. (a) The dipole of a peptide unit. Values in boxes give the approximate fractional charges of the atoms of the peptide unit. (b) The dipoles of peptide units are aligned along the α-helical axis, which creates an overall dipole moment of the α helix, positive at the amino end and negative at the carboxy end. (c) A phosphate group hydrogen-bonded to the NH end of an α helix. Nitrogen atoms are blue; oxygen atoms are red; main-chain carbon atoms are black; and phosphorus is green.

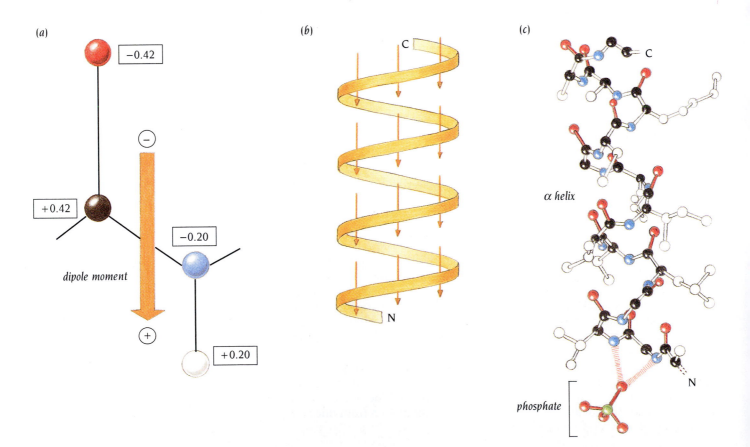

(a)

−0.42

+0.42

−0.20

dipole moment

+0.20

(b)

C

N

(c)

α helix

phosphate

N

C

Table 2.1 Amino acid sequences of three α helices

1. - Leu - Ser - Phe - Ala - Ala - Ala - Met - Asn - Gly - Leu - Ala -
2. - Ile - Asn - Glu - Gly - Phe - Asp - Leu - Leu - Arg - Ser - Gly -
3. - Lys - Glu - Asp - Ala - Lys - Gly - Lys - Ser - Glu - Glu - Glu -

The first sequence is from the enzyme citrate synthase, residues 260–270, which form a buried helix; the second sequence is from the enzyme alcohol dehydrogenase, residues 355–365, which form a partially exposed helix; and the third sequence is from troponin-C, residues 87–97, which form a completely exposed helix. Charged residues are colored red, polar residues are blue, and hydrophobic residues are green.

chain is bonded to the main-chain N atom, which forms a ring structure, C_α–CH_2–CH_2–CH_2–N (see Panel 1.1, p. 7). This prevents the N atom from participating in hydrogen-bonding and also provides some steric hindrance to the α-helical conformation. Proline fits very well in the first turn of an α helix, but it usually produces a significant bend if it is anywhere else in the helix. Such bends occur in many α helices, not just in those few that contain a proline in the middle. Therefore, although we can predict that a proline residue may cause a bend in an α helix, it does not follow that all bends result from the presence of proline.

Different side chains have been found to have weak but definite preferences either for or against being in α helices. Thus Ala (A), Glu (E), Leu (L), and Met (M) are good α-helix formers, while Pro (P), Gly (G), Tyr (Y), and Ser (S) are very poor. Such preferences were central to all early attempts to predict secondary structure from amino acid sequence, but they are not strong enough to give accurate predictions.

The most common location for an α helix in a protein structure is along the outside of the protein, with one side of the helix facing the solution and the other side facing the hydrophobic interior of the protein. Therefore, with 3.6 residues per turn, there is a tendency for side chains to change from hydrophobic to hydrophilic with a periodicity of three to four residues. Although this trend can sometimes be seen in the amino acid sequence, it is not strong enough for reliable structural prediction by itself, because residues that face the solution can be hydrophobic and, furthermore, α helices can be either completely buried within the protein or completely exposed. Table 2.1 shows examples of the amino acid sequences of a totally buried, a partially buried, and a completely exposed α helix.

A convenient way to illustrate the amino acid sequences in helices is the **helical wheel** or spiral. Since one turn in an α helix is 3.6 residues long, each residue can be plotted every 360/3.6 = 100° around a circle or a spiral, as shown in Figure 2.4. Such a plot shows the projection of the position of the

Figure 2.4 The helical wheel or spiral. Amino acid residues are plotted every 100° around the spiral, following the sequences given in Table 2.1. The following color code is used: green is an amino acid with a hydrophobic side chain, blue is a polar side chain, and red is a charged side chain. The first helix is all hydrophobic, the second is polar on one side and hydrophobic on the other side, and the third helix is all polar.

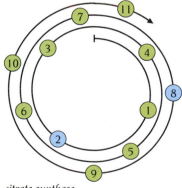

citrate synthase

1	2	3	4	5	6	7	8	9	10	11
L	S	F	A	A	A	M	N	G	L	A

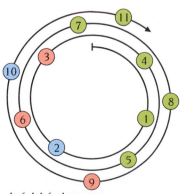

alcohol dehydrogenase

1	2	3	4	5	6	7	8	9	10	11
I	N	E	G	F	D	L	L	R	S	G

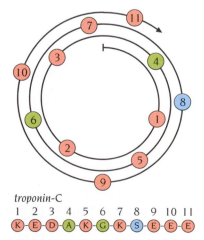

troponin-C

1	2	3	4	5	6	7	8	9	10	11
K	E	D	A	K	G	K	S	E	E	E

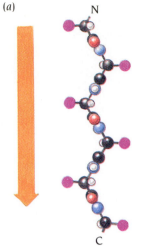

(a)

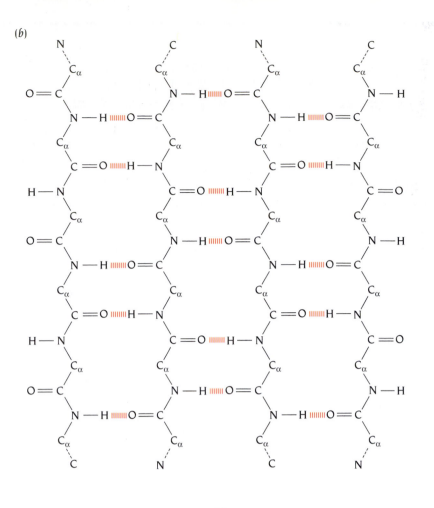

(b)

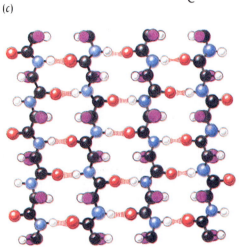

(c)

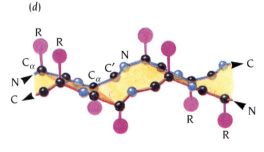

(d)

Figure 2.5 Schematic illustrations of antiparallel β sheets. Beta sheets are the second major element of secondary structure in proteins. The β strands are either all antiparallel as in this figure or all parallel or mixed as illustrated in following figures. (a) The extended conformation of a β strand. Side chains are shown as purple circles. The orientation of the β strand is at right angles to those of (b) and (c). A β strand is schematically illustrated as an arrow, from N to C terminus. (b) Schematic illustration of the hydrogen bond pattern in an antiparallel β sheet. Main-chain NH and O atoms within a β sheet are hydrogen bonded to each other. (c) A ball-and-stick version of (b). Oxygen atoms are red; nitrogen atoms are blue. The hydrogen atom in N–H···O is white. The carbon atom in the main chain, C_α, is black. Side chains are illustrated by one purple atom. The orientation of the β strands is different from that in (a). (d) Illustration of the pleat of a β sheet. Two antiparallel β strands are viewed from the side of the β sheet. Note that the directions of the side chains, R (purple), follow the pleat, which is emphasized in yellow.

residues onto a plane perpendicular to the helical axis. Residues on one side of the helix are plotted on one side of the spiral. The three helices whose sequences are given in Table 2.1 are plotted in this way in Figure 2.4, using a color code for hydrophobic, polar, and charged residues. It is immediately obvious that one side of the helix from alcohol dehydrogenase is hydrophilic and the other side hydrophobic.

Alpha helices that cross membranes are in a hydrophobic environment. Therefore, most of their side chains are hydrophobic. Long regions of hydrophobic residues in the amino acid sequence of a protein that is membrane-bound can therefore be predicted with a high degree of confidence to be transmembrane helices, as will be discussed in Chapter 12.

(a)

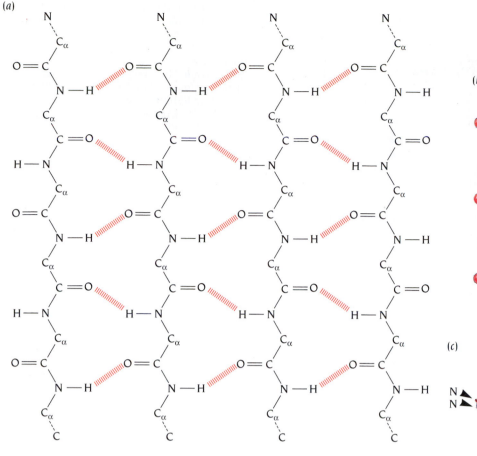

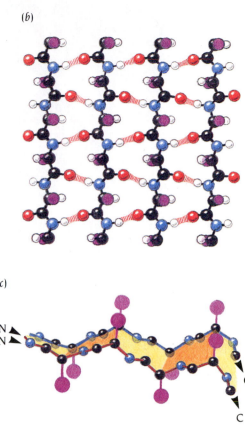

(b)

(c)

Beta (β) sheets usually have their β strands either parallel or antiparallel

The second major structural element found in globular proteins is the β sheet. This structure is built up from a combination of several regions of the polypeptide chain, in contrast to the α helix, which is built up from one continuous region. These regions, β strands, are usually from 5 to 10 residues long and are in an almost fully extended conformation with ϕ, ψ angles within the broad structurally allowed region in the upper left quadrant of the Ramachandran plot (see Figure 1.7). These β strands are aligned adjacent to each other (see Figures 2.5 and 2.6) such that hydrogen bonds can form between C'=O groups of one β strand and NH groups on an adjacent β strand and vice versa. The β sheets that are formed from several such β strands are "**pleated**" with C_α atoms successively a little above and below the plane of the β sheet. The side chains follow this pattern such that within a β strand they also point alternately above and below the β sheet.

Beta strands can interact in two ways to form a pleated sheet. Either the amino acids in the aligned β strands can all run in the same biochemical direction, amino terminal to carboxy terminal, in which case the sheet is described as **parallel,** or the amino acids in successive strands can have alternating directions, amino terminal to carboxy terminal followed by carboxy terminal to amino terminal, followed by amino terminal to carboxy terminal, and so on, in which case the sheet is called **antiparallel.** Each of the two forms has a distinctive pattern of hydrogen-bonding. The antiparallel β sheet (Figure 2.5) has narrowly spaced hydrogen bond pairs that alternate with widely spaced pairs. Parallel β sheets (Figure 2.6) have evenly spaced hydrogen bonds that bridge the β strands at an angle. Within both types of β sheets all possible main-chain hydrogen bonds are formed, except for the two flanking strands of the β sheet that have only one neighboring β strand.

Figure 2.6 Parallel β sheet. (a) Schematic diagram showing the hydrogen bond pattern in a parallel β sheet. (b) Ball-and-stick version of (a). The same color scheme is used as in Figure 2.5c. (c) Schematic diagram illustrating the pleat of a parallel β sheet.

(a)

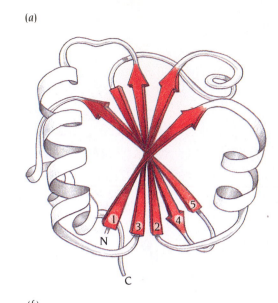

Figure 2.7 (a) Illustration of the twist of β sheets. Beta strands are drawn as arrows from the amino end to the carboxy end of the β strand in this schematic drawing of the protein thioredoxin from *E. coli,* the structure of which was determined in the laboratory of Carl Branden, Uppsala, Sweden, to 2.8 Å resolution. The mixed β sheet is viewed from one of its ends. (b) The hydrogen bonds between the β strands in the mixed β sheet of the same protein. [(a) Adapted from B. Furugren.]

(b)

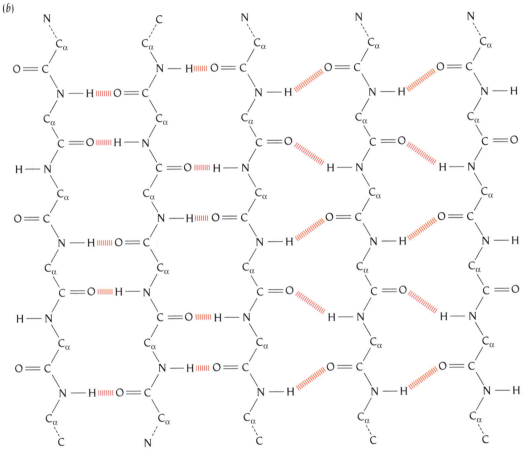

Beta strands can also combine into mixed β sheets with some β strand pairs parallel and some antiparallel. There is a strong bias against mixed β sheets; only about 20% of the strands inside the β sheets of known protein structures have parallel bonding on one side and antiparallel bonding on the other. Figure 2.7 illustrates how the hydrogen bonds between the β strands are arranged in a mixed β sheet.

As they occur in known protein structures, almost all β sheets—parallel, antiparallel, and mixed—have twisted strands. This twist always has the same handedness as that shown in Figure 2.7, which is defined as a right-handed twist.

Loop regions are at the surface of protein molecules

Most protein structures are built up from combinations of secondary structure elements, α helices and β strands, which are connected by **loop regions** of various lengths and irregular shape. A combination of secondary structure elements forms the stable hydrophobic core of the molecule. The loop regions are at the surface of the molecule. The main-chain C'=O and NH groups of these loop regions, which in general do not form hydrogen bonds to each other, are exposed to the solvent and can form hydrogen bonds to water molecules.

Loop regions exposed to solvent are rich in charged and polar hydrophilic residues. This has been used in several prediction schemes, and it has proved possible to predict loop regions from an amino acid sequence with a higher degree of confidence than α helices or β strands, which is ironic since the loops have irregular structures.

When homologous amino acid sequences from different species are compared, it is found that insertions and deletions of a few residues occur almost exclusively in the loop regions. During evolution, cores are much more stable than loops. Intron positions are also often found at sites in structural genes that correspond to loop regions in the protein structure. Since proteins that exhibit sequence homology in general have similar core structures, it is apparent that the specific arrangement of secondary structure elements in the core is rather insensitive to the lengths of the loop regions. In addition to their function as connecting units between secondary structure elements, loop regions frequently participate in forming binding sites and enzyme active sites. Thus antigen-binding sites in antibodies are built up from six loop regions, which vary both in length and in amino acid sequence between different antibodies. Modeling antigen-binding sites from a known antibody sequence (discussed in Chapter 17) is thus essentially a problem of modeling the three-dimensional structures of loop regions since the core structures of all antibodies are very similar. Such model building has been facilitated by the recent findings that loop regions have preferred structures. Surveys of known three-dimensional structures of loops have shown that they fall into a rather limited set of structures and are not a random collection of possible structures (Figure 2.8). Loop regions that connect two adjacent antiparallel β strands are called **hairpin loops**. Short hairpin loops are usually called **reverse turns** or simply turns. Figure 2.8b shows two of the most frequently

Figure 2.8 Adjacent antiparallel β strands are joined by hairpin loops. Such loops are frequently short and do not have regular secondary structure. Nevertheless, many loop regions in different proteins have similar structures. (a) Histogram showing the frequency of hairpin loops of different lengths in 62 different proteins. (b) The two most frequently occurring two-residue hairpin loops; Type I turn to the left and Type II turn to the right. Bonds within the hairpin loop are green. [(a) Adapted from B.L. Sibanda and J.M. Thornton, *Nature* 316: 170–174, 1985.]

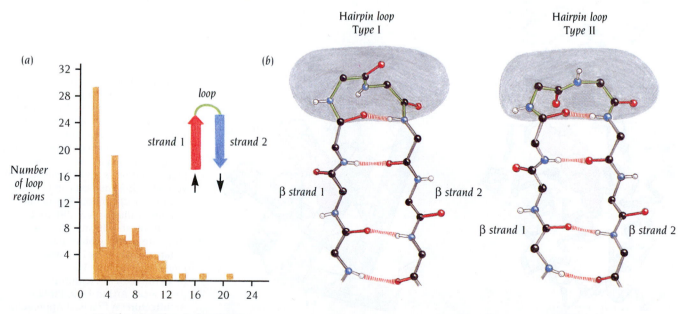

21

occurring turns; the Type I turn and the Type II turn. The Type II turn usually has a glycine residue as the second of the two residues in the turn.

Long loop regions are often flexible and can frequently adopt several different conformations, making them "invisible" in x-ray structure determinations and undetermined in NMR studies. Such loops are frequently involved in the function of the protein and can switch from an "open" conformation, which allows access to the active site, to a "closed" conformation, which shields reactive groups in the active site from water.

Long loops are in many cases susceptible to proteolytic degradation. One specific type of long loop, the omega loop, is compact with good internal packing interactions and is therefore quite stable. Other long loops, which by themselves would be attacked by proteolytic enzymes, are stabilized and protected by binding metal ions, especially calcium.

Schematic pictures of proteins highlight secondary structure

All pictorial representations of molecules are simplified versions of our current model of real molecules, which are quantum mechanical, probabilistic collections of atoms as both particles and waves. These are difficult to illustrate. Therefore we use different types of simplified representations, including space-filling models; ball-and-stick models, where atoms are spheres and bonds are sticks; and models that illustrate surface properties. The most detailed representation is the ball-and-stick model. However, a model of a protein structure where all atoms are displayed is confusing because of the sheer amount of information present (Figure 2.9a).

A two-dimensional picture of such a model is impossible to interpret. Even if the side-chain atoms are stripped off, it is still difficult to extract

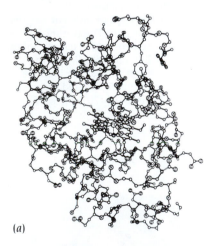

(a)

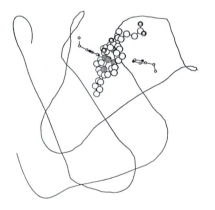

(b)

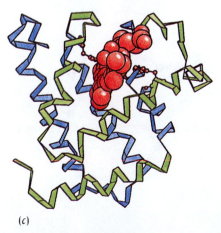

(c)

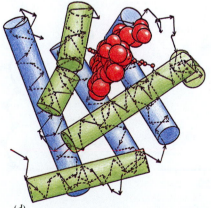

(d)

Figure 2.9 (a) The structure of myoglobin displaying all atoms as small circles connected by straight lines. Even though only side chains at the surface of the molecule are shown, the picture contains so many atoms that such a two-dimensional representation is very confusing and very little information can be gained from it. (b–d) Computer-generated schematic diagrams at different degrees of simplification of the structure of myoglobin. [(a) Half of a stereo diagram by H.C. Watson, *Prog. Stereochem.* 4: 299–333, 1969, by permission of Plenum Press. (b–d) From Arthur Lesk, Protein Architecture: A Practical Approach, 1991, Oxford University Press.]

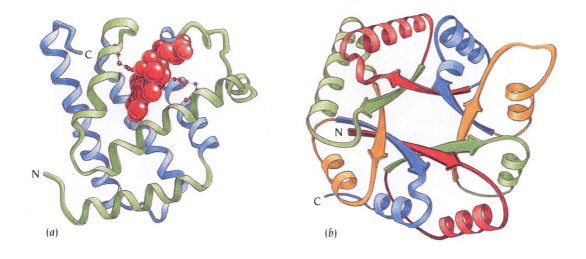

(a)   (b)

meaningful information from a flat picture of such a model, mainly because it is difficult to see the relationships between the secondary structural elements. Since these elements dominate the structure, the picture becomes clearer if they are simplified and highlighted in some way. This is usually done by representing the path of the polypeptide chain by three different symbols: cylinders for α helices; arrows for β strands, which give the direction of the strands from amino to carboxy end; and ribbons for the remaining parts. Such schematic diagrams give good and very useful overall views of protein structures but, of course, give no detailed information. Details are best studied on graphic display systems where one can manipulate the computer-generated model on the screen; in this way the power of the graphics device makes it possible for the viewer to study much greater detail intelligibly.

The Kinemage Supplement, produced by Jane Richardson at Duke University, is available from the publisher as a complement to this book for readers with access to a personal computer. The system is easy to use and provides interactive three-dimensional viewing of many of the structures discussed in various chapters.

Jane Richardson has also made a very popular collection of **schematic diagrams** of various protein molecules with an artistic touch that gives an aesthetic impression without losing too much accuracy. Arthur Lesk at the MRC Laboratory of Molecular Biology in Cambridge, UK, and Karl Hardman at IBM pioneered the use of computer programs to generate schematic diagrams on a computer display from a list of atomic coordinates of the main-chain atoms. Figures 2.9b–d and 2.10 show representative examples of both Lesk-type and Richardson-type diagrams.

Topology diagrams are useful for classification of protein structures

It is very convenient for some purposes to have an even more simplified schematic representation of the secondary structure elements, especially for β sheets. The most characteristic features of a β sheet are the number of strands, their relative directions (parallel or antiparallel), and how the strands are connected along the polypeptide chain (the strand order). This information can be easily conveyed through simple diagrams of connected arrows like those in Figure 2.11, where such simple **topology diagrams** are compared to the more elaborate Richardson diagrams. The twist of the β sheet is not represented in these topology diagrams. They are, nevertheless, very helpful when used to compare β structures and to analyze and present data in computerized database searches of similar structures. Such topology diagrams will be used frequently in this book.

Figure 2.10 Examples of schematic diagrams of the type pioneered by Jane Richardson. Diagram (a) illustrates the structure of myoglobin in the same orientation as the computer-drawn diagrams of Figures 2.9b–d. Diagram (b), which is adapted from J. Richardson, illustrates the structure of the enzyme triosephosphate isomerase, determined to 2.5 Å resolution in the laboratory of David Phillips, Oxford University. Such diagrams can easily be obtained from databases of protein structures, such as PDB, SCOP or CATH, available on the World Wide Web.

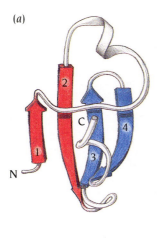

(a)

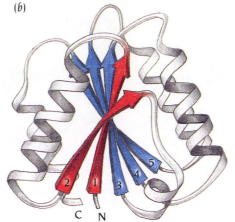

(b)

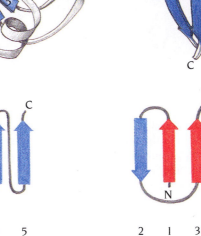

(c)

Secondary structure elements are connected to form simple motifs

Simple combinations of a few secondary structure elements with a specific geometric arrangement have been found to occur frequently in protein structures. These units have been called either supersecondary structures or **motifs.** We will use the term "motif" throughout this book. Some of these motifs can be associated with a particular function such as DNA binding; others have no specific biological function alone but are part of larger structural and functional assemblies.

The simplest motif with a specific function consists of two α helices joined by a loop region. Two such motifs, each with its own characteristic geometry and amino acid sequence requirements, have been observed as parts of many protein structures (Figure 2.12).

One of these motifs, called the helix-turn-helix motif, is specific for DNA binding and is described in detail in Chapters 8 and 9. The second motif is specific for calcium binding and is present in parvalbumin, calmodulin, troponin-C, and other proteins that bind calcium and thereby regulate cellular activities. This calcium-binding motif was first found in 1973 by Robert Kretsinger, University of Virginia, when he determined the structure of parvalbumin to 1.8 Å resolution.

Parvalbumin is a muscle protein with a single polypeptide chain of 109 amino acids. Its function is uncertain, but calcium binding to this protein probably plays a role in muscle relaxation. The helix-loop-helix motif appears three times in this structure, in two of the cases there is a calcium-binding site. Figure 2.13 shows this motif which is called an **EF hand** because the fifth and sixth helices from the amino terminus in the structure of parvalbumin, which were labeled E and F, are the parts of the structure that were originally used to illustrate calcium binding by this motif. Despite this trivial origin, the name has remained in the literature.

Figure 2.11 Beta sheets are usually represented simply by arrows in topology diagrams that show both the direction of each β strand and the way the strands are connected to each other along the polypeptide chain. Such topology diagrams are here compared with more elaborate schematic diagrams for different types of β sheets. (a) Four strands. Antiparallel β sheet in one domain of the enzyme aspartate transcarbamoylase. The structure of this enzyme has been determined to 2.8 Å resolution in the laboratory of William Lipscomb, Harvard University. (b) Five strands. Parallel β sheet in the redox protein flavodoxin, the structure of which has been determined to 1.8 Å resolution in the laboratory of Martha Ludwig, University of Michigan. (c) Eight strands. Antiparallel barrel in the electron carrier plastocyanin. This is a closed barrel where the sheet is folded such that β strands 2 and 8 are adjacent. The structure has been determined to 1.6 Å resolution in the laboratory of Hans Freeman in Sydney, Australia. (Adapted from J. Richardson.)

24

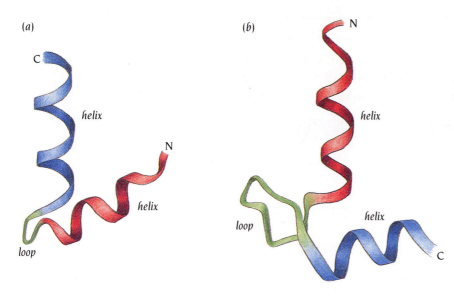

(a)

C

helix

N

helix

loop

(b)

N

helix

loop

helix

C

Figure 2.12 Two α helices that are connected by a short loop region in a specific geometric arrangement constitute a helix-turn-helix motif. Two such motifs are shown: the DNA-binding motif (a), which is further discussed in Chapter 8, and the calcium-binding motif (b), which is present in many proteins whose function is regulated by calcium.

The loop region between the two α helices binds the calcium atom. Carboxyl side chains from Asp and Glu, main-chain C'=O and H_2O form the ligands to the metal atom (see Figure 2.13b). Thus both the specific main-chain conformation of the loop and specific side chains are required to provide the function of this motif. The helix-loop-helix motif provides a scaffold that holds the calcium ligands in the proper position to bind and release calcium.

When Kretsinger analyzed in detail the structure of the calcium-binding motif in parvalbumin in 1973, he deduced a set of constraints that an amino acid sequence must conform to in order to form such a motif, one of the first to be recognized in protein structures. The motif comprises two α helices, E and F, that flank a loop of 12 contiguous residues. Five of the loop residues are calcium ligands, and their side chains should contain an oxygen atom and preferably be Asp or Glu (Table 2.2). Residue 6 of the loop must be a glycine because the side chain of any other residue would disturb the structure of the motif. Finally, a number of side chains form a hydrophobic core between the α helices and thus must be hydrophobic. Applying these constraints, Kretsinger predicted that several different calcium-binding proteins could form this motif. Among these proteins were the important muscle

Figure 2.13 Schematic diagrams of the calcium-binding motif. (a) The calcium-binding motif is symbolized by a right hand. Helix E (red) runs from the tip to the base of the forefinger. The flexed middle finger corresponds to the green loop region of 12 residues that binds calcium (pink). Helix F (blue) runs to the end of the thumb. (b) The calcium atom is bound to one of the motifs in the muscle protein troponin-C through six oxygen atoms: one each from the side chains of Asp (D) 9, Asn (N) 11, and Asp (D) 13; one from the main chain of residue 15; and two from the side chain of Glu (E) 20. In addition, a water molecule (W) is bound to the calcium atom. (c) Schematic diagram illustrating that the structure of troponin-C is built up from four EF motifs—colored as in (a). Two of these bind Ca (pink balls) in the molecules that were used for the structure determination. (Adapted from a diagram by J. Richardson in O. Herzberg and M. James, *Nature* 313: 653–659, 1985.)

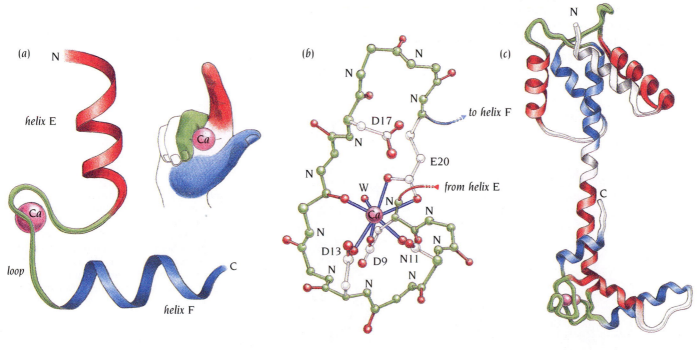

(a)

N

helix E

Ca

loop

Ca

helix F

C

(b)

N

N

N

D17

N

to helix F

N

E20

W

from helix E

N

N

N

D13

N

N

N

D9 N11

N

N

(c)

N

C

Table 2.2 Amino acid sequences of calcium-binding EF motifs in three different proteins

Parvalbumin	V K K A F A I I D Q D K S G F I E E D E L K L F L Q N F
Calmodulin	F K E A F S L F D K D G D G T I T T K E L G T V M R S L
Troponin-C	L A D C F R I F D K N A D G F I D I E E L G E I L R A T

E helix	loop	F helix

Calcium-binding residues are orange, and residues that form the hydrophobic core of the motif are light green. The helix-loop-helix region shown underneath is colored as in Figure 2.13.

protein troponin-C as well as calmodulin, which regulates a variety of cellular functions by calcium binding. Subsequent structure determinations of these two proteins have shown that Kretsinger's prediction was correct. The structure of one of these, troponin-C, which was determined by Osnat Herzberg in the laboratory of Michael James in Edmonton, Canada, to 2.0 Å resolution, is shown in Figure 2.13. Functional aspects of calcium binding to calmodulin and the dramatic structural change of this molecule when it binds to target peptides are discussed in Chapter 6.

The hairpin β motif occurs frequently in protein structures

The simplest motif involving β strands, simpler than the α-helical calcium-binding motif, is two adjacent antiparallel strands joined by a loop. This motif, which is called either a hairpin or a β-β unit, occurs quite frequently; it is present in most antiparallel β structures both as an isolated ribbon and as part of more complex β sheets. There is a strong preference for β strands to be adjacent in β sheets when they are adjacent in the amino acid sequence and thus to form a **hairpin β** motif (β hairpin for short). The lengths of the loop regions between the β strands vary but are generally from two to five residues long (see Figure 2.8). There is no specific function associated with this motif.

Figure 2.14 shows examples of both cases, an isolated ribbon and a β sheet. The isolated ribbon is illustrated by the structure of bovine trypsin inhibitor (Figure 2.14a), a small, very stable polypeptide of 58 amino acids that inhibits the activity of the digestive protease trypsin. The structure has been determined to 1.0 Å resolution in the laboratory of Robert Huber in Munich, Germany, and the folding pathway of this protein is discussed in Chapter 6. Hairpin motifs as parts of a β sheet are exemplified by the structure of a snake venom, erabutoxin (Figure 2.14b), which binds to and inhibits

(a)

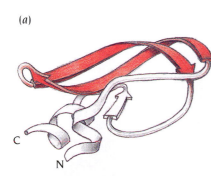

(b)

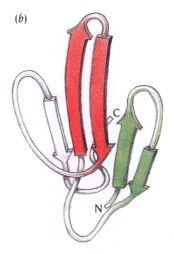

Figure 2.14 The hairpin motif is very frequent in β sheets and is built up from two adjacent β strands that are joined by a loop region. Two examples of such motifs are shown. (a) Schematic diagram of the structure of bovine trypsin inhibitor. The hairpin motif is colored red. (b) Schematic diagram of the structure of the snake venom erabutoxin. The two hairpin motifs within the β sheet are colored red and green. (Adapted from J. Richardson.)

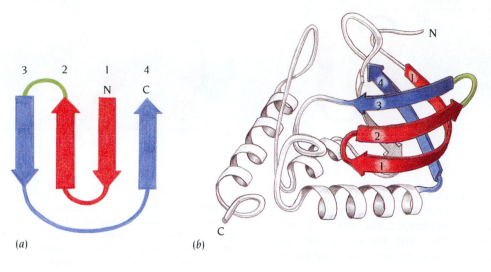

(a)

(b)

Figure 2.15 The Greek key motif is found in antiparallel β sheets when four adjacent β strands are arranged in the pattern shown as a topology diagram in (a). The motif occurs in many β sheets and is exemplified here by the enzyme *Staphylococcus* nuclease (b). The four β strands that form this motif are colored red and blue. The structure of this enzyme was determined to 1.5 Å resolution in the laboratory of Al Cotton at MIT. (Adapted from J. Richardson.)

the acetylcholine receptor in nerve cells. The structure has been determined to 1.4 Å resolution in the laboratory of Barbara Low at Columbia University. The core of this structure is a β sheet of five strands that contains two hairpin motifs and one additional β strand.

The Greek key motif is found in antiparallel β sheets

Four adjacent antiparallel β strands are frequently arranged in a pattern similar to the repeating unit of an ornamental pattern, or fret, used in ancient Greece, which is now called a Greek key. In proteins the motif is therefore called a **Greek key motif**, and Figure 2.15 shows an example of such a motif in the structure of *Staphylococcus* nuclease, an enzyme that degrades DNA. The Greek key motif is not associated with any specific function, but it occurs frequently in protein structures.

The Greek key motifs have been analyzed in detail, and it has been suggested that their frequent occurrence compared with other arrangements of four antiparallel β strands is based on an initial formation of one long antiparallel structure with loops in the middle of both β strands, as shown in Figure 2.16. By structural changes in the loop regions between β strands 1 and 2 and between β strands 3 and 4, the top part folds down so that β strand 2 associates with β strand 1. Beta strands 1 and 2 then form hydrogen bonds, and the Greek key motif is thus formed.

The β-α-β motif contains two parallel β strands

The hairpin motif is a simple and frequently used way to connect two antiparallel β strands, since the connected ends of the β strands are close together at the same edge of the β sheet. How are parallel β strands connected? If two adjacent strands are consecutive in the amino acid sequence, the two ends that must be joined are at opposite edges of the β sheet. The polypeptide chain must cross the β sheet from one edge to the other and connect the next β strand close to the point where the first β strand started. Such crossover connections are frequently made by α helices. The polypeptide chain must turn twice using loop regions, and the motif that is formed is thus a β strand followed by a loop, an α helix, another loop, and, finally, the second β strand.

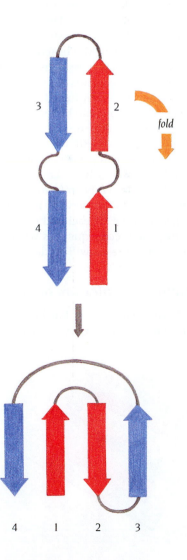

Figure 2.16 Suggested folding pathway from a hairpinlike structure to the Greek key motif. Beta strands 2 and 3 fold over such that strand 2 is aligned adjacent and antiparallel to strand 1. The topology diagram of the Greek key shown here is the same as in Figure 2.15a but rotated 180° in the plane of the page.

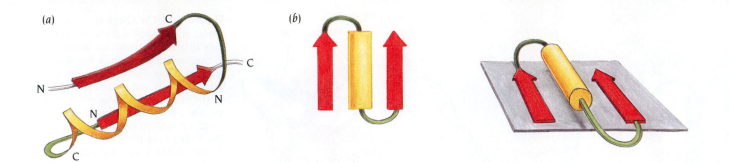

(a) (b)

This motif is called a **beta-alpha-beta motif** (Figure 2.17) and is found as part of almost every protein structure that has a parallel β sheet. For example, the molecule shown in Figure 2.10b, triosephosphate isomerase, is entirely built up by repeated combinations of this motif, where two successive motifs share one β strand. Alternatively, it can be regarded as being built up from four consecutive β-α-β-α motifs.

The α helix in the β-α-β motif connects the carboxy end of one β strand with the amino end of the next β strand (see Figure 2.17) and is usually oriented so that the helical axis is approximately parallel to the β strands. The α helix packs against the β strands and thus shields the hydrophobic residues of the β strands from the solvent. The β-α-β motif thus consists of two parallel β strands, an α helix, and two loop regions (except in a few cases where the connection between the two parallel β strands is not an α helix but a polypeptide chain of irregular structure). The loop regions can be of very different lengths, from one or two residues to over a hundred. The two loops have different functions. The loop (dark green in Figure 2.17) that connects the carboxy end of the β strand with the amino end of the α helix is often involved in forming the functional binding site, or active site, of these structures. These loop regions thus usually have conserved amino acid sequences in homologous proteins. In contrast, the other loop (light green in Figure 2.17) has not yet been found to contribute to an active site.

The β-α-β motif can be regarded as a loose helical turn from one β strand, around the connection, and into the next β strand. The motif can thus in principle have two different "hands" (Figure 2.18). Essentially every β-α-β motif in the known protein structures has been found to have the same hand as a right-handed α helix and therefore is called right-handed. No convincing explanation has been found for this regularity, even though it is the only general rule that describes how three secondary structure elements are arranged relative to each other. This handedness has important structural and functional consequences when several of these motifs are linked into a domain structure, as will be described in Chapter 4.

Protein molecules are organized in a structural hierarchy

The Danish biochemist Kai Linderstrøm-Lang coined the terms "primary," "secondary," and "tertiary" structure to emphasize the structural hierarchy in

Figure 2.17 Two adjacent parallel β strands are usually connected by an α helix from the C-terminus of strand 1 to the N-terminus of strand 2. Most protein structures that contain parallel β sheets are built up from combinations of such β-α-β motifs. Beta strands are red, and α helices are yellow. Arrows represent β strands, and cylinders represent helices. (a) Schematic diagram of the path of the main chain. (b) Topological diagrams of the β-α-β motif.

(a) (b)

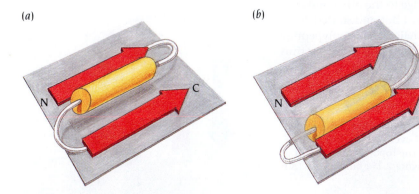

Figure 2.18 The β-α-β motif can in principle have two "hands." (a) This connection with the helix above the sheet is found in almost all proteins and is called right-handed because it has the same hand as a right-handed α helix. (b) The left-handed connection with the helix below the sheet.

proteins (see Figure 1.1). **Primary structure** is the amino acid sequence, or, in other words, the arrangement of amino acids along a linear polypeptide chain. Two different proteins that have significant similarities in their primary structures are said to be homologous to each other, and since their corresponding DNA sequences also are significantly similar, it is generally assumed that the two proteins are evolutionarily related, that they have evolved from a common ancestral gene.

Secondary structure occurs mainly as α helices and β strands. The formation of secondary structure in a local region of the polypeptide chain is to some extent determined by the primary structure. Certain amino acid sequences favor either α helices or β strands; others favor formation of loop regions. Secondary structure elements usually arrange themselves in simple motifs, as described earlier. Motifs are formed by packing side chains from adjacent α helices or β strands close to each other.

Several motifs usually combine to form compact globular structures, which are called **domains.** In this book we will use the term **tertiary structure** as a common term both for the way motifs are arranged into domain structures and for the way a single polypeptide chain folds into one or several domains. In all cases examined so far it has been found that if there is significant amino acid sequence homology in two domains in different proteins, these domains have similar tertiary structures.

Protein molecules that have only one chain are called **monomeric** proteins. But a fairly large number of proteins have a **quaternary structure**, which consists of several identical polypeptide chains (subunits) that associate into a **multimeric** molecule in a specific way. These subunits can function either independently of each other or cooperatively so that the function of one subunit is dependent on the functional state of other subunits. Other protein molecules are assembled from several different subunits with different functions; for example, RNA polymerase from *E. coli* contains five different polypeptide chains.

Large polypeptide chains fold into several domains

The fundamental unit of tertiary structure is the domain. A domain is defined as a polypeptide chain or a part of a polypeptide chain that can fold independently into a stable tertiary structure. Domains are also units of function. Often, the different domains of a protein are associated with different functions. For example, in the lambda repressor protein, discussed in Chapter 8, one domain at the N-terminus of the polypeptide chain binds DNA, while a second domain at the C-terminus contains a site necessary for the dimerization of two polypeptide chains to form the dimeric repressor molecule.

Proteins may comprise a single domain or as many as several dozen domains (Figure 2.19). There is no fundamental structural distinction

Figure 2.19 Organization of polypeptide chains into domains. Small protein molecules like the epidermal growth factor, EGF, comprise only one domain. Others, like the serine proteinase chymotrypsin, are arranged in two domains that are required to form a functional unit (see Chapter 11). Many of the proteins that are involved in blood coagulation and fibrinolysis, such as urokinase, factor IX, and plasminogen, have long polypeptide chains that comprise different combinations of domains homologous to EGF and serine proteinases and, in addition, calcium-binding domains and Kringle domains.

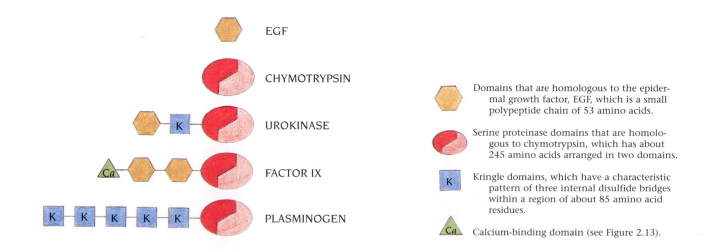

Domains that are homologous to the epidermal growth factor, EGF, which is a small polypeptide chain of 53 amino acids.

Serine proteinase domains that are homologous to chymotrypsin, which has about 245 amino acids arranged in two domains.

Kringle domains, which have a characteristic pattern of three internal disulfide bridges within a region of about 85 amino acid residues.

Calcium-binding domain (see Figure 2.13).

between a domain and a subunit, there are many known examples where several biological functions that are carried out by separate polypeptide chains in one species are performed by domains of a single protein in another species. For example, synthesis of fatty acids requires catalysis of seven different chemical reactions. In plant chloroplasts these reactions are catalyzed by seven different proteins, whereas in mammals they are performed by one polypeptide chain arranged in seven domains with short linker regions between the domains. Such differences thus reflect the organization of the genome rather than the dictates of structure.

Domains are built from structural motifs

Domains are formed by different combinations of secondary structure elements and motifs. The α helices and β strands of the motifs are adjacent to each other in the three-dimensional structure and connected by loop regions. Sequentially adjacent motifs, or motifs that are formed from consecutive regions of the primary structure of a polypeptide chain, are usually close together in the three-dimensional structure (Figure 2.20). Thus to a first approximation a polypeptide chain can be considered as a sequential arrangement of these simple motifs. The number of such combinations found in proteins is limited, and some combinations seem to be structurally favored. Thus similar domain structures frequently occur in different proteins with different functions and with completely different amino acid sequences.

Simple motifs combine to form complex motifs

Figure 2.21 illustrates the 24 possible ways in which two adjacent β hairpin motifs, each consisting of two antiparallel β strands connected by a loop region, can be combined to make a more complex motif.

A survey of all known structures in 1991 showed that only those eight arrangements shown in Figure 2.21a occurred either as a complete β sheet or as a fragment of a β sheet with more than four strands. The number of times that these complex motifs occurred were 65, 29, 23, 11, 9, 3, 2, 1 for (i) to

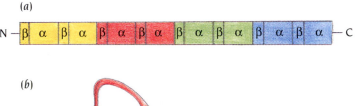

(a)

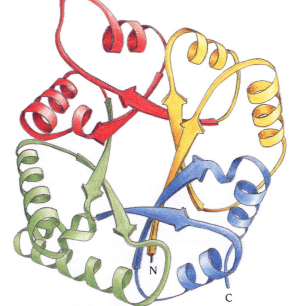

(b)

Figure 2.20 Motifs that are adjacent in the amino acid sequence are also usually adjacent in the three-dimensional structure. Triosephosphate isomerase is built up from four β-α-β-α motifs that are consecutive both in the amino acid sequence (a) and in the three-dimensional structure (b).

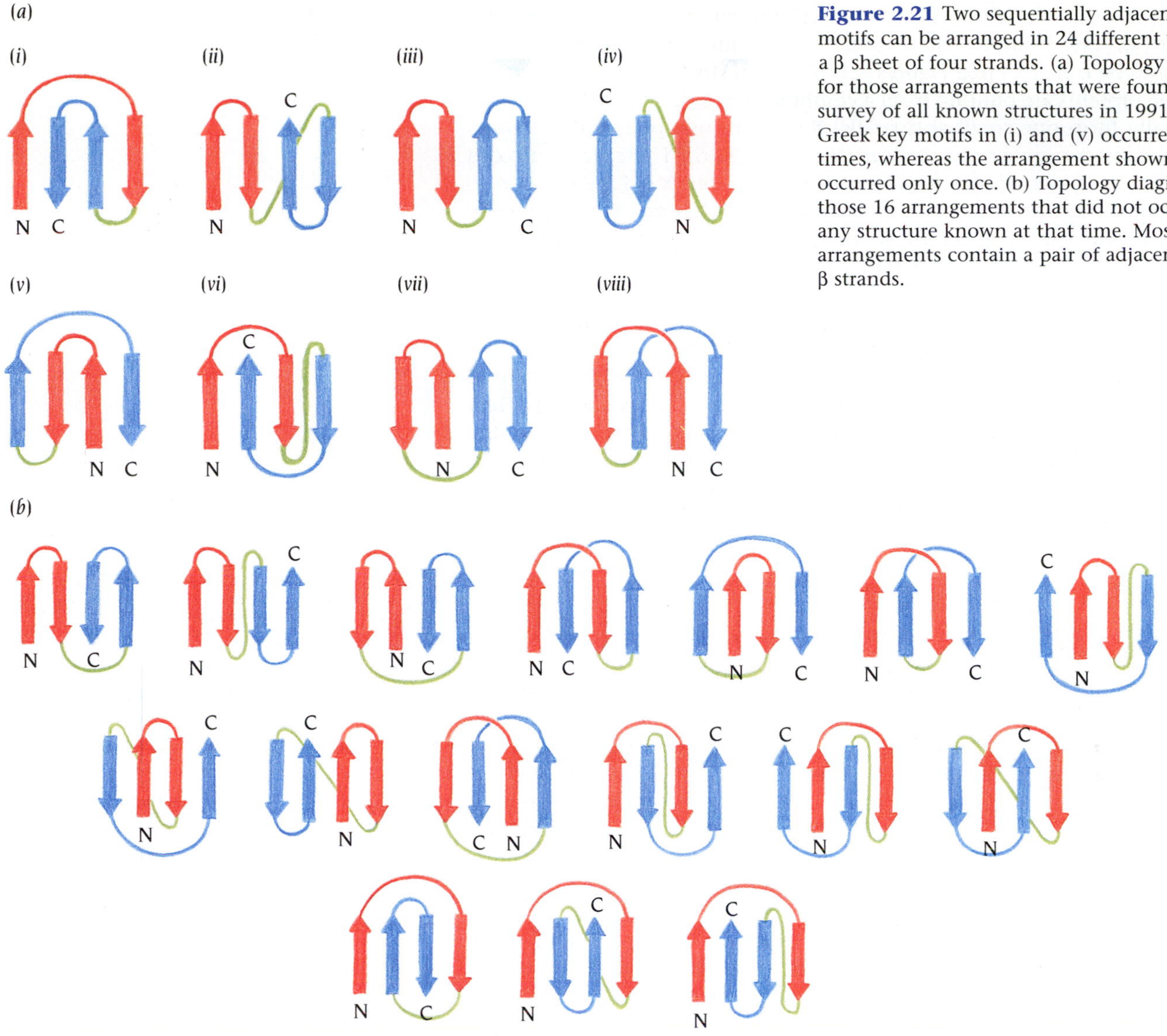

(a)

(i) (ii) (iii) (iv)

(v) (vi) (vii) (viii)

(b)

Figure 2.21 Two sequentially adjacent hairpin motifs can be arranged in 24 different ways into a β sheet of four strands. (a) Topology diagrams for those arrangements that were found in a survey of all known structures in 1991. The Greek key motifs in (i) and (v) occurred 74 times, whereas the arrangement shown in (viii) occurred only once. (b) Topology diagrams for those 16 arrangements that did not occur in any structure known at that time. Most of these arrangements contain a pair of adjacent parallel β strands.

(viii), respectively. The Greek key motifs shown in (i) and (v) thus occur more frequently than all the other possible motifs together. It is also apparent that two hairpin motifs strongly prefer to combine in such a way that all four strands become antiparallel, rather than being arranged with two adjacent parallel β strands. Of 143 β sheets with 4 strands, 138 were antiparallel and only 5 contained 2 adjacent parallel β strands.

The 16 possible motifs shown in Figure 2.21b never occur. Even if this database is limited compared with the universe of existing proteins, the survey clearly demonstrates that a few topological arrangements occur much more frequently than others, and that most possible complex motifs never occur or occur only in a few cases. In 1995 a preliminary survey of a much larger database of known structures yielded the same basic conclusions.

Protein structures can be divided into three main classes

On the basis of simple considerations of connected motifs, Michael Levitt and Cyrus Chothia of the MRC Laboratory of Molecular Biology derived a taxonomy of protein structures and have classified domain structures into three main groups: α domains, β domains, and α/β domains. In α structures the core is built up exclusively from α helices (see Figure 2.9); in β structures the core comprises antiparallel β sheets and are usually two β sheets packed against each other (see Figure 2.11c). The α/β structures are made from combinations of β-α-β motifs that form a predominantly parallel β sheet surrounded by α helices (see Figures 2.10b and 2.11b).

Some proteins are built up from a combination of discrete α and β motifs and usually form one small antiparallel β sheet in one part of the domain packed against a number of α helices (see Figure 2.15). These structures can be considered to belong to a small fourth group called α + β. In addition to these groups, there are a number of small proteins that are rich in disulfide bonds or metal atoms and form a special group. The structures of these proteins seem to be strongly influenced by the presence of these metals or disulfides and often look like distorted versions of more regular proteins.

In this book the domains of the known protein structures are classified according to Levitt and Chothia's scheme. The three main classes—α, β, and α/β—will be examined in more detail in Chapters 3, 4, and 5. The group of Cyrus Chothia has constructed a database in which all available protein structures are arranged according to this hierarchy. Within each class the structures are arranged in superfamilies according to their tertiary structure and, within the superfamilies, in families according to function and sequence homology. This database is freely available on the World Wide Web.

Conclusion

The interiors of protein molecules contain mainly hydrophobic side chains. The main chain in the interior is arranged in secondary structures to neutralize its polar atoms through hydrogen bonds. There are two main types of secondary structure, α helices and β sheets. Beta sheets can have their strands parallel, antiparallel, or mixed.

Protein structures are built up by combinations of secondary structural elements, α helices, and β strands. These form the core regions—the interior of the molecule—and they are connected by loop regions at the surface. Schematic and simple topological diagrams where these secondary structure elements are highlighted are very useful and are frequently used. Alpha helices or β strands that are adjacent in the amino acid sequence are also usually adjacent in the three-dimensional structure. Certain combinations, called motifs, occur very frequently, including the helix-loop-helix motif and the hairpin motif. A DNA-binding helix-loop-helix motif and a calcium-binding helix-loop-helix motif, each with its own specific geometry and amino acid sequence requirements, are used in many different proteins.

The β-α-β motif, which consists of two parallel β strands joined by an α helix, occurs in almost all structures that have a parallel β sheet. Four antiparallel β strands that are arranged in a specific way comprise the Greek key motif, which is frequently found in structures with antiparallel β sheets.

Polypeptide chains are folded into one or several discrete units, domains, which are the fundamental functional and three-dimensional structural units. The cores of domains are built up from combinations of small motifs of secondary structure, such as α-loop-α, β-loop-β, or β-α-β motifs. Domains are classified into three main structural groups: α structures, where the core is built up exclusively from α helices; β structures, which comprise antiparallel β sheets; and α/β structures, where combinations of β-α-β motifs form a predominantly parallel β sheet surrounded by α helices.

Selected readings

General

Chothia, C. Principles that determine the structure of proteins. *Annu. Rev. Biochem.* 53: 537–572, 1984.

Doolittle, R.F. Proteins. *Sci. Am.* 253: 88–99, 1985.

Hardie, D.G., Coggins, J.R. *Multidomain Proteins: Structure and Evolution.* Amsterdam: Elsevier, 1986.

Janin, J., Chothia, C. Domains in proteins: definitions, location and structural principles. *Methods Enzymol.* 115: 420–430, 1985.

Klotz, I.M., et al. Quaternary structure of proteins. *Annu. Rev. Biochem.* 39: 25–62, 1970.

Lesk, A.M. Themes and contrasts in protein structures. *Trends Biochem. Sci.* 9: June V, 1984.

Levitt, M., Chothia, C. Structural patterns in globular proteins. *Nature* 261: 552–558, 1976.

Matthews, B.W., Bernhard, S.A. Structure and symmetry of oligomeric enzymes. *Annu. Rev. Biophys. Bioeng.* 2: 257–317, 1973.

Richardson, J.S. Describing patterns of protein tertiary structure. *Methods Enzymol.* 115: 349–358, 1985.

Richardson, J.S. Schematic drawings of protein structures. *Methods Enzymol.* 115: 359–380, 1985.

Richardson, J.S. The anatomy and taxonomy of protein structure. *Adv. Prot. Chem.* 34: 167–339, 1981.

Rossmann, M.G., Argos, P. Protein folding. *Annu. Rev. Biochem.* 50: 497–532, 1981.

Schulz, G.E. Protein differentiation: emergence of novel proteins during evolution. *Angew. Chem.,* int. ed. 20: 143–151, 1981.

Schulz, G.E. Structural rules for globular proteins. *Angew. Chem.,* int. ed. 16: 23–33, 1977.

Strynadka, N.C.J., James, M.N.G. Crystal structures of the helix-loop-helix calcium-binding proteins. *Annu. Rev. Biochem.* 58: 951–998, 1989.

Specific structures

Adams, M.J., et al. Structure of lactate dehydrogenase at 2.8 Å resolution. *Nature* 227: 1098–1103, 1970.

Baba, Y.S., et al. Three-dimensional structure of calmodulin. *Nature* 315: 37–40, 1985.

Banner, B.W., et al. Structure of chicken muscle triose phosphate isomerase determined crystallographically at 2.5 Å resolution using amino acid sequence data. *Nature* 255: 609–614, 1975.

Bourne, P.E., et al. Erabutoxin b. Initial protein refinement and sequence analysis at 0.140 nm resolution. *Eur. J. Biochem.* 153: 521–527, 1985.

Burnett, R.M., et al. The structure of oxidized form of clostridial flavodoxin at 1.9 Å resolution. *J. Biol. Chem.* 249: 4383–4392, 1974.

Chothia, C. Structural invariants in protein folding. *Nature* 254: 304–308, 1975.

Chothia, C., Levitt, M., Richardson, D. Structure of proteins: packing of α-helices and pleated sheets. *Proc. Natl. Acad. Sci. USA* 74: 4130–4134, 1977.

Chou, P.Y., Fasman, G.D. β-turns in proteins. *J. Mol. Biol.* 115: 135–175, 1977.

Colman, P., et al. X-ray crystal structure analysis of plastocyanin at 2.7 Å resolution. *Nature* 272: 319–324, 1978.

Crawford, J.L., Lipscomb, W.N., Schellmann, C.G. The reverse turn as a polypeptide conformation in globular proteins. *Proc. Natl. Acad. Sci. USA* 70: 538–542, 1973.

Efimov, A.V. Stereochemistry of α-helices and β-sheet packing in compact globule. *J. Mol. Biol.* 134: 23–40, 1979.

Eklund, H., et al. Three-dimensional structure of horse liver alcohol dehydrogenase at 2.4 Å resolution. *J. Mol. Biol.* 102: 27–59, 1976.

Gouaux, J.E., Lipscomb, W.N. Crystal structures of phosphonoacetamide ligated T and phosphono-acetamide and malonate ligated R states of aspartate carbamoyltransferase at 2.8 Å resolution and neutral pH. *Biochemistry* 29: 389–402, 1990.

Herzberg, O., James, M.N.G. Structure of the calcium regulatory muscle protein troponin-C at 2.8 Å resolution. *Nature* 313: 653–659, 1985.

Hol, W.G.J., van Duijnen, P.T., Berendsen, H.J.C. The α-helix dipole and the properties of proteins. *Nature* 273: 443–446, 1978.

Holmgren, A., et al. Three-dimensional structure of *E. coli* thioredoxin-S_2 to 2.8 Å resolution. *Proc. Natl. Acad. Sci. USA* 72: 2305–2309, 1975.

Jones, A., Thirup, S. Using known substructures in protein model building and crystallography. *EMBO J.* 5: 819–822, 1986.

Kendrew, J.C. The three-dimensional structure of a protein molecule. *Sci. Am.* 205: 96–110, 1961.

Kendrew, J.C., et al. A three-dimensional model of the myoglobin molecule obtained by x-ray analysis. *Nature* 181: 662–666, 1958.

Kendrew, J.C., et al. Structure of myoglobin. *Nature* 185: 422–427, 1960.

Koch, I., Kaden, F., Selbig, J. Analysis of protein sheet topologies by graph theoretical methods. *Prot: Struc. Func. Genet.* 12: 314–323, 1992.

Kretsinger, R.H. Structure and evolution of calcium-modulated proteins. *CRC Crit. Rev. Biochem.* 8: 119–174, 1980.

Lesk, A.M., Hardman, K.D. Computer-generated pictures of proteins. *Methods Enzymol.* 115: 381–390, 1985.

Levitt, M. Conformational preferences of amino acids in globular proteins. *Biochemistry* 17: 4277–4285, 1978.

Matthews, B.W., Rossmann, M.G. Comparison of protein structures. *Methods Enzymol.* 115: 397–420, 1985.

Milner-White, E.J., Poet, R. Loops, bulges, turns and hairpins in proteins. *Trends Biochem. Sci.* 12: 189–192, 1987.

Moews, P.C., Kretsinger, R.H. Refinement of the structure of carp muscle calcium-binding parvalbumin by model building and difference Fourier analysis. *J. Mol. Biol.* 91: 201–228, 1975.

Park, C.H., Tulinsky, A. Three-dimensional structure of the Kringle sequence: structure of prothrombin fragment 1. *Biochemistry* 25: 3977–3982, 1986.

Pauling, L., Corey, R.B. Configurations of polypeptide chains with favored orientations around single bonds: two new pleated sheets. *Proc. Natl. Acad. Sci. USA* 37: 729–740, 1951.

Pauling, L., Corey, R.B., Branson, H.R. The structure of proteins: two hydrogen-bonded helical configurations of the polypeptide chain. *Proc. Natl. Acad. Sci. USA* 37: 205–211, 1951.

Perutz, M.F. Electrostatic effects in proteins. *Science* 201: 1187–1191, 1978.

Perutz, M.F. New x-ray evidence on the configuration of polypeptide chains. Polypeptide chains in poly-g-benzyl-t-glutamate, keratin and haemoglobin. *Nature* 167: 1053–1054, 1951.

Perutz, M.F., et al. Structure of haemoglobin. A three-dimensional Fourier synthesis at 5.5 Å resolution, obtained by x-ray analysis. *Nature* 185: 416–422, 1960.

Rao, S.T., Rossmann, M.G. Comparison of supersecondary structures in proteins. *J. Mol. Biol.* 76: 241–256, 1973.

Remington, S.J., Matthews, B.W. A systematic approach to the comparison of protein structures. *J. Mol. Biol.* 140: 77–99, 1980.

Richards, F.M. Calculation of molecular volumes and areas for structures of known geometry. *Methods Enzymol.* 115: 440–464, 1985.

Rose, G.D. Automatic recognition of domains in globular proteins. *Methods Enzymol.* 115: 430–440, 1985.

Rose, G.D. Prediction of chain turns in globular proteins on a hydrophobic basis. *Nature* 272: 586–590, 1978.

Rose, G.D., Roy, S. Hydrophobic basis of packing in globular proteins. *Proc. Natl. Acad. Sci. USA* 77: 4643–4647, 1980.

Rose, G.D., Young, W.B., Gierasch, L.M. Interior turns in globular proteins. *Nature* 304: 654–657, 1983.

Sibanda, B.L., Thornton, J.M. β-hairpin families in globular proteins. *Nature* 316: 170–174, 1985.

Tucker, P.W., Hazen, E.E., Cotton, F.A. Staphylococcal nuclease reviewed: a prototypic study in contemporary enzymology. III. Correlation of the three-dimensional structure with the mechanisms of enzymatic action. *Mol. Cell. Biochem.* 23: 67–86, 1979.

Venkatachalam, C.M. Stereochemical criteria for polypeptides and proteins. V. Conformation of a system of three linked peptide units. *Biopolymers* 6: 1425–1436, 1968.

Wiegand, G., et al. Crystal structure analysis and molecular model of a complex of citrate synthase with oxaloacetate and S-acetonyl-coenzyme A. *J. Mol. Biol.* 174: 205–219, 1984.

Wlodawer, A., Deisenhofer, J., Huber, R. Comparison of two highly refined structures of bovine pancreatic trypsin inhibitor. *J. Mol. Biol.* 193: 145–156, 1987.

Wright, C.S., Alden, R., Kraut, J. Structure of subtilisin BPN' at 2.5 Å resolution. *Nature* 221: 235–242, 1969.

Alpha-Domain Structures

The first globular protein structure that was determined, myoglobin, belongs to the class of alpha- (α-) domain structures. The structure illustrated in Figure 2.9 is called the globin fold and is a representative example of one class of α domains in proteins; short α helices, the building blocks, are connected by loop regions and packed together to produce a hydrophobic core. Packing interactions within the core hold the helices together in a stable globular structure, while the hydrophilic residues on the surface make the protein soluble in water. In this chapter we will describe some of the different α-domain structures in soluble proteins.

Alpha helices are sufficiently versatile to produce many very different classes of structures. In membrane-bound proteins, the regions inside the membranes are frequently α helices whose surfaces are covered by hydrophobic side chains suitable for the hydrophobic environment inside the membranes. Membrane-bound proteins are described in Chapter 12. Alpha helices are also frequently used to produce structural and motile proteins with various different properties and functions. These can be typical fibrous proteins such as keratin, which is present in skin, hair, and feathers, or parts of the cellular machinery such as fibrinogen or the muscle proteins myosin and dystrophin. These α-helical proteins will be discussed in Chapter 14.

Coiled-coil α helices contain a repetitive heptad amino acid sequence pattern

Despite its frequent occurrence in proteins an isolated α helix is only marginally stable in solution. Alpha helices are stabilized in proteins by being packed together through hydrophobic side chains. The simplest way to achieve such stabilization is to pack two α helices together. As early as 1953 Francis Crick showed that the side-chain interactions are maximized if the two α helices are not straight rods but are wound around each other in a supercoil, a so-called **coiled-coil** arrangement (Figure 3.1). Coiled-coils are the basis for some of the fibrous proteins we shall discuss in Chapter 14. Coiled-coils in fibers can extend over many hundreds of amino acid residues

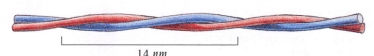

14 nm

Figure 3.1 Schematic diagram of the coiled-coil structure. Two α helices are intertwined and gradually coil around each other.

$$
\begin{array}{ccccccc}
a & b & c & d & e & f & g \\
\end{array}
$$

NH$_2$ - Met - Lys - Gln - **Leu** - Glu - Asp - Lys -
Val - Glu - Glu - **Leu** - Leu - Ser - Lys -
Asn - Tyr - His - **Leu** - Glu - Asn - Glu -
Val - Ala - Arg - **Leu** - Lys - Lys - Leu - COOH

(a)

(b)

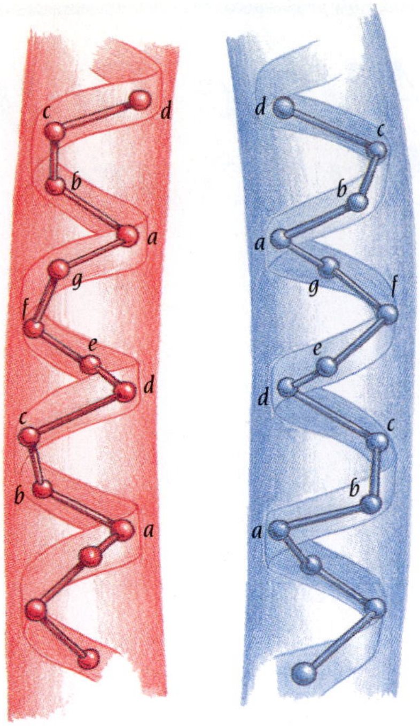

Figure 3.2 Repetitive pattern of amino acids in a coiled-coil α helix. (a) The amino acid sequence of the transcription factor GCN4 showing a heptad repeat of leucine residues. Within each heptad the amino acids are labeled a–g. (b) Schematic diagram of one heptad repeat in a coiled-coil structure showing the backbone of the polypeptide chain. The α helices in the coiled-coil are slightly distorted so that the helical repeat is 3.5 residues rather than 3.6, as in a regular helix. There is therefore an integral repeat of seven residues along the helix.

to produce long, flexible dimers that contribute to the strength and flexibility of the fibers. Much shorter coiled-coils are used in some transcription factors to promote or prevent formation of homo- and heterodimers, as we shall discuss in Chapter 10.

Crick showed that a left-handed supercoil of two right-handed α helices reduces the number of residues per turn in each helix from 3.6 to 3.5 so that the pattern of side-chain interactions between the helices repeats every seven residues, that is, after two turns. This is reflected in the amino acid sequences of polypeptide chains that form α-helical coiled-coils. Such sequences are repetitive with a period of seven residues, the **heptad repeat**. The amino acid residues within one such heptad repeat are usually labeled a–g (Figure 3.2a), and one of these, the d-residue, is hydrophobic, usually a leucine or an isoleucine. When two α helices form a coiled-coil structure the side chains of these d-residues pack against each other every second turn of the α helices (Figure 3.2b). The hydrophobic region between the α helices is completed by the a-residues, which are frequently hydrophobic and also pack against each other (Figure 3.3). Residues "e" and "g," which border the hydrophobic core (see Figure 3.2b), frequently are charged residues. The side chains of these residues provide ionic interactions (salt bridges) between the α helices that define the relative chain alignment and orientation (Figure 3.4).

The repetitive heptad amino acid sequence pattern required for a coiled-coil structure can be identified in computer searches of amino acid sequence databases. Heptad repeats provide strong indications of α-helical coiled-coil structures, and they have been found in a number of different proteins with very diverse functions. Fibrinogen, which plays an essential role in blood coagulation; some RNA- and DNA-binding proteins; the class of cell-surface recognition proteins called collectins; both spectrin and dystrophin, which link actin molecules; and the muscle protein myosin all contain heptad repeats and therefore coiled-coil α helices. An illustrative example is provided by GCN4, a DNA-binding protein. GCN4 contains one region of α helix, the leucine zipper region, and its dimerization is accomplished by the formation of an α-helical coiled-coil with the leucine zipper regions of two subunits. The structure and DNA-binding function of this protein are described in Chapter 10.

Detailed structure determinations of GCN4 and other coiled-coil proteins have shown that the α helices pack against each other according to the "knobs in holes" model first suggested by Francis Crick (Figure 3.5). Each side chain in the hydrophobic region of one of the α helices can contact four side chains from the second α helix. The side chain of a residue in position "d"

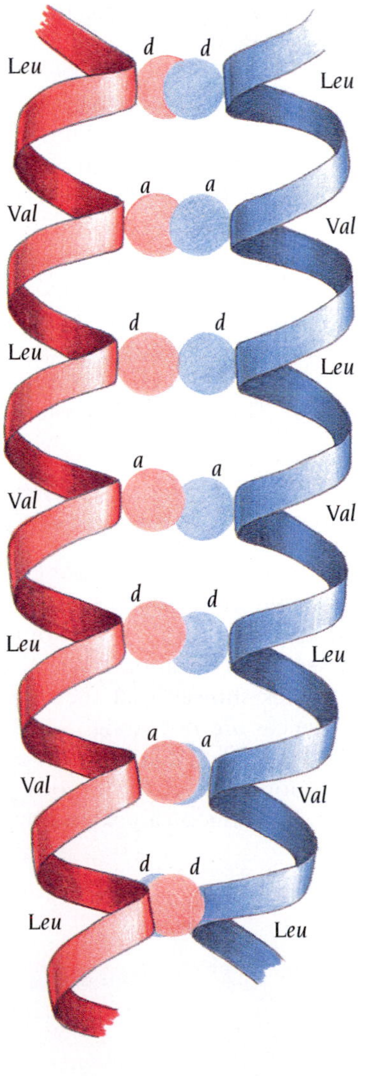

Figure 3.3 Schematic diagram showing the packing of hydrophobic side chains between the two α helices in a coiled-coil structure. Every seventh residue in both α helices is a leucine, labeled "d." Due to the heptad repeat, the d-residues pack against each other along the coiled-coil. Residues labeled "a" are also usually hydrophobic and participate in forming the hydrophobic core along the coiled-coil.

36

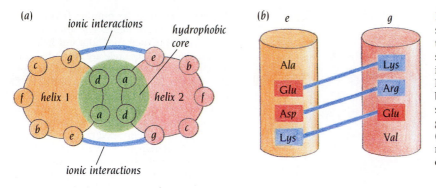

(a) ionic interactions
hydrophobic core
helix 1
helix 2
ionic interactions

(b) e
g

Figure 3.4 Salt bridges can stabilize coiled-coil structures and are sometimes important for the formation of heterodimeric coiled-coil structures. The residues labeled "e" and "g" in the heptad sequence are close to the hydrophobic core and can form salt bridges between the two α helices of a coiled-coil structure, the e-residue in one helix with the g-residue in the second and vice versa. (a) Schematic view from the top of a heptad repeat. (b) Schematic view from the side of a coiled-coil structure.

in one helix is directed into a hole at the surface of the second helix surrounded by one d-residue, two a-residues, and one e-residue, with numbers n, $n - 3$, $n + 4$, and $n + 1$, respectively. The two helices are aligned in such a way that the two d-residues, frequently leucines or isoleucines, face each other (see Figure 3.3).

The four-helix bundle is a common domain structure in α proteins

Two α helices packed together into a coiled-coil are building blocks within a domain or a fiber but are not sufficient to form a complete domain. The simplest and most frequent α-helical domain consists of four α helices arranged in a bundle with the helical axes almost parallel to each other. A schematic representation of the structure of the **four-helix bundle** is shown in Figure 3.6a. The side chains of each helix in the four-helix bundle are arranged so that hydrophobic side chains are buried between the helices and hydrophilic side chains are on the outer surface of the bundle (Figure 3.6b). This arrangement creates a hydrophobic core in the middle of the bundle along its length, where the side chains are so closely packed that water is excluded.

The four-helix bundle occurs in several widely different proteins, such as myohemerythrin (an oxygen-transport protein in marine worms that does not contain heme iron), cytochrome c' and cytochrome b562 (heme-containing electron carriers) (Figure 3.7a), ferritin (a storage molecule for iron atoms in eucaryotic cells), and the coat protein of tobacco mosaic virus. In these examples, sequentially adjacent α helices are always antiparallel. However, four-helix bundles can also be formed with different topological arrangements of the α helices. In human growth hormone (Figure 3.7b), a four-helix bundle is formed from two pairs of parallel α helices that are joined in an

Figure 3.5 Schematic diagram of packing side chains in the hydrophobic core of coiled-coil structures according to the "knobs in holes" model. The positions of the side chains along the surface of the cylindrical α helix is projected onto a plane parallel with the helical axis for both α helices of the coiled-coil. (a) Projected positions of side chains in helix 1. (b) Projected positions of side chains in helix 2. (c) Superposition of (a) and (b) using the relative orientation of the helices in the coiled-coil structure. The side-chain positions of the first helix, the "knobs," superimpose between the side-chain positions in the second helix, the "holes." The green shading outlines a d-residue (leucine) from helix 1 surrounded by four side chains from helix 2, and the brown shading outlines an a-residue (usually hydrophobic) from helix 1 surrounded by four side chains from helix 2.

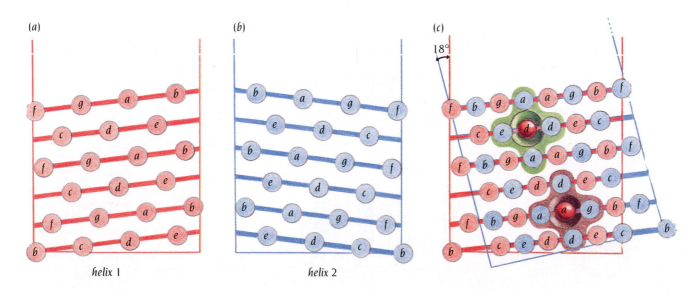

(a) helix 1

(b) helix 2

(c)

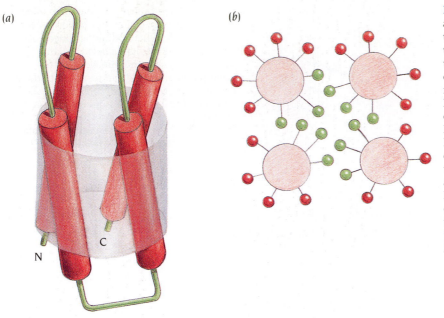

(a)

(b)

Figure 3.6 Four-helix bundles frequently occur as domains in α proteins. The arrangement of the α helices is such that adjacent helices in the amino acid sequence are also adjacent in the three-dimensional structure. Some side chains from all four helices are buried in the middle of the bundle, where they form a hydrophobic core. (a) Schematic representation of the path of the polypeptide chain in a four-helix-bundle domain. Red cylinders are α helices. (b) Schematic view of a projection down the bundle axis. Large circles represent the main chain of the α helices; small circles are side chains. Green circles are the buried hydrophobic side chains; red circles are side chains that are exposed on the surface of the bundle, which are mainly hydrophilic. [(a) Adapted from P.C. Weber and F.R. Salemme, *Nature* 287: 82–84, 1980.]

antiparallel fashion. The interaction of this hormone with its receptor is described in Chapter 13.

In most four-helix bundle structures, including those shown in Figure 3.7, the α helices are packed against each other according to the "ridges in grooves" model discussed later in this chapter. However, there are also examples where coiled-coil dimers packed by the "knobs in holes" model participate in four-helix bundle structures. A particularly simple illustrative example is the Rop protein, a small RNA-binding protein that is encoded by certain plasmids and is involved in plasmid replication. The monomeric subunit of Rop is a polypeptide chain of 63 amino acids built up from two

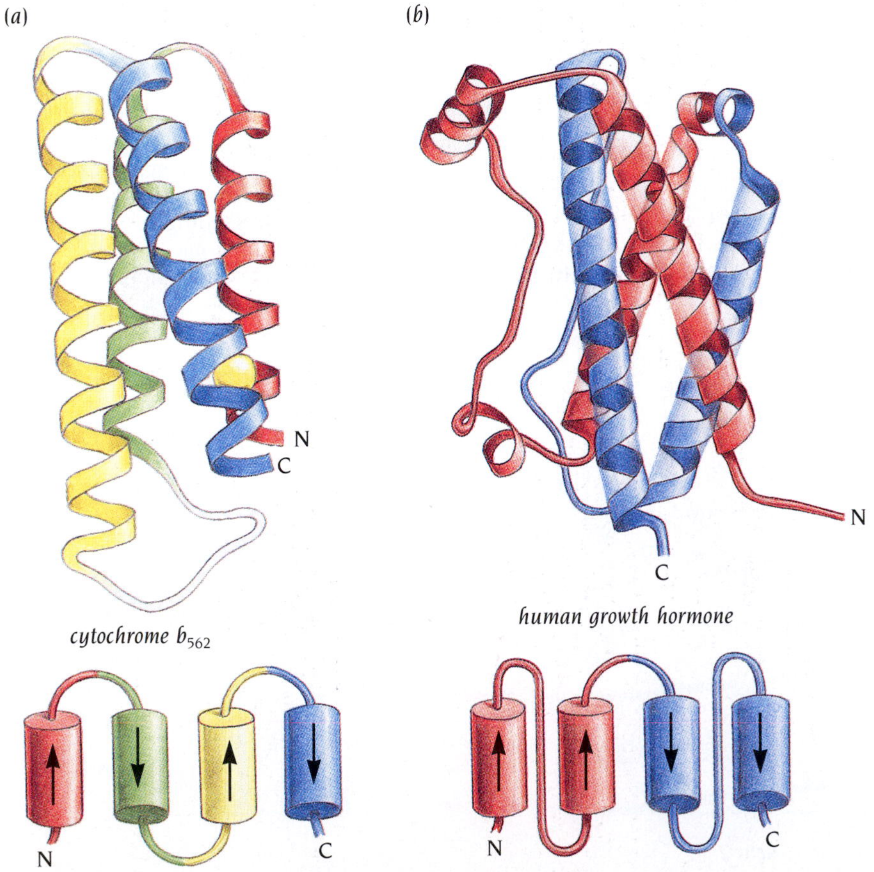

(a)

(b)

cytochrome b₅₆₂

human growth hormone

Figure 3.7 The polypeptide chains of cytochrome b₅₆₂ and human growth hormone both form four-helix-bundle structures. In cytochrome b₅₆₂ (a) adjacent helices are antiparallel, whereas the human growth hormone (b) has two pairs of parallel α helices joined in an antiparallel fashion.

antiparallel α helices joined by a short loop of three amino acids. The structure of Rop was determined by David Banner at EMBL, Heidelberg, Germany.

The two α helices of the Rop subunit are arranged as an antiparallel coiled-coil in which the hydrophobic side chains are packed against each other according to the "knobs in holes" model. Two such subunits, each with the same structure, form the dimeric Rop molecule in which the subunits are arranged as a bundle of four α helices with their long axes aligned (Figure 3.8). The two dimers pack against each other according to the "ridges in grooves" model. The helix-loop-helix (HLH) family of transcription factors, discussed in Chapter 10, is another example of a four-helix bundle structure involving coiled-coil helices.

Alpha-helical domains are sometimes large and complex

The structures of several enzymes are known in which a long polypeptide chain of 300–400 amino acids is arranged in more than 20 α helices packed together in a complex pattern to form a globular domain. One such enzyme is a bacterial muramidase that is involved in the metabolism of peptidoglycans, which form part of the bacterial cell wall. The structure of this enzyme was determined by Bauke Dijkstra and colleagues in Groningen, Netherlands, as a basis for the design of specific inhibitors to the enzyme, which might lead eventually to novel types of antibacterial drugs.

The polypeptide chain of this monomeric enzyme has 618 amino acids, of which the N-terminal 450 residues form one α-helical domain. This domain is built up from 27 α helices arranged in a two-layered ring with a right-handed superhelical twist (Figure 3.9). The ring has a large central hole, like in a doughnut, with a diameter of about 30 Å. The remaining residues form the catalytic domain that lies on top of the ring. The function of the

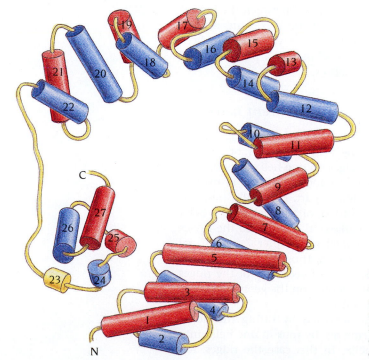

Figure 3.9 Schematic diagram of the structure of one domain of a bacterial muramidase, comprising 450 amino acid residues. The structure is built up from 27 α helices arranged in a two-layered ring. The ring has a large central hole, like a doughnut, with a diameter of about 30 Å.

doughnut-shaped domain is not known, but its shape may be required for the specificity of the catalytic reaction *in vivo*.

The globin fold is present in myoglobin and hemoglobin

One of the most important α structures is the **globin fold**. This fold has been found in a large group of related proteins, including myoglobin, hemoglobins, and the light-capturing assemblies in algae, the phycocyanins. The functional and evolutionary aspects of these structures will not be discussed in this book; instead, we will examine some features that are of general structural interest.

The pairwise arrangements of the sequential α helices in the globin fold are quite different from the antiparallel organization found in the four-helix-bundle α structures. The globin structure is a bundle of eight α helices, usually labeled A–H, connected by rather short loop regions and arranged so that the helices form a pocket for the active site, which in myoglobin and the hemoglobins binds a heme group (Figure 3.10). The lengths of the α helices vary considerably, from 7 residues in the shortest helix (C) to 28 in the longest helix (H) in myoglobin. In the globin fold the α helices wrap around the core in different directions so that sequentially adjacent α helices are usually not adjacent to each other in the structure. The only exceptions are the last two α helices (G and H), which form an antiparallel pair with extensive packing interactions between them. All other packing interactions are formed between pairs of α helices that are not sequentially adjacent. Because the globin fold is not built up from an assembly of smaller motifs, it is quite difficult to visualize conceptually in spite of its relatively small size and simplicity.

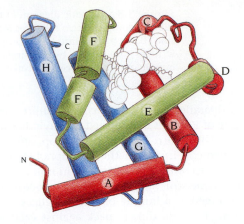

Figure 3.10 Schematic diagram of the globin domain. The eight α helices are labeled A–H. A–D are red, E and F green, and G and H blue. The heme group is shown in white. (Adapted from originals provided by A. Lesk.)

Geometric considerations determine α-helix packing

When we compare the arrangements of the α helices in coiled-coil structures (see Figure 3.1), in the four-helix-bundle structure (see Figure 3.8), and in the globin fold (see Figure 3.10), it is obvious that the geometry of α-helix packing is quite different. We described earlier the way that the side chains of coiled-coil α helices pack according to the "knobs in holes" model. In contrast, other α-helical structures pack their α helices according to a "ridges in grooves" model. In the four-helix bundle the α helices pack almost parallel, or antiparallel, to each other, with an angle of about 20° between the helical axes. In the globin fold the angles between the helical axes are usually larger, in most cases around 50°. These are the two main ways that α helices pack against each other in the "ridges in grooves" model, a packing motif dictated by the geometry of the surfaces of α helices.

Ridges of one α helix fit into grooves of an adjacent helix

Since the side chains of an α helix are arranged in a helical row along the surface of the helix, they form ridges separated by shallow furrows, or grooves, on the surface. Alpha helices pack with the ridges on one helix packing into the grooves of the other and vice versa. The ridges and grooves are formed by amino acids that are usually three or four residues apart. This is illustrated in Figure 3.11, which shows slices through the surface of a polyalanine α helix on which the directions of the ridges are marked. In contrast to the ridges and grooves of the DNA double helix described in Chapter 7, which are formed by the sugar-phosphate main-chain atoms, those of an α helix are formed by the amino acid side chains. The detailed geometry of the ridges and grooves of an α helix is thus dependent not only on the geometry of the helix but also on the actual amino acid sequence.

The most common way of packing α helices is by fitting the ridges formed by a row of residues separated in sequence by four in one helix into the same type of grooves in the other helix. In this case the ridges and

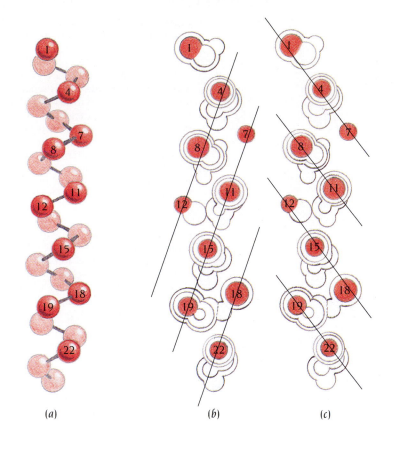

Figure 3.11 The side chains on the surface of an α helix form ridges separated by grooves, as schematically illustrated here. (a) An α helix with each residue represented by the first atom in the side chain, $C_β$. (b) The surface relief of a polyalanine α helix in the orientation shown in (a). Sections are cut through a space-filling model and superimposed. The residue numbers are placed on the side-chain atom. The ridges caused by the side chains separated by four residues are shown as lines. (c) The same as (b), but here the ridges are caused by side chains separated by three residues.

(a)   (b)   (c)

grooves form an angle of about 25° to the helical axis. In order to pack the two helices shown in Figure 3.12a (red and blue) against each other, one of these (the blue in Figure 3.12a) must be turned around 180° out of the plane of the paper and placed on top of the other (red). In the interface between the two α helices the directions of the ridges and grooves are then on opposite sides of the vertical axis, as illustrated in Figure 3.12a. The α helices must thus be inclined by an angle of about 50° (25° + 25°) in order for the ridges of one helix to fit into the grooves of the other and vice versa. This is the type of packing of several of the helix-helix interactions in the globin fold, and in many other helical structures.

In the second frequently occurring packing mode the ridges formed by amino acids three residues apart fit into the grooves of amino acids four residues apart and vice versa. The direction of the first type of ridge forms an angle of about 45° to the helical axis, whereas the other type makes an angle of about 25° to the axis in the opposite direction (Figure 3.12b). In the interface, however, after one helix has been rotated 180°, these directions are on the same side of the helical axis. Thus an inclination of about 20° (45°–25°) between the two α helices will fit these ridges and grooves into each other. Some four-helix-bundle structures (see Figure 3.8b) have this mode of packing.

These two rules for fitting ridges into grooves are quite general: they apply to most packing interactions between α helices, and they explain the geometrical arrangements of adjacent α helices observed in many protein structures.

The globin fold has been preserved during evolution

The three-dimensional structures of globin domains from many diverse sources, including mammals, insects, and plant root nodules, have been determined independently of each other. All these domains have amino acid

(a)

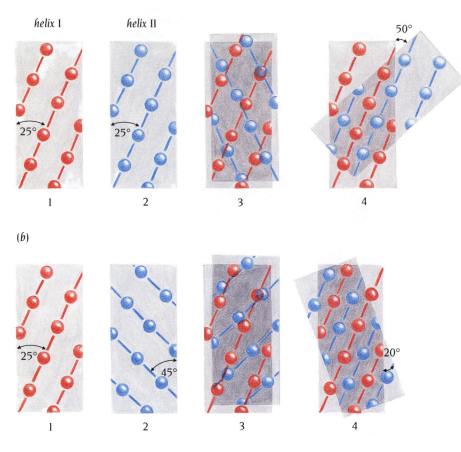

helix I   helix II

25°   25°   50°

1   2   3   4

(b)

25°   45°   20°

1   2   3   4

Figure 3.12 By fitting the ridges of side chains from one helix into the grooves between side chains of the other helix and vice versa, α helices pack against each other. (a) Two α helices, I and II, with ridges from side chains separated by four residues marked in red and blue, respectively. Panels 1 and 2 are the same view of the two α helices. In panel 3 the blue α helix is turned over through 180° in order to form an interface with the red α helix. In panel 4 the orientation of the helices has been rotated 50° in order to pack the ridges of one α helix into the grooves of the other. (b) In the red α helix the ridges are formed by side chains separated by four residues and in the blue α helix by three residues. The α helices are rotated 20° in order to pack ridges into grooves, in a direction opposite that in (a). (Adapted from C. Chothia et al., *Proc. Natl. Acad. Sci. USA* 74: 4130–4134, 1977.)

sequence homologies that range from 99% to 16% in pairwise comparisons, but they all share the same essential features of the globin fold. This family of structures is thus the prime example of a situation where natural selection has produced proteins whose amino acid sequences have diverged widely (although some homology is usually still recognizable) but whose three-dimensional structure has been essentially preserved.

Arthur Lesk and Cyrus Chothia at the MRC Laboratory of Molecular Biology in Cambridge, UK, compared the family of globin structures with the aim of answering two general questions: How can amino acid sequences that are very different form proteins that are very similar in their three-dimensional structure? What is the mechanism by which proteins adapt to mutations in the course of their evolution?

The hydrophobic interior is preserved

To answer the first question, Lesk and Chothia examined in detail residues at structurally equivalent positions that are involved in helix-heme contacts and in packing the α helices against each other. After comparing the nine globin structures then known, the 59 positions they found that fulfilled these criteria were divided into 31 positions buried in the interior of the protein and 28 in contact with the heme group. These positions are the principal determinants of both the function and the three-dimensional structure of the globin family.

One might expect these positions to exhibit a higher degree of amino acid conservation and hence sequence identity than the rest of the molecule. This is not, however, the case for distantly related molecules that have low sequence identity and derive from distantly related species. The sequence identity of these residues is no greater than in the rest of the molecule. Since the important residues involved in packing the α helices are not conserved, we used to assume that the changes which have occurred compensate each other in size. This is not the case either. The volumes occupied by the 31 buried residues vary considerably between individual members. Thus neither conserved sequence nor size-compensatory mutations in the hydrophobic core are important factors in preserving three-dimensional structure during evolution. We now know that this is also true for other proteins, such as the immunoglobulins.

Lesk and Chothia did find, however, that there is a striking preferential conservation of the hydrophobic character of the amino acids at the 59 buried positions, but that no such conservation occurs at positions exposed on the surface of the molecule. With a few exceptions on the surface, hydrophobic residues have replaced hydrophilic ones and vice versa. However, the case of sickle-cell hemoglobin, which is described below, shows that a charge balance must be preserved to avoid hydrophobic patches on the surface. In summary, the evolutionary divergence of these nine globins has been constrained primarily by an almost absolute conservation of the hydrophobicity of the residues buried in the helix-to-helix and helix-to-heme contacts.

Helix movements accommodate interior side-chain mutations

Lesk and Chothia also found a simple answer to the question of how proteins adapt to changes in size of buried residues. The mode of packing the α helices is the same in all the globin structures: the same types of packing of ridges into grooves occur in corresponding α helices in all these structures. However, the relative positions and orientations of the α helices change to accommodate changes in the volume of side chains involved in the packing.

The proteins thus adapt to mutations of buried residues by changing their overall structure, which in the globins involves movements of entire α helices relative to each other. The structure of loop regions changes so that the movement of one α helix is not transmitted to the rest of the structure. Only movements that preserve the geometry of the heme pocket are accepted. Mutations that cause such structural shifts are tolerated because many different combinations of side chains can produce well-packed helix-helix interfaces of similar but not identical geometry and because the shifts are coupled so that the geometry of the active site is retained.

Sickle-cell hemoglobin confers resistance to malaria

Sickle-cell anemia is the classic example of an inherited disease that is caused by a change in a protein's amino acid sequence. Linus Pauling proposed in 1949 that it was caused by a defect in the hemoglobin molecule; he thus coined the term **molecular disease.** Seven years later Vernon Ingram showed that the disease was caused by a single mutation, a change in residue 6 of the β chain of hemoglobin from Glu to Val.

Hemoglobin is a tetramer built up of two copies each of two different polypeptide chains, α- and β-globin chains in normal adults. Each of the four chains has the globin fold with a heme pocket. Residue 6 in the β chain is on the surface of α helix A, and it is also on the surface of the tetrameric molecule (Figure 3.13).

The hemoglobin concentration in red blood cells, erythrocytes, is extremely high, 340 mg/ml. This is almost as high as in the crystalline state: the hemoglobin molecules, which are spheroids of dimension $50 \times 55 \times 65$ Å, are on average only 10 Å apart in the cells. It is thus surprising that they can nevertheless rotate and flow past one another. The mutation in sickle-cell hemoglobin converts a charged residue to a hydrophobic residue, and as a result, it produces a hydrophobic patch on the surface. This patch happens to fit and bind a hydrophobic pocket in the deoxygenated form of another

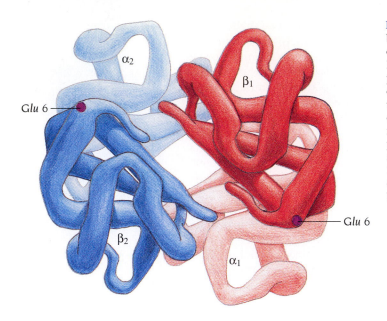

Figure 3.13 The hemoglobin molecule is built up of four polypeptide chains: two α chains and two β chains. Compare this with Figure 1.1 and note that for purposes of clarity parts of the α chains are not shown here. Each chain has a three-dimensional structure similar to that of myoglobin: the globin fold. In sickle-cell hemoglobin Glu 6 in the β chain is mutated to Val, thereby creating a hydrophobic patch on the surface of the molecule. The structure of hemoglobin was determined in 1968 to 2.8 Å resolution in the laboratory of Max Perutz at the MRC Laboratory of Molecular Biology, Cambridge, UK.

hemoglobin molecule (Figure 3.14a). In the oxygenated form of hemoglobin the shape of this pocket is slightly different, and thus no interaction occurs between hemoglobin molecules in the lungs. However, when hemoglobin in the blood capillaries delivers its oxygen, these hydrophobic interactions can occur and the highly concentrated hemoglobin in the cells polymerizes into fibers (Figure 3.14b and c). These fibers stiffen the erythrocytes and deform them into a sickle shape: hence the name sickle-cell anemia. In heterozygotes, where only one β-globin allele is mutated, this occurs only to a minor extent, whereas it is lethal for homozygotes, all of whose hemoglobin molecules carry the mutation.

We thus have here a case where a mutation on the surface of the globin fold, replacing a hydrophilic residue with a hydrophobic one, changes important properties of the molecule and produces a lethal disease. Why has the

Figure 3.14 Sickle-cell hemoglobin molecules polymerize due to the hydrophobic patch introduced by the mutation Glu 6 to Val in the β chain. The diagram (a) illustrates how this hydrophobic patch (green) interacts with a hydrophobic pocket (red) in a second hemoglobin molecule, whose hydrophobic patch interacts with the pocket in a third molecule, and so on. Electron micrographs of sickle-cell hemoglobin fibers are shown in cross-section in (b) and along the fibers in (c). [(b) and (c) from J.T. Finch et al., *Proc. Natl. Acad. Sci. USA* 70: 718–722, 1973.]

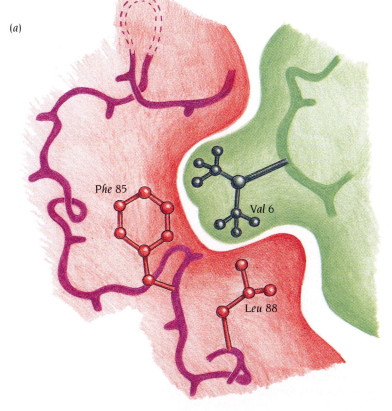

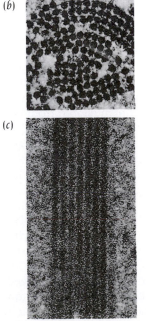

mutation survived during evolution? It turns out that the disease gives increased resistance to malaria which has had a high survival value for heterozygotes, especially in Africa. In evolutionary terms, the death of homozygotes has been an acceptable price to pay for increased survival of heterozygotes in a malarial environment.

Conclusion

Coiled-coil α-helical structures are found both in fibrous proteins and as parts of smaller domains in many globular proteins. Alpha- (α-) domain structures consist of a bundle of α helices that are packed together to form a hydrophobic core. A common motif is the four-helix bundle structure, where four helices are pairwise arranged in either a parallel or an antiparallel fashion and packed against each other. The most intensively studied α structure is the globin fold, which has been found in a large group of related proteins, including myoglobin and hemoglobin. This structure comprises eight α helices that wrap around the core in different directions and form a pocket where the heme group is bound.

Rules have been derived that explain the different geometrical arrangements of α helices observed in α-domain structures. The helix packing in coiled-coil structures is determined by fitting the knobs of side chains in the first helix into holes between side chains in the second helix. For other α-helical structures the helix packing is determined by fitting ridges of side chains along one α helix into grooves between side chains of another helix.

The globin fold has been used to study evolutionary constraints for maintaining structure and function. Evolutionary divergence is primarily constrained by conservation of the hydrophobicity of buried residues. In contrast, neither conserved sequence nor size-compensatory mutations in the hydrophobic core are important. Proteins adapt to mutations in buried residues by small changes of overall structure that in the globins involve movements of entire helices relative to each other.

Selected readings

General

Chothia, C., Lesk, A.M. Helix movements in proteins. *Trends Biochem. Sci.* 13: 116–118, 1985.

Chothia, C., Levitt, M., Richardson, D. Helix-to-helix packing in proteins. *J. Mol. Biol.* 145: 215–250, 1981.

Dickerson, R.E., Geis, I. *Hemoglobin: Structure, Function, Evolution and Pathology.* Menlo Park, CA: Benjamin/Cummings, 1983.

Lesk, A.M., Chothia, C. How different amino acid sequences determine similar protein structures: the structure and evolutionary dynamics of the globins. *J. Mol. Biol.* 136: 225–270, 1980.

Murzin, A.G., Finkelstein, A.V. General architecture of the α-helical globule. *J. Mol. Biol.* 204: 749–769, 1988.

Pauling, L., et al. Sickle cell anemia: a molecular disease. *Science* 110: 543–548, 1949.

Perutz, M.F. Hemoglobin structure and respiratory transport. *Sci. Am.* 239(6): 92–125, 1978.

Perutz, M.F. *Protein Structure: New Approaches to Disease and Therapy.* New York: Freeman, 1992.

Specific structures

Argos, P., Rossmann, M.G., Johnsson, J.E. A four-helical super-secondary structure. *Biochem. Biophys. Res. Comm.* 75: 83–86, 1977.

Banner, D.W., Kokkinidis, M., Tsernoglou, D. Structure of the col1 rop protein at 1.7 Å resolution. *J. Mol. Biol.* 196: 657–675, 1987.

Bashford, D., Chothia, C., Lesk, A.M. Determinants of a protein fold. Unique features of the globin amino acid sequences. *J. Mol. Biol.* 196: 199–216, 1987.

Bloomer, A.C., et al. Protein disk of tobacco mosaic virus at 2.8 Å resolution showing the interactions within and between subunits. *Nature* 276: 362–368, 1978.

Blundell, T., et al. Solvent-induced distortions and the curvature of α-helices. *Nature* 306: 281–283, 1983.

Clegg, G.A., et al. Helix packing and subunit conformation in horse spleen apoferritin. *Nature* 288: 298–300, 1980.

Cohen, C., Parry, D.A.D. Alpha-helical coiled coils—a widespread motif in proteins. *Trends Biochem. Sci.* 11: 245–248, 1986.

Crick, F.H.C. The packing of α-helices: simple coiled coils. *Acta Cryst.* 6: 689–697, 1953.

de Vos, A.M., Ultsch, M., Kossiakoff, A.A. Human growth hormone and extracellular domain of its receptor: crystal structure of the complex. *Science* 255: 306–312, 1992.

Embury, S.H. The clinical pathophysiology of sickle-cell disease. *Annu. Rev. Med.* 37: 361–376, 1986.

Fermi, G., et al. The crystal structure of human deoxy-haemoglobin at 1.74 Å resolution. *J. Mol. Biol.* 175: 159–174, 1984.

Fermi, G., Perutz, M.F. *Atlas of Molecular Structures in Biology. 2. Haemoglobin and Myoglobin.* Oxford, UK: Clarendon Press, 1981.

Finch, J.T., et al. Structure of sickled erythrocytes and of sickle-cell hemoglobin fibers. *Proc. Natl. Acad. Sci. USA* 70: 718–722, 1973.

Finzel, B.C., et al. Structure of ferricytochrome c′ from *Rhodospirillum molischianum* at 1.67 Å resolution. *J. Mol. Biol.* 186: 627–643, 1985.

Ingram, V.M. Gene mutation in human haemoglobin: the chemical difference between normal and sickle cell haemoglobin. *Nature* 180: 326–328, 1957.

Lederer, F., et al. Improvement of the 2.5 Å resolution model of cytochrome b_{562} by redetermining the primary structure and using molecular graphics. *J. Mol. Biol.* 148: 427–448, 1981.

Nordlund, P., Sjoberg, B.-M., Eklund, H. Three-dimensional structure of the free radical protein of ribonucleotide reductase. *Nature* 345: 593–598, 1990.

Pastore, A., et al. Structural alignment and analysis of two distantly related proteins: *Aplysia limacina* myoglobin and sea lamprey globin. *Proteins* 4: 240–250, 1988.

Pastore, A., Lesk, A.M. Comparison of the structures of globins and phycocyanins: evidence for evolutionary relationship. *Proteins* 8: 133–155, 1990.

Phillips, S.E.V. Structure and refinement of oxymyoglobin at 1.6 Å resolution. *J. Mol. Biol.* 142: 531–554, 1980.

Presnell, S.R., Cohen, F.E. Topological distribution of four-α-helix bundles. *Proc. Natl. Acad. Sci. USA* 86: 6592–6596, 1989.

Richmond, T.J., Richards, F.M. Packing of α-helices: geometrical constraints and contact areas. *J. Mol. Biol.* 119: 537–555, 1978.

Sheriff, S., Hendrickson, W.A., Smith, J.L. Structure of myohemerythrin in the azidomet state at 1.7/1.3 Å resolution. *J. Mol. Biol.* 197: 273–296, 1987.

Thunissen, A.-M., et al. Doughnut-shaped structure of a bacterial muramidase revealed by x-ray crystallography. *Nature* 367: 750–753, 1994.

Watson, H.C. The stereochemistry of the protein myoglobin. *Progr. Stereochem.* 4: 299–333, 1969.

Weber, P.C., Salemme, F.R. Structural and functional diversity in 4-α-helical proteins. *Nature* 287: 82–84, 1980.

Alpha/Beta Structures

The most frequent of the domain structures are the alpha/beta (α/β) domains, which consist of a central parallel or mixed β sheet surrounded by α helices. All the glycolytic enzymes are α/β structures as are many other enzymes as well as proteins that bind and transport metabolites. In α/β domains, binding crevices are formed by loop regions. These regions do not contribute to the structural stability of the fold but participate in binding and catalytic action.

Parallel β strands are arranged in barrels or sheets

There are three main classes of α/β proteins. In the first class there is a core of twisted parallel β strands arranged close together, like the staves of a barrel. The α helices that connect the parallel β strands are on the outside of this barrel (Figure 4.1a). This domain structure is often called the **TIM barrel** from the structure of the enzyme triosephosphate isomerase, where it was first observed. The second class contains an open twisted β sheet surrounded by α helices on both sides. A typical example is shown in Figure 4.1b, a nucleotide-binding domain sometimes called the **Rossman fold** after Michael Rossman, Purdue University, who first discovered this fold in the enzyme lactate dehydrogenase in 1970. The third class is formed by amino acid sequences that contain repetitive regions of a specific pattern of leucine residues, so-called **leucine-rich motifs**, which form α helices and β strands. The β strands form a curved parallel β sheet with all the α helices on the outside. The structure of one member of this class, a ribonuclease inhibitor (illustrated in Figure 4.11), is shaped like a horseshoe, and consequently this class is called the **horseshoe fold.**

Barrels, open sheets, and horseshoe structures are all built up from β-α-β motifs. To illustrate how they differ, let us consider two β-α-β motifs: β_1-$\alpha_{1,2}$-β_2 and β_3-$\alpha_{3,4}$-β_4 linked together by helix $\alpha_{2,3}$. There are two fundamentally different ways these two motifs can be connected into a β sheet of four parallel strands, as shown in Figure 4.2. Strand β_3 can be aligned adjacent either to strand β_2, giving the strand order 1 2 3 4, or to strand β_1, giving the strand order 4 3 1 2. In the first case the two β-α-β motifs are joined with the same orientation. Since the β-α-β unit is almost always a right-handed structure, all three α helices (one from each motif and the joining helix) are on the same side, above the β sheet (Figure 4.2a). In barrel and horseshoe structures the β-α-β motifs are linked in this way and consist of consecutive β-α-β units, all in the same orientation.

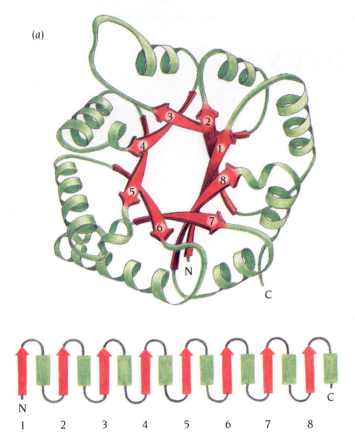

(a)

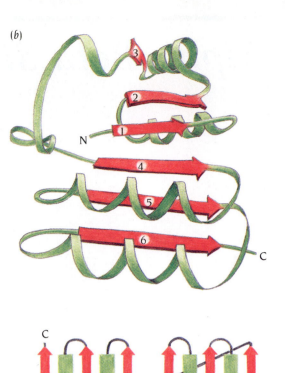

(b)

Figure 4.1 Alpha/beta domains are found in many proteins. They occur in different classes, two of which are shown here: (a) a closed barrel exemplified by schematic and topological diagrams of the enzyme triosephosphate isomerase and (b) an open twisted sheet with helices on both sides, as in the coenzyme-binding domain of some dehydrogenases. Both classes are built up from β-α-β motifs that are linked such that the β strands are parallel. Rectangles represent α helices, and arrows represent β strands in the topological diagrams. [(a) Adapted from J. Richardson. (b) Adapted from B. Furugren.]

In the second case we must turn the second motif around in order to align β strands 1 and 3. As a result of the right-handed structure of the β-α-β motif, its α helix is on the other side of the β sheet (Figure 4.2b). In open twisted β-sheet structures there are always one or more such alignments and therefore there are α helices on both sides of the β sheet. These geometric rules apply because virtually all β-α-β motifs are right-handed. As pointed out in Chapter 2, this is an empirical rule that almost always applies, although no convincing explanation has been found. `

Alpha/beta barrels occur in many different enzymes

In α/β structures where the strand order is 1 2 3 4, all connections are on the same side of the β sheet. An open twisted β sheet of this sort with four or more parallel β strands would leave one side of the parallel β sheet exposed to the solvent and the other side shielded by the α helices. Such a domain structure is rarely observed, except in the horseshoe structure or as part of more complex structures where loop regions, extra α helices, or additional β sheets cover the exposed side of the β sheet. Instead, a closed barrel of twisted β strands is formed with all the connecting α helices on the outside of the barrel, as shown in Figure 4.1a. However, more than four β strands are needed to provide enough staves to form a closed barrel, and almost all the closed α/β barrels observed to date have eight parallel β strands. These are arranged such that β strand 8 is adjacent and hydrogen-bonded to β strand 1. In a few cases the barrels do not have eight parallel β strands; there are also barrels that contain ten parallel β strands and some that contain eight parallel and two antiparallel β strands. In almost all cases the cross-connections between the parallel β strands are α helices; in addition, there is usually an α helix after the last β strand.

The eight-stranded α/β-barrel structure is one of the largest and most regular of all domain structures. A minimum of about 200 residues are required to form this structure. It has been found in many different proteins, most of which are enzymes, with completely different amino acid sequences and

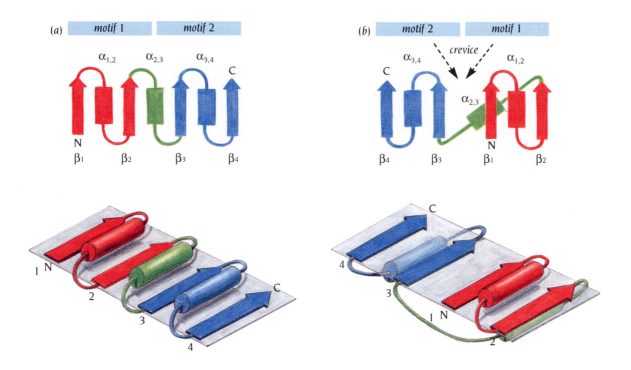

different functions. Superimposing the structures of these proteins shows that around 160 residues are structurally equivalent. These residues form the β strands and α helices. The remaining residues form the loop regions that connect the β strands with the α helices. These loops have quite different lengths and conformations in the different proteins. This reflects the fact that the β strands and α helices form the structural framework of the enzyme, whereas the loops contain the amino acids responsible for its catalytic chemistry. In some cases the loops are very long and form independent domains in the overall subunit structure.

Branched hydrophobic side chains dominate the core of α/β barrels

In barrels the hydrophobic side chains of the α helices are packed against hydrophobic side chains of the β sheet. The α helices are antiparallel and adjacent to the β strands that they connect. Thus the barrel is provided with a shell of hydrophobic residues from the α helices and the β strands.

Since the side chains of consecutive amino acids of a β strand are on opposite sides of the β sheet, every second residue of the β strands contributes to this hydrophobic shell. The other side chains of the β strands point inside the barrel to form a hydrophobic core; this core is therefore comprised exclusively of side chains of β-strand residues (Figure 4.3).

The packing interactions between α helices and β strands are dominated by the residues Val (V), Ile (I), and Leu (L), which have branched hydrophobic side chains. This is reflected in the amino acid composition: these three amino acids comprise approximately 40% of the residues of the β strands in parallel β sheets. The important role that these residues play in packing α helices against β sheets is particularly obvious in α/β-barrel structures, as shown in Table 4.1.

Figure 4.2 A β-α-β motif is a right-handed structure. Two such motifs can be joined into a four-stranded parallel β sheet in two different ways. They can be aligned with the α helices either on the same side of the β sheet (a) or on opposite sides (b). In case (a) the last β strand of motif 1 (red) is adjacent to the first β strand of motif 2 (blue), giving the strand order 1 2 3 4. The motifs are aligned in this way in barrel structures (see Figure 4.1a) and in the horseshoe fold (see Figure 4.11). In case (b) the first β strands of both motifs are adjacent, giving the strand order 4 3 1 2. Open twisted sheets (see Figure 4.1b) contain at least one motif alignment of this kind. In both cases the motifs are joined by an α helix (green).

Figure 4.3 In most α/β-barrel structures the eight β strands of the barrel enclose a tightly packed hydrophobic core formed entirely by side chains from the β strands. The core is arranged in three layers, with each layer containing four side chains from alternate β strands. The schematic diagram shows this packing arrangement in the α/β barrel of the enzyme glycolate oxidase, the structure of which was determined by Carl Branden and colleagues in Uppsala, Sweden.

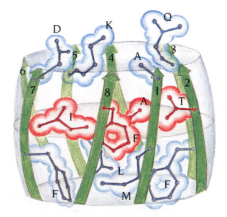

Table 4.1 The amino acid residues of the eight parallel β strands in the barrel structure of the enzyme triosephosphate isomerase from chicken muscle

		Positions				
Strand no.	Residue no.	1	2	3	4	5
1	6–10	Phe	Val	Gly	Gly	Asn
2	37–41	Glu	Val	Val	Cys	Gly
3	59–63	Gly	Val	Ala	Ala	Gln
4	89–93	Trp	Val	Ile	Leu	Gly
5	121–125	Gly	Val	Ile	Ala	Cys
6	158–162	Lys	Val	Val	Leu	Ala
7	204–208	Arg	Ile	Ile	Tyr	Gly
8	227–231	Gly	Phe	Leu	Val	Gly

The sequences are aligned so that residues in positions 1, 3, and 5 point into the barrel and residues in positions 2 and 4 point toward the α helices on the outside and are involved in the hydrophobic interactions between the β strands and the α helices.

Bulky hydrophobic residues from positions 1, 3, and 5 of the β strands fill the interior of the barrel and form a tightly packed hydrophobic core (see Figure 4.3). Note from Table 4.1 that some of these residues are Lys, Arg, or Gln, which have a polar end group (see Panel 1.1, pp. 6–7) terminating a chain of hydrophobic –CH_2 groups. These chains are in the hydrophobic interior and traverse part of the barrel; their polar end groups are on the top or bottom surface of the barrel and are in contact with the aqueous environment. By this arrangement even amino acids that are classified as polar can participate in the formation of hydrophobic cores of compact globular domains through the hydrophobic parts of their side chains.

There is one exception to the rule that requires bulky hydrophobic residues to fill the interior of eight-stranded α/β barrels in order to form a tightly packed hydrophobic core. The coenzyme B_{12}–dependent enzyme methylmalonyl–coenzyme A mutase, the x-ray structure of which was determined by Phil Evans and colleagues at the MRC Laboratory of Molecular

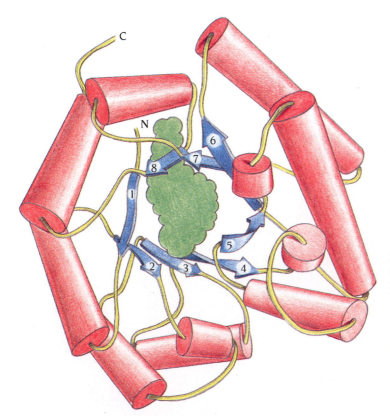

Figure 4.4 Schematic diagram of the structure of the α/β-barrel domain of the enzyme methylmalonyl–coenzyme A mutase. Alpha helices are red, and β strands are blue. The inside of the barrel is lined by small hydrophilic side chains (serine and threonine) from the β strands, which creates a hole in the middle where one of the substrate molecules, coenzyme A (green), binds along the axis of the barrel from one end to the other. (Adapted from a computer-generated diagram provided by P. Evans.)

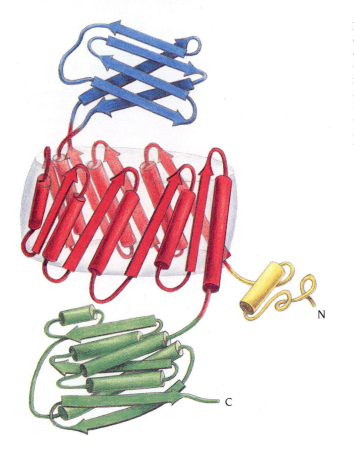

Figure 4.5 The polypeptide chain of the enzyme pyruvate kinase folds into several domains, one of which is an α/β barrel (red). One of the loop regions in this barrel domain is extended and comprises about 100 amino acid residues that fold into a separate domain (blue) built up from antiparallel β strands. The C-terminal region of about 140 residues forms a third domain (green), which is an open twisted α/β structure.

Biology, in Cambridge, UK, has a hole in the middle of its α/β-barrel domain (Figure 4.4). The serine and threonine side chains that project from the β strands into the interior of the barrel are small and polar, and therefore do not fill up the space available inside the barrel. The resulting tunnel through the barrel provides an ideal environment for the catalytic reaction and is sufficiently large for the substrate molecule, methylmalonyl–coenzyme A, to bind. Many enzyme-catalyzed reactions, including the reaction catalyzed by this mutase, require that the reactive part of the substrate molecule be shielded from solvent during the catalytic reaction. When the substrate is bound in the tunnel inside the barrel, it is shielded from the outside world. In α/β barrels with a hydrophobic core, the substrate binds at the surface of the barrel and conformational changes of loop regions shield the substrate from the solvent.

Pyruvate kinase contains several domains, one of which is an α/β barrel

All known eight-stranded α/β-barrel domains have enzymatic functions that include isomerization of small sugar molecules, oxidation by flavin coenzymes, phosphate transfer, and degradation of sugar polymers. In some of these enzymes the barrel domain comprises the whole subunit of the protein; in others the polypeptide chain is longer and forms several additional domains. An enzymatic function in these multidomain subunits, however, is always associated with the barrel domain.

For example, each subunit of the dimeric glycolytic enzyme triosephosphate isomerase (see Figure 4.1a) consists of one such barrel domain. The polypeptide chain has 248 residues in which the first β strand of the barrel starts at residue 6 and the last α helix of the barrel ends at residue 246. In contrast, the subunit of the glycolytic enzyme pyruvate kinase (Figure 4.5), which was solved at 2.6 Å resolution in the laboratory of Hilary Muirhead, Bristol University, UK, is folded into four different domains. The polypeptide chain of this cat muscle enzyme has 530 residues. In Figure 4.5, residues 1–42

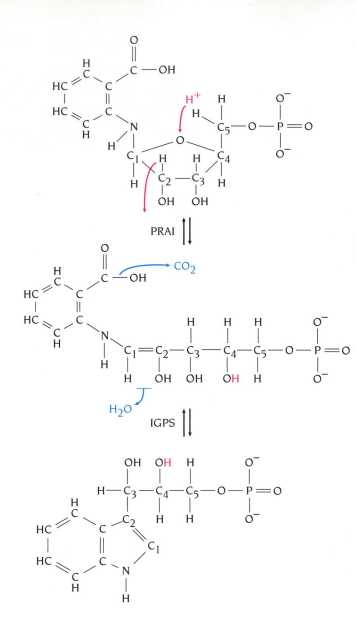

Figure 4.6 The bifunctional enzyme PRA-isomerase (PRAI):IGP-synthase (IGPS) catalyzes two sequential reactions in the biosynthesis of tryptophan. In the first reaction (top half), which is catalyzed by the C-terminal PRAI domain of the enzyme, the substrate N-(5′-phosphoribosyl) anthranilate (PRA) is converted to 1-(o-carboxyphenylamino)-1-deoxyribulose 5-phosphate (CdRP) by a rearrangement reaction. The succeeding step (bottom half), a ring closure reaction from CdRP to indole-3-glycerol phosphate (IGP), is catalyzed by the N-terminal IGPS domain.

form a small domain (yellow) involved in subunit contacts in the tetrameric molecule; residues 43–115 and 224–387 form an α/β-barrel domain (red) that binds substrate and provides the catalytic groups; residues 116–223 loop out from the end of β-strand number 3 in the barrel domain and are folded into a separate domain consisting of an antiparallel β sheet (blue); and finally, residues 388–530 form an open twisted α/β domain (green). This structure illustrates perfectly how a long polypeptide chain can be arranged in domains of different structural types.

Double barrels have occurred by gene fusion

PRA-isomerase:IGP-synthase, a bifunctional enzyme from *E. coli* that catalyzes two reactions in the synthesis of tryptophan (Figure 4.6), has a polypeptide chain that forms two α/β barrels. The structure of this enzyme, solved at 2.8 Å in the laboratory of Hans Jansonius in Basel, Switzerland, showed that residues 48–254 form one barrel with IGP-synthase activity, while residues 255–450 form the second barrel with PRA-isomerase activity (Figure 4.7).

In *Bacillus subtilis* these two reactions are catalyzed by two separate enzymes that have amino acid sequences homologous to the corresponding regions of the bifunctional enzyme from *E. coli,* and thus each forms a barrel

Figure 4.7 Two of the enzymatic activities involved in the biosynthesis of tryptophan in *E. coli,* phosphoribosyl anthranilate (PRA) isomerase and indoleglycerol phosphate (IGP) synthase, are performed by two separate domains in the polypeptide chain of a bifunctional enzyme. Both these domains are α/β-barrel structures, oriented such that their active sites are on opposite sides of the molecule. The two catalytic reactions are therefore independent of each other. The diagram shows the IGP-synthase domain (residues 48–254) with dark colors and the PRA-isomerase domain with light colors. The α helices are sequentially labeled a–h in both barrel domains. Residue 255 (arrow) is the first residue of the second domain. (Adapted from J.P. Priestle et al., *Proc. Natl. Acad. Sci. USA* 84: 5690–5694, 1987.)

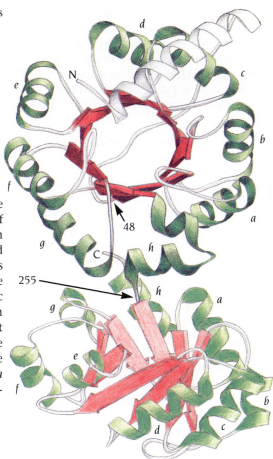

structure. There is no obvious functional advantage to *E. coli* in having these two enzymatic activities in one polypeptide chain, since the active sites of the two barrels are on opposite sides of the molecule facing away from each other and the two reactions are thus independent of each other. A third organism, *Neurospora crassa,* has an enzyme with three catalytic activities within the same polypeptide chain; here two domains similar to those of the *E. coli* enzyme are linked to a third domain that has yet another enzymatic function in the same biosynthetic pathway. These differences between species reflect different ways to organize the genome. DNA sequences that code for protein domains with different functions are organized into separate genes in one organism and fused into a single gene in another. Although the three-dimensional structures of these enzymes in *B. subtilis* and *N. crassa* have not been solved by crystallography, we can be certain that they are α/β-barrel domains because of their sequence homologies to the *E. coli* proteins.

The active site is formed by loops at one end of the α/β barrel

In all these α/β-barrel domains the active site is in a very similar position. It is situated in the bottom of a funnel-shaped pocket created by the eight loops that connect the carboxy end of the β strands with the amino end of the α helices (Figure 4.8). Residues that participate in binding and catalysis are in

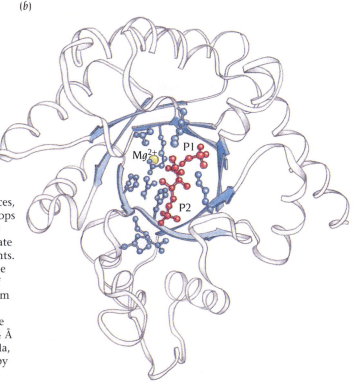

(a)

(b)

Figure 4.8 The active site in all α/β barrels is in a pocket formed by the loop regions that connect the carboxy ends of the β strands with the adjacent α helices, as shown schematically in (a), where only two such loops are shown. (b) A view from the top of the barrel of the active site of the enzyme RuBisCo (ribulose bisphosphate carboxylase), which is involved in CO_2 fixation in plants. A substrate analog (red) binds across the barrel between the two phosphate groups, P1 and P2, on opposite sides of the pocket. A number of charged side chains (blue) from different loops as well as a Mg^{2+} ion (yellow) form the substrate-binding site and provide catalytic groups. The structure of this 500 kD enzyme was determined to 2.4 Å resolution in the laboratory of Carl Branden, in Uppsala, Sweden. (Adapted from an original drawing provided by Bo Furugren.)

these loop regions. In other words, these enzymes are modeled on a common stable scaffold of eight parallel β strands surrounded by eight α helices. In each case the specific enzymatic activity is determined by the eight loop regions at the carboxy end of the β strands, which do not contribute to the structural stability of the scaffold. In some cases an additional loop region from a second domain or a different subunit comes close to this active site and also participates in binding and catalysis.

Alpha/beta barrels provide examples of evolution of new enzyme activities

How do new enzyme activities evolve? Are new enzymes formed from random sequences generated by recombination and other genetic rearrangements or do they arise by divergent evolution from a preexisting set of enzymes. Greg Petsko at Brandeis University has provided strong evidence for the latter case from studies of α/β-barrel enzymes in a rare metabolic pathway, conversion of mandelate to benzoate. This rare metabolic pathway is thought to be of recent evolutionary origin, since it is present in only a few pseudomonad species.

The first enzyme in this pathway, mandelate racemase, catalyzes the interconversion of the two optical isomers of mandelate (Figure 4.9a). The key step in this reaction is proton abstraction from a carbon atom, producing an enolic intermediate. Petsko found that the three-dimensional structure of this enzyme, including its α/β barrel, is very similar to that of a quite different enzyme, muconate lactonizing enzyme, which catalyzes a different chemical reaction (Figure 4.9b) but which also involves the formation of an intermediate by proton abstraction. The amino acid sequences of the 350 residues of these enzymes showed 26% sequence identity, which clearly demonstrates that they are evolutionarily related. By comparing these two structures in detail Petsko found significant similarities in the region of the active site that catalyzes proton abstraction and intermediate formation but substantial differences in those regions of the active site that confer substrate specificity.

These results are compatible with an evolutionary history in which the new enzyme activity of mandelate racemase has evolved from a preexisting enzyme that catalyzes the basic chemical reaction of proton abstraction and formation of an intermediate. Subsequent mutations have modified the

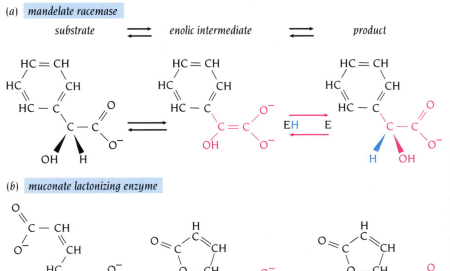

Figure 4.9 Mechanisms of the reactions catalyzed by the enzymes mandelate racemase (a) and muconate lactonizing enzyme (b). The two overall reactions are quite different: a change of configuration of a carbon atom for mandelate racemase versus ring closure for the lactonizing enzyme. However, one crucial step (pink) in the two reactions is the same: addition of a proton (blue) to an intermediate of the substrate (pink) from a lysine residue of the enzyme (E) or, in the reverse direction, formation of an intermediate by proton abstraction from the carbon atom adjacent to the carboxylate group.

←β strand →   loop   ←── α helix ──→

                2   5   7   12   17 20   24
(type A) NH₂ -X-L-E-X-L-X-L-X-X-C-X-L-T-X-X-X-C-X-X-L-X-X-*a*-L-X-X-X-X-

(type B) NH₂ -X-L-**R**-**E**-**L**-X-**L**-X-X-**N**-X-**L**-**G**-**D**-X-**G**-*a*-X-X-**L**-X-X-X-**L**-X-X-**P**-X-X-

Figure 4.10 Consensus amino acid sequence and secondary structure of the leucine-rich motifs of type A and type B. "X" denotes any amino acid; "a" denotes an aliphatic amino acid. Conserved residues are shown in bold in type B.

substrate specificity while preserving the ability to catalyze the basic chemical reaction. Chemistry is the important factor to preserve during evolution of new enzymes, while specificity can be modified. It would therefore seem that relatively nonspecific enzymes, which may have existed earlier in evolution or which may arise occasionally through random genetic rearrangements, are the clay from which nature sculpts new enzymes. To preserve the original enzyme activity and at the same time allow divergence, the precursor gene for the enzyme must first be duplicated at some point.

Further strong evidence that proteins with new functions evolve by the process of gene duplication and subsequent modification by mutation has recently been found by examination of the genome sequence of the bacterium *Haemophilus influenzae*, the first free-living organism to be completely sequenced. Sequence comparisons showed that, of the 1680 identified proteins coded by this genome, at least one third are related to one or more other proteins within the genome and therefore have arisen from processes that involve gene duplications.

Alpha/beta barrels are particularly well suited to such an evolutionary strategy, since substrate specificity and catalytic function reside in loop regions which are separated from the residues of the α helices and β strands that contribute to the structural stability of these domain structures. Such enzymes should also be excellent targets for genetic redesign *in vitro*. By changing the lengths and specific residues of the active-site loop regions, it might be possible to produce novel substrate specificities without affecting the stability of the structural framework and therefore the enzyme.

Leucine-rich motifs form an α/β-horseshoe fold

Leucine-rich motifs—tandem homologous amino acid sequences of about 20–30 residues—have been identified from sequence studies in over 60 different proteins, including receptors, cell adhesion molecules, bacterial virulence factors, and molecules involved in RNA splicing and DNA repair. The x-ray structure of one member of this class of proteins, a ribonuclease inhibitor, has been determined by Johann Deisenhofer and colleagues at the University of Texas, Dallas. The 456 amino acids of the polypeptide chain are arranged in 15 tandem leucine-rich motifs of two types that alternate along the chain: type A, with 29 residues, and type B, with 28 residues. In addition there are two short regions with nonhomologous sequences at the termini of the chain. The consensus sequence of these homologous repeats (Figure 4.10) indicates that both types of repeat contain a characteristic pattern of leucine residues that play an important structural role, as we will see.

Each repeat forms a right-handed β-loop-α structure similar to those found in the two other classes of α/β structures described earlier. Sequential β-loop-α repeats are joined together in a similar way to those in the α/β-barrel structures. The β strands form a parallel β sheet, and all the α helices are on one side of the β sheet. However, the β strands do not form a closed barrel; instead they form a curved open structure that resembles a horseshoe with α helices on the outside and a β sheet forming the inside wall of the horseshoe (Figure 4.11). One side of the β sheet faces the α helices and participates in a hydrophobic core between the α helices and the β sheet; the other side of the β sheet is exposed to solvent, a characteristic other α/β structures do not have.

The leucine residues in this leucine-rich motif form a hydrophobic core between the β sheet and the α helices. Leucine residues 2, 5, and 7 (see Figure

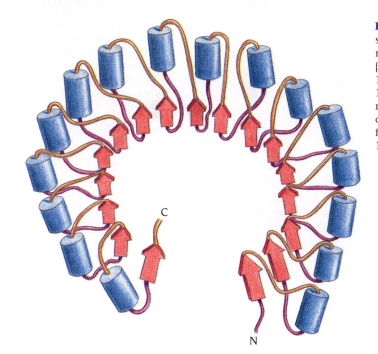

Figure 4.11 Schematic diagram of the structure of the ribonuclease inhibitor. The molecule, which is built up by repetitive β-loop-α motifs, resembles a horseshoe with a 17-stranded parallel β sheet on the inside and 16 α helices on the outside. The β sheet is light red, α helices are blue, and loops that are part of the β-loop-α motifs are orange. (Adapted from B. Kobe et al., *Nature* 366: 751–756, 1993.)

4.10) from the β strand of the motif pack against leucine residues 20 and 24 of the α helix to form the main part of the hydrophobic region. Leucine residue 12 from the loop region is also part of this hydrophobic core, as is residue 17 from the α helix, which is usually hydrophobic (Figure 4.12).

Leucine residues 2, 5, 7, 12, 20, and 24 of the motif are invariant in both type A and type B repeats of the ribonuclease inhibitor. An examination of more than 500 tandem repeats from 68 different proteins has shown that residues 20 and 24 can be other hydrophobic residues, whereas the remaining four leucine residues are present in all repeats. On the basis of the crystal structure of the ribonuclease inhibitor and the important structural role of these leucine residues, it has been possible to construct plausible structural models of several other proteins with leucine-rich motifs, such as the extracellular domains of the thyrotropin and gonadotropin receptors.

Alpha/beta twisted open-sheet structures contain α helices on both sides of the β sheet

In the next class of α/β structures there are α helices on both sides of the β sheet. This has at least three important consequences. First, a closed barrel cannot be formed unless the β strands completely enclose the α helices on one side of the β sheet. Such structures have never been found and are very unlikely to occur, since a large number of β strands would be required to enclose even a single α helix. Instead, the β strands are arranged into an open twisted β sheet such as that shown in Figure 4.1b.

Second, there are always two adjacent β strands (β_1 and β_3 in Figure 4.2b) in the interior of the β sheet whose connections to the flanking β strand are on opposite sides of the β sheet. One of the loops from one of these two β strands goes above the β sheet, whereas the other loop goes below. This creates a crevice outside the edge of the β sheet between these two loops (Figure 4.13). Almost all binding sites in this class of α/β proteins are located in crevices of this type at the carboxy edge of the β sheet, as we discuss in detail

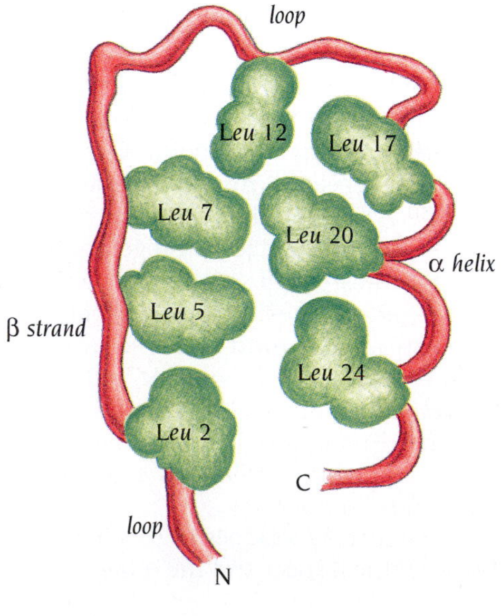

Figure 4.12 Schematic diagram illustrating the role of the conserved leucine residues (green) in the leucine-rich motif in stabilizing the β-loop-α structural module. In the ribonuclease inhibitor, leucine residues 2, 5, and 7 from the β strand pack against leucine residues 17, 20, and 24 from the α helix as well as leucine residue 12 from the loop to form a hydrophobic core between the β strand and the α helix.

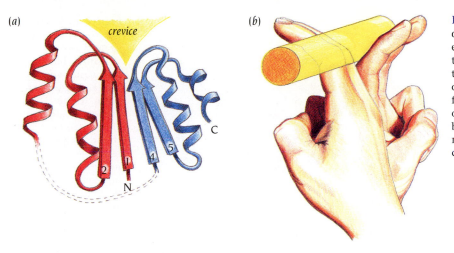

(a)

crevice

2 1 4 5

N

C

(b)

Figure 4.13 (a) The active site in open twisted α/β domains is in a crevice outside the carboxy ends of the β strands. This crevice is formed by two adjacent loop regions that connect the two strands with α helices on opposite sides of the β sheet. This is illustrated by the curled fingers of two hands (b), where the top halves of the fingers represent loop regions and the bottom halves represent the β strands. The rod represents a bound molecule in the binding crevice.

later. (We define the carboxy edge of the sheet as the edge that is formed by the carboxy ends of the parallel β strands in the sheet.)

Third, in open-sheet structures the α helices are packed against both sides of the β sheet. Each β strand thus contributes hydrophobic side chains to pack against α helices in two similar hydrophobic core regions, one on each side of the β sheet.

Open β-sheet structures have a variety of topologies

We have seen that all members of the α/β-barrel domain structures have the same basic arrangement of eight α helices and eight β strands. Within the open α/β sheets, however, there is much more variation in structure, as is obvious from purely geometric considerations. Since the β strands form an open β sheet, there are no geometric restrictions on the number of strands involved. In fact, the number varies from four to ten. Furthermore, the two β strands joined by a crossover connection need not be adjacent in the β sheet, although the β-α-β motif where the two β strands are adjacent is a preferred structural building block. In addition, there can be mixed β sheets in which hairpin connections give rise to some antiparallel β strands mixed with the parallel β strands. All these variations occur in actual structures, some of which are illustrated in Figure 4.14a–d. There are thus many variations on the regular arrangement of six parallel β strands (see Figure 4.1b).

The positions of active sites can be predicted in α/β structures

We have described a general relationship between structure and function for the α/β-barrel structures. They all have the active site at the same position with respect to their common structure in spite of having different functions as well as different amino acid sequences. We can now ask if similar relationships also occur for the open α/β-sheet structures in spite of their much greater variation in structure. Can the position of the active sites be predicted from the structures of many open-sheet α/β proteins?

In almost every one of the more than 100 different known α/β structures of this class the active site is at the carboxy edge of the β sheet. Functional residues are provided by the loop regions that connect the carboxy end of the β strands with the amino end of the α helices. In this one respect a fundamental similarity therefore exists between the α/β-barrel structures and the open α/β-sheet structures.

The general shapes of the active sites are quite different, however. Open α/β structures cannot form funnel-shaped active sites like the barrel structures. Instead, they form crevices at the edge of the β sheet. Such crevices occur when there are two adjacent connections that are on opposite sides of the β sheet. One of the loop regions in these two connections goes out from

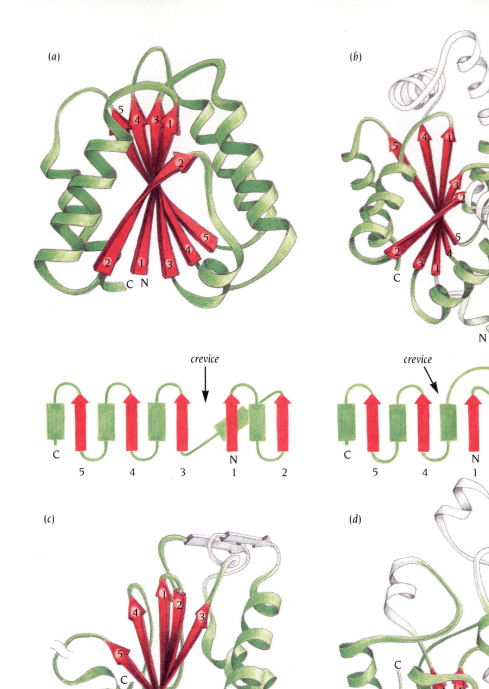

(a)

(b)

crevice

crevice   crevice

C   C

5   4   3   1   2   5   4   1   3   2

(c)

(d)

crevice   crevice

C   C

5   4   1   2   3   6   5   1   4   2   3

Figure 4.14 Examples of different types of open twisted α/β structures. Both schematic and topological diagrams are given. In the topological diagrams, arrows denote strands of β sheet and rectangles denote α helices. (a) The FMN-binding redox protein flavodoxin. (b) The enzyme adenylate kinase, which catalyzes the reaction AMP + ATP ⇌ 2 ADP. The structure was determined to 3.0 Å resolution in the laboratory of Georg Schulz in Heidelberg, Germany. (c) The ATP-binding domain of the glycolytic enzyme hexokinase, which catalyzes the phosphorylation of glucose. The structure was determined to 2.8 Å resolution in the laboratory of Tom Steitz, Yale University. (d) The glycolytic enzyme phosphoglycerate mutase, which catalyzes transfer of a phosphoryl group from carbon 3 to carbon 2 in phosphoglycerate. The structure was determined to 2.5 Å resolution in the laboratory of Herman Watson, Bristol University, UK. (Adapted from J. Richardson.)

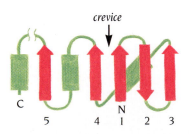

its β strand above the β sheet and the other below, creating a crevice between them (see Figure 4.13). The active site or part of it is usually found in such a crevice. The position of such crevices is determined by the topology of the β sheet and can be predicted from a topology diagram. The crevices occur when the strand order is reversed, and can be easily identified in a topology diagram as the place where connections from the carboxy ends of two adjacent β strands go in opposite directions, one to the left and one to the right. Let us examine the first two diagrams given in Figure 4.14.

The first structure, flavodoxin (Figure 4.14a), has one such position, between strands 1 and 3. The connection from strand 1 goes to the right and that from strand 3 to the left. In the schematic diagram in Figure 4.14a we can see that the corresponding α helices are on opposite sides of the β sheet. The loops from these two β strands, 1 and 3, to their respective α helices form the major part of the binding cleft for the coenzyme FMN (flavin mononucleotide).

The second structure, adenylate kinase (Figure 4.14b), has two such positions, one on each side of β strand 1. The connection from strand 1 to strand 2 goes to the right, whereas the connection from the flanking strands 3 and 4 both go to the left. Crevices are formed between β strands 1 and 3 and between strands 1 and 4. One of these crevices forms part of an AMP-binding site, and the other crevice forms part of an ATP-binding site that catalyzes the formation of ADP from AMP and ATP.

Such positions in a topology diagram are called topological switch points. It was postulated in 1980 by Carl Branden, in Uppsala, Sweden, that the position of active sites could be predicted from such switch points. Since then at least one part of the active site has been found in crevices defined by such switch points in almost all new α/β structures that have been determined. Thus we can predict the approximate position of the active site and possible loop regions that form this site in α/β proteins. This is in contrast to proteins of the other two main classes—α-helical proteins and antiparallel β proteins—where no such predictive rules have been found. We will now examine a few examples that illustrate the relationship between the topology diagrams of some α/β proteins, their switch points, and the active-site residues. These examples have been chosen because they represent different types of α/β open-sheet structures.

Tyrosyl-tRNA synthetase has two different domains (α/β + α)

One of the crucial steps in protein synthesis is performed by the group of enzymes called aminoacyl-tRNA synthetases. These enzymes connect each amino acid with its specific transfer RNA molecule in a two-step reaction. First the amino acid is activated by ATP to give an enzyme-bound amino acid adenylate; then this complex is attacked by the tRNA to give the aminoacyl-tRNA.

The structure of the synthetase specific for the amino acid tyrosine was determined to 2.7 Å resolution in the laboratory of David Blow in London. Figure 4.15 shows a schematic diagram of the first 320 residues of a single subunit of this dimeric molecule. The last 100 residues are disordered in the crystal and are not visible. There are essentially two different domains, one

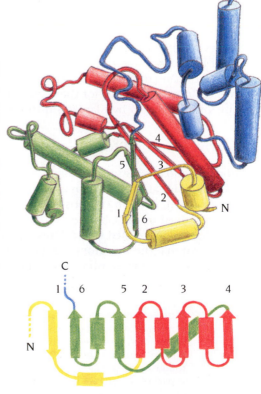

Figure 4.15 Schematic diagram of the enzyme tyrosyl-tRNA synthetase, which couples tyrosine to its cognate transfer RNA. The central region of the catalytic domain (red and green) is an open twisted α/β structure with five parallel β strands. The active site is formed by the loops from the carboxy ends of β strands 2 and 5. These two adjacent strands are connected to α helices on opposite sides of the β sheet. Where more than one α helix connects two β strands (for example, between strands 4 and 5), they are represented as one rectangle in the topology diagram. (Adapted from T.N. Bhat et al., *J. Mol. Biol.* 158: 699–709, 1982.)

Figure 4.16 A schematic view of the active site of tyrosyl-tRNA synthetase. Tyrosyl adenylate, the product of the first reaction catalyzed by the enzyme, is bound to two loop regions: residues 38–47, which form the loop after β strand 2, and residues 190–193, which form the loop after β strand 5. The tyrosine and adenylate moieties are bound on opposite sides of the β sheet outside the carboxy ends of β strands 2 and 5.

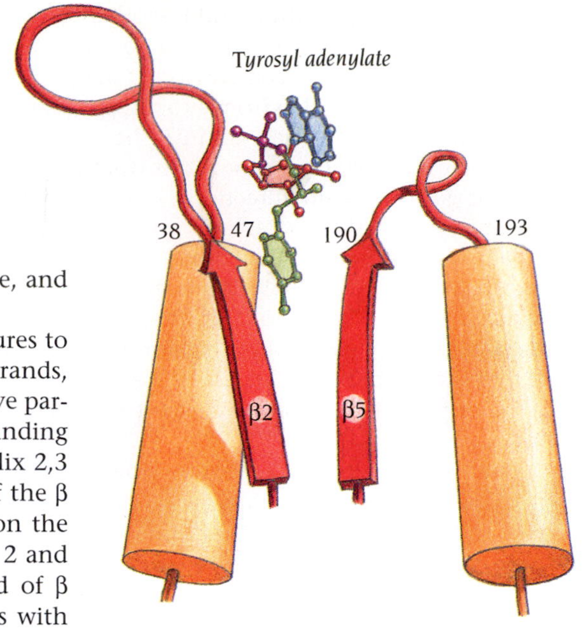

α/β domain (red and green in Figure 4.15) that binds ATP and tyrosine, and one α-helical domain (blue), the function of which is not known.

Let us now apply the rules for predicting active sites of α/β structures to the topological diagram shown in Figure 4.15. The β sheet has six strands, one of which, number 1, is antiparallel to the others. The remaining five parallel β strands are arranged in a way rather similar to the nucleotide-binding fold (see Figure 4.1b), but here the strand order is 6 5 2 3 4. Alpha helix 2,3 (which connects β strands 2 and 3) and α helix 3,4 are on one side of the β sheet (red helices in Figure 4.15), whereas α helices 4,5 and 5,6 are on the other side (green helices). The switch point is thus between β strands 2 and 5. We would predict that the active site is outside the carboxy end of β strands 2 and 5 and that the loop regions that connect these strands with their respective α helices participate in binding the substrates. These loop regions comprise residues 38–47 and 190–193, respectively. The active site has been identified in the crystal structure by diffusing tyrosine and ATP into the crystals. The enzyme molecules in the crystals are active, so tyrosyl adenylate is formed, but because no tRNA is present, it stays bound to the enzyme.

The position of this bound tyrosyl adenylate was determined from an electron density map of the complex just after the predictive rules were formulated. The region where it binds proved to be as predicted. Loop regions 38–47, after β strand 2, and 190–193, after β strand 5, line a cleft where the substrate binds (Figure 4.16). The phosphate and the sugar moieties are hydrogen-bonded to the main-chain nitrogen atoms of residues 38 and 192, respectively. This part of the substrate is thus very close to the switch point. The substrate straddles the edge of the β sheet so that the tyrosine and adenine ends are on opposite sides of the β sheet. The substrate also interacts with some of the other regions at this end of the β sheet, especially residues 173–177 of the α helix that connects β strands 4 and 5 and, in addition, some residues within β strand 2. Figure 4.17 shows a schematic diagram of the position of bound tyrosine (red) in relation to these regions of the protein, and Figure 4.18 gives the important hydrogen bonds to the substrate, knowledge of which formed the basis for the beautiful site-directed mutagenesis experiments on this system by Alan Fersht in London.

Carboxypeptidase is an α/β protein with a mixed β sheet

Carboxypeptidases are zinc-containing enzymes that catalyze the hydrolysis of polypeptides at the C-terminal peptide bond. The bovine enzyme form A is a monomeric protein comprising 307 amino acid residues. The structure was determined in the laboratory of William Lipscomb, Harvard University, in 1970 and later refined to 1.5 Å resolution. Biochemical and x-ray studies have shown that the zinc atom is essential for catalysis by binding to the carbonyl oxygen of the substrate. This binding weakens the C'=O bond by

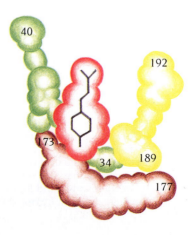

Figure 4.17 Schematic diagram of bound tyrosine to tyrosyl-tRNA synthetase. Colored regions correspond to van der Waals radii of atoms within a layer of the structure through the tyrosine ring. Red is bound tyrosine; green is the end of β strand 2 and the beginning of the following loop region; yellow is the loop region 189–192; and brown is part of the α helix in loop region 173–177.

60

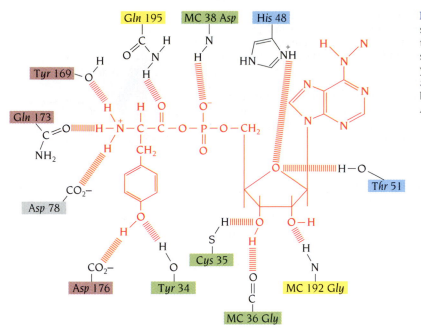

Figure 4.18 Side chains of the tyrosyl-tRNA synthetase that form hydrogen bonds to tyrosyl adenylate. Green residues are from β strand 2 and the following loop regions, yellow residues are from the loop after β strand 5, and brown residues are from the α helix before β strand 5. (Adapted from T. Wells and A. Fersht, *Nature* 316: 656–657, 1985.)

abstracting electrons from the carbon atom and thus facilitates cleavage of the adjacent peptide bond. Carboxypeptidase is a large single domain structure comprising a mixed β sheet of eight β strands (Figure 4.19) with α helices on both sides. Some of the loop regions are very long and curl around the central theme of the structure.

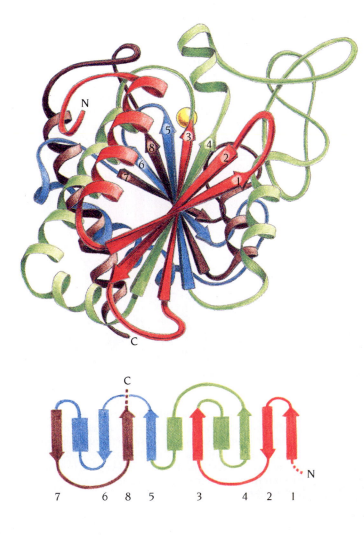

Figure 4.19 Schematic and topological diagrams for the structure of the enzyme carboxypeptidase. The central region of the mixed β sheet contains four adjacent parallel β strands (numbers 8, 5, 3, and 4), where the strand order is reversed between strands 5 and 3. The active-site zinc atom (yellow circle) is bound to side chains in the loop regions outside the carboxy ends of these two β strands. The first part of the polypeptide chain is red, followed by green, blue, and brown. (Adapted from J. Richardson.)

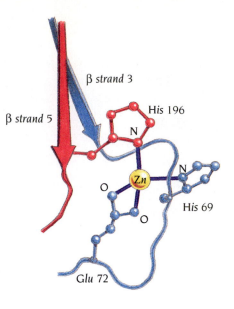

Figure 4.20 Detailed view of the zinc environment in carboxy-peptidase. The active-site zinc atom is bound to His 69 and Glu 72, which are part of the loop region outside β strand 2. In addition, His 196, which is the last residue of β strand 5, also binds the zinc.

The four central strands of the β sheet are parallel and have the strand order 8 5 3 4 (see Figure 4.19). The strand order is thus reversed once, and there is a switch point in the middle of this β sheet between β strands 5 and 3 where we would expect the active site to be located.

This is precisely where the catalytically essential zinc atom is found. This zinc atom is located precisely at this switch point, where it is firmly anchored to the protein by three side-chain ligands, His 69, Glu 72, and His 196 (Figure 4.20). The last residue of β strand 3 is residue 66, so the two zinc ligands His 69 and Glu 72 are at the beginning of the loop region that connects this β strand with its corresponding α helix. The last residue of β strand 5 is the third zinc ligand, His 196.

In this structure the loop regions adjacent to the switch point do not provide a binding crevice for the substrate but instead accommodate the active-site zinc atom. The essential point here is that this zinc atom and the active site are in the predicted position outside the switch point for the four central parallel β strands, even though these β strands are only a small part of the total structure. This sort of arrangement, in which an active site formed from parallel β strands is flanked by antiparallel β strands, has been found in a number of other α/β proteins with mixed β sheets.

Arabinose-binding protein has two similar α/β domains

The arabinose-binding protein is one of the group of proteins that occur in the periplasmic space between the inner and outer cell membranes of Gram-negative bacteria such as *E. coli*. These proteins are components of active transport systems for various sugars, amino acids, and ions. Arabinose-binding protein is involved in arabinose transport. It is a single polypeptide chain of 306 amino acids folded into two domains of similar structure and topology (Figure 4.21), as was ascertained in the laboratory of Florante Quiocho in Houston, Texas, by a structure determination of the protein with bound

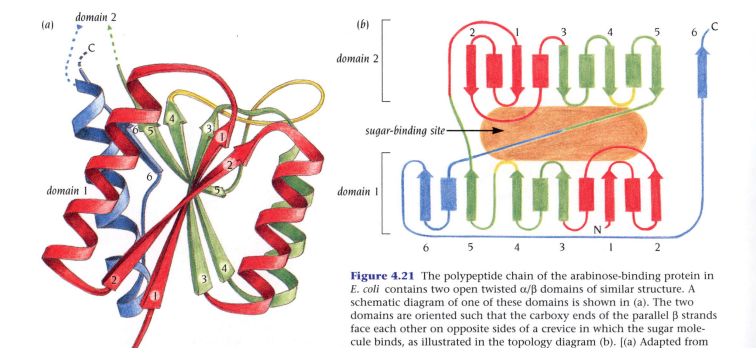

Figure 4.21 The polypeptide chain of the arabinose-binding protein in *E. coli* contains two open twisted α/β domains of similar structure. A schematic diagram of one of these domains is shown in (a). The two domains are oriented such that the carboxy ends of the parallel β strands face each other on opposite sides of a crevice in which the sugar molecule binds, as illustrated in the topology diagram (b). [(a) Adapted from J. Richardson.]

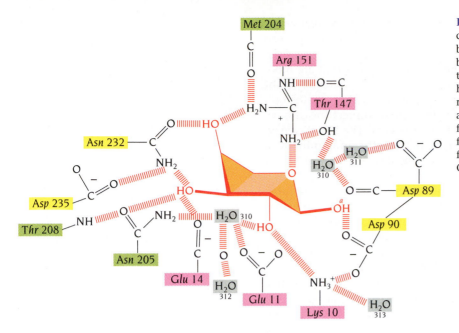

Figure 4.22 Schematic diagram of the complex networks of hydrogen bonds formed by polar side chains from the arabinose-binding protein and L-arabinose. The residues that interact with the sugar are in turn hydrogen-bonded to each other or to other residues or isolated water molecules. The pink and green residues are in loop regions that, from the topology diagram, are predicted to form the binding site. The yellow residues are from adjacent loop regions. (Adapted from C.F. Sams et al., *Nature* 310: 429–430, 1984.)

arabinose at 1.7 Å resolution. The first five β strands in both domains are parallel; the strand order is reversed once, and the switch points are between β strands 1 and 3 in both domains.

A schematic diagram relating binding to the topology of the domains is shown in Figure 4.21b. Twelve amino acid residues from both domains are involved in forming a complicated network of hydrogen bonds to the oxygen atoms of the bound arabinose (Figure 4.22). We predict from the topology of these domains that loop residues from β strands 1 and 3 of both domains should participate in these interactions. Five residues in loop regions after β strand 1 in both domains (pink residues in Figure 4.22) and three residues after β strand 3 in the second domain (green residues in Figure 4.22) participate in binding. The remaining four residues (yellow residues in Figure 4.22) are all from loop regions after β strand 4 in both domains. These are one β strand removed from the switch point. Both domains thus participate approximately equally in binding the sugar at the carboxy edge of the β sheets, outside the switch point, where the strand order is reversed.

A number of proteins consist of two α/β open-sheet domains formed from a single polypeptide chain, as in arabinose-binding protein. In almost all these cases the active sites are found in cleft regions between these two domains. The domains are oriented in such a way that the carboxy edge of both β sheets points toward the active site. Loop regions adjacent to the switch points of both domains participate in forming the active site. In enzymatic reactions where two different substrates participate, they are bound to different domains and are brought together for catalytic reactions by the orientation of these domains. In other proteins the two domains bind different regions of the same ligand. Sugar-binding proteins from bacteria are examples of this second case.

Conclusion

Alpha/beta (α/β) structures are the most frequent and most regular of the protein structures. They fall into three classes: the first class comprises a central core of usually eight parallel β strands arranged close together like the staves of a barrel, surrounded by α helices; the second class comprises an open twisted parallel or mixed β sheet with α helices on both sides of the β sheet; and the third class is formed by leucine-rich motifs in which a large number of parallel β strands form a curved β sheet with all the α helices on the outside of this sheet.

The α/β-barrel structure is one of the largest and most regular of all domain structures, comprising about 250 amino acids. It has so far been found in more than 20 different proteins, with completely different amino acid sequences and different functions. They are all enzymes that are modeled on this common scaffold of eight parallel β strands surrounded by eight α helices. They all have their active sites in very similar positions, at the bottom of a funnel-shaped pocket created by the loops that connect the carboxy end of the β strands with the amino end of the α helices. The specific enzymatic activity is, in each case, determined by the lengths and amino acid sequences of these loop regions, which do not contribute to the stability of the fold.

The horseshoe structure is formed by homologous repeats of leucine-rich motifs, each of which forms a β-loop-α unit. The units are linked together such that the β strands form an open curved β sheet, like a horseshoe, with the α helices on the outside of the β sheet and the inside exposed to solvent. The invariant leucine residues of these motifs form the major part of the hydrophobic region between the α helices and the β sheet.

The open α/β-sheet structures vary considerably in size, number of β strands, and strand order. Independent of these variations, they all have their active sites at the carboxy edge of the β strands, and these active sites are lined by the loop regions that connect the β strands with the α helices. In this respect, they are similar to the α/β-barrel structures. However, the active-site regions are created differently in open structures. They are formed in those regions outside the carboxy edge of the β sheet, where two adjacent loops are on opposite sides of the β sheet. The positions of these regions can be predicted from topology diagrams. The rules that relate the general position of functional binding sites to the overall structure of the protein are thus known for β barrels and open-sheet α/β proteins.

Selected readings

General

Babbit, P.C., et al. A functionally diverse enzyme superfamily that abstracts the α protons of carboxylic acids. *Science* 267: 1159–1161, 1995.

Brändén, C.-I. Relation between structure and function of α/β proteins. *Q. Rev. Biophys.* 13: 317–338, 1980.

Brenner, S.E. Gene duplications in *H. influenzae*. *Nature* 378: 140, 1995.

Cohen, F.E., Sternberg, M.J.E., Taylor, W.R. Analysis and prediction of the packing of α-helices against a β-sheet in the tertiary structure of globular proteins. *J. Mol. Biol.* 156: 821–862, 1982.

Farber, G., Petsko, G.A. The evolution of α/β-barrel enzymes. *Trends Biochem. Sci.* 15: 228–234, 1990.

Janin, J., Chothia, C. Packing of α-helices onto β-pleated sheets and the anatomy of α/β-proteins. *J. Mol. Biol.* 143: 95–128, 1980.

Kajava, A.V., Vassart, G., Wodak, S.J. Modelling of the three-dimensional structure of proteins with the typical leucine-rich repeats. *Structure* 3: 867–877, 1995.

Lasters, I., et al. Structural principles of parallel β barrels in proteins. *Proc. Natl. Acad. Sci. USA* 85: 3338–3342, 1988.

Lasters, I., Wodak, S.J., Pio, F. The design of idealized α/β-barrels: analysis of β-sheet closure requirements. *Proteins* 7: 249–256, 1990.

Lesk, A.M., Branden, C.-I., Chothia, C. Structural principles of α/β barrel proteins: the packing of the interior of the sheet. *Proteins* 5: 139–148, 1989.

Murzin, A.G., Lesk, A.M., Chothia, C. Principles determining the structure of β sheet barrels in proteins. *J. Mol. Biol.* 236: 1369–1400, 1994.

Ohlsson, I., Nordström, B., Brändén, C.-I. Structural and functional similarities within the coenzyme binding domains of dehydrogenases. *J. Mol. Biol.* 89: 339–354, 1974.

Richardson, J.S. β-sheet topology and the relatedness of proteins. *Nature* 268: 495–500, 1977.

Richardson, J.S. Handedness of crossover connections in β sheets. *Proc. Natl. Acad. Sci. USA* 73: 2619–2623, 1976.

Sternberg, M.J.E., et al. Analysis and prediction of structural motifs in the glycolytic enzymes. *Phil. Trans. R. Soc. Lond.* B293: 177–189, 1981.

Sternberg, M.J.E., Thornton, J.M. On the conformation of proteins: the handedness of the connection between parallel β-strands. *J. Mol. Biol.* 110: 269–283, 1977.

Specific structures

Brick, P., Bhat, T.N., Blow, D.M. Structure of tyrosyl-tRNA synthetase refined at 2.3 Å resolution. Interaction of the enzyme with the tyrosyl adenylate intermediate. *J. Mol. Biol.* 208: 83–98, 1988.

Campbell, J.W., Watson, H.C., Hodgson, G.I. Structure of yeast phosphoglycerate mutase. *Nature* 250: 301–303, 1974.

Carrel, H.L., et al. X-ray structure of D-xylose isomerase from *Streptomyces rubiginosus* at 4 Å resolution. *J. Biol. Chem.* 259: 3230–3236, 1984.

Dreusicke, D., Karplus, P.A., Schulz, G.E. Refined structure of porcine adenylate kinase at 2.1 Å resolution. *J. Mol. Biol.* 199: 359–371, 1988.

Eklund, H., et al. Three-dimensional structure of horse liver alcohol dehydrogenase at 2.4 Å resolution. *J. Mol. Biol.* 102: 27–59, 1976.

Gilliland, G.L., Quiocho, F.A. Structure of the L-arabinose-binding protein from *Escherichia coli* at 2.4 Å resolution. *J. Mol. Biol.* 146: 341–362, 1981.

Goldman, A., Ollis, D.L., Steitz, T.A. Crystal structure of muconate lactonizing enzyme at 3 Å resolution. *J. Mol. Biol.* 194: 143–153, 1987.

Hofmann, B.E., Bender, H., Schulz, G.E. Three-dimensional structure of cyclodextrin glycosyltransferase from *Bacillus circulans* at 3.4 Å resolution. *J. Mol. Biol.* 209: 793–800, 1989.

Hyde, C.C., et al. Three-dimensional structure of the tryptophan synthase $\alpha_2\beta_2$ multienzyme complex from *Salmonella typhimurium*. *J. Biol. Chem.* 263: 17857–17871, 1988.

Knight, S., Andersson, I., Brändén, C.-I. Crystallographic analysis of ribulose-1,5-bisphosphate carboxylase from spinach at 2.4 Å resolution. Subunit interactions and active site. *J. Mol. Biol.* 215: 113–160, 1990.

Kobe, B., Deisenhofer, J. Crystal structure of porcine ribonuclease inhibitor, a protein with leucine-rich repeats. *Nature* 366: 751–756, 1993.

Lebioda, L., Stec, B., Brewer, J.M. The structure of yeast enolase at 2.5 Å resolution. An 8-fold $\beta + \alpha$ barrel with a novel $\beta\beta\alpha\alpha(\beta\alpha)_6$ topology. *J. Biol. Chem.* 264: 3685–3693, 1989.

Lim, L.W., et al. Three-dimensional structure of the iron-sulfur flavoprotein trimethylamine dehydrogenase at 2.4 Å resolution. *J. Biol. Chem.* 261: 15140–15146, 1986.

Lindqvist, Y. Refined structure of spinach glycolate oxidase at 2 Å resolution. *J. Mol. Biol.* 209: 151–166, 1989.

Mancia, F., et al. How coenzyme B_{12} radicals are generated: the crystal structure of methylmalonyl–coenzyme A mutase at 2 Å resolution. *Structure* 4: 339–350, 1996.

Matsuura, Y., et al. Structure and possible catalytic residues of taka-amylase A. *J. Biochem.* 95: 697–702, 1984.

Mavridis, I.M., et al. Structure of 2-keto-3-deoxy-6-phosphogluconate aldolase at 2.8 Å resolution. *J. Mol. Biol.* 162: 419–444, 1982.

Muirhead, H., et al. The structure of cat muscle pyruvate kinase. *EMBO J.* 5: 475–481, 1986.

Neidhart, D.J., et al. Mandelate racemase and muconate lactonizing enzyme are mechanistically distinct and structurally homologous. *Nature* 347: 692–694, 1990.

Priestle, J.P., et al. Three-dimensional structure of the bifunctional enzyme N-(5′-phosphoribosyl) anthranilate isomerase-indole-3-glycerol-phosphate synthase from *Escherichia coli*. *Proc. Natl. Acad. Sci. USA* 84: 5690–5694, 1987.

Rees, D.C., Lewis, M., Lipscomb, W.N. Refined crystal structure of carboxypeptidase A at 1.54 Å resolution. *J. Mol. Biol.* 168: 367–387, 1983.

Rouvinen, J., et al. Three-dimensional structure of cellobiohydrolase II from *Trichoderma resei*. *Science* 249: 380–386, 1990.

Schneider, G., Lindqvist, Y., Lundqvist, T. Crystallographic refinement and structure of ribulose-1,5-bisphosphate carboxylase from *Rhodospirillum rubrum* at 1.7 Å resolution. *J. Mol. Biol.* 211: 989–1008, 1990.

Steitz, T.A., et al. High resolution x-ray structure of yeast hexokinase, an allosteric protein exhibiting a non-symmetric arrangement of subunits. *J. Mol. Biol.* 104: 197–222, 1976.

Sygusch, J., Beaudry, D., Allaire, M. Molecular architecture of rabbit skeletal muscle aldolase at 2.7 Å resolution. *Proc. Natl. Acad. Sci. USA* 84: 7846–7850, 1987.

Uhlin, U., Eklund, H. Structure of ribonucleotide reductase protein R1. *Nature* 370: 553–559, 1994.

Xia, Z.-X., et al. Three-dimensional structure of flavocytochrome b_2 from baker's yeast at 3.0 Å resolution. *Proc. Natl. Acad. Sci. USA* 84: 2629–2633, 1987.

Beta Structures

5

Antiparallel beta (β) structures comprise the second large group of protein domain structures. Functionally, this group is the most diverse; it includes enzymes, transport proteins, antibodies, cell surface proteins, and virus coat proteins. The cores of these domains are built up by β strands that can vary in number from four or five to over ten. The β strands are arranged in a predominantly antiparallel fashion and usually in such a way that they form two β sheets that are joined together and packed against each other.

The β sheets have the usual twist, and when two such twisted β sheets are packed together, they form a barrel-like structure (Figure 5.1). Antiparallel β structures, therefore, in general have a core of hydrophobic side chains inside the barrel provided by residues in the β strands. The surface is formed by residues from the loop regions and from the strands. The aim of this chapter is to examine a number of antiparallel β structures and demonstrate how these rather complex structures can be separated into smaller comprehensible motifs.

Figure 5.1 The enzyme superoxide dismutase (SOD). SOD is a β structure comprising eight antiparallel β strands (a). In addition, SOD has two metal atoms, Cu and Zn (yellow circles), that participate in the catalytic action: conversion of a superoxide radical to hydrogen peroxide and oxygen. The eight β strands are arranged around the surface of a barrel, which is viewed along the barrel axis in (b) and perpendicular to this axis in (c). [(a) Adapted from J.S. Richardson. The structure of SOD was determined in the laboratory of J.S. and D.R. Richardson, Duke University.]

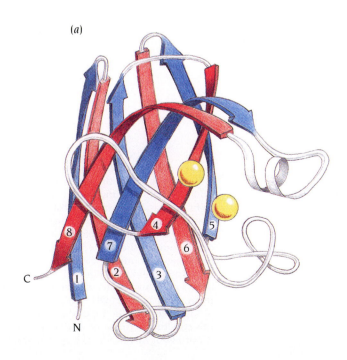

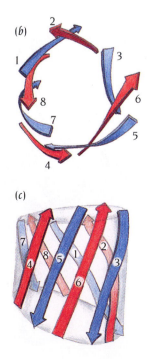

In Chapter 2 we described the 24 different ways that two β-loop-β units can form a four-stranded β sheet. The number of possible ways to form antiparallel β-sheet structures rapidly increases as the number of strands increases. It is thus surprising, but reassuring, that the number of topologies actually observed is small and that most β structures fall into a few groups of common or similar topology. The three most frequently occurring groups—up-and-down barrels, Greek keys, and jelly roll barrels—can all be related to simple ways of connecting antiparallel β strands arranged in a barrel structure.

Up-and-down barrels have a simple topology

The simplest topology is obtained if each successive β strand is added adjacent to the previous strand until the last strand is joined by hydrogen bonds to the first strand and the barrel is closed (Figure 5.2). These are called **up-and-down β sheets** or **barrels**. The arrangement of β strands is similar to that in the α/β-barrel structures we have just described in Chapter 4, except that here the strands are antiparallel and all the connections are hairpins. The structural and functional versatility of even this simple arrangement will be illustrated by two examples.

The retinol-binding protein binds retinol inside an up-and-down β barrel

The first example is the plasma-borne **retinol-binding protein**, RBP, which is a single polypeptide chain of 182 amino acid residues. This protein is responsible for transporting the lipid alcohol **vitamin A** (retinol) from its storage site in the liver to the various vitamin-A-dependent tissues. It is a disposable package in the sense that each RBP molecule transports only a single retinol molecule and is then degraded.

RBP is synthesized in the hepatocytes, where it picks up one molecule of retinol in the endoplasmic reticulum. Both its synthesis and its secretion from the hepatocytes to the plasma are regulated by retinol. In plasma, the

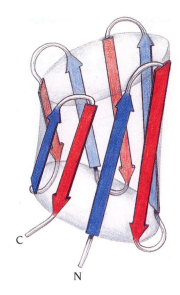

Figure 5.2 Schematic and topological diagrams of an up-and-down β barrel. The eight β strands are all antiparallel to each other and are connected by hairpin loops. Beta strands that are adjacent in the amino acid sequence are also adjacent in the three-dimensional structure of up-and-down barrels.

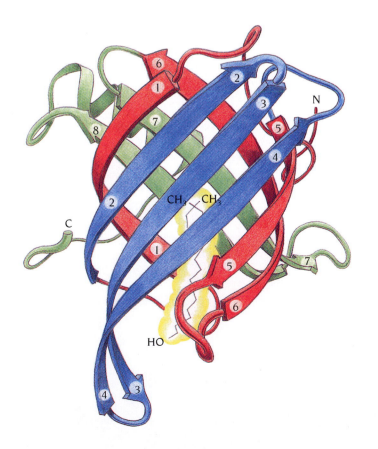

Figure 5.3 Schematic diagram of the structure of human plasma retinol-binding protein (RBP), which is an up-and-down β barrel. The eight antiparallel β strands twist and curl such that the structure can also be regarded as two β sheets (green and blue) packed against each other. Some of the twisted β strands (red) participate in both β sheets. A retinol molecule, vitamin A (yellow), is bound inside the barrel, between the two β sheets, such that its only hydrophilic part (an OH tail) is at the surface of the molecule. The topological diagram of this structure is the same as that in Figure 5.2. (Courtesy of Alwyn Jones, Uppsala, Sweden.)

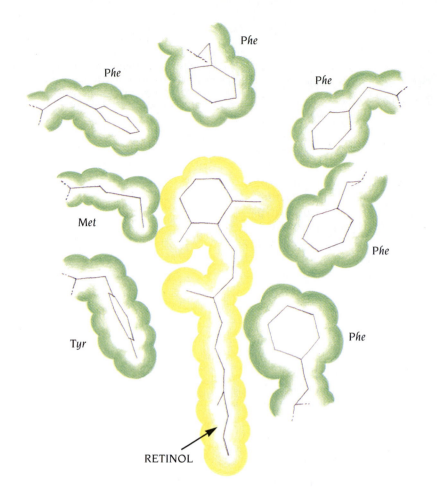

Phe

Phe

Phe

Met

Phe

Tyr

Phe

RETINOL

Figure 5.4 The binding site for retinol inside the RBP barrel is lined with hydrophobic residues. They provide a hydrophobic surrounding for the hydrophobic part of the retinol molecule.

RBP-retinol complex binds to a larger protein molecule, prealbumin, which further stabilizes it and prevents its loss via the kidney. Recognition of this complex by a cell-surface receptor causes RBP to release the retinol and, as a result, to undergo a conformational change that drastically reduces its affinity for prealbumin. The free RBP molecule is then excreted through the kidney glomerus, reabsorbed in the proximal tubule cells, and degraded.

The structure of RBP with bound retinol has been determined in the laboratory of Alwyn Jones in Uppsala, Sweden to 2.0 Å resolution. Its most striking feature is a β-barrel core consisting of eight up-and-down antiparallel β strands as shown in Figure 5.3. In addition, there is a four-turn α helix at the carboxy end of the polypeptide chain that is packed against the outside of the β barrel. The β strands are curved and twisted, and the barrel is wrapped around the retinol molecule. One end of the barrel is open to the solvent whereas the other end is closed by tight side-chain packing: the tail of the retinol molecule is at the open end of the barrel.

The hydrophobic retinol molecule is packed against hydrophobic side chains from the β strands in the barrel's core (Figure 5.4). The structure of the apo-form where the retinol molecule is removed is also known. Surprisingly, there is almost no change in the barrel structure, and a large hole is left inside the barrel.

Amino acid sequence reflects β structure

On a large part of the surface of RBP (the front face in Figure 5.3), side chains from residues in the β strands are exposed to the solvent. This is achieved by alternating hydrophobic with polar or charged hydrophilic residues in the

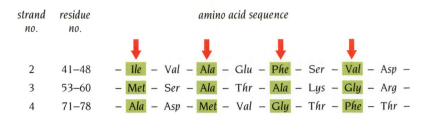

strand no.	residue no.	amino acid sequence
2	41–48	– Ile – Val – Ala – Glu – Phe – Ser – Val – Asp –
3	53–60	– Met – Ser – Ala – Thr – Ala – Lys – Gly – Arg –
4	71–78	– Ala – Asp – Met – Val – Gly – Thr – Phe – Thr –

Figure 5.5 Amino acid sequence of β strands 2 3 4 in human plasma retinol-binding protein. The sequences are listed in such a way that residues which point into the barrel are aligned. These hydrophobic residues are arrowed and colored green. The remaining residues are exposed to the solvent.

amino acid sequences of the β strands; in other words, side chains of the β strands form the hydrophobic core of the barrel as well as part of the hydrophilic outer surface. Strands 2 3 4 of RBP clearly illustrate this arrangement (Figure 5.5, where the core residues are colored green). This structure also very clearly illustrates that an antiparallel β barrel is built up from two β sheets that are packed against each other (see Figure 5.3). Beta strands 1 2 3 4 5 6 (blue and red color) form one sheet, and strands 1 8 7 6 5 (green and red color) form the second sheet. Strands 1 5 6 thus contribute to both sheets by having sharp corners where they can turn over from one sheet to the other.

The retinol-binding protein belongs to a superfamily of protein structures

RBP is one member of a **superfamily** of proteins with different functions, marginally homologous amino acid sequences, but similar three-dimensional structures. This superfamily also includes an insect protein that binds a blue pigment, biliverdin, and β-lactoglobulin, a protein that is abundant in milk. All have polypeptide chains of approximately the same lengths that are wrapped into very similar up-and-down eight-stranded antiparallel β barrels. They all tightly bind hydrophobic ligands inside this barrel.

There is a second family of small lipid-binding proteins, the **P2 family**, which include among others cellular retinol- and fatty acid-binding proteins as well as a protein, P2, from myelin in the peripheral nervous system. However, members of this second family have ten antiparallel β strands in their barrels compared with the eight strands found in the barrels of the RBP superfamily. Members of the P2 family show no amino acid sequence homology to members of the RBP superfamily. Nevertheless, their three-dimensional structures have similar architecture and topology, being up-and-down β barrels.

Neuraminidase folds into up-and-down β sheets

A second example of up-and-down β sheets is the protein neuraminidase from influenza virus. Here the packing of the sheets is different from that in RBP. They do not form a simple barrel but instead six small sheets, each with four β strands, which are arranged like the blades of a six-bladed propeller. Loop regions between the β strands form the active site in the middle of one side of the propeller. Other similar structures are known with different numbers of the same motif arranged like propellers with different numbers of blades such as the G-proteins discussed in Chapter 13.

Influenza virus is an RNA virus with an outer lipid envelope. There are two viral proteins anchored in this membrane, neuraminidase and hemagglutinin. They are both transmembrane proteins with a few residues inside the membrane and a transmembrane region followed by a stalk and a headpiece outside the membrane. The heads are exposed on the surface of the virion and thus provide the antigenic determinants of this epidemic virus. The function of hemagglutinin, which is glycosylated, is to mediate the binding of virus particles to host cells by recognizing and binding to sialic acid residues on glycoproteins of the cell membrane, as we shall discuss in more detail at the end of this chapter.

The role of the viral neuraminidase, conversely, seems to be to facilitate the release of progeny virions from infected cells by cleaving sialic acid residues from the carbohydrate side chains both of the viral hemagglutinin and of the glycosylated cellular membrane proteins. This helps prevent progeny virions from binding to and reinfecting the cells from whose surface they have just budded. From the point of view of the viruses, reinfecting an already infected cell is, of course, a waste of time.

The neuraminidase molecule is a homotetramer made up of four identical polypeptide chains, each of around 470 amino acids; the exact number varies depending on the strain of the virus. If influenza virus is treated with the proteolytic enzyme pronase, the head of the neuraminidase, which is soluble, is cleaved off from the stalk projecting from the viral envelope. The soluble head, comprising four subunits of about 400 amino acids each, can be crystallized.

Folding motifs form a propeller-like structure in neuraminidase

The structure of these tetrameric neuraminidase heads was determined in the laboratory of Peter Colman in Parkville, Australia to 2.9 Å resolution. Each of the four subunits of the tetramer is folded into a single domain built up from six closely packed, similarly folded motifs. The motif is a simple up-and-down antiparallel β sheet of four strands (Figure 5.6). The strands have a rather large twist such that the directions of the first and the fourth strands differ by 90°. To a first very rough approximation the six motifs are arranged within each subunit with an approximate sixfold symmetry around an axis through the center of the subunit (Figure 5.7a). These six β sheets are arranged like six blades of a propeller.

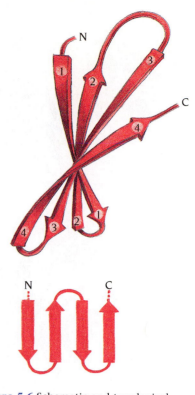

Figure 5.6 Schematic and topological diagrams of the folding motif in neuraminidase from influenza virus. The motif is built up from four antiparallel β strands joined by hairpin loops, an up-and-down open β sheet.

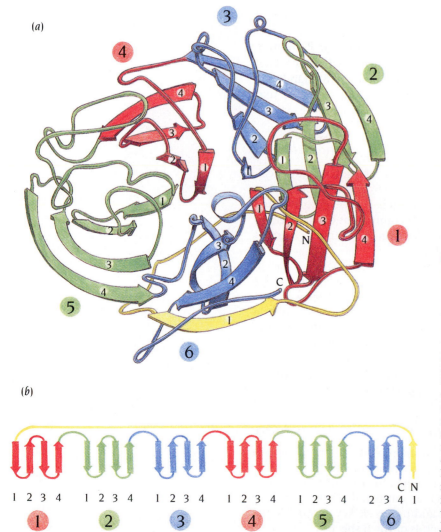

Figure 5.7 The subunit structure of the neuraminidase headpiece (residues 84–469) from influenza virus is built up from six similar, consecutive motifs of four up-and-down antiparallel β strands (Figure 5.6). Each such motif has been called a propeller blade and the whole subunit structure a six-blade propeller. The motifs are connected by loop regions from β strand 4 in one motif to β strand 1 in the next motif. The schematic diagram (a) is viewed down an approximate sixfold axis that relates the centers of the motifs. Four such six-blade propeller subunits are present in each complete neuraminidase molecule (see Figure 5.8). In the topological diagram (b) the yellow loop that connects the N-terminal β strand to the first β strand of motif 1 is not to scale. In the folded structure it is about the same length as the other loops that connect the motifs. (Adapted from J. Varghese et al., *Nature* 303: 35–40, 1983.)

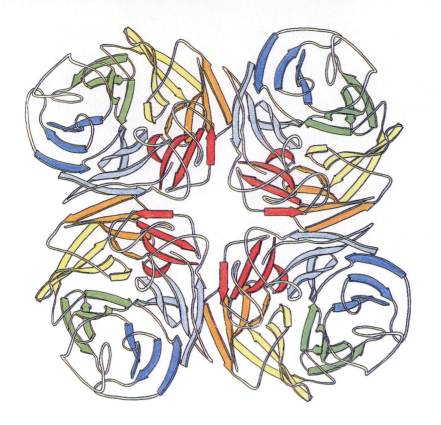

Figure 5.8 Schematic view down the fourfold axis of the tetrameric molecule of neuraminidase as it appeared on the cover of *Nature*, May 5, 1983.

In summary, the whole molecule has almost 1600 amino acid residues. It is composed of four identical polypeptide chains, each of which is folded into a superbarrel with 24 β strands (Figure 5.8). These 24 β strands are arranged in six similar motifs, each of which contains four β strands that form the blades of a propeller-like structure.

The active site is in the middle of one side of the propeller

Not only are the topologies within the six β sheets in each subunit identical, but so are their connections to each other, with the exception of the last β sheet (see Figure 5.7b). The fourth strand of each β sheet is connected across the top of the subunit (seen in Figure 5.7a coming out of the page) to the first strand of the next sheet. The loop that connects strands 2 and 3 within the sheet is also at the top of the subunit.

Furthermore, because of the approximate sixfold symmetry of the β-sheet motifs, these 12-loop regions, derived from the six β sheets, are on the same side of the molecule, as can be seen in Figure 5.9a, where we see a single polypeptide chain (one of the four subunits) from the side of the propeller. The β sheets are arranged cyclically around an axis through the center of the molecule. The loop regions at the top of this barrel are extensive (Figure 5.9a) and together they form a wide funnel-shaped pocket containing the active site (Figure 5.9b). This is analogous to the active site formed by the loop regions at the top of the α/β-barrel structures.

Greek key motifs occur frequently in antiparallel β structures

We saw in Chapter 2 that the **Greek key motif** provides a simple way to connect antiparallel β strands that are on opposite sides of a barrel structure. We will now look at how this motif is incorporated into some of the simple antiparallel β-barrel structures and show that an antiparallel β sheet of eight strands can be built up only by hairpin and/or Greek key motifs, if the connections do not cross between the two ends of the β sheet.

(a)

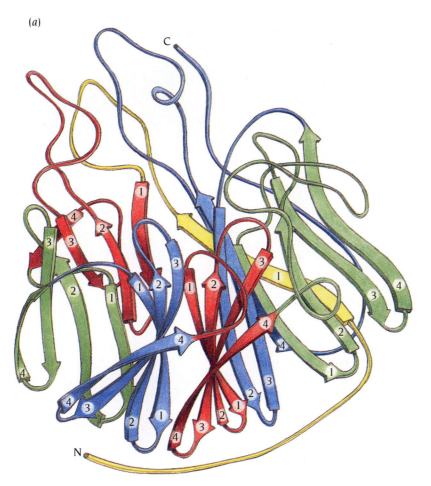

C

N

(b)

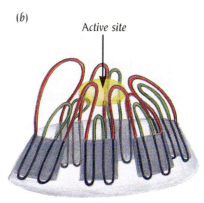

Active site

Figure 5.9 The six four-stranded motifs in a single subunit of neuraminidase form the six blades of a propeller-like structure. A schematic diagram of the subunit structure shows the propeller viewed from its side (a). An idealized propeller structure viewed from the side to highlight the position of the active site is shown in (b). The loop regions that connect the motifs (red in b) in combination with the loops that connect strands 2 and 3 within the motifs (green in b) form a wide funnel-shaped active site pocket. [(a) Adapted from P. Colman et al., *Nature* 326: 358–363, 1987.]

Assume that we have eight antiparallel β strands arranged in a barrel structure. We decide that we want to connect strand number *n* to an antiparallel strand at the same end of the barrel. We do not want to connect it to strand number *n* + 1 as in the up-and-down barrels just described, nor do we want to connect it to strand number *n* − 1 which is equivalent to turning the up-and-down barrel in Figure 5.2 upside down. What alternatives remain?

It is easy to see from Figure 5.10 that there are only two alternatives. We can connect it either to strand number *n* + 3 or to *n* − 3. Both cases require only short loop regions that traverse the end of the barrel. How do we now continue the connections? The simplest way to connect the strands that were skipped over is to join them by up-and-down connections, as illustrated in Figure 5.10.

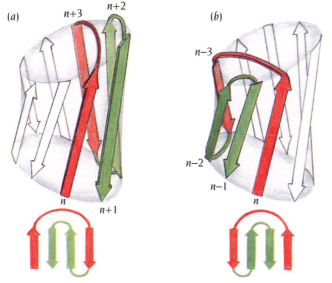

(a)   *n*+3   *n*+2   (b)   *n*−3   *n*−2   *n*−1   *n*   *n*+1

Figure 5.10 Idealized diagrams of the Greek key motif. This motif is formed when one of the connections of four antiparallel β strands is not a hairpin connection. The motif occurs when strand number *n* is connected to strand *n* + 3 (a) or *n* - 3 (b) instead of *n* + 1 or *n* - 1 in an eight-stranded antiparallel β sheet or barrel. The two different possible connections give two different hands of the Greek key motif. In all protein structures known so far only the hand shown in (a) has been observed.

We have now connected four adjacent strands of the barrel in a simple and logical fashion requiring only short loop regions. The result is the Greek key motif described in Chapter 2, which is found in the large majority of antiparallel β structures. The two cases represent the two possible different hands, but in all structures known to us the hand that corresponds to the case where β strand *n* is linked to β strand *n* + 3 as in Figure 5.10a is present.

The remaining four strands of the barrel can be joined either by up-and-down connections before and after the motif or by another Greek key motif. We will examine examples of both cases.

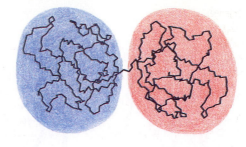

Figure 5.11 A computer-generated diagram of the structure of γ crystallin comprising one polypeptide chain of 170 amino acid residues. The diagram illustrates that the polypeptide chain is arranged in two domains (blue and red). Only main chain (N, C′, C$_\alpha$) atoms and no side chains are shown.

The γ–crystallin molecule has two domains

The transparency and refractive power of the lenses of our eyes depend on a smooth gradient of refractive index for visible light. This is achieved partly by a regular packing arrangement of the cells in the lens and partly by a smoothly changing concentration gradient of lens-specific proteins, the crystallins.

There are at least three different classes of crystallins. The α and β are heterogeneous assemblies of different subunits specified by different genes, whereas the **gamma (γ) crystallins** are monomeric proteins with a polypeptide chain of around 170 amino acid residues. The structure of one such γ crystallin was determined in the laboratory of Tom Blundell in London to 1.9 Å resolution. A picture of this molecule generated from a graphics display is shown in Figure 5.11.

Let us now examine this molecule and dissect it into its structural components to see if we can understand how these are put together. We will reduce this rather complex, and at first sight bewildering, structure to its simplest representation as a series of motifs. This will help us to understand the structure and see its relationships to other structures.

We can immediately discern from Figure 5.11 that the molecule is divided into two clearly separated domains that seem to be of similar size. For the next step we would need a stereopicture of the model or, much better, a graphics display where we could manipulate the model and look at it from different viewpoints. Here instead we have made a schematic diagram of one domain (Figure 5.12), which is normally not done until the analysis is completed and the structural principles are clear.

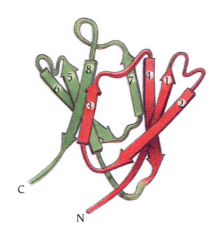

Figure 5.12 Schematic diagram of the path of the polypeptide chain in one domain (the blue region in Figure 5.11) of the γ-crystallin molecule. The domain structure is built up from two β sheets of four antiparallel β strands, sheet 1 from β strands 1, 2, 4, and 7 and sheet 2 from strands 3, 5, 6, and 8.

The domain structure has a simple topology

We will now follow the main polypeptide chain and trace out a topological diagram for this domain. We can immediately see from Figure 5.13 that the only secondary structure in the molecule is made up of β strands, which are arranged in an antiparallel fashion into two separate β sheets. Beta strands 1, 2, 4, and 7 form one antiparallel β sheet with the strand order 2 1 4 7. We thus draw the left four arrows in Figure 5.13 and connect strands 1 and 2. Similarly, we see that β strands 3, 5, 6, and 8 form another antiparallel β sheet with the strand order 6 5 8 3. We notice that strands 7 and 6 are adjacent although not hydrogen bonded to each other on the back side of the domain. We thus position strand 6 adjacent to strand 7 in the topology diagram but make a space between them to indicate that they belong to different β sheets. Alternatively, we could have positioned strands 2 and 3 adjacent to each other, which would have given a topologically identical diagram. We then connect the strands in consecutive order along the polypeptide chain.

Two Greek key motifs form the domain

The topological diagram of Figure 5.13 has been drawn to reflect the observation that the two β sheets are separate: β strands 2 and 3 are not hydrogen bonded to each other, nor are strands 6 and 7. The connections

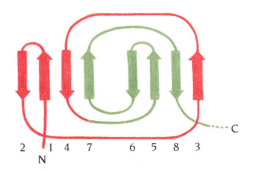

Figure 5.13 A preliminary topological diagram of the structure of one domain of γ crystallin shown in Figure 5.12, illustrating that the two β sheets are separate within the domain.

look unnecessarily complicated, but notice from the schematic diagram of the domain in Figure 5.12 that the two β sheets are packed against each other so that they form a distorted barrel. To see if the diagram can be simplified, we idealize the barrel and plot the strands along the surface of the barrel as shown in Figure 5.14. It is then immediately obvious that strands 1, 2, 3, and 4 form a Greek key motif, as do strands 5, 6, 7, and 8. These two motifs are joined by a loop across the bottom of the barrel, between strands 4 and 5.

On the basis of this new insight we can draw the topology diagram shown in the left half of Figure 5.15b. What is the difference between this and the previous topological diagram we made? The only changes we have made are to move β strand 3 from the right edge to the left edge of the domain topology and to close the gap between strands 7 and 6. We have changed neither the strand order nor the connections between the strands; thus the two diagrams are topologically identical.

The two domains have identical topology

Using a graphics display, we could do the same thing for the second domain and arrive at the full topology diagram in Figure 5.15b. From this diagram it is obvious that the two domains have identical topology and thus in all probability similar structures. This realization is not at all trivial. To be able

Figure 5.14 The eight β strands in one domain of the crystallin structure in this idealized diagram are drawn along the surface of a barrel. From this diagram it is obvious that the β strands are arranged in two Greek key motifs, one (red) formed by strands 1–4 and the other (green) by strands 5–8. Notice that the β strands that form one motif contribute to both β sheets as shown in Figure 5.12.

(a)

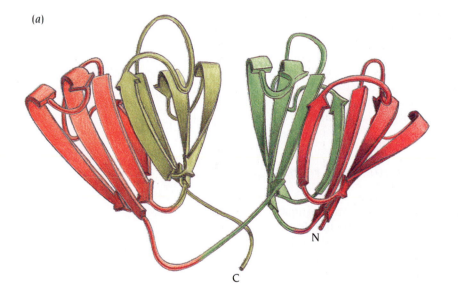

(b)

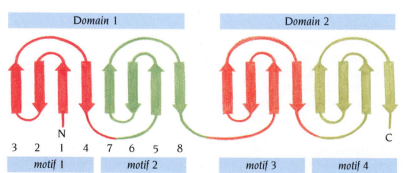

Domain 1   Domain 2

N

C

3 2 1 4   7 6 5 8

motif 1   *motif 2*   *motif 3*   *motif 4*

Figure 5.15 Schematic diagram (a) and topology diagram (b) for the γ-crystallin molecule. The two domains of the complete molecule have the same topology; each is composed of two Greek key motifs that are joined by a short loop region. [(a) Adapted from T. Blundell et al., *Nature* 289: 771–777, 1981.]

to see it when one looks at the structure on the display requires considerable experience because the two domains are in different orientations in the molecule. The brain therefore has to store the image of one domain while examining different orientations of the second domain. A topology diagram, on the other hand, immediately reveals similarities in domain structures. This illustrates one very important use of topology diagrams—namely, to reduce a complicated pattern to a simpler one, from which conclusions can be drawn that are also valid for the complicated pattern.

The two domains have similar structures

A relevant question to ask at this stage is, do the topological identities displayed in the diagram reflect structural similarity? We can now see that topologically the polypeptide chain is divided into four consecutive Greek key motifs arranged in two domains. How similar are the domain structures to each other, and how similar are the two motifs within each domain?

Tom Blundell has answered these questions by superposing the C_α atoms of the two motifs within a domain with each other and by superposing the C_α atoms of the two domains with each other. As a rule of thumb, when two structures superpose with a mean deviation of less than 2 Å they are considered structurally equivalent. For each pair of motifs Blundell found that 40 C_α atoms superpose with a mean distance of 1.4 Å. These 40 C_α atoms within each motif are therefore structurally equivalent. Since each motif comprises only 43 or 44 amino acid residues in total, these comparisons show that the structures of the complete motifs are very similar. Not only are the individual motifs similar in structure, but they are also pairwise arranged into the two domains in a similar way since superposition of the two domains showed that about 80 C_α atoms of each domain were structurally equivalent.

This structural similarity is also reflected in the amino acid sequences of the domains, which show 40% identity. They are thus clearly homologous to each other. The motif structures within the domains superpose equally well but their sequence homology is less, being around 30% between motifs 1 and 2 and 20% between 3 and 4. This study, however, clearly shows that the topological description in terms of four Greek key motifs is also valid at the structural and amino acid sequence levels.

The Greek key motifs in γ crystallin are evolutionarily related

These comparisons strongly suggest that the four Greek key motifs are evolutionarily related. We can guess from the amino acid sequence comparison that this protein evolved in two stages, beginning with the duplication of a primordial gene coding for one motif of about 40 amino acid residues, followed by fusion of the duplicated genes to give a single gene encoding one domain. The gene for this domain, we may imagine, later duplicated in turn and fused to give the full gene for the present-day γ-crystallin polypeptide. The evidence that this was the second step lies in the fact that the amino acid sequence homology is greater between the domains than between the motifs within each domain.

There is some circumstantial evidence in the organization of the crystallin gene for the evolutionary history that we have reconstructed. The amino acid sequence of a mouse β crystallin is homologous to that of γ crystallin and shows the same four homologous motifs. Its coding sequence is in separate DNA sequences (exons) interrupted by noncoding DNA sequences (introns). Walter Gilbert at Harvard University suggested in 1978 that genes for large proteins might have evolved by the accidental juxtaposition of exons coding for specific functions. In β crystallin the three introns are positioned at the junctions between the four motifs, supporting Gilbert's ideas. These introns could, therefore, be evolutionary remnants of the gene duplication and fusion events.

The Greek key motifs can form jelly roll barrels

In antiparallel barrel structures with the Greek key motif one of the connections in the motif is made across one end of the barrel. Such connections can be made several times in a β barrel giving different variations and combinations of the Greek key motif. In the structure of crystallin there are two consecutive Greek key motifs that form a barrel with two such connections. There is a different but frequently occurring motif, the **jelly roll motif**, in which there are four connections of this type. It is called jelly roll, or Swiss roll, because the polypeptide chain is wrapped around a barrel core like a jelly roll. This motif has been found in a variety of different structures including the coat proteins of most of the spherical viruses examined thus far by x-ray crystallography, the plant lectin concanavalin A and the hemagglutinin protein from the influenza virus.

The jelly roll motif is wrapped around a barrel

To illustrate how this rather complicated structure is built up, we will start by wrapping a piece of string around a barrel as shown in Figure 5.16. The string goes up and down the barrel four times, crosses over once at the bottom and twice at the top of the barrel. This configuration is the basic pattern for the jelly roll motif.

Let us do the same thing with a strip of paper the width of which is approximately one-eighth of the circumference of the top of the barrel. We imagine that a polypeptide chain follows the edges of this strip, starting at the bottom right corner of the strip and ending at the bottom left corner (Figure 5.17a). The polypeptide chain has eight straight sections, β strands, interrupted by loop regions. The β strands are arranged in a long antiparallel hairpin such that strand 1 is hydrogen bonded to strand 8, strand 2 to strand 7, and so on.

We now wrap the strip around the barrel following the path of the string in Figure 5.16 and in such a way that the β strands go along the sides of the barrel and the loop regions form the connections at the top and bottom of the barrel (Figure 5.17b).

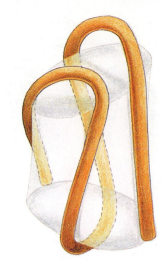

Figure 5.16 A diagram of a piece of string wrapped around a barrel to illustrate the basic pattern of a jelly roll motif.

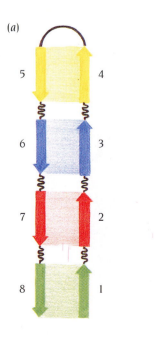

Figure 5.17 A simple illustration of the way eight β strands are arranged in a jelly roll motif. (a) The eight β strands are drawn as arrows along two edges of a strip of paper. The strands are arranged such that strand 1 is opposite strand 8, etc. The β strands are separated by loop regions. (b) The strip of paper in (a) is wrapped around a barrel in the same way as the string in Figure 5.16, such that the β strands follow the surface of the barrel and the loop regions (gray) provide the connections at both ends of the barrel. The β strands are now arranged in a jelly roll motif.

(a)

(b)

Figure 5.18 Topological diagrams of the jelly roll structure. The same color scheme is used as in Figure 5.17.

The hydrogen-bonded antiparallel β strand pairs 1:8, 2:7, 3:6, and 4:5 are now arranged such that β strand 1 is adjacent to strand 2, 7 is adjacent to 4, 5 to 6, and 3 to 8. These can also form hydrogen bonds to each other. All adjacent β strands are antiparallel. This is the basic jelly roll β-barrel structure for eight β strands (Figure 5.18a). Most such barrels have eight strands, but any even number of strands greater than four can form a jelly roll barrel. In eight-stranded barrels there are two connections across the top of the barrel and two across the bottom. In addition, there are two connections between adjacent β strands at the top and one at the bottom. A topological diagram of this fold is given in Figure 5.18b.

The jelly roll barrel is thus conceptually simple, but it can be quite puzzling if it is not considered in this way. Discussion of these structures will be exemplified in this chapter by hemagglutinin and in Chapter 16 by viral coat proteins.

The jelly roll barrel is usually divided into two sheets

The barrels we have used to illustrate both the Greek key and the jelly roll structures provide topological descriptions, as defined in Chapter 2. A topological description accurately represents the connectivity and the strand order around the barrel and thus is very useful in the same way that a subway map tells you how stations are interconnected. However, when one analyzes the pattern of hydrogen bonds between the β strands of such barrels, one finds that they usually form two sheets with few if any hydrogen bonds between strands that belong to the different β sheets, as we saw in the crystallin structure. The barrel is distorted and adjacent β strands are separated from each other in two places across the barrel. The division of β strands into these two sheets does not necessarily follow the division into topological motifs. The β strands in jelly roll barrels are also usually arranged in two sheets that are packed against each other. This does not, however, change either the topology or the usefulness of the description of these structures as barrels as long as one keeps in mind that these barrels are distorted and flattened.

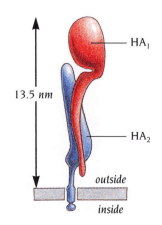

Figure 5.19 Schematic picture of a single subunit of influenza virus hemagglutinin. The two polypeptide chains HA1 and HA2 are held together by disulfide bridges.

The functional hemagglutinin subunit has two polypeptide chains

We have already discussed one envelope protein of influenza virus, neuraminidase, as an example of an up-and-down antiparallel β motif. In the second envelope protein, hemagglutinin, one domain of the polypeptide chain is folded into a jelly roll motif. We shall now look at some other features of hemagglutinin that are important for its biological function.

The hemagglutinin polypeptide chain is synthesized on membrane-bound ribosomes of the rough endoplasmic reticulum and then cotranslationally inserted into the membrane. The polypeptide chain is proteolytically cleaved to yield two chains of 328 and 221 amino acids called HA$_1$ and HA$_2$, respectively, which are held together by disulfide bonds (Figure 5.19). Three hemagglutinin monomers, each with one HA$_1$ and one HA$_2$ chain, trimerize in the rough endoplasmic reticulum and are transported from there through the Golgi apparatus to the plasma membrane, where the functional part of the molecule is now outside the cell anchored to the cell membrane via the HA$_2$ tails.

Progeny virus particles then bud from patches of the infected cell's plasma membrane that contain both the viral hemagglutinin and neuraminidase. The viral envelopes therefore contain both viral membrane proteins but no cellular membrane proteins.

Protease treatment of influenza virus particles cleaves the three HA$_2$ chains of the trimeric hemagglutinin molecules. Such cleavage leaves three C-terminal chains, 47 amino acids long, still inserted into the viral envelope and releases a second much larger soluble fragment. The soluble fragment consists of three complete HA$_1$ chains disulfide bonded to three HA$_2$ chains complete except for their membrane anchor regions. This soluble trimeric fragment has been crystallized at neutral pH and its structure was determined at 3 Å resolution and subsequently refined to a high resolution in the laboratory of Don Wiley at Harvard University. The influenza virus that he used was the Hong Kong 1968 strain, which caused the "Asian flu" pandemic disease.

The subunit structure is divided into a stem and a tip

The monomeric subunit is divided into a long, fibrous stemlike region extending outward from the membrane with a globular region at its tip (Figure 5.20). The globular region contains only residues of HA$_1$, while the stem contains some residues of HA$_1$ and all of HA$_2$.

The amino terminus of HA$_1$ is found at the base of the stem close to the viral membrane. The first 63 amino acids of HA$_1$ (pale red in Figure 5.20) reach, in an almost fully extended structure, nearly 100 Å along the length of the molecule before the first compact fold. These 63 residues form part of the stem region of the subunit. The globular tip is an eight-stranded distorted jelly roll structure comprising residues 116 to 261, which are folded into a distorted barrel. The remaining 70 residues of HA$_1$ return to the stem region, running nearly antiparallel to the initial stretch of 63 residues.

The major structural feature of the HA$_2$ chain (blue in Figure 5.20) is a hairpin loop of two α helices packed together. The second α helix is 50 amino acids long and reaches back 76 Å toward the membrane. At the bottom of the stem there is a β sheet of five antiparallel strands. The central β strand is from HA$_1$, and this is flanked on both sides by hairpin loops from HA$_2$. About 20 residues at the amino terminal end of HA$_2$ are associated with the activity by which the virus penetrates the host cell membrane to initiate infection. This region, which is quite hydrophobic, is called the fusion peptide.

The hemagglutinin trimer molecule is 135 Å long (from membrane to tip) and varies in cross-section between 15 Å and 40 Å. It is thus an unusually

Figure 5.20 Schematic diagram of the subunit structure of hemagglutinin from influenza virus. The structure comprises about 550 amino acids arranged in two chains HA$_1$ (red) and HA$_2$ (blue). The first half of each chain has a lighter color in the diagram. The subunit is very elongated with a long stemlike region built up by residues from both chains and includes one of the longest α helices known in a globular structure, about 75 Å long. The globular head is formed by residues only from HA$_1$. (Courtesy of Don Wiley, Harvard University.)

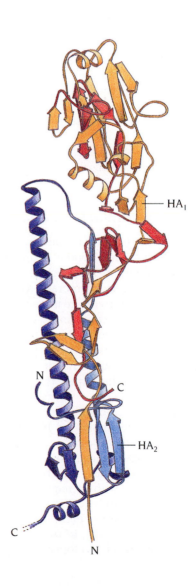

elongated molecule. The long fibrous stems of each subunit form the major subunit contacts (Figure 5.21). In particular, the three long HA2 helices, one from each subunit, intertwine by forming a coiled-coil structure (see Chapter 3) and pack against each other for part of their lengths forming a core 40 Å long stabilized by both hydrophobic residues and internal salt bridges. In addition the three heads of the subunits are close together and interact with each other, further stabilizing the trimer.

The receptor binding site is formed by the jelly roll domain

To initiate infection, the virus hemagglutinin binds to the sialic acid residues of glycosylated receptor proteins on the target cell surface. Once bound to the receptor, the virus is then taken into the cell by endocytosis. The receptor binding site on the hemagglutinin molecule has been determined from experiments with mutants of hemagglutinin and from the determination of structures with bound inhibitors (modified sialic acid molecules with substituents at two different positions of the sugar ring, Figure 5.22).

The binding site is located at the tip of the subunit within the jelly roll structure (Figure 5.23). The sialic acid moiety of the hemagglutinin inhibitors binds in the center of a broad pocket on the surface of the barrel (Figure 5.24). In addition to this groove there is a hydrophobic channel that can accomodate large hydrophobic substituents at the C_2 position of sialic acid (Figures 5.22 and 5.24).

Antibodies in our immune system bind to the receptor binding site, so preventing the virus from entering a cell. The virus can escape this neutralization through mutations in residues that form this binding site. Such mutations, however, are found only at the rim of the sialic acid binding pocket, presumably because mutation of residues inside this pocket would prevent the virus from binding to the cell surface receptor and consequently prevent viral propagation. The receptor binding site is therefore an ideal target for drug design, and studies of inhibitor binding to this site have provided valuable clues and ideas for the design of molecules that are candidate drug therapies for influenza virus infections.

Hemagglutinin acts as a membrane fusogen

In addition to binding to sialic acid residues of the carbohydrate side chains of cellular proteins that the virus exploits as receptors, hemagglutinin has a second function in the infection of host cells. Viruses, bound to the plasma membrane via their membrane receptors, are taken into the cells by endocytosis. Proton pumps in the membrane of endocytic vesicles that now contain the bound viruses cause an accumulation of protons and a consequent lowering of the pH inside the vesicles. The acidic pH (below pH 6) allows hemagglutinin to fulfill its second role, namely, to act as a membrane fusogen by inducing the fusion of the viral envelope membrane with the membrane of the endosome. This expels the viral RNA into the cytoplasm, where it can begin to replicate.

This fusogenic activity of influenza hemagglutinin is frequently exploited in the laboratory. If, for example, the virus is bound to cells at a temperature too low for endocytosis and then the pH of the external medium is lowered, the hemagglutinin causes direct fusion of the viral envelope with the plasma membrane; infection is achieved without endocytosis. Similarly, artificial vesicles with hemagglutinin in their membrane and other molecules in their lumen can be caused to fuse with cells by first allowing the vesicles to bind to the plasma membrane via the hemagglutinin and then lowering the pH of the medium. In this way the contents of the vesicles are delivered to the recipient cell's cytoplasm.

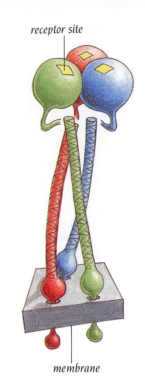

Figure 5.21 The hemagglutinin molecule is formed from three subunits. Each of these subunits is anchored in the membrane of the influenza virus. The globular heads contain the receptor sites that bind to sialic acid residues on the surface of eukaryotic cells. A major part of the subunit interface is formed by the three long intertwining helices, one from each subunit. (Adapted from I. Wilson et al., *Nature* 289: 366–373, 1981.)

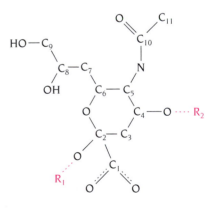

Figure 5.22 Chemical formula for sialic acid (α–5–*n*–acetylneuraminic acid) drawn in approximately the same orientation as the ball and stick models in Figure 5.24. R_1 and $R_{2'}$ which are H atoms in sialic acid, denote substituents introduced to design tightly bound inhibitors. These are large and hydrophobic as shown in Figure 5.24.

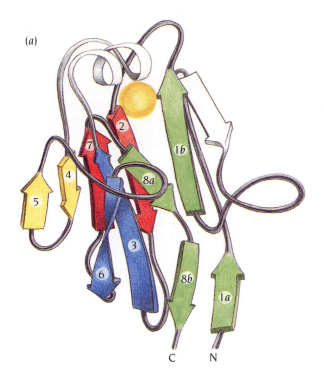

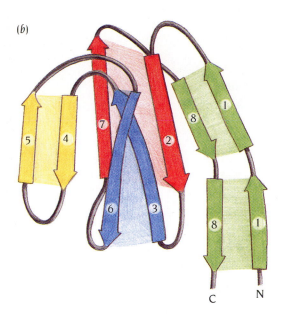

The structure of hemagglutinin is affected by pH changes

The structure of the soluble trimeric hemagglutinin fragment above pH 6 gave no indication of how hemagglutinin induces membrane fusion. When the soluble trimers were exposed to pH below 6 they aggregated and therefore could not be crystallized. However, by using monoclonal antibodies to specific epitopes on the two chains, HA$_1$ and HA$_2$, it was shown that lowering the pH causes a massive conformational change in hemagglutinin. As a result of this induced conformational change the hemagglutinin becomes highly susceptible to proteolytic cleavage. This property was used in the laboratory of Don Wiley to produce a soluble fragment of the low pH form comprising

Figure 5.23 The globular head of the hemagglutinin subunit is a distorted jelly roll structure (a). β strand 1 contains a long insertion, and β strand 8 contains a bulge in the corresponding position. Each of these two strands is therefore subdivided into shorter β strands. The loop region between β strands 3 and 4 contains a short α helix, which forms one side of the receptor binding site (yellow circle). A schematic diagram (b) illustrates the organization of the β strands into a jelly roll motif.

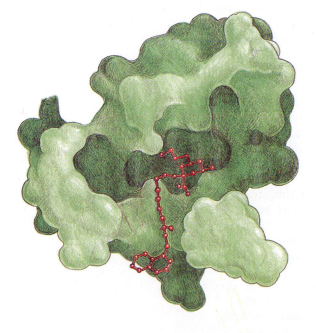

Figure 5.24 Space-filling model (green) of the sialic acid binding domain of hemagglutinin with a bound inhibitor (red) illustrating the different binding grooves. The sialic acid moiety of the inhibitor binds in the central groove. A large hydrophobic substituent, R$_1$, at the C$_2$ position of sialic acid binds in a hydrophobic channel that runs from the central groove to the bottom of the domain. (Adapted from S.J. Watowich et al., *Structure* 2: 719–731, 1994.)

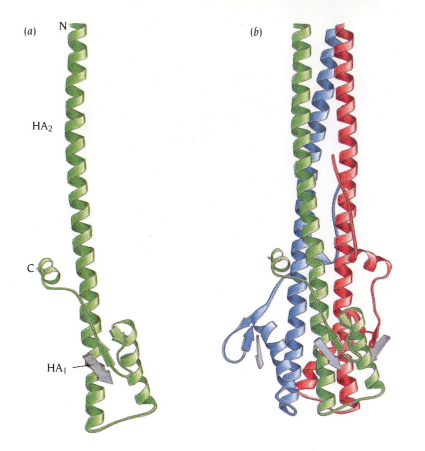

(a) N

HA₂

C

HA₁

(b)

Figure 5.25 Schematic diagrams of the structure of a proteolytic fragment of hemagglutinin at low pH where the molecule induces membrane fusion. (a) Residues 38–175 of the HA₂ polypeptide chain (green) form a 100-Å long α helix starting at the N-terminus followed by a loop, a β hairpin and finally a short C-terminal helix. In addition the diagram shows a β strand from the HA₁ polypeptide chain (gray) which participates with the β hairpin to form a three-stranded antiparallel β sheet. (b) The proteolytic fragment forms a trimer like the intact hemagglutinin molecule. The three long α helices of the three subunits intertwine to form a three-stranded coiled coil. (Adapted from P.A. Bullough et al., *Nature* 371: 37–43, 1994.)

residues 1–27 of the HA₁ chain and residues 38–175 of the HA₂ chain that could be crystallized; its structure was determined in 1994 (Figure 5.25).

The structure of this fragment, which like the high pH fragment is a trimer, confirms that the HA₂ subunit undergoes major structural changes at low pH. Most of the secondary structure elements are essentially preserved but there are two important exceptions (Figure 5.26). First, the loop region B between the two long α helices A and C + D in the high pH structure changes into an α helix. Second, an α-helical region in the middle of helix C + D changes into a loop region (Figure 5.26). The resulting helix A + B + C comprises 65 residues and is about 100 Å long. Previous examination of the amino acid sequence of this whole region, residues 40–106, had shown a very clear heptad repeat typical of coiled-coil structures (see Chapter 3). It was therefore surprising that in the high pH structure only a part of this region, helix C, actually has a coiled-coil structure. Reassuringly, in the low pH structure the whole region is a coiled coil which is involved in trimerization (Figure 5.25).

The new loop region between helices C and D changes completely the position and orientation of helix D so that instead of being continuous with helix C it is packed sideways against helix C in an antiparallel fashion (Figure 5.26). The small α helices and β strands at the C-terminal region of the chain collectively follow the realignment of helix D and occupy totally different positions (Figure 5.26).

The consequences for a possible fusion mechanism of this structural change are significant. In the high pH structure, the fusion peptide is attached to the N-terminus of helix A and is about 100 Å away from the receptor binding site. At low pH the N-terminus of helix A moves about 100 Å and is brought to the same region of the molecule as the receptor binding site (Figure 5.27). Even though the picture is incomplete, since only a proteolytic fragment of the low pH form was studied, it is quite clear that after these structural changes the likely positions of the receptor binding site and the fusion peptide are compatible with a fusion mechanism whereby hemagglutinin brings the viral and cellular membranes close together.

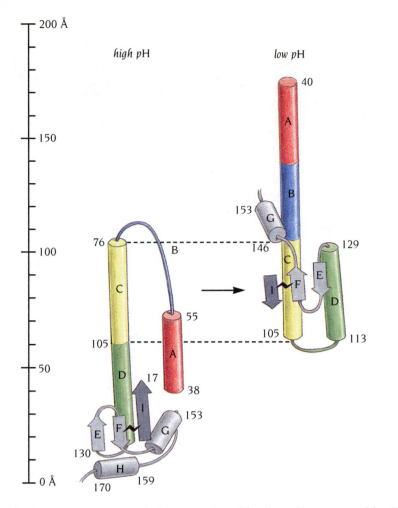

Figure 5.26 Schematic diagram illustrating the large conformational differences between the high and low pH forms of hemagglutinin. The loop region B (blue) in the high pH form has changed into an α helix producing a continuous 100-Å-long helix composed of regions A (red), B (blue) and C (yellow) at low pH. Furthermore, residues 105–113, which in the high pH form are in the middle of helix C–D, form a loop in the low pH form causing helix D (green) to be at a very different position. Consequently the β-hairpin E–F and the C-terminal helix G as well as the β strand I from HA$_1$ occupy very different positions in the two forms even though they have the same internal structure. The squiggle between β strands F and I denotes the S-S bond that joins HA$_1$ to HA$_2$. (Adapted from P.A. Bullough et al., *Nature* 371: 37–43, 1994.)

The large conformational change induced by low pH is irreversible. The low pH form is more thermostable than the high pH form and does not revert to the initial state when pH is raised. This suggests that the low pH form is lower in energy than the high pH form and that the energy required for fusion is stored up during the formation of the hemagglutinin molecule. Each of the three subunits of the trimeric hemagglutinin starts out as a single

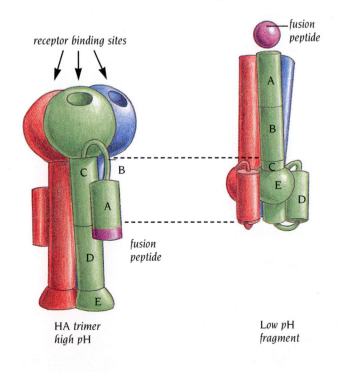

Figure 5.27 Schematic representation of a model for the conformational change of hemagglutinin that at low pH brings the fusion peptide to the same end of the molecule as the receptor binding site. The fusion peptide (purple) is at the end of helix A about 100 Å away from the receptor binding site in the high pH form. In the low pH fragment this region of helix A has moved about 100 Å towards the area where the receptor binding sites are expected to be in the intact hemagglutinin molecule. (Adapted from D. Stuart, *Nature* 371: 19–20, 1994.)

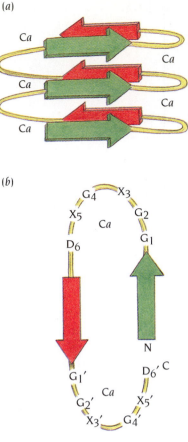

Figure 5.28 Schematic diagrams of the two-sheet β helix. Three complete coils of the helix are shown in (a). The two parallel β sheets are colored green and red, the loop regions that connect the β strands are yellow. (b) Each structural unit is composed of 18 residues forming a β-loop-β-loop structure. Each loop region contains six residues of sequence Gly-Gly-X-Gly-X-Asp where X is any residue. Calcium ions are bound to both loop regions. (Adapted from F. Jurnak et al., *Curr. Opin. Struct. Biol.* 4: 802–806, 1994.)

polypeptide chain, folded into a stable conformation. On cleavage to form the HA_1 and HA_2 chains of the mature molecules the free ends snap 20 Å apart to give the metastable high pH structure. This molecule is like a set trap: lowering the endosomal pH springs the trap, setting in motion a series of events which starts with a large conformational change to bring the fusion peptide into a position where it can engage the target membrane.

Parallel β–helix domains have a novel fold

In the first edition of this book this chapter was entitled "Antiparallel Beta Structures" but we have had to change this because an entirely unexpected structure, the β helix, was discovered in 1993. The β helix, which is not related to the numerous antiparallel β structures discussed so far, was first seen in the bacterial enzyme pectate lyase, the structure of which was determined by the group of Frances Jurnak at the University of California, Riverside. Subsequently several other protein structures have been found to contain β helices, including extracellular bacterial proteinases and the bacteriophage P22 tailspike protein.

In these β-helix structures the polypeptide chain is coiled into a wide helix, formed by β strands separated by loop regions. In the simplest form, the two-sheet β helix, each turn of the helix comprises two β strands and two loop regions (Figure 5.28). This structural unit is repeated three times in extracellular bacterial proteinases to form a right-handed coiled structure which comprises two adjacent three-stranded parallel β sheets with a hydrophobic core in between.

This structural organization has striking similarities to that of α/β proteins, the difference being that the loop-α helix-loop that connects the parallel β strands in α/β structures is substituted by a loop-β strand-loop in these β-helix structures. Instead of having one β sheet adjacent to a stack of α helices as in the α/β structures described in the previous chapter, β-helix structures have two parallel β sheets. In α/β structures a twist of about 20° between adjacent β strands is imposed by the packing requirements of the α helices: in order to pack ridges into grooves, as described in Chapter 3, the α helices have to be twisted with respect to each other and this forces the β strands also to be twisted. In β-helix structures no such constraint is present and therefore the sheets are almost planar and form straight walls (Figure 5.28a).

The basic structural unit of these two-sheet β helix structures contains 18 amino acids, three in each β strand and six in each loop. A specific amino acid sequence pattern identifies this unit; namely a double repeat of a nine-residue consensus sequence Gly-Gly-X-Gly-X-Asp-X-U-X where X is any amino acid and U is large, hydrophobic and frequently leucine. The first six residues form the loop and the last three form a β strand with the side chain of U involved in the hydrophobic packing of the two β sheets. The loops are stabilized by calcium ions which bind to the Asp residue (Figure 5.28). This sequence pattern can be used to search for possible two-sheet β structures in databases of amino acid sequences of proteins of unknown structure.

A more complex β helix is present in pectate lyase and the bacteriophage P22 tailspike protein. In these β helices each turn of the helix contains three short β strands, each with three to five residues, connected by loop regions. The β helix therefore comprises three parallel β sheets roughly arranged as the three sides of a prism. However, the cross-section of the β helix is not quite triangular because of the arrangement of the β sheets. Two of the sheets are arranged adjacent to each other as in the two-sheet β helix, and the third β sheet is almost perpendicular to the other two (Figure 5.29). The β strands in these three parallel sheets are connected by three loop regions. One loop (loop *a* in Figure 5.29) is short and always formed by only two residues which have invariant conformations. The other two loops are much longer and vary in size and conformation. These long loops protrude from the β sheets and probably form the active site regions on the external surface of the protein. Since the long loop regions vary in size, the number and type of amino acids in each turn of the β-helix structures vary. Consequently, no specific amino acid sequence pattern has been identified for the three-sheet β-helix structures.

The number of helical turns in these structures is larger than those found so far in two-sheet β helices. The pectate lyase β helix consists of seven complete turns and is 34 Å long and 17–27 Å in diameter (Figure 5.30) while the β-helix part of the bacteriophage P22 tailspike protein has 13 complete turns. Both these proteins have other structural elements in addition to the β-helix moiety. The complete tailspike protein contains three intertwined, identical subunits each with the three-sheet β helix and is about 200 Å long and 60 Å wide. Six of these trimers are attached to each phage at the base of the icosahedral capsid.

The interior of the β-helix structure in pectate lyase is completely filled with side chains leaving no room for a channel. Interestingly, these side chains are not limited to hydrophobic groups but include polar and charged groups which are all neutralized either by hydrogen bonding or by electrostatic interactions. All the side chains in the interior of the helix are stacked such that the side chains of adjacent turns of the helix have a linear arrangement parallel to the helix axis. These internal stacks fall into different classes; polar stacks of Asn or Ser side chains, aliphatic stacks of Ala, Val, Leu or Ile side chains and aromatic stacks of Phe or Tyr side chains. Pectate lyase is an unusually stable protein and this stacking arrangement no doubt contributes to its stability.

Conclusion

Antiparallel β structures comprise the second major class of protein conformations. In these the antiparallel β strands are usually arranged in two β sheets that pack against each other and form a distorted barrel structure, the core of the molecule. Depending on the way the β strands around the barrel are connected along the polypeptide chain—in other words, depending on the topology of the barrels—they are divided into three main groups: up-and-down barrels, Greek key barrels, and jelly roll barrels.

The number of possible ways to form antiparallel β structures is very large. The number of topologies actually observed is small, and most β structures fall into these three major groups of barrel structures. The last two groups—the Greek key and jelly roll barrels—include proteins of quite diverse function, where functional variability is achieved by differences in the loop regions that connect the β strands that build up the common core region.

Up-and-down barrels are the simplest structures. Each β strand is connected to the next strand by a short loop region. Eight β strands arranged

(a)

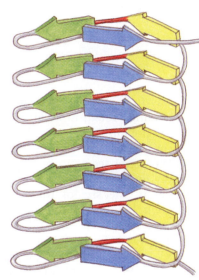

(b)

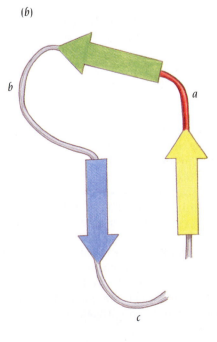

Figure 5.29 Schematic diagrams of the three-sheet β helix. (a) The three sheets of parallel β strands are colored green, blue and yellow. Seven complete coils are shown in this diagram but the number of coils varies in different structures. Two of the β sheets (blue and yellow) are parallel to each other and are perpendicular to the third (green). (b) Each structural unit is composed of three β strands connected by three loop regions (labeled *a*, *b* and *c*). Loop *a* (red) is invariably composed of only two residues, whereas the other two loop regions vary in length. (Adapted from F. Jurnak et al., *Curr. Opin. Struct. Biol.* 4: 802–806, 1994.)

(a)   (b)

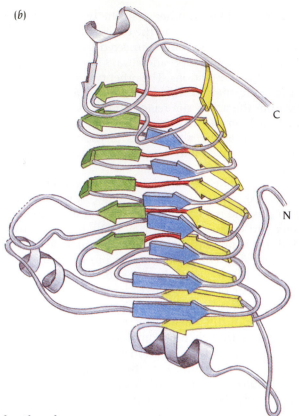

in this way form the core of a family of proteins that includes the plasma retinol-binding protein in mammals, biliverdin-binding proteins in insects, and β-lactoglobulin from milk. Members of this family, as well as of the related P2 family with 10 β strands, bind large hydrophobic ligands inside the barrel. The barrel seems to be particularly suited to act as a container for chemically quite diverse ligands. Diversity in ligand binding is achieved by differences in the size of the barrel and in the amino acids that also participate in building up the common core.

Most of the known antiparallel β structures, including the immunoglobulins and a number of different enzymes, have barrels that comprise at least one Greek key motif. An example is γ crystallin, which has two consecutive Greek key motifs in each of two barrel domains. These four motifs are homologous in terms of both their three-dimensional structure and amino acid sequence and are thus evolutionarily related.

The jelly roll barrels are found in a variety of protein molecules, including viral coat proteins and hemagglutinin from influenza virus. This structure looks complicated but, in principle, is very simple if one thinks of the analogy of wrapping a strip of paper around a barrel, like a jelly roll. The hemagglutinin receptor-binding domain forms such a jelly roll barrel of eight β strands, where the receptor binding site is at one end of the barrel. During the membrane fusion process of influenza virus infection the hemagglutinin molecule undergoes a major structural change in which the fusion peptide moves about 100 Å to a position close to the receptor binding site.

The second protein in the membrane of influenza virus, neuraminidase, does not belong to any of these three groups of barrel structures. Instead, it forms a propeller-like structure of 24 β strands, arranged in six similar motifs that form the six blades of the propeller. Each motif is a β sheet of 4 up-and-down-connected β strands. The enzyme active site is formed by loop regions on one side of the propeller.

In addition to the antiparallel β-structures, there is a novel fold called the β helix. In the β-helix structures the polypeptide chain is folded into a wide helix with two or three β strands for each turn. The β strands align to form either two or three parallel β sheets with a core between the sheets completely filled with side chains.

Figure 5.30 Schematic diagrams of the structure of the enzyme pectate lyase C, which has a three-sheet parallel β-helix topology.
(a) Idealized diagram highlighting the helical nature of the path of the polypeptide chain which comprises eight helical turns. Dotted regions indicate positions where large external loops have been removed for clarity.
(b) Ribbon diagram of the polypeptide chain. The predominant secondary structural elements are three parallel β sheets which are colored green, blue and yellow. Each β sheet is composed of 7–10 parallel β strands with an average length of four to five residues in each strand. The short loop regions of two residues length are shown in red. (Adapted from M.D. Yoder et al., *Science* 260: 1503–1507, 1993.)

Selected readings

General

Chothia, C. Conformation of twisted β-pleated sheets in proteins. *J. Mol. Biol.* 75: 295–302, 1973.

Chothia, C., Janin, J. Orthogonal packing of β-pleated sheets in proteins. *Biochemistry* 21: 3955–3965, 1982.

Chothia, C., Janin, J. Relative orientation of close-packed β-pleated sheets in proteins. *Proc. Natl. Acad. Sci. USA* 78: 4146–4150, 1981.

Cohen, F.E., Sternberg, M.J.E., Taylor, W.R. Analysis of the tertiary structure of protein β-sheet sandwiches. *J. Mol. Biol.* 148: 253–272, 1981.

Edison, A.S. Propagation of an error: β-sheet structures. *Trends Biochem. Sci.* 15: 216–217, 1990.

Efimov, A.E. Favoured structural motifs in globular proteins. *Structure* 2: 999–1002, 1995.

Gilbert, W. Why genes in pieces? *Nature* 271: 501, 1978.

Godovac-Zimmerman, J. The structural motif of β-lactoglobulin and retinol-binding protein: a basic framework for binding and transport of small hydrophobic molecules? *Trends Biochem. Sci.* 13: 64–66, 1988.

Jurnak, F., et al. Parallel β domains: a new fold in protein structures. *Curr. Opin. Struct. Biol.* 4: 802–806, 1994.

Lifson, S., Sander, C. Antiparallel and parallel β-strands differ in amino acid residue preference. *Nature* 282: 109–111, 1979.

Lifson, S., Sander, C. Specific recognition in the tertiary structure of β-sheets of proteins. *J. Mol. Biol.* 139: 627–639, 1980.

Ptitsyn, O.B., Finkelstein, A.V. Similarities of protein topologies: evolutionary divergence, functional convergence or principles of folding? *Q. Rev. Biophys.* 13: 339–386, 1980.

Ptitsyn, O.B., Finkelstein, A.V., Falk, P. Principal folding pathway and topology of all-β proteins. *FEBS Lett.* 101: 1–5, 1979.

Richardson, J.S. β-sheet topology and the relatedness of proteins. *Nature* 268: 495–500, 1977.

Richardson, J.S. Handedness of crossover connections in β-sheets. *Proc. Natl. Acad. Sci. USA* 72: 1349–1353, 1975.

Richardson, J.S., Getzoff, E.D., Richardson, D.C. The β-bulge: a common small unit of nonrepetitive protein structure. *Proc. Natl. Acad. Sci. USA* 75: 2574–2578, 1978.

Sawyer, L. Protein structure. One fold among many. *Nature* 327: 659, 1987.

Sibanda, B.L., Blundell, T.L., Thornton, J. Conformation of β-hairpins in protein structures. A systematic classification with applications to modelling by homology, electron density fitting and protein engineering. *J. Mol. Biol.* 206: 759–777, 1989.

Wilmot, C.M., Thornton, J.M. Analysis and prediction of the different types of β-turns in proteins. *J. Mol. Biol.* 203: 221–232, 1988.

Specific structures

Boumann, U., et al. Three-dimensional structure of the alkaline protease of *Pseudomonas aeruginosa*, a two-domain protein with a calcium binding parallel beta roll motif. *EMBO J.* 12: 3357–3364, 1993.

Bullough, P.A., et al. Structure of influenza haemagglutinin at the pH of membrane fusion. *Nature* 371: 37–43, 1994.

Colman, P.M., Varghese, J.N., Laver, W.G. Structure of the catalytic and antigenic sites in influenza virus neuraminidase. *Nature* 303: 41–44, 1983.

Daniels, R.S., et al. Fusion mutants of the influenza virus hemagglutinin glycoprotein. *Cell* 40: 431–439, 1985.

Jones, T.A., et al. The three-dimensional structure of P2 myelin protein. *EMBO J.* 7: 1597–1604, 1988.

McLachlan, A.D. Repeated folding pattern in copper-zinc superoxide dismutase. *Nature* 285: 267–268, 1980.

Newcomer, M.E., et al. The three-dimensional structure of retinol-binding protein. *EMBO J.* 3: 1451–1454, 1984.

Papiz, M.Z., et al. The structure of β-lactoglobulin and its similarity to plasma retinol-binding protein. *Nature* 324: 383–385, 1986.

Richardson, J.S., et al. Similarity of three-dimensional structure between the immunoglobulin domain and the copper, zinc superoxide dismutase subunit. *J. Mol. Biol.* 102: 221–235, 1976.

Sacchettini, J.C., et al. Refined apoprotein structure of rat intestinal fatty acid binding protein produced in *Escherichia coli. Proc. Natl. Acad. Sci. USA* 86: 7736–7740, 1989.

Steinbacher, S., et al. Crystal structure of P22 tailspike protein: interdigitated subunits in a thermostable trimer. *Science* 265: 383–386, 1994.

Summers, L., et al. X-ray studies of the lens specific proteins. The crystallins. *Pept. Protein Rev.* 3: 147–168, 1984.

Tainer, J.A., et al. Determination and analysis of the 2 Å structure of copper, zinc superoxide dismutase. *J. Mol. Biol.* 160: 181–217, 1982.

Varghese, J.N., Laver, W.G., Colman, P.M. Structure of the influenza virus glycoprotein antigen neuraminidase at 2.9 Å resolution. *Nature* 303: 35–40, 1983.

Watowich, S.I., et al. Crystal structures of influenza virus haemagglutinin in complex with high affinity receptor analogs. *Structure* 2: 719–731, 1994.

Weiss, W., et al. Structure of the influenza virus haemagglutinin complexed with its receptor, sialic acid. *Nature* 333: 426–431, 1988.

Wiley, D.C., Skehel, J.J. The structure and function of the hemagglutinin membrane glycoprotein of influenza virus. *Annu. Rev. Biochem.* 56: 365–394, 1987.

Wiley, D.C., Wilson, I.A., Skehel, J.J. Structural identification of the antibody-binding sites of Hong Kong influenza haemagglutinin and their involvement in the antigenic variation. *Nature* 289: 373–378, 1981.

Wilson, I.A., Skehel, J.J., Wiley, D.C. Structure of the haemagglutinin membrane glycoprotein of influenza virus at 3 Å resolution. *Nature* 289: 366–373, 1981.

Wistow, G., et al. X-ray analysis of the eye lens protein gamma-crystallin at 1.9 Å resolution. *J. Mol. Biol.* 170: 175–202, 1983.

Yoder, M.D., et al. New domain motifs: the structure of pectate lyase C, a secreted plant virulence factor. *Science* 260: 1503–1507, 1993.

Folding and Flexibility

6

A protein, as we have seen, is a polypeptide chain folded into one or more domains, each of which is made up of α helices, β sheets and loops. The process by which a polypeptide chain acquires its correct three-dimensional structure to achieve the biologically active native state is called protein folding. Although some polypeptide chains spontaneously fold into the native state, others require the assistance of enzymes, for example, to catalyze the formation and exchange of disulfide bonds; and many require the assistance of a class of proteins called chaperones. A chaperone binds to a partly folded polypeptide chain and prevents it from making illicit associations with other folded or partly folded proteins, hence the name chaperone. A chaperone also promotes the folding of the polypeptide chain it holds. After a polypeptide has acquired most of its correct secondary structure, with the α-helices and β-sheets formed, it has a looser tertiary structure than the native state and is said to be in the molten globular state. The compaction that is necessary to go from the molten globular state to the final native state occurs spontaneously.

Protein folding generates a particular three-dimensional structure from an essentially linear, one-dimensional structure—a polypeptide chain with a particular sequence of amino acid residues. How to predict the three-dimensional structure of a protein from its amino acid sequence is the major unsolved problem in structural molecular biology. If we had a general solution to the protein folding problem, it would be possible to write a computer program to simulate protein folding and generate the precise three-dimensional structure of any protein from its amino acid sequence. However, a general solution to the folding problem is still not in sight, even though the number of proteins whose three-dimensional structure has been solved experimentally, in other words, the database of known protein structures, is doubling every 2 years.

A protein in its native state is not static. The secondary structural elements of the domains as well as the entire domains continually undergo small movements in space, either fluctuations of individual atoms or collective motions of groups of atoms. Furthermore, the functional activities of many proteins depend upon large conformational changes triggered by ligand binding. In this chapter, after discussing protein folding, we shall examine some examples of functionally important conformational changes of proteins.

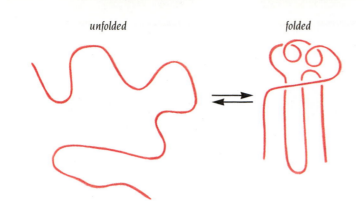

unfolded   folded

Figure 6.1 A polypeptide chain is extended and flexible in the unfolded, denatured state whereas it is globular and compact in the folded, native state.

Globular proteins are only marginally stable

Every biochemist or molecular biologist who has worked with proteins knows by experience that they are unstable. Slight changes in pH or temperature can convert a solution of biologically active protein molecules in their **native state** to a biologically inactive **denatured state**. The energy difference between these two states in physiological conditions is quite small, about 5–15 kcal/mol, not much more than the energy contribution of a single hydrogen bond, which is of the order of 2–5 kcal/mol.

There are two major contributors to the energy difference between the folded and the denatured state: enthalpy and entropy. **Enthalpy** derives from the energy of the noncovalent interactions within the polypeptide chain— the hydrophobic interactions, hydrogen bonds and ionic bonds. The covalent bonds within and between the amino acid residues in the polypeptide chain are the same in the native and denatured states, with the exceptions of disulfide bonds in those proteins where these form between cysteine residues. The noncovalent interactions on the other hand differ significantly between the two states. In the native state these interactions are maximized to produce a compact globular molecule with a tightly packed hydrophobic core whereas the denatured state is more open and the side chains are more loosely packed (Figure 6.1). These noncovalent interactions are therefore stronger and more frequent in the native state and hence their energy contribution, enthalpy, is much larger. The enthalpy difference between native and denatured states can reach several hundred kcal/mol.

Entropy derives from the second law of thermodynamics which states that energy is required to create order. Proteins in the native state are highly ordered in one main conformation whereas the denatured state is highly disordered, with the protein molecules in many different conformations. A typical experimental preparation of unfolded protein (a solution in 6 M guanidinium chloride or 8 M urea) contains 10^{15}–10^{20} protein molecules, each of which will have a unique conformation. In the absence of compensating factors it would therefore be entropically much more favorable for the protein to be in the disordered denatured state. The energy difference due to entropy between the native ordered state and the denatured state can also reach several hundred kcal/mole but in the opposite direction to the enthalpy difference. The total energy difference between the native and the denatured state of 5–15 kcal/mol, which is called the **free energy** difference, is thus a difference between two large numbers, the enthalpy difference and the entropy difference. The fact that this difference is very small is a severe complicating factor both for predictions of possible native states and for interpretation of factors responsible for the stability or instability of protein molecules, because our knowledge about the denatured state is very incomplete.

We know much more about factors that influence the stability of the native state, mainly from experiments using directed mutations in proteins of known three-dimensional structure. Such experiments have yielded precise information about energy contributions to the stability of the native state from close packing of hydrophobic side chains in the interior of the protein, and from the presence of disulfide bridges and interior hydrogen bonds and salt bridges, as well as from side chains that compensate the dipole moment of α helices (see Chapter 17).

The marginal stability of the native state over the denatured state is biologically very important. Living cells need globular proteins in correct quantities at appropriate times. It is therefore as important to be able easily to degrade these proteins as it is to be able to synthesize them. Globular proteins in living cells usually have a rather rapid turnover and their native states have therefore evolved to be only marginally stable. Moreover, the catalytic activities of enzymes, and other important functions of proteins, generally require some structural flexibility, which would be inconsistent with a rigidly stabilized structure.

Kinetic factors are important for folding

High resolution x-ray structure determinations of several hundred proteins have shown that in each case the specific sequence of a polypeptide chain appears to yield only a single, compact, biologically active fold in the native state. This fold generally has many substates with minor structural differences between them, as will be discussed later in this chapter, but all of these substates have the same general fold. Comparisons with structure determinations in solution by NMR show that the same fold also prevails in solution. In other words, under physiological conditions there appears to be one conformation for a given amino acid sequence that has a significantly lower free energy than any other. How is this folded state reached?

Intuitively one might imagine that all protein molecules search through all possible conformations in a random fashion until they are frozen at the lowest energy in the conformation of native state. The biophysicist Cyrus Levinthal showed in 1968 by a simple calculation that this is impossible. Assume as a gross simplification that each peptide group has only three possible conformations, the allowed regions α, β and L in the Ramachandran diagram (see Figure 1.7), and that it converts one conformation into another in the shortest possible time, one picosecond (10^{-12} seconds). A polypeptide chain of 150 residues would then have $3^{150} = 10^{68}$ possible conformations. To search all these conformations would require 10^{48} years (10^{56} seconds)—an astronomical number compared with the actual folding time, which is between 0.1 and 1000 seconds both *in vivo* and *in vitro*. To occur on this short time scale, the folding process must be directed in some way through a kinetic pathway of unstable intermediates to escape sampling a large number of irrelevant conformations.

Such a folding mechanism raises several important questions that are difficult to examine experimentally, since the possible intermediates have a very short lifetime. If kinetic factors are important for the folding process it is possible that the observed folded conformation is not the one with the lowest free energy but rather the most stable of those conformations that are kinetically accessible. The protein might be kinetically trapped in a local low energy state with a high energy barrier that prevents it from reaching the global energy minimum which might have a different fold. In such a case structure prediction by energy calculations would give the wrong structure even if such calculations could be made with great accuracy. One important question therefore is how a living cell can prevent the folding pathway from becoming blocked at an intermediate stage. The most common obstacles to correct folding seem to be (1) aggregation of the intermediates through exposed hydrophobic groups, (2) formation of incorrect disulfide bonds, and (3) isomerization of proline residues. To circumvent these three obstacles cells produce special proteins that assist the folding process, as we shall discuss later in this chapter.

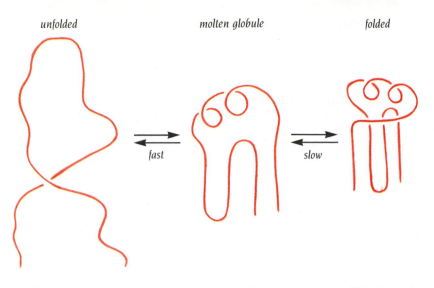

unfolded   molten globule   folded

fast   *slow*

Figure 6.2 The molten globule state is an important intermediate in the folding pathway when a polypeptide chain converts from an unfolded to a folded state. The molten globule has most of the secondary structure of the native state but it is less compact and the proper packing interactions in the interior of the protein have not been formed.

An alternative way to remove kinetic barriers is exemplified by α-lytic protease, a bacterial enzyme which belongs to the serine protease superfamily of enzymes (Chapter 11). Like many other proteases it is synthesized and folded *in vivo* as an inactive precursor protein with a prosegment of 77 residues. This segment is excised after folding to produce the active enzyme. Unfolded precursor protein refolds easily *in vitro* but unfolded α-lytic protease lacking the prosegment does not refold. However, a solution of unfolded enzyme can be induced to refold by adding the excised prosegment. The capacity for folding obviously exists in the unfolded enzyme but there is a barrier present somewhere in the folding pathway that prevents folding. The prosegment removes this kinetic barrier, presumably by interacting with the enzyme in the unfolded state and thereby lowering the free energy of the transition states for folding; just as enzymes lower the free energy of transition states for chemical reactions and thereby increase the rates of the reactions (see Chapter 11).

Molten globules are intermediates in folding

The first observable event in the folding pathway of at least some proteins is a collapse of the flexible disordered unfolded polypeptide chain into a partly organized globular state, which is called the **molten globule** (Figure 6.2). This event is fast, usually within the deadtime of the experimental observation, which is a few milliseconds. We therefore know almost nothing about the process that leads to the molten globule, but we know some of the properties of this state. The molten globule has most of the secondary structure of the native state and in some cases even native-like positions of the α helices and β strands. It is less compact than the native structure and the proper packing interactions in the interior of the protein have not been formed. The interior side chains may be mobile, more closely resembling a liquid than the solid-like interior of the native state. Also loops and other elements of surface structure remain largely unfolded, with different conformations. The molten globule should, therefore, not be viewed as a single structural entity but as an ensemble of related structures that are rapidly interconverting (see Figure 6.3a).

In a second step, which can last up to 1 second, persistent native-like elements of tertiary structure begin to develop, possibly in the form of subdomains that are not yet properly docked. The ensemble of conformations is much reduced compared with those of the molten globule but it is still far from a single form. The single native form is reached in the final stage of folding, which involves the formation of native interactions throughout the protein, including hydrophobic packing in the interior as well as the fixation of surface loops.

Burying hydrophobic side chains is a key event

The collapse of the unfolded state to generate the molten globule embodies the main mystery of protein folding. What is the driving force behind the choice of native tertiary fold from a randomly oriented polypeptide chain?

There is very little change in free energy by forming the internal hydrogen bonds that are characteristic of α helices and β sheets because in the unfolded state equally stable hydrogen bonds can be formed to water molecules. Secondary structure formation therefore cannot be the thermodynamic driving force of protein folding. On the other hand there is a large free energy change by bringing hydrophobic side chains out of contact with water and into contact with each other in the interior of a globular entity. Thus the most likely scenario is that the polypeptide chain begins to form a compact shape with hydrophobic side chains at least partially buried very early in the folding process. This scenario has several important consequences. It vastly reduces the number of possible conformations that need to be searched because only those that are sterically accessible within this shape can be sampled. Second, when some of the side chains are partly buried, their polar backbone -NH and -CO groups are also buried in a hydrophobic environment unable to form hydrogen bonds to water. This is energetically unfavorable unless they form hydrogen bonds to each other, which they can only do if they are close together. The simplest way to form such bonds is by forming elements of secondary structure: α helices and β sheets. The formation of secondary structure early in the folding process can therefore be regarded as a consequence of burying hydrophobic side chains and not as a driving force for the formation of the molten globule.

Looking at the amino acid sequence of a globular protein one finds that hydrophobic side chains are usually scattered along the entire sequence in a seemingly random manner. In the native state of the folded protein about half of these side chains are buried in the interior and the rest are scattered on the surface of the protein, surrounded by hydrophilic side chains. The buried hydrophobic side chains are not clustered in the sequence but are scattered along the entire polypeptide chain. What causes these residues to be selectively buried during the early and rapid formation of the molten globule? This question must be answered before one can solve the folding problem and be able to predict the fold of a protein from its amino acid sequence.

Both single and multiple folding pathways have been observed

In order to understand fully any folding pathway, all states of the pathway must be characterized both structurally and energetically. The simplified diagram in Figure 6.3 illustrates that during the folding process the protein proceeds from a high energy unfolded state to a low energy native state through metastable intermediate states with local low energy minima separated by unstable transition states of higher energy. The characterization of these states is not trivial and many different experimental techniques are employed, including NMR, hydrogen exchange, spectroscopy and thermochemistry.

Recently Alan Fersht, Cambridge University, has developed a protein engineering procedure for such studies. The technique is based on investigation of the effects on the energetics of folding of single-site mutations in a protein of known structure. For example, if minimal mutations such as Ala to Gly in the solvent-exposed face of an α helix, destabilize both an intermediate state and the native state, as well as the transition state between them, it is likely that the helix is already fully formed in the intermediate state. If on the other hand the mutations destabilize the native state but do not affect the energy of the intermediate or transition states at all, it is likely that the helix is not formed until after the transition state.

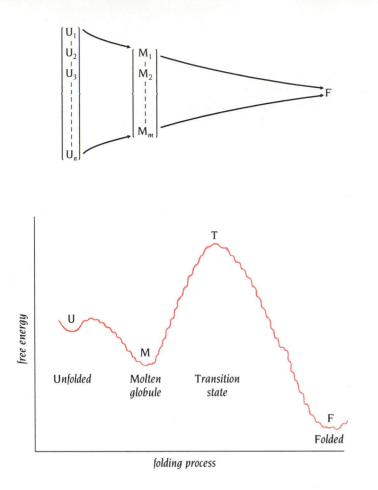

Figure 6.3 The unfolded state is an ensemble of a large number of conformationally different molecules, $U_1...U_n$, which undergo rapid interconversions. The molten globule is an ensemble of structurally related molecules, $M_1...M_m$, which are rapidly interconverting and which slowly change to a single unique conformation, the folded state F. During the folding process the protein proceeds from a high energy unfolded state to a low energy native state. The conversion from the molten globule state to the folded state is slow and passes through a high energy transition state, T.

The small bacterial ribonuclease, **barnase**, is a single chain protein with 110 amino acids and no disulfide bridges. Its three-dimensional structure was determined by the group of Guy Dodson, York University, and comprises three amino terminal α helices and a carboxy terminal five-stranded antiparallel β sheet (Figure 6.4). The group of Alan Fersht have examined the effects of mutations all along the structure and have made a detailed residue by residue characterization of its folding intermediate and transition states. They have concluded from their results that the intermediate molten globule state already has not only most of the native secondary structure elements but also the native-like relative positions of the α helix and β sheet as well as

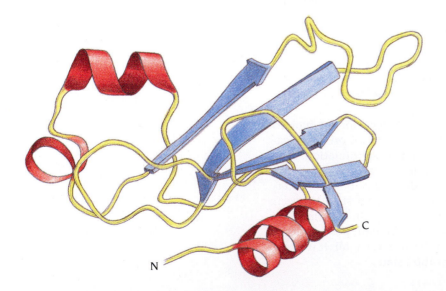

Figure 6.4 Schematic diagram of the structure of the enzyme barnase which is folded into a five stranded antiparallel β sheet (blue) and two α helices (red).

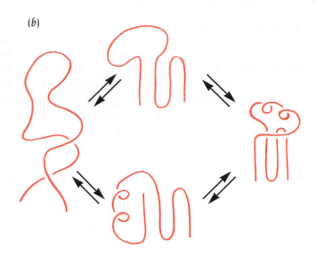

the relative positions of the β strands within the sheet. These results are consistent with the notion that the folding of barnase proceeds through a single major transition state and consequently through one major pathway (Figure 6.5a).

In contrast, folding of the enzyme **lysozyme** involves parallel pathways and distinct folding domains. Hen egg-white lysozyme was the first enzyme to have its structure determined crystallographically, in the laboratory of David Phillips then at the Royal Institution, London in 1965. The native structure consists of two lobes separated by a cleft (Figure 6.6). The first lobe comprises five α helices and the second is predominantly a three-stranded antiparallel β sheet. The folding of lysozyme has been studied extensively by a variety of complementary techniques (NMR, circular dichroism, fluorescence, hydrogen–deuterium exchange) to follow the development of different aspects of the structure such as formation of secondary structure, burial of hydrophobic aromatic groups and formation of hydrogen bonds. The group of Christopher Dobson, Oxford University, has used pulsed amide hydrogen–deuterium exchange to follow secondary structure formation. Amide hydrogen atoms are readily exchanged with the solvent in unfolded proteins, but this exchange is often strongly inhibited in a folded protein, especially for those amide groups that are hydrogen bonded in secondary structure elements. As a result, by measuring the rate of amide–hydrogen exchange as a function of folding time it is possible to monitor the formation of structure during the folding reaction. At 20 milliseconds, two major intermediate stages of lysozyme were detected: one in which the α-helical domain

Figure 6.5 (a) Some proteins such as barnase fold through one major pathway whereas others fold through multiple pathways. (b) The folding of the enzyme lysozyme proceeds through at least two different pathways.

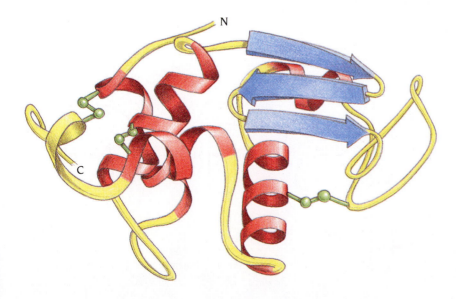

Figure 6.6 Schematic diagram of the structure of the enzyme lysozyme which folds into two domains. One domain is essentially α-helical whereas the second domain comprises a three stranded antiparallel β sheet and two α helices. There are three disulfide bonds (green), two in the α-helical domain and one in the second domain.

had achieved a high degree of secondary structure while the β-sheet domain contained no detectable structure, and a second state in which no stable structure was observed in either the α or the β domain. In addition, a third, less populated state was observed in which both the α domain and the β domain had significant structure. Therefore in this case it is likely that the folding follows different pathways, two major and one minor, with different molten globule states which in a later stage of the folding process converge to one final native state (see Figure 6.5b).

Enzymes assist formation of proper disulfide bonds during folding

The formation of correct disulfide bonds during the folding process poses special problems for cells. In the denatured, unfolded state of proteins there are no disulfide bridges; the cysteine residues are reduced. Formation of disulfide bonds requires oxidation of the cysteine residues. In bacteria this oxidation occurs mainly in the periplasmic space and is catalyzed by a family of enzymes called disulfide bridge-forming enzymes, **Dsb**. In bacteria proteins with disulfide bridges are therefore essentially only found in the periplasmic space and in the outer membrane or they are secreted. In eucaryotic cells disulfide bond formation occurs in the endoplasmic recticulum before proteins are exported to the cell surface. Here an enzyme called **protein disulfide isomerase, PDI**, catalyzes internal disulfide exchange to remove folding intermediates with incorrectly formed disulfide bridges. Again, proteins with disulfide bonds are not found in the cytosol but are located in the plasma membrane or are secreted.

The influence of disulfide bond formation on folding *in vitro* has been extensively studied by Thomas Creighton, EMBL, Heidelberg, and more recently by Peter Kim at MIT. Creighton's pioneering work, which introduced the trapping of disulfide-bonded intermediates as a method for studying the folding pathways of proteins, as well as recent experiments, have shown that these *in vitro* results are also relevant for the folding process *in vivo*. Both Creighton and Kim have studied a small eucaryotic protein, **bovine pancreatic trypsin inhibitor, BPTI**, which has six cysteine residues that form three disulfide bonds within its polypeptide chain of 58 residues (Figure 6.7a).

Figure 6.7 (a) Schematic diagram of the structure of the small protein bovine pancreatic trypsin inhibitor, BPTI. The three disulfide bonds are green, β strands are blue and α helices are red. (b) The folding pathway of BPTI according to Creighton. The unfolded protein has six cysteine residues in their reduced state. The major single disulfide bond intermediate has a disulfide bond between residues 30 and 51. This intermediate forms double disulfide bond intermediates which contain non-native disulfide bonds. According to Creighton these intermediates are essential for the formation of the native double disulfide bond intermediate 30–51, 5–55 which rapidly forms the third native disulfide bond 14–38.

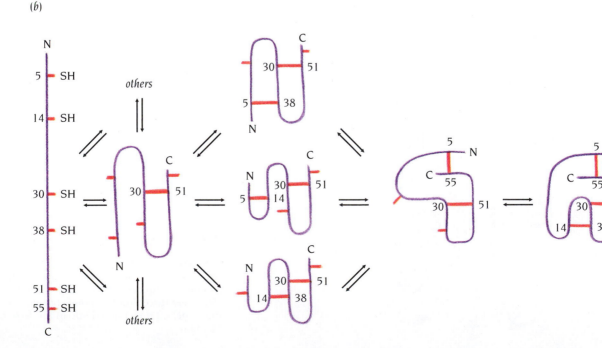

The fully reduced protein is largely unfolded and does not fold until the cysteine residues are oxidized to disulfide bridges. In the native state these bonds are between cysteine residues 30–51, 5–55 and 14–38. During the folding process formation of the first disulfide bond is almost random and the 15 possible single-disulfide species are in rapidly interchanging equilibrium. However, the intermediate with the disulfide bond at 30–51 is more stable than the other 14 and is present in about 60% of the molecules. The productive folding pathway goes through this stable intermediate, consequently all other intermediates must eventually rearrange their disulfide bond in order to fold properly. It has a partly folded conformation comprising the native-like α helix and β sheet linked by the 30–51 disulfide bond. The remaining part of the polypeptide chain is disordered or unfolded and includes cysteines 5, 14 and 38 but not cysteine 55 which is within the folded α helix.

The second disulfide bonds formed in the 30–51 intermediate are between all three possible pairs of flexible cysteines, to form 5–14, 5–38 or 14–38 disulfide bonds (Figure 6.7b). The first two are non-native disulfide bonds, which are not stabilized to any significant extent, and occur primarily because they are in flexible parts of the polypeptide chain. In contrast, upon forming the 14–38 disulfide bond, there are now two native disulfide bonds, and the molecule adopts a very native-like, folded conformation. However, formation of the last disulfide bond between cysteines 5 and 55 occurs only very slowly in this intermediate because these cysteine residues are buried inside the folded intermediate and not accessible to oxidizing agents or disulfide rearrangement. The productive precursor to the native state is instead the intermediate with the 30–51 and 5–55 disulfide bonds since the remaining cysteines 14 and 38 are in good proximity on the surface of the folded protein and there is no barrier to them forming the final disulfide bond. This intermediate can be reached either by rearrangement of the non-native disulfide bonds formed in the 30–51 intermediate or by unfolding the 30–51, 14–38 intermediate.

The folding pathway of BPTI illustrates clearly the importance of disulfide rearrangements for the folding process and hence the advantage for a cell of having enzymes that increase the rate of this reaction. Adding the enzyme protein disulfide isomerase significantly increases the rate of folding of BPTI *in vitro*. The three-dimensional structure of a eucaryotic protein disulfide isomerase has not yet been determined but the structure of DsbA from *Escherichia coli*, a member of the Dsb family of enzymes, has been solved by John Kuriyan, Rockefeller University. This monomeric enzyme has 189 amino acid residues that fold into two domains; one comprises five tightly packed α helices while the second domain has a structure very similar to that of **thioredoxin** (Figure 6.8). Thioredoxin, the x-ray structure of which was determined by the group of Carl Branden, Uppsala, is a ubiquitous protein that functions as a general protein disulfide oxido-reductase and in bacteria is responsible for keeping disulfide bridges reduced. The well-characterized mechanism of thioredoxin is based on reversible oxidation of two cysteine thiol groups to a disulfide, accompanied by the transfer of two electrons and two protons. Presumably DsbA functions in a similar way since the redox-active disulfide bridges in these two proteins are very similar.

The thioredoxin domain (see Figure 2.7) has a central β sheet surrounded by α helices. The active part of the molecule is a βαβ unit comprising β strands 2 and 3 joined by α helix 2. The redox-active disulfide bridge is at the amino end of this α helix and is formed by a Cys-X-X-Cys motif where X is any residue in DsbA, in thioredoxin, and in other members of this family of redox-active proteins. The α-helical domain of DsbA is positioned so that this disulfide bridge is at the center of a relatively extensive hydrophobic protein surface. Since disulfide bonds in proteins are usually buried in a hydrophobic environment, this hydrophobic surface in DsbA could provide an interaction area for exposed hydrophobic patches on partially folded protein substrates.

Disulfide bonds in proteins are generally stable and nonreactive, acting like bolts in the structure. However, oxidized DsbA is less stable than the reduced form and its disulfide bond is very reactive. DsbA is thus a strong

Figure 6.8 Schematic diagram of the enzyme DsbA which catalyzes disulfide bond formation and rearrangement. The enzyme is folded into two domains, one domain comprising five α helices (green) and a second domain which has a structure similar to the disulfide-containing redox protein thioredoxin (violet). The N-terminal extension (blue) is not present in thioredoxin. (Adapted from J.L. Martin et al., *Nature* 365: 464–468, 1993.)

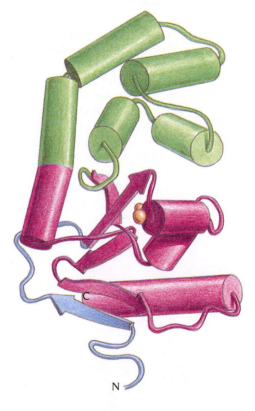

oxidizing agent which is consistent with its ability to catalyze the exchange of disulfide bonds in other proteins. Although the structure of eucaryotic protein disulfide isomerase is unknown, sequence alignment shows that its 189 amino acid polypeptide chain contains two domains that are homologous to thioredoxin. This indicates that both the bacterial and the eucaryotic proteins that catalyze the formation of disulfide bonds in proteins during folding use thioredoxin-like domains and presumably are evolutionarily related.

Isomerization of proline residues can be a rate-limiting step in protein folding

In Chapter 1 we presented a picture of the peptide as a planar unit with the C=O and N-H groups pointing in opposite directions in the plane (see Figure 1.5). This so called *trans*-peptide is the most stable form of the peptide group. However, there is another possible form, the *cis*-peptide, which is also planar but in which the C=O and N-H groups point in the same direction (Figure 6.9a). For most peptides the *cis*-form is about 1000 times less stable than the *trans*-form and consequently *cis*-peptides are rarely found in native proteins. However, when the second residue (R_2 in Figure 6.9a) is a proline the *cis*-form is only about four times less stable than the *trans*-form (Figure 6.9b) and some *cis*-proline peptides occur in many proteins. Most proline residues in proteins are of course in the *trans*-configuration because in *trans* there are fewer steric collisions, but *cis*-prolines are found in tight bends of the polypeptide chain and are sometimes essential for activity or for conformational flexibility.

In the native protein these less stable *cis*-proline peptides are stabilized by the tertiary structure but in the unfolded state these constraints are relaxed and there is an equilibrium between *cis*- and *trans*-isomers at each peptide bond. When the protein is refolded a substantial fraction of the molecules have one or more proline-peptide bonds in the incorrect form and the greater the number of proline residues the greater the fraction of such molecules. *Cis–trans* isomerization of proline peptides is intrinsically a slow process and *in vitro* it is frequently the rate-limiting step in folding for those molecules that have been trapped in a folding intermediate with the wrong isomer.

In vivo the rate of this process is enhanced by enzymes initially called **peptidyl prolyl isomerases**, which are found in both procaryotic and eucaryotic organisms. Surprisingly these enzymes were later found to be involved in immunosuppression by inhibiting T cell proliferation after binding of immunosuppressive drugs; this medically important activity of peptidyl prolyl isomerases is unrelated to their isomerase activity in folding. The first peptidyl prolyl isomerase to be discovered is now called **cyclophilin** because of its role as target for the most frequently used therapeutic agent for prevention of graft rejection, cyclosporin A. Cyclophilin is an abundant soluble protein consisting of a single polypeptide chain of 165 amino acids. Because it is a target for cyclosporin A the determination of its structure was pursued in the early 1990s by several pharmaceutical companies intent

Figure 6.9 (a) Peptide units can adopt two different conformations, *trans* and *cis*. In the *trans*-form the C=O and the N-H groups point in opposite directions whereas in the *cis*-form they point in the same direction. For most peptides the *trans*-form is about 1000 times more stable than the *cis*-form. (b) When the second residue in a peptide is proline the *trans*-form is only about four times more stable than the *cis*-form. *Cis*-proline peptides are found in many proteins.

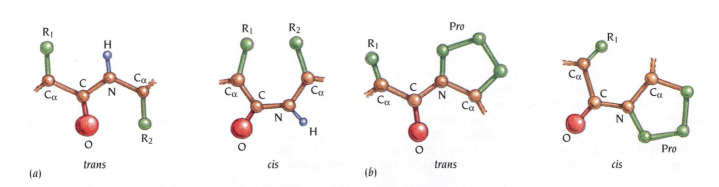

(a) trans   cis   (b) trans   cis

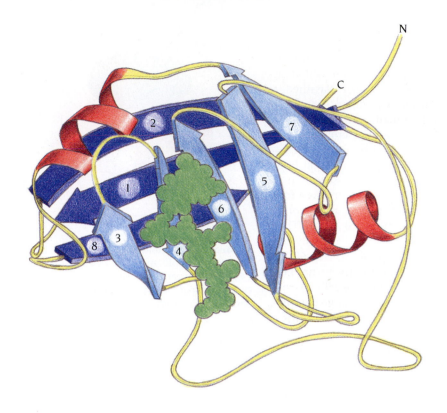

Figure 6.10 Schematic diagram of the structure of the protein cyclophilin, a prolyl peptide isomerase that catalyzes the conversion between *cis*- and *trans*-proline peptides. The protein is folded into an eight-stranded antiparallel β barrel (blue) with two α helices (red) on the outside. The active site, which has been located by the binding of a proline containing tetrapeptide (green), is on the outside of the β barrel. (Adapted from J. Kallen et al., *Nature* 353: 276–279, 1991.)

upon designing better immunosuppressive drugs, and a group from Sandoz, Basel, in collaboration with Kurt Würthrich, ETH, Zurich, published the first structure of cyclophilin with a bound proline-containing peptide using a combination of x-ray crystallography and NMR.

Their work showed that the structure of cyclophilin complexed with a proline-containing tetrapeptide (Figure 6.10) is essentially an eight-stranded β barrel with an unusual topology, different from both the up-and-down and the Greek key β barrels discussed in Chapter 5. The interior of the barrel is tightly packed with hydrophobic residues. In contrast to the up-and-down β barrels that bind ligands inside the barrel, the tetrapeptide ligand of cyclophilin is bound on the outside of the barrel in a deep groove between one face of the β sheet and a long loop region (see Figure 6.10).

Cyclophilin enhances the rate of *cis–trans* isomerization of proline peptides by a factor of one million over the nonenzymatic rate. The detailed mechanism of the enzymatic process is still unclear but mechanisms based on distortion or desolvation of the peptide group are possible. The proline residue is tightly bound in a hydrophobic pocket of the binding groove and the carbonyl oxygen atom is hydrogen-bonded to basic side chains. The binding of a peptide segment into a hydrophobic environment may promote *cis–trans* isomerization by decreasing the charge separation in the peptide group (see Figure 2.3a) and thus creating a peptide bond with a more single-bond character. Alternatively, the tight hydrophobic interactions with cyclophilin might distort the geometry of the peptide group toward the transition state for isomerization. Both mechanisms or a combination of them will decrease the energy barrier for rotation around the peptide bond, which is required for the *cis–trans* isomerization.

Proteins can fold or unfold inside chaperonins

Before protein molecules attain their native folded state they may expose hydrophobic patches to the solvent. Isolated purified proteins will aggregate during folding even at relatively low protein concentrations. Inside cells, where there are high concentrations of many different proteins, aggregation could therefore occur during the folding process. This is prevented by

molecular chaperones, ubiquitous and abundant families of proteins that assist the folding of both nascent polypeptides still attached to ribosomes and released completed polypeptide chains. Several chaperones are induced by heat shock, because protein unfolding and aggregation is increased at elevated temperatures; and consequently chaperones were first classified as **heat-shock proteins** (Hsps). Several of these heat-shock proteins, such as the 70 kDa protein **DnaK** (**Hsp 70**) are also expressed normally and participate in various other cellular processes including assembly and disassembly of multimeric proteins and membrane translocation of secreted proteins. The Hsp 70 polypeptide chain is divided into two functional regions, one that binds and hydrolyses ATP and a second that binds hydrophobic segments of unfolded polypeptide chains. The N-terminal ATP-binding region has a four-domain structure very similar to that of actin (Chapter 14); the polypeptide-binding domain is an antiparallel β sandwich in the C-terminal region. There is as yet no structural information on the complete Hsp 70 molecule. By contrast, the role during folding of two other heat-shock proteins, Hsp 60 and Hsp 10, with molecular weights of 60 kDa and 10 kDa, respectively, has been studied extensively, especially in *Escherichia coli* where they are called **GroEL** and **GroES**, respectively. These proteins function together as a large complex, called a **chaperonin**, consisting of 14 subunits of GroEL and 7 subunits of GroES and requiring ATP to function.

Chaperonins bind unfolded, partly folded and incorrectly folded protein molecules but not proteins in their native state. They are promiscuous in that they bind to and assist the folding of a large number of different proteins independent of the latter's amino acid sequences. How can these chaperonins distinguish between correctly and incorrectly folded versions of almost any water-soluble polypeptide chain and how can they mediate the efficient conversion of unfolded or misfolded proteins to their native form? The x-ray structure determinations of GroEL and GroES, in combination with electron microscopic studies of GroEL–GroES–polypeptide complexes, have made possible major steps towards understanding these processes.

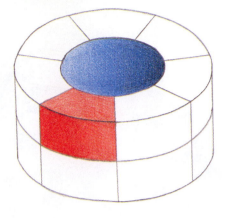

Figure 6.11 Schematic diagram of the chaperonin GroEL molecule as a cylinder with 14 subunits arranged in two rings of 7 subunits each. The space occupied by one subunit is red and the hole inside the cylinder is blue.

GroEL is a cylindrical structure with a central channel in which newly synthesized polypeptides bind

The x-ray structure of GroEL was determined in 1994 by the groups of Paul Sigler and Arthur Horwich, Yale University. The 14 subunits of GroEL, each comprising 547 amino acids, form two rings in which the 7 subunits of each ring are arranged with nearly exact sevenfold rotational symmetry. The rings are arranged back-to-back, forming an extensive interface with one another across an almost flat equatorial plane. The whole structure resembles a porous thick-walled cylinder that is about 150 Å long and 140 Å in diameter and contains a large central cavity or channel (Figure 6.11). Each subunit has three distinct domains, equatorial, intermediate and apical, that are arranged in the cylinder as shown in Figure 6.12. The equatorial domain which is largest and comprises 243 residues is mainly α-helical. It serves as the foundation of the GroEL structure providing all of the contacts between the two sevenmembered rings across the equatorial plane. In addition it provides most of the contacts between the subunits within each ring as well as the ATP binding site, which is essential for function. The equatorial domain contains both the N-terminus and the C-terminus of the polypeptide chain. About 30 residues at these ends are not visible in the x-ray structure; either they are disordered, or they occupy differently ordered positions in different molecules. The visible ends point into the central cavity close to the equatorial plane and presumably the nonvisible residues partially block the channel in this region; three-dimensional electron microscopic reconstruction of individual chaperonin particles shows that the channel is narrow in the center of the molecule (Figure 6.13).

The apical domain (residues 191–376) is essentially a four-layer structure comprising two β sheets sandwiched between α helices. One β sheet has

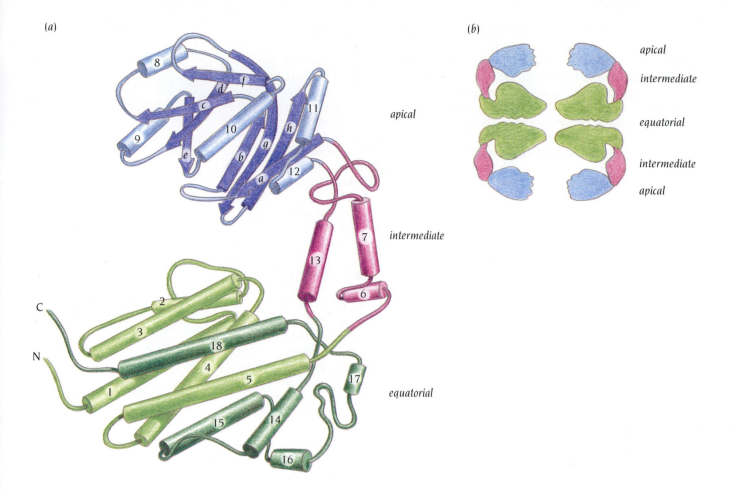

Figure 6.12 (a) Schematic diagram of one subunit of GroEL. The polypeptide chain is folded into three domains. The equatorial domain (green) is the largest domain, comprising 10 α helices, and is built up from both the N-terminal and the C-terminal regions. The apical domain (blue), which is a β sandwich flanked by α helices, is formed by the middle region of the polypeptide chain. The two linker regions between the equatorial and the apical domains form a small intermediate domain (purple) comprising three α helices. (b) Schematic diagram illustrating the domain arrangement of four subunits in the GroEL molecule, two in each of the seven membered rings. The equatorial domains form the middle part of the molecule and interact with each other both within each ring and between the rings. The apical domains are at the top and the bottom of the cylinder and form an opening to the interior of the molecule. The small intermediate domains form the thinnest part of the cylinder wall in the middle of each ring. [(a) Adapted from K. Braig et al., *Nature* 371: 578–586, 1994.]

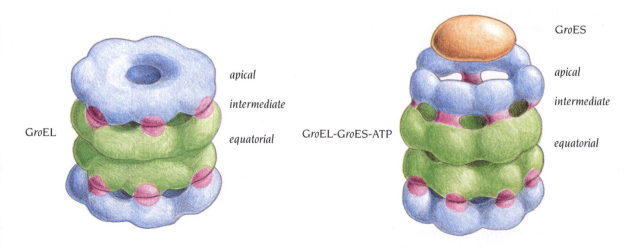

Figure 6.13 Models of the GroEL molecule in two different functional states based on three-dimensional reconstruction from electron microscopy pictures. A large conformational change of GroEL occurs when GroES and ATP are bound. The GroES molecule binds to one of the GroEL rings and closes off the central cavity. The GroEL ring becomes larger and the cavity inside that part of the cylinder becomes wider. (Adapted from S. Chen et al., *Nature* 371: 261–264, 1994.)

101

antiparallel β strands whereas the other is part of an α/β domain with four parallel β strands. The apical domains form the opening to the solvent of the central channel. The segments of this domain that form the top surface of the molecule as well as those that face the upper regions of the channel are flexible and not very well ordered. These segments are rich in hydrophobic residues and they are involved in binding to the hydrophobic areas exposed by non-native folds of polypeptide chains. Mutating these hydrophobic residues to charged ones abolishes both GroES and polypeptide binding whereas mutations conserving hydrophobic residues, such as Phe to Val, have no functional effects. The flexibility of these segments probably accounts for the promiscuous binding of a wide range of unfolded polypeptide sequences.

The equatorial and apical domains are linked by a small intermediate domain which forms part of the outer wall and extends only about 25 Å in the radial direction; consequently the internal cavity is wider in this region, up to 90 Å in diameter. In addition there are holes in the wall between the intermediate domains in adjacent subunits. These seven holes are large enough to permit ATP and ADP to diffuse in and out. The intermediate domain is connected to the other two domains through short antiparallel segments that could easily serve as hinges during conformational changes. Electron microscopic studies of GroEL with different ligands bound have shown that substantial changes in the orientations of the domains and in the size of the central cavity occur during the functional cycle of the chaperonin (see Figure 6.13).

GroES closes off one end of the GroEL cylinder

GroES binds to the apical domain of GroEL, closing off the central cavity (see Figure 6.13). Once GroES has bound to one of the rings in the GroEL molecule a conformational change occurs which decreases the affinity of the second GroEL ring for GroES. The predominant functional state of the GroEL–GroES complex is, therefore, asymmetric with GroES bound to only one end of the GroEL cylinder. Obviously there is strong structural communication between the halves of the GroEL molecule since GroES binding to one half affects the properties of the other half, despite a distance of about 150 Å between the two GroES binding sites.

The GroES molecule comprises seven subunits, each of 97 amino acids linked together around a sevenfold rotation axis, the same symmetry arrangement as in one GroEL ring. The x-ray structure of GroES was determined in 1996 by the group of Johann Deisenhofer, University of Texas, Dallas. The GroES molecule is shaped like a dome, about 75 Å in diameter and 30 Å high with a small hole in the middle (Figure 6.14a). The core of the subunit structure is a β barrel comprising two antiparallel β sheets packed against each other (Figure 6.14b). Two large loop regions protrude from this core, one of which extends above the plane of the ring creating a loosely packed top of the dome that covers the central hole. The other loop region, which is rich in hydrophobic residues, extends below the dome and presumably interacts with the apical domain in the GroES–GroEL complex. This loop is disordered in the x-ray structure of GroES but NMR studies have shown that the loop becomes ordered when GroES binds to GroEL and mutational studies have shown that the hydrophobic residues in this loop are important for chaperonin function.

The GroEL–GroES complex binds and releases newly synthesized polypeptides in an ATP-dependent cycle

How does the GroEL–GroES complex function as a chaperone to assist protein folding? Although several aspects of the mechanism are not clear, the main features of the functional cycle are known. The first step is the

(a)

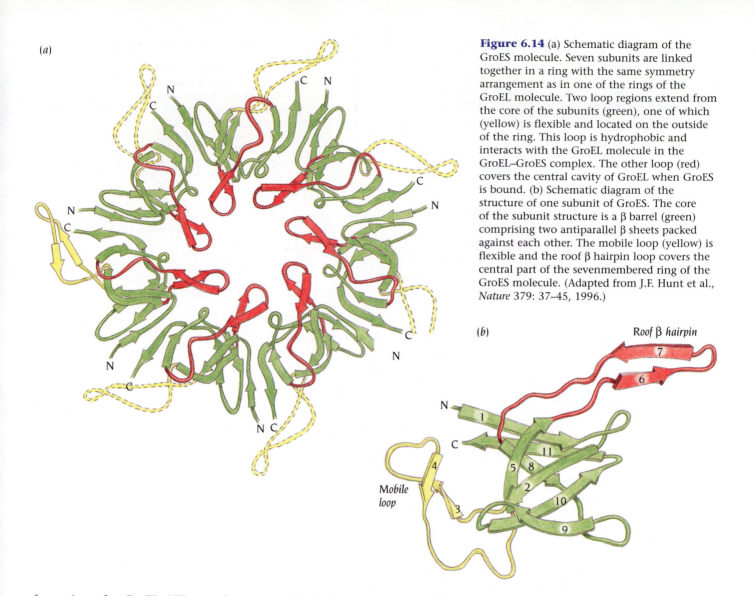

Figure 6.14 (a) Schematic diagram of the GroES molecule. Seven subunits are linked together in a ring with the same symmetry arrangement as in one of the rings of the GroEL molecule. Two loop regions extend from the core of the subunits (green), one of which (yellow) is flexible and located on the outside of the ring. This loop is hydrophobic and interacts with the GroEL molecule in the GroEL–GroES complex. The other loop (red) covers the central cavity of GroEL when GroES is bound. (b) Schematic diagram of the structure of one subunit of GroES. The core of the subunit structure is a β barrel (green) comprising two antiparallel β sheets packed against each other. The mobile loop (yellow) is flexible and the roof β hairpin loop covers the central part of the sevenmembered ring of the GroES molecule. (Adapted from J.F. Hunt et al., *Nature* 379: 37–45, 1996.)

formation of a GroEL–ATP complex, one end of which which then binds one molecule of GroES, with the hydrolysis of ATP. The resulting GroEL–ADP–GroES is a stable complex the halves of which have very different properties (Figure 6.15a). The GroEL ring where GroES is bound (*cis*-position) has undergone a large structural change forming a wide internal cavity whose walls are formed from both the apical and intermediate domains. This cavity is partly closed off from the solvent by the GroES dome. The other ring (*trans*-position) has a smaller cavity which is open to the solvent. Unfolded proteins can bind in both the *cis*- and the *trans*-positions but only those that are bound in the *cis*-position undergo subsequent folding rearrangements (Figure 6.15c–e). Binding and release of polypeptides at the *trans*-position seem to be functionally unimportant.

Release of bound polypeptides from the closed cavity in the *cis*-position requires that GroES is first released. The release of GroES, like its binding, requires ATP hydrolysis, but this time by ATP molecules bound to the distant GroEL ring in the *trans*-position (Figure 6.15e,f). This is another example of the strong structural communication between the halves of the complex since ATP hydrolysis in the *trans*-GroEL ring affects the GroES binding site almost 100 Å away. Once GroES is released the polypeptide chain is released and it can bind to another GroEL–GroES complex to repeat the cycle. The native state is reached after iterative rounds involving multiple binding and release to GroEL–GroES complex.

The crucial question that remains to be answered is, of course, what happens to the polypeptide chain inside the closed *cis*-cavity. Two different

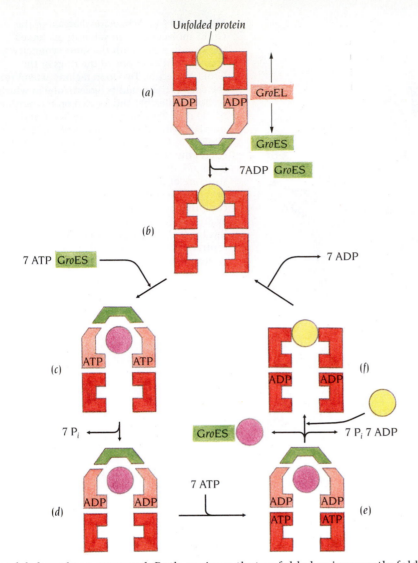

Figure 6.15 Possible functional cycle of the GroEL–GroES molecule. (a) An unfolded protein molecule (yellow) binds to one end of the GroEL–ADP complex (red) with bound GroES (green) at the other end. (b and c) GroES is released from the *trans*-position and rebound together with ATP at the *cis*-position (light red) of GroEL. (d) ATP hydrolysis occurs as the protein is folding or unfolding inside the central cavity. (e) ATP binding and hydrolysis in the *trans*-position is required for release of GroES and the protein molecule. (f) A new unfolded protein molecule can now bind to GroEL. (Adapted from M. Mayhew et al., *Nature* 379: 420–426, 1996.)

models have been proposed. Both envisage that unfolded or incorrectly folded proteins are recognized by their exposed hydrophobic areas and bound to hydrophobic regions inside the GroEL cavity. In one of these models the subsequent function of the cavity is to unfold unproductive intermediates and then eject them in the unfolded state into the bulk solution for spontaneous folding, giving them another chance to reach the folded state. In this model folding would occur in the solvent during jumps of the polypeptide between GroEL–GroES complexes. In the second model folding occurs inside the *cis*-cavity of the GroEL–GroES complex, either to the native state or to an intermediate state along one of the productive pathways to the native state. Experiments by the group of Ulrich Hartl, Memorial Sloan-Kettering Cancer Center, New York, have provided support, at least for some proteins, for the second model, which has the attractive feature that folding occurs in a closed environment preventing aggregation with other unfolded proteins during the folding process. However, GroEL-assisted folding of large protein molecules must be different since they are too large to fit inside the cavity. In addition, assisted folding by Hsp 70 does not occur inside a closed cavity.

The folded state has a flexible structure

For simplicity we have so far described a native folded protein molecule as being in one single state. However, within this state, the protein molecule does not have a static rigid structure at normal temperatures. Instead, all the atoms are subject to small temperature-dependent fluctuations. The molecule as a whole undergoes **breathing** and every atom is constantly in motion. These atomic movements are in general random, but sometimes they can be collective and cause groups of atoms to move in the same direction. Side chains can flip from one conformation to another, some loop regions may not be fixed in one single conformation, helices may slide relative to each other and entire domains can change their packing contacts and open or close the distance between them. Usually these motions are small, a few tenths of an Ångstrom; but occasionally the collective motions can be large and very significant.

Such large collective movements are reflected in x-ray studies as a low level of electron density and in some cases no electron density at all. The regions that undergo these movements are usually described by crystallographers as flexible or disordered, but in order to distinguish between collective movements and a few discrete and well-ordered but different conformations of these regions, x-ray studies must be made at different temperatures. In NMR studies the experimental data for these regions are compatible with many different conformations. Insight into these individual and collective motions has been obtained by theoretical studies, called molecular dynamics simulations, which use classical Newtonian descriptions of atomic movements.

These calculations have shown that collective movements occur on the picosecond time scale for individual residues, and in nanoseconds for loop regions. Such movements are very important for the function of many protein molecules. Reactions such as electron transfer or ligand binding and release occur on these time scales and usually require movements of protein atoms. As soon as the structure of myoglobin was determined it was immediately apparent that the static picture of the myoglobin molecule seen in the crystal did not allow oxygen atoms to enter its binding site or diffuse out. We now know that while the myoglobin molecule is breathing, pathways are opened up between the solvent and the buried binding site to allow oxygen binding or release on a time scale of nanoseconds.

In addition to these small breathing movements of protein atoms there can also be larger conformational changes between different functional states of the molecule. Small differences in the environment such as different pH or the presence or absence of ligands can stabilize different conformational states of the protein. These conformational changes can vary from adjustments of side chain orientations in the active site to movements of loop regions, differences in relative orientation of domains or changes in the quaternary structure of oligomeric proteins. Such movements are usually essential for function, for enzyme catalysis, binding of antigens to antibodies, receptor–ligand interactions, muscle action and energy transduction and so on. We have already discussed two extreme examples, the effect of pH change on hemagglutinin (Chapter 5) and ATP hydrolysis in GroEL–GroES. We will here give a few further examples of striking conformational transitions.

Conformational changes in a protein kinase are important for cell cycle regulation

The cell division cycle of all eucaryotic cells can be divided into four major phases: G_1, S, G_2, and M (Figure 6.16a). In S phase (DNA synthesis), the entire DNA content of the nucleus is duplicated and the number of chromosomes doubled; in M phase (mitosis), the duplicated chromosomes of the parental cell are segregated using the microtubules of the mitotic spindle, such that each daughter cell receives the same set of chromosomes. After mitosis, including cytokinesis which divides the cytoplasm between the two daughter cells, and before the onset of the next S phase there is an interval designated Gap 1 or G_1. After the completion of DNA replication and chromosome duplication, in other words after S phase is completed, there is a second interval designated Gap 2, or G_2, before the onset of mitosis. The

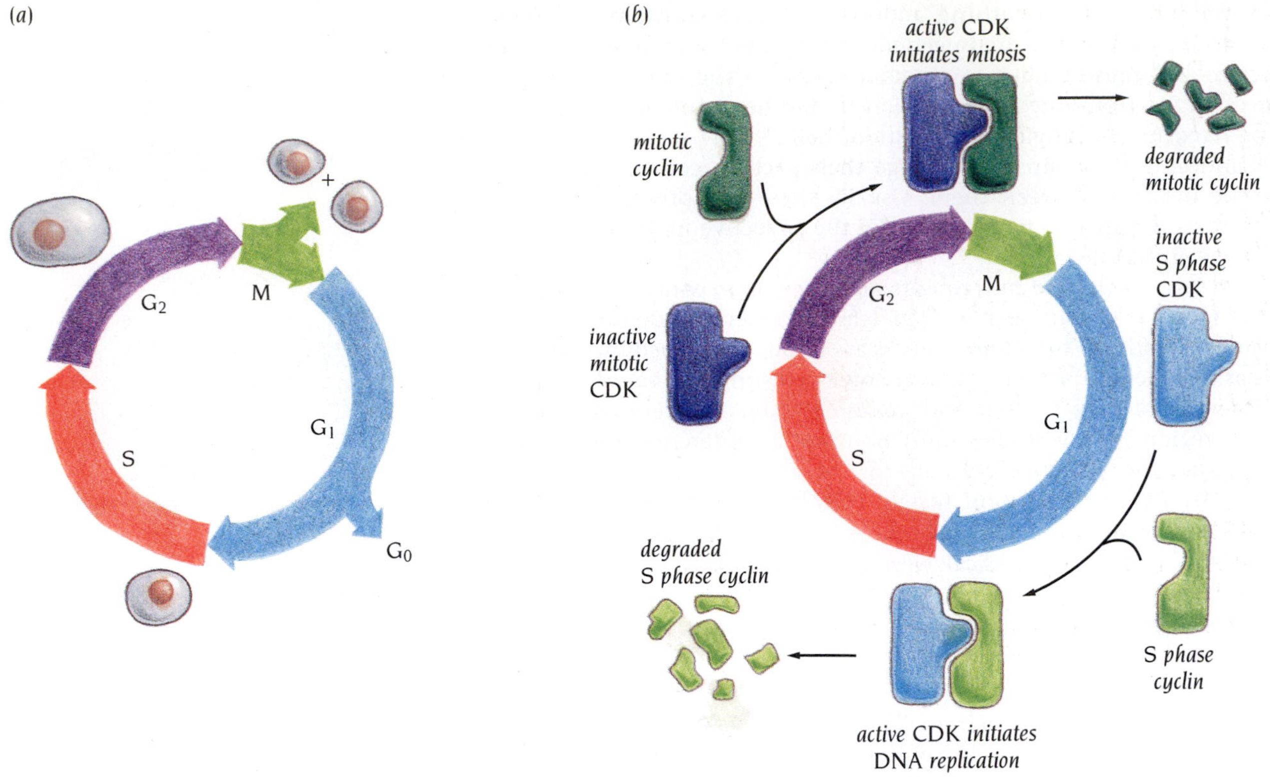

(a)

(b)

Figure 6.16 (a) The five phases of a standard eucaryotic cell cycle. During M phase growth stops and the cell then divides. DNA replication is confined to the S phase; G_1 phase is the gap between M phase and S phase; G_2 phase is the gap between S phase and M phase. Cells which are not dividing enter the stationary phase, G_0. (b) The regulation of CDKs by cyclin degradation. Only two types of cyclin–CDK complexes are shown, one that triggers S phase and one that triggers M phase. In both cases the activation of CDK requires cyclin binding and its inactivation depends on cyclin degradation. (Adapted from B. Alberts et al, *Essential Cell Biology,* New York: Garland Publishing, 1998.)

complete cycle therefore is M, G_1, S, G_2 and M again. Throughout G_1, S and G_2 the cell's protein-synthesising machinery, macromolecules and organelles are built up, and the cell increases in volume. During mitosis the chromosomes and the cytoplasm are divided into two essentially equal parts. An additional stationary phase G_0 occurs in cells which are not actively dividing. How is the progression of a dividing cell through the consecutive phases of the cell cycle regulated? What triggers DNA synthesis in a G_1 cell and mitosis in a G_2 cell? Why don't the two daughter cells emerging from mitosis immediately begin a new round of DNA synthesis?

A combination of biochemical and genetic studies have shown that the progression through the cell cycle is dependent upon the successive activation of a series of enzymes called **cyclin-dependent protein kinases, CDKs** (Figure 6.16b). Each CDK during its transient existence phosphorylates target proteins that then directly or indirectly activate the next set of events of the cell cycle. Each CDK is a heterodimer comprising a catalytic subunit, the protein kinase, complexed with a **cyclin** molecule which activates the kinase. In vertebrate cells there are at least four different CDKs involved in control of passage through the cell cycle, as well as other CDKs with other functions. The different catalytic subunits are products of the genes of a closely related gene family. The different cyclins, one or more for each sort of catalytic subunit, are also members of a gene family. The cyclin components of the CDKs undergo sequential programmed synthesis, accumulation and then degradation; the short half-lives of the cyclins ensure that the CDKs of which they are part are active kinases only for short periods and at the correct time in the cell cycle. The CDKs can therefore act as a relay of switches, governing passage from G_1 to S phase, from G_2 to M phase and all other steps that constitute the cell cycle.

Although there is still a very great deal to learn about the physiological substrates of the CDKs and how the enzymes locate their targets, detailed structural information, including the activation of the kinase by cyclin, is available for CDK2-cyclin A, which regulates DNA replication in human cells. The x-ray structure of a functional fragment of cyclin A was determined by the group of Louise Johnson, Oxford University, that of inactive CDK2 by the group of Sung-Ho Kim, University of California, Berkeley, and that of the

(a)

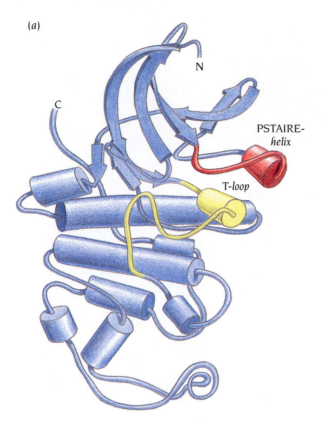

(b)

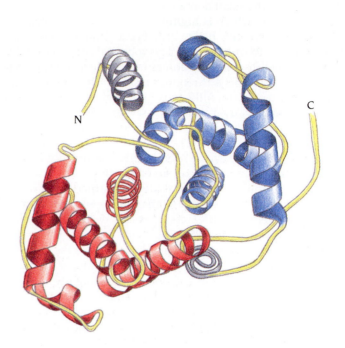

(c)

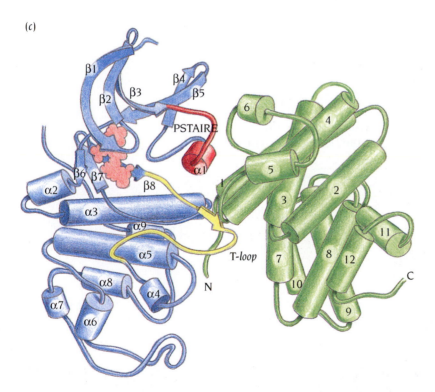

Figure 6.17 (a) Schematic diagram of the structure of the cyclin-dependent kinase CDK2. The molecule is folded into two domains. The N-terminal domain comprises a five stranded β sheet and an α helix which has an amino acid sequence PSTAIRE (red) that is highly conserved in all protein kinases. The C-terminal domain is mainly α-helical and has a flexible loop region called the T-loop (yellow). This loop contains a threonone residue which is phosphorylated in the fully active enzyme. (b) Schematic diagram of the structure of a fragment of cyclin A. The structure comprises two very similar domains (red and blue). This domain structure is called the cyclin fold and contains five α helices. The N-terminal domain (red) has an amino acid sequence that is conserved in all cyclins and which is called the cyclin box. (c) Schematic diagram of the structure of the active complex between CDK2 (blue) and the fragment of cyclin A (green). The cyclin box (helices 2–6) interacts with the PSTAIRE helix (dark red) and the T-loop (yellow) of CDK2. The structure of cyclin A in this complex is the same as that of cyclin A whereas that of CDK2 has undergone major conformational changes involving the PSTAIRE helix, the T-loop and the ATP binding site (light red). (Adapted from P.D. Jeffrey et al., *Nature* 376: 313–320, 1995.)

active cyclin A fragment-CDK2 complex by the group of Nicola Pavletich, Memorial Sloan-Kettering Cancer Center, New York. Comparison of these structures reveals how cyclin A binding to CDK2 causes large conformational changes in the active site of CDK2, converting the protein from an inactive to an active kinase. The structure of cyclin A, in contrast, is not changed.

CDK2 has two domains, a small (85 residue) amino-terminal domain comprising a single α helix and a five-stranded β sheet and a larger (213 residues) domain that is mainly α-helical (Figure 6.17a). The cofactor in the phosphorylating reaction, ATP, is bound in a cleft between the two domains. The α helix in the small domain contains a sequence of residues, -Pro-Ser-Thr-Ala-Ile-Arg-Glu-, which is highly conserved in all protein kinases and which is called the PSTAIRE helix. Mutational studies have shown that the Glu residue in the PSTAIRE helix plays a crucial role for activity of the enzyme. The large, mainly α-helical domain has a flexible loop region, called the T-loop, which contains a threonine residue, Thr 160 in human CDK2, that is phosphorylated in the fully active enzyme. The cyclin A fragment (residues 173–432) that was used in these x-ray studies is built up from two domains with a very similar structure that is now called the cyclin fold (Figure 6.17b) and which comprises five α helices. The fragment activates CDK2 almost as well as the complete cyclin A molecule. The first domain has an amino acid sequence that is strongly conserved in all cyclins and which is called the cyclin box. In spite of the almost identical structures of the two domains their amino acid sequences are not similar, only one has the cyclin box.

In the cyclin A–CDK2 complex the cyclin box domain interacts with CDK2, mainly with the PSTAIRE helix and the T-loop (Figure 6.17c). The structure of cyclin A in the complex is virtually the same as that of cyclin A alone whereas that of CDK2 has undergone major conformational changes. The whole of the N-terminal domain has slightly changed its orientation relative to the C-terminal domain. In addition the PSTAIRE helix has moved closer to the active site cleft of CDK2 and rotated about 90° so that the catalytically essential residue Glu 51 points into the cleft instead of away from it in free CDK2 (Figure 6.18). Some of the main chain atoms of this helix have moved up to 8 Å due to these concerted movements. Coupled to the structural change of the PSTAIRE helix there is a major rearrangement of the T-loop with some residues moving up to 20 Å (Figure 6.19). As well as adopting a completely new position, the T loop undergoes a structural transformation. The part of the T-loop that is α-helical in free CDK2 melts away, and instead a β strand appears in the complex.

In free CDK2 the active site cleft is blocked by the T-loop and Thr 160 is buried (Figure 6.20a). Substrates cannot bind and Thr 160 cannot be phosphorylated; consequently free CDK2 is inactive. The conformational changes induced by cyclin A binding not only expose the active site cleft so that ATP and protein substrates can bind but also rearrange essential active site residues to make the enzyme catalytically competent (Figure 6.20b). In addition Thr

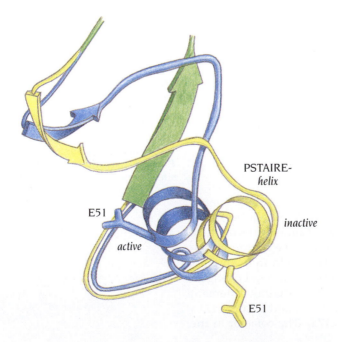

Figure 6.18 The PSTAIRE helix undergoes a major conformational change when CDK2 binds to cyclin A. In the inactive free CDK2 (yellow) the active site residue Glu 51 is far from the active site. Upon binding of cyclin A to CDK2 the PSTAIRE helix (blue) rotates 90° and changes its position so that Glu 51 becomes positioned into the active site. (Adapted from P.D. Jeffry et al., *Nature* 376: 313–320, 1995.)

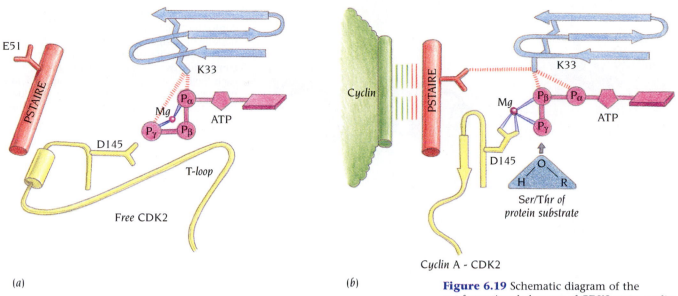

(a)   (b)

Cyclin A - CDK2

160 is exposed and ready to become phosphorylated, which enhances the catalytic activity. In short, spectacular flexibility of protein structure is essential for regulating the CDK family of enzymes and hence for controlling the cell cycle.

Peptide binding to calmodulin induces a large interdomain movement

Calmodulin is a ubiquitous calcium-binding protein of 148 amino acid residues that is involved in a range of calcium-dependent signaling pathways. Calmodulin binds to a variety of proteins such as kinases, calcium pumps and proteins involved in motility, thereby regulating their activities. The calmodulin-binding regions of these proteins, comprising about 20 sequentially adjacent residues, vary in their amino acid sequences but they all have a strong propensity to form α helices. Structure determinations of calmodulin alone and of complexes with peptides have shown that peptide binding induces a large conformational change in the calmodulin molecule.

The x-ray structure of free calmodulin was determined by the group of Charles Bugg, University of Alabama. It is a dumbbell-shaped molecule

Figure 6.19 Schematic diagram of the conformational changes of CDK2 upon cyclin binding. (a) In the inactive form the PSTAIRE helix (red) is oriented such that Glu 51 points away from the ATP binding site (purple) and the T-loop (yellow) blocks the substrate binding site and prevents proteins from binding to CDK2. (b) In the active cyclin–CDK2 complex the PSTAIRE helix is reoriented so that Glu 51 points into the active site and forms a salt bridge to another residue involved in catalysis, Lys 33. The T-loop has drastically changed its conformation and one of its residues, Asp 145, forms ligands to the Mg atom in the active site. The substrate-binding site is now open, proteins can bind and the cyclin-CDK2 complex can phosphorylate Ser/Thr residues and thereby activate the bound proteins.

Figure 6.20 Space-filling diagram illustrating the structural changes of CDK2 upon cyclin binding. (a) The active site is in a cleft between the N-terminal domain (blue) and the C-terminal domain (purple). In the inactive form this site is blocked by the T-loop. (b) In the active cyclin bound form of CDK2 the T-loop has changed its structure, the active site is open and available and Thr 160 is available for phosphorylation.

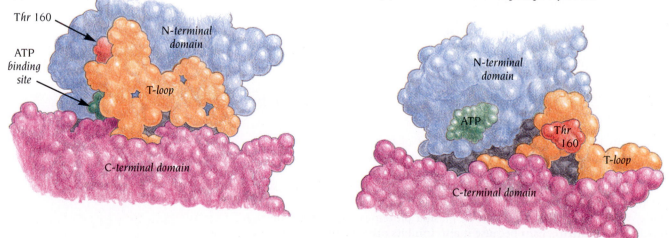

(a)   (b)

(a)   (b)

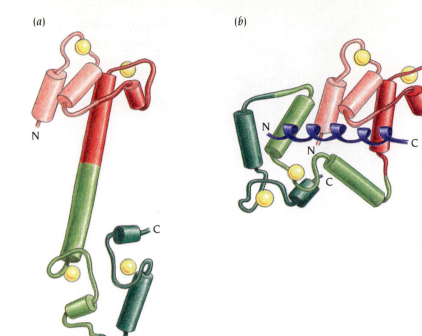

Figure 6.21 Schematic diagram of the conformational changes of calmodulin upon peptide binding. (a) In the free form the calmodulin molecule is dumbbell-shaped comprising two domains (red and green), each having two EF hands with bound calcium (yellow). (b) In the form with bound peptides (blue) the α helix linker has been broken, the two ends of the molecule are close together and they form a compact globular complex. The internal structure of each domain is essentially unchanged. The bound peptide binds as an α helix.

(Figure 6.21a) comprising two domains separated by a long straight α helix, similar in shape to troponin-C described in Chapter 2 (see Figure 2.13c). Each domain comprises two EF hands (see Figure 2.13a), each of which binds a calcium atom. The two domains are clearly separated in space at the two ends of the α helix linker.

The structures of two different complexes of calmodulin with binding peptides have been determined, one by the group of Ad Bax, National Institutes of Health, Bethesda, using NMR and the second by the group of Florante Quiocho, Baylor College of Medicine, Houston, using x-ray crystallography. These two structures are quite similar but the molecular shape of calmodulin in these complexes is very different from that of free calmodulin. The internal structures of the two domains have not changed but the α helix linker has been broken into two helices that are oriented in different directions so that the relative positions of the domains have changed (Figure 6.21b). They are now close together forming a compact molecule of ellipsoidal shape. The bound peptide is in a wide cleft between the two domains and adopts an α-helical conformation.

When calmodulin binds a ligand, only five groups actually change their conformation. They are five consecutive residues in the linker helix, which unwind and turn into a loop region. The α helix continues after this loop but now in an entirely new direction, which positions the second domain close to the first and in a different orientation. This rather small local change in peptide conformation causes one of the largest ligand-induced interdomain motions known in a protein, comparable to the large repositioning of domains during the pH-induced conformational change of hemagglutinin discussed in Chapter 5.

Serpins inhibit serine proteinases with a spring-loaded safety catch mechanism

Infections in the lung elicit an accumulation of activated leucocytes that secrete enzymes involved in removing the damage done by the infection. The most important of these enzymes is neutrofil elastase, which belongs to the serine proteinase family of enzymes described in Chapter 11. The health of the lung depends to a large extent on proper control of the activity of this enzyme, which is achieved by a blood plasma proteinase inhibitor named, misleadingly, **α1-antitrypsin** because it also inhibits other serine proteinases, among them trypsin. Alpha₁-antitrypsin belongs to a family of serine pro-teinase inhibitors found in blood plasma that are collectively called **serpins**. Other members of this family are **antithrombin** and **plasminogen activator inhibitor**, **PAI**, both of which are essential regulators of the blood coagula-tion cascade of reactions. All serpin molecules are homologous with very similar three-dimensional structures.

Serpins form very tight complexes with their corresponding serine pro-teinases, thereby inhibiting the latter. A flexible loop region of the serpin binds to the active site of the proteinases. Upon release of the serpin from the complex its polypeptide chain is cleaved by the proteinase in the middle of this loop region and the molecule is subsequently degraded. In addition to the active and cleaved states of the serpins there is also a latent state with an intact polypeptide chain that is functionally inactive and does not bind to the proteinase.

The structures of all three states of the serpins have been determined by x-ray crystallography, the cleaved form of α₁-antitrypsin by the group of Robert Huber, Max-Planck Institute for Biochemistry, Munich, the latent form of PAI by the group of Elizabeth Goldsmith, University of Texas, Dallas, and the active form of antithrombin by the groups of Wim Hol, University of Washington, Seattle, and of Robin Carrel, Cambridge University. In addi-tion, the group of Robin Carrel has determined the structure of another member of the serpin family, **ovalbumin**, which is present in uncleaved form in egg white. The general folds of these serpin molecules in their different states are the same, but the positions of the flexible loop regions vary in a novel and intriguing way.

The serpin fold comprises a compact body of three antiparallel β sheets, A, B and C, which are partly covered by α helices (Figure 6.22). In the struc-ture of the uncleaved form of ovalbumin, which can be regarded as the canonical structure of the serpins, sheet A has five strands. The flexible loop starts at the end of strand number 5 of β sheet A (β15 in Figure 6.22), then

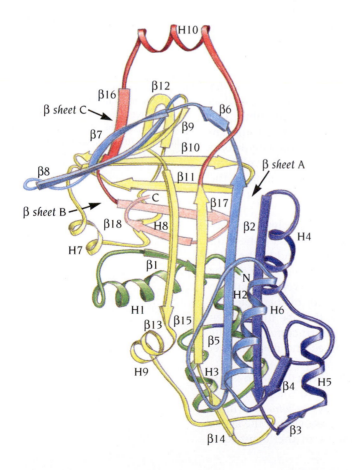

Figure 6.22 Schematic diagram of the structure of ovalbumin which illustrates the serpin fold. The structure is built up of a compact body of three antiparallel β sheets, A, B, and C, surrounded by α helices. The polypeptide chain is colored in sections from the N-terminus to facilitate following the chain tracing in the order green, blue, yellow, red and pink. The red region corresponds to the active site loop in the serpins which in ovalbumin is protruding like a handle out of the main body of the structure. (Adapted from R.W. Carrell et al., *Structure* 2: 257–270, 1994.)

111

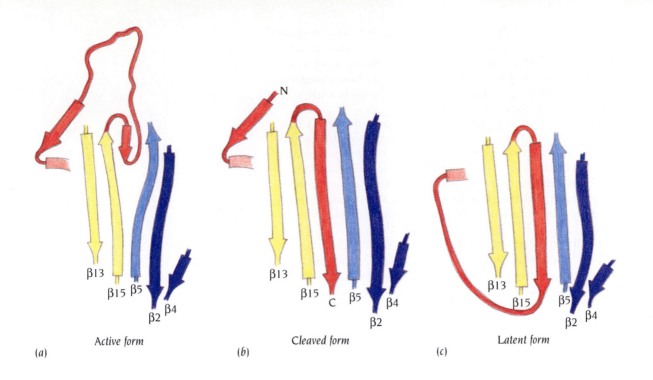

β13	β13	β13
β15 β5	β15	β15 β5
β2 β4	C β5 β4	β2 β4
	β2	
Active form	*Cleaved form*	*Latent form*
(a)	(b)	(c)

forms an α helix outside the top of the molecule followed by an edge strand of β sheet C (β16) and finally ends at the beginning of one of the strands in β sheet B (β17). The central α-helical region of the loop, which contains the cleavage site for the serpins, is extended like a handle on the outside of the molecule.

The flexible loop region in the active form of antithrombin (Figure 6.23a) is in the same general position as in ovalbumin but the first few residues form a short sixth β strand in β sheet A inserted between strands β5 and β15. Furthermore there is no α helix in the loop which is extended outside the main body of the molecule, ready to be inserted into the active site of thrombin.

In the cleaved form of α1-antitrypsin the first half of the loop region up to the cleavage site forms a complete β strand inserted between strands β5 and β15 in β sheet A (Figure 6.23b). The other half of the loop region has approximately the same position as in the active form of antithrombin. The two new ends of the polypeptide, which are joined in the active form, are here at opposite ends of the molecule, 70 Å apart. Finally, in the latent form of PAI the additional β strand in β sheet A is present as in the cleaved form of α1-antitrypsin, but the rest of the flexible loop region makes a loop on the outside of the molecule and enters the β strand of β sheet B without forming an edge strand of β sheet C (Figure 6.23c).

The conversion of the active form to the latent form involves the conversion of a loop into a long β strand inserted in the middle of a β sheet. To achieve the remarkable structural change of inserting a β strand in the middle of a stable preformed β sheet, adjacent strands in the β sheet must first be separated. This involves breaking many hydrogen bonds as well as changing a number of hydrophobic packing contacts in the interior of the molecule. New hydrogen bonds and new packing contacts must then be made when the extra β strand is inserted. Such major changes in a β structure were quite unforseen before these serpin structures were determined and have not yet been observed in any other system.

Which of these forms is most stable? Surprisingly, the active form is less stable than the latent form. Conversion from the active to the latent form can occur spontaneously over a period of hours or days *in vitro* and more quickly under mild denaturing conditions. In contrast, recovery of the active form from the latent form requires complete unfolding of the latent form

Figure 6.23 Schematic diagram illustrating the active site loop regions (red) in three forms of the serpins. (a) In the active form the loop protrudes from the main part of the molecule poised to interact with the active site of a serine proteinase. The first few residues of the loop form a short β strand inserted between β5 and β15 of sheet A. (b) As a result of inhibiting proteases, the serpin molecules are cleaved at the tip of the active site loop region. In the cleaved form the N-terminal part of the loop inserts itself between β strands 5 and 15 and forms a long β strand (red) in the middle of the β sheet. (c) In the most stable form, the latent form, which is inactive, the N-terminal part of the loop forms an inserted β strand as in the cleaved form and the remaining residues form a loop at the other end of the β sheet. (Adapted from R.W. Carrell et al., *Structure* 2: 257–270, 1994.)

under strong denaturing conditions, and subsequent refolding. Refolding does, however, produce the less stable active form in preference to the more stable latent form. This is one of the few pieces of direct experimental evidence that the folding process can be kinetically controlled through intermediates that produce a native state that is not the thermodynamically most stable state.

In vivo PAI and antithrombin are stabilized in their active forms by binding to vitronectin and heparin, respectively. These two serpins seem to have evolved what Max Perutz has called "a spring-loaded safety catch" mechanism that makes them revert to their latent, stable, inactive form unless the catch is kept in a loaded position by another molecule. Only when the safety catch is in the loaded position is the flexible loop of these serpins exposed and ready for action; otherwise it snaps back and is buried inside the protein. This remarkable biological control mechanism is achieved by the flexibility that is inherent in protein structures.

Emphysema is often associated with a specific mutation of the serpin antitrypsin. The mutant serpin molecules form aggregates in the liver, causing a deficiency of antitrypsin in the blood plasma and consequently increased proteolytic degradation of elastin fibers in the lung by the enzyme elastase. It has been shown that the formation of aggregates *in vivo* is due to an extremely slow folding process of the mutant antitrypsin leading to accumulation of a folding intermediate that aggregates. This is one example of aggregation of incompletely folded or misfolded molecules that can lead to pathologic consequences or even severe disease. Other examples involve the formation of large aggregates, plaques, of proteins in amyloid structures associated with Alzheimer's disease and spongiform encephalopathies such as scrapie, BSE "mad cow disease" and Creutzfeld-Jacob disease (see Chapter 14). A better understanding of the folding and misfolding processes might, therefore, open up new approaches to drugs for these diseases.

Effector molecules switch allosteric proteins between R and T states

In 1963 Jaques Monod, Jean-Pierre Changeaux and Francois Jacob published a theory that radically changed our views on the control of protein function and which still influences protein biochemists as well as structural biologists. Their theory of **allosteric control** provided a unifying theme for such diverse concepts as **feedback inhibition** of enzymes, repressor/corepressor binding (see Chapter 8) and **cooperative binding** of ligand by proteins, with oxygen binding to hemoglobin as the prime example. The allosteric theory they proposed has the following main features. Cooperative substrate binding and modification of a protein's activity by allosteric effector molecules may arise in proteins with two or more preformed structural states that are in equilibrium. Substrates and effector molecules bind at different sites on the protein and therefore need no stereochemical relationship to each other, hence the name allostery (different shapes).

The theory predicts that such proteins are built up of several subunits which are symmetrically arranged and that the two states differ by the arrangements of the subunits and the number of bonds between them. In one state the subunits are constrained by strong bonds that would resist the structural changes needed for substrate binding, and this state would consequently bind substrates weakly; they called it the tense or T state. In the other state, called the R state, these constraints are relaxed.

This **concerted model** assumes furthermore that the symmetry of the molecule is conserved so that the activity of all its subunits is either equally low or equally high, that is, all structural changes are concerted. Subsequently Daniel Koshland, University of California, Berkeley, postulated a **sequential model** in which each subunit is allowed independently to change its tertiary structure on substrate binding. In this model tertiary structural changes in the subunit with bound ligand alter the interactions of this subunit with its neighbours and this leads sequentially to changes in the latter's reactive sites. Koshland's model was based on his theories of induced fit; ligand binding induces a conformational change that converts an enzyme from an inactive to an active state.

For many years hemoglobin was the only allosteric protein whose stereochemical mechanism was understood in detail. However, more recently detailed structural information has been obtained for both the R and the T states of several enzymes as well as one genetic repressor system, the trp-repressor, described in Chapter 8. We will here examine the structural differences between the R and the T states of a key enzyme in the glycolytic pathway, phosphofructokinase.

X-ray structures explain the allosteric properties of phosphofructokinase

Phosphofructokinase, PFK, is the key regulatory enzyme in glycolysis, the breakdown of glucose to generate ATP which occurs in most cells (Figure 6.24a). The enzyme catalyzes one of the early steps in this pathway, phosphorylation by ATP of fructose-6-phosphate, F6P, to fructose-1,6-bisphosphate (Figure 6.24b). Binding to PFK of one of the substrate molecules, F6P, is highly cooperative whereas binding of the second substrate ATP is

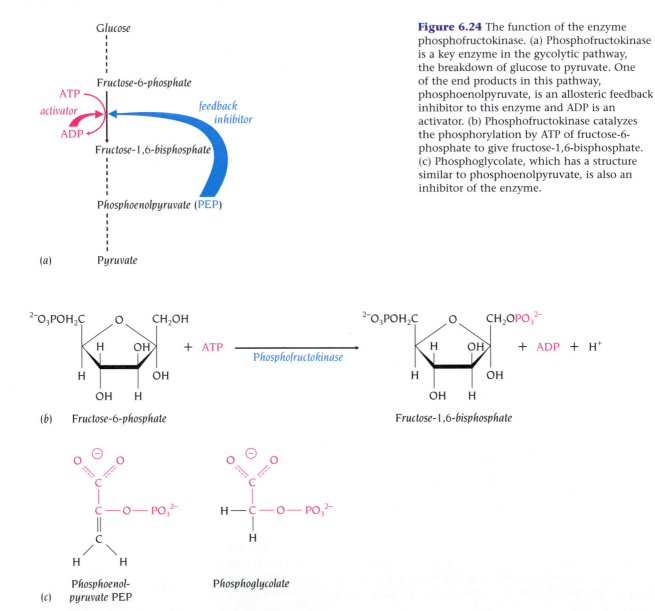

Figure 6.24 The function of the enzyme phosphofructokinase. (a) Phosphofructokinase is a key enzyme in the glycolytic pathway, the breakdown of glucose to pyruvate. One of the end products in this pathway, phosphoenolpyruvate, is an allosteric feedback inhibitor to this enzyme and ADP is an activator. (b) Phosphofructokinase catalyzes the phosphorylation by ATP of fructose-6-phosphate to give fructose-1,6-bisphosphate. (c) Phosphoglycolate, which has a structure similar to phosphoenolpyruvate, is also an inhibitor of the enzyme.

noncooperative. Phosphofructokinase is inhibited by phosphoenolpyruvate, PEP, one of the products of a late step in the glycolytic pathway, and by chemical analogs of PEP, for example 2-phosphoglycolate (Figure 6.24c). By contrast the reaction rate is enhanced by ADP, an allosteric effector molecule. The regulation of PFK by effector molecules is the main way that the glucose degradation by glycolysis is controlled in cells.

Because of the crucial role of this enzyme in one of the most important biochemical pathways in the cell, its allosteric properties have been studied extensively in solution. Interpretation of these studies in terms of the theory of allosteric enzymes led Monod and coworkers to conclude that:

1. The enzyme is made of four identical subunits each having a single binding site for each ligand.
2. The subunits can switch between two distinct conformational states, R and T, which are in equilibrium.
3. The transitions between these states in each tetrameric molecule are concerted, in other words all four subunits of each molecule are in the same state, either R or T.
4. The two states have the same affinity for ATP but differ with respect to their affinity for the substrate F6P, the allosteric effector ADP and the inhibitor PEP. Because of these differences in affinity, ligand binding can shift the equilibrium between the R and T states to favor one or the other state depending on which ligand is bound.

The group of Phil Evans, MRC Laboratory of Molecular Biology, Cambridge, UK, has determined x-ray structures of bacterial PFK both in the R and the T states. These studies have confirmed the above conclusions and given insight into how an allosteric enzyme accomplishes its complex behavior.

Each subunit of the homotetrameric PFK of *Escherichia coli* comprises 320 amino acids arranged in two domains, one large and one smaller, both of which have an α/β structure reminiscent of the Rossman fold (Figure 6.25).

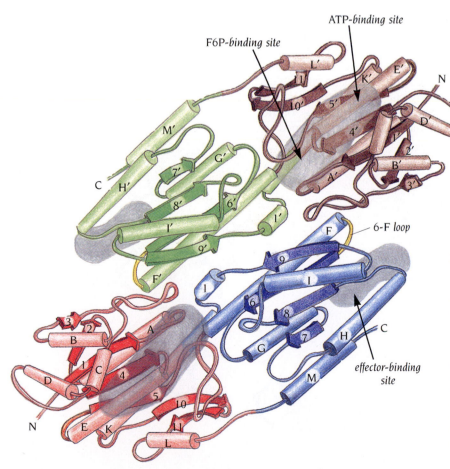

Figure 6.25 Schematic diagram of the structure of one dimer of phosphofructokinase. Each polypeptide chain is folded into two domains (blue and red, and green and brown), each of which has an α/β structure. Helices are labeled A to M and β strands 1 to 11 from the amino terminus of one polypeptide chain, and respectively A' to M' and 1' to 11' for the second polypeptide chain. The binding sites of substrate and effector molecules are schematically marked in gray. The effector site of one subunit is linked to the active site of the other subunit of the dimer through the 6-F loop between helix F and strand 6. (Adapted from T. Schirmer and P.R. Evans, *Nature* 343: 140–145, 1990.)

The subunits are pairwise linked into two dimers (A-B and C-D in Figure 6.26a) with extensive close contacts between the subunits within the dimers. The two dimers are loosely packed against each other into a symmetrical tetramer. The close contacts between the two subunits of the dimer are the same in the R and the T states but the interactions between the dimers are quite different. The orientation of the dimers with respect to each other in the R and T states differs by a rotation of 7°. This difference affects the packing of the dimers against each other and hence the quaternary structure. In the T state the dimers are close together and there are direct hydrogen bonds between two β strands, one from each dimer (Figure 6.26b). In the R state these two β strands are further apart and the gap between them is filled with water molecules that form hydrogen bond bridges (Figure 6.26c). The inclusion or exclusion of water molecules between the dimers is an all-or-none effect that acts like a two-way switch. As we shall see this change in quaternary structure of the tetrameric molecule is intimately linked to differences in tertiary structure of the subunits in the R and T states.

Each of the four subunits contains three binding sites (see Figure 6.25). There is one site for each of the substrate molecules, ATP and F6P, which together form the active site facing a cleft between the two domains. The third binding site of each subunit is the regulator binding site to which both the inhibitor PEP and the allosteric activator ADP can bind; this site is distant from the active site. Evans has studied crystals of the catalytically competent R state of three types: (1) with the subunits complexed with substrates, (2) subunits complexed with ADP, and (3) subunits unliganded. He also studied crystals of the T state in which the subunits are complexed with the inhibitor 2-phosphoglycolate.

The catalytic site of each subunit is in the cleft between the two domains. The large domain binds ATP with the terminal phosphate pointing into the cleft. The main binding site for the second substrate molecule, F6P, is in the smaller domain and the phosphate group of F6P interacts with a neighboring subunit affecting subunit interactions that are crucial for catalytic activity. In the active R state this phosphate group forms hydrogen bonds to an arginine residue, Arg 162, of a small α helix, called the 6-F helix, in the neighboring subunit (Figure 6.27a). By contrast in the T form this helix is unwound and instead forms a loop with Arg 162 pointing away from the F6P molecule (Figure 6.27b). In the T state a negatively charged glutamate side chain, Glu 161, occupies the same position as the positively charged Arg 162 in the R state. This negatively charged glutamate 161 repels the negative charge of the phosphate group of F6P. Consequently, when the R state is transformed to

Figure 6.26 The quaternary structure of phosphofructokinase. (a) The four subunits are pairwise arranged in two dimers A-B (blue) and C-D (red or green). The subunit interactions within the dimers are extensive and tight whereas the two dimers are loosely packed against each other and the packing contacts are different in the R and the T states. The orientation of the dimers with respect to each other in the T (red contours of the C-D dimer) and R (green contours) differs by a rotation of 7°. (b) The dimers are close together in the T state and there are direct hydrogen bonds between two β strands, one from the A-B dimer (blue) and one from the C-D dimer (green). Hydrogen bonds are shown in orange. (c) The dimers are further apart in the R state and there is a gap between the two β strands from the two dimers which is filled by water molecules (red). These water molecules form bridges between the dimers by making hydrogen bonds to the C=O and N-H groups of the two β strands.

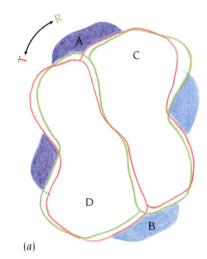

(a)

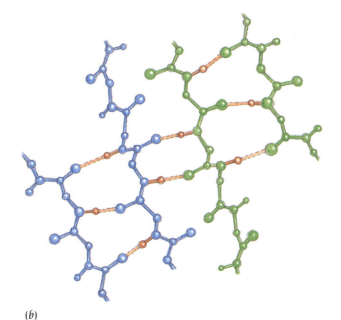

(b)

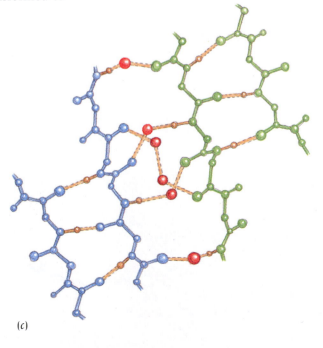

(c)

Figure 6.27 Conformational changes in the active site of phosphofructokinase. (a) In the active R state the phosphate group of the substrate fructose-1-phosphate, F6P, (red) forms a salt bridge to an arginine residue, Arg 162, of a small α helix (orange). This salt bridge contributes substantially to promote binding of the substrate to the enzyme. (b) In the inactive T state the helix has been partially unwound and changed its orientation so that Arg 162 points away from the substrate binding site. Instead a negatively charged glutamate residue, Glu 161, points towards the phosphate binding site of the substrate molecule. Repulsive forces between the negative charges of Glu 161 and the phosphate of F6P prevent binding and result in a thousandfold lower affinity for F6P when the enzyme is in the T state compared with the R state.

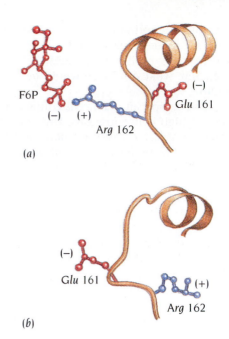

the T state the change in the F6P binding site results in a thousandfold lower affinity for F6P. In addition, catalytically important residues are properly arranged for catalysis in the R state but not in the T state. Since the binding site of ATP is virtually identical in the R and T states the binding of ATP is not affected by the structural differences between them.

The change in the quaternary structure and the structural change in the 6-F helix as the molecule moves from one state to the other are intimately related. The dimer interactions in the T state are not compatible with the presence of the 6-F helix, which would, if present, clash with the neighbouring dimer. The quaternary structure of the T state requires that the 6-F helix be unwound. Conversely the R state quaternary structure depends on the presence of the 6-F helix.

The basic kinetic properties of this allosteric enzyme are clearly explained by combining Monod's theory and these structural results. The tetrameric enzyme exists in equilibrium between a catalytically active R state and an inactive T state. There is a difference in the tertiary structure of the subunits in these two states, which is closely linked to a difference in the quaternary structure of the molecule. The substrate F6P binds preferentially to the R state, thereby shifting the equilibrium to that state. Since the mechanism is concerted, binding of one F6P to the first subunit provides an additional three subunits in the R state, hence the cooperativity of F6P binding and catalysis. ATP binds to both states, so there is no shift in the equilibrium and hence there is no cooperativity of ATP binding. The inhibitor PEP preferentially binds to the effector binding site of molecules in the T state and as a result the equilibrium is shifted to the inactive state. By contrast the activator ADP preferentially binds to the effector site of molecules in the R state and as a result shifts the equilibrium to the R state with its four available, catalytically competent, active sites per molecule.

Conclusion

The thermodynamic stability of a protein in its native state is small and depends on the differences in entropy and enthalpy between the native state and the unfolded state. From the biological point of view it is important that this free energy difference is small because cells must be able to degrade proteins as well as synthesize them, and the functions of many proteins require structural flexibility.

When a fully extended unfolded polypeptide chain begins to fold, hydrophobic residues tend to be buried in the interior, greatly restricting the number of possible conformations the chain can assume, and therefore allowing proteins to fold in seconds rather than years. Within milliseconds the polypeptide chain achieves the molten globule state, a term used to describe a set of structures that have in common a loosely packed hydrophobic core and some secondary structure. Some proteins have one preferred folding pathway, while others seem to have multiple parallel pathways to the native state. There are certain high energy barriers to folding such as, for example, the formation of correct disulfide bonds and the isomerization of proline residues. Cells contain enzymes such as protein disulfide isomerases and *cis–trans*-proline isomerases that catalyze these reactions and therefore overcome what otherwise would be an insuperable energy barrier to rapid folding.

The cytoplasm of all cells contains folded proteins and folding polypeptides at high concentrations. Unfolded proteins with exposed hydrophobic patches aggregate easily by non-specific hydrophobic interactions. To circumvent this problem a class of proteins called chaperones have evolved to sequester unfolded polypeptides. The complex structure of one class of multimeric chaperones, the chaperonins GroEL and GroES, has been elucidated and has shed light on how chaperones function. These chaperonins are short cylinders which, because they have hydrophobic residues in the interior, can bind to any unfolded polypeptide that has large exposed hydrophobic patches, regardless of its amino acid sequence. Once the polypeptide chain is shielded inside the chaperonins it is protected from aggregation with other protein molecules. During folding some polypeptide chains go through many cycles of binding and release from the chaperonins.

Even inside crystals all atoms in protein molecules undergo small oscillations. Protein structures determined by x-rays are an average of these breathing structures. In addition to breathing, some proteins undergo large conformational changes in response to ligand binding or to changes in their environment, and these conformational changes are essential for function. The switches that control the successful passage through the eucaryotic cell cycle depend on changes in the conformation of an α helix and a flexible loop region in the cyclin-dependent kinases. In the case of calmodulin, structural changes play a crucial role in calcium signaling pathways. The dumbbell-shaped inactive molecule collapses into a globular structure that binds to regulatory proteins in such pathways. The serpins, a class of specific serine proteinase inhibitors, undergo an extraordinary conformational change: activation of the inactive form involves conversion of a β strand in the middle of a β sheet into a flexible loop and the converse occurs when the active form changes into the latent form.

Many multimeric enzymes and some other multimeric proteins, the classic examples being hemoglobin and phosphofructokinase, PFK, are subject to allosteric control. Allosteric proteins exist in two states, classically known as the R (relaxed) and the T (tense) states. Effector molecules have a high affinity for only one of these states. Therefore when an effector molecule is present it shifts the equilibrium to favor the high affinity state. The binding site for effector molecules is unrelated to and distinct from the active site. In the case of PFK there are two effector molecules, the activator ADP, which shifts the four identical subunits of the enzyme to the enzymatically active R state, and the inhibitor PEP, which shifts the four subunits to the inactive T state. The active site in the R state has a thousandfold higher affinity for the substrate F6P than does the active site of the T state, due to structural differences between the active sites in these two states. In the cell the activity of PFK is controlled through this allosteric mechanism by the relative concentrations of the two effectors. The inhibitor or negative effector is PEP, which is the product of an enzyme downstream of PFK in the glycolytic pathway. As the concentration of PEP increases it inhibits PFK and downregulates the pathway. This is the classic case of feedback inhibition.

Selected readings

General

Cohen, F.E., et al. Structural clues to prion replication. *Science* 264: 530–531, 1994.

Creighton, T.E. Up the kinetic pathway. *Nature* 356: 194–195, 1992.

Dobson, C.M. Finding the right fold. *Nature (Struct. Biol.)* 2: 513–517, 1995.

Fersht, A.R. Characterizing transition states in protein folding: an essential step in the puzzle. *Curr. Opin. Struct. Biol* 5: 79–84, 1994.

Finn, B.E., Forsen, S. The evolving model of calmodulin structure, function and activation. *Structure* 3: 7–11, 1995.

Freedman, R.B. The formation of protein disulfide bonds. *Curr. Opin. Struct. Biol.* 5: 85–91, 1995.

Goldsmith, E.J, Mottonen, J. Serpins: the uncut version. *Structure* 2: 241–244, 1994.

Hartl, F.U. Molecular chaperones in cellular protein folding. *Nature* 381: 571–580, 1996.

Lorimer, G.H. GroEL structure: a new chapter on assisted folding. *Structure* 2: 1125–1128, 1994.

Matthews, C.R. Pathways of protein folding. *Annu. Rev. Biochem.* 62: 653–683, 1993.

Miranker, A., Dobson, C.M. Collapse and cooperativity in protein folding. *Curr. Opin. Struct. Biol.* 6: 31–42, 1996.

Morgan, D.O. Principles of CDK regulation. *Nature* 374: 131–134, 1995.

Perutz, M. *Mechanisms of cooperativity and allosteric regulation in proteins.* Cambridge: Cambridge University Press, 1990.

Pines, J. Conformational change. *Nature* 376: 294–295, 1995.

Ptitsyn, O.B. Structures of folding intermediates. *Curr. Opin. Struct. Biol.* 5: 74–78, 1995.

Radzio-Andzelm, E.R., Lew, J., Taylor, S. Bound to activate: conformational consequences of cyclin binding to CDK2. *Structure* 3: 1135–1141, 1995.

Saibil, H.R. The lid that shapes the pot: structure and function of the chaperonin GroES. *Structure* 4: 1–4, 1996.

Weissman, J.S. All the roads lead to Rome? The multiple pathways of protein folding. *Chem. Biol.* 2: 255–260, 1995.

Specific structures

Babu, Y.S., et al. Three-dimensional structure of calmodulin. *Nature* 315: 37–40, 1985.

Baker, D., Sohl, J.L., Agard, D.A. A protein-folding reaction under kinetic control. *Nature* 356: 263–265, 1992.

Braig, K., et al. The crystal structure of the bacterial chaperonin GroEL at 2.8 Å. *Nature* 371: 578–586, 1994.

Brown, N.R., et al. The crystal structure of cyclin A. *Structure* 3: 1235–1247, 1995.

Carrell, R.W., et al. Biological implications of a 3 Å structure of dimeric antithrombin. *Structure* 2: 257–270, 1994.

Carrell, R.W., Evans, D.L., Stein, P.E. Mobile reactive centre of serpins and the control of thrombosis. *Nature* 353: 576–578, 1991.

Chen, S. et al. Location of a folding protein and shape changes in GroEL–GroES complexes imaged by cryo-electron microscopy. *Nature* 371: 261–264, 1994.

Creighton, T.E. The disulfide folding pathway of BPTI. *Science* 256: 111–114, 1992.

De Bondt, H.L., et al. Crystal structure of cyclin-dependent kinase 2. *Nature* 363: 595–602, 1993.

Dobson, C.M., Evans, P.A., Radford, S.E. Understanding protein folding: the lysozyme story so far. *Trends Biol. Sci.* 19: 31–37, 1994.

Fenton, W.A., et al. Residues in chaperonin GroEL required for polypeptide binding and release. *Nature* 371: 614–619, 1994.

Hunt, J.F., et al. The crystal structure of the GroES co-chaperonin at 2.8 Å resolution. *Nature* 379: 37–45, 1996.

Ikura, M., et al. Solution structure of a calmodulin-target peptide complex by multidimensional NMR. *Science* 256: 632–638, 1992.

Jeffrey, P.D., et al. Mechanism of CDK activation revealed by the structure of a cyclinA-CDK2 complex. *Nature* 376: 313–320, 1995.

Kabsch, W., et al. Atomic structure of the actin:DNAse I complex. *Nature* 347: 37–44, 1990.

Kallen, J., et al. Structure of human cyclophilin and its binding site for cyclosporin A determined by x-ray crystallography and NMR spectroscopy. *Nature* 353: 276–279, 1991.

Loebermann, H., et al. Human α_1-proteinase inhibitor. *J. Mol. Biol.* 177: 531–556, 1984.

Martin, J.L., Bardwell, J.C.A., Kuriyan, J. Crystal structure of the DsbA protein required for disulphide bond formation *in vivo*. *Nature* 365: 464–468, 1993.

Mauguen, Y., et al. Molecular structure of a new family of ribonucleases. *Nature* 297: 3162–3164, 1982.

Mayhew, M., et al. Protein folding in the central cavity of the GroEL–GroES chaperonin complex. *Nature* 379: 420–426, 1996.

Meador, W.E., Means, A.R., Quiocho, F.A. Target enzyme recognition by calmodulin: 2.4 Å structure of a calmodulin–peptide complex. *Science* 257: 1251–1255, 1992.

Mottonen, J., et al. Structural basis of latency in plasminogen activator inhibitor-1. *Nature* 355: 270–273, 1992.

Schirmer, T., Evans, P.R. Structural basis of the allosteric behaviour of phosphofructokinase. *Nature* 343: 140–145, 1990.

Schreuder, H.A., et al. The intact and cleaved human antithrombin III complex as a model for serpin–proteinase interactions. *Nature (Struct. Biol.)* 1: 48–54, 1994.

Stein, P.E., et al. Crystal structure of ovalbumin as a model for the reactive centre of serpins. *Nature* 347: 99–102, 1990.

Takahashi, N., Hayano, T., Suzuki, M. Peptidyl-prolyl *cis–trans* isomerase is the cyclosporin A binding protein cyclophilin. *Nature* 337: 473–475, 1989.

Weissman, J.S., Kim, P.S. Kinetic role of non-native species in the folding of bovine pancreatic trypsin inhibitor. *Proc. Natl. Acad. Sci. USA* 89: 9900–9904, 1992.

Weissman, J.S., Kim, P.S. Re-examination of the folding of BPTI: predominance of native intermediates. *Science* 253: 1386–1393, 1992.

Yu, M.-H., Lee, K.N., Kim, J. The Z type variation of human α_1-antitrypsin causes a protein folding defect. *Nature (Struct. Biol.)* 2: 363–367, 1995.

Zhu, X., et al. Structural analysis of substrate binding by the molecular chaperone DnaK. *Science* 272: 1606–1614, 1996.

DNA Structures

7

DNA replication, DNA transcription, and the regulation of gene expression all depend upon the recognition of DNA by proteins. By a powerful combination of structural and genetic studies, in recent years we have begun to understand how these functions are achieved. These insights will be described in Chapters 8 to 10, with emphasis on the contribution of structural studies. Before tackling the structures of DNA-binding proteins and the complexes they form with DNA, however, we need to understand the structure of the **double-stranded, base-paired helical DNA** molecule on its own in order to see what possibilities it offers for the recognition of specific sequences by proteins.

The DNA double helix is different in A- and B-DNA

The DNA molecules in each cell of an organism contain all the genetic information necessary to ensure the normal development and function of that organism. This genetic information is encoded in the precise linear sequence of the nucleotide bases from which the DNA is built. DNA is a linear molecule: while its diameter is only about 20 Å, if stretched out its length can reach many millimeters. This means that concentrated solutions of DNA can be pulled into fibers in which the long thin DNA molecules are oriented with their long axes parallel.

Early diffraction photographs of such DNA fibers taken by Rosalind Franklin and Maurice Wilkins in London and interpreted by James Watson and Francis Crick in Cambridge revealed two types of DNA structures: **A-DNA** and **B-DNA**. The B-DNA form is obtained when DNA is fully hydrated as it is *in vivo*. A-DNA is obtained under dehydrated nonphysiological conditions. Improvements in the methods for the chemical synthesis of DNA have recently made it possible to study crystals of short DNA molecules of any selected sequence. These studies have essentially confirmed the refined fiber diffraction models for A- and B-DNA and in addition have given details of small structural variations for different DNA sequences. Furthermore, a new structural form of DNA, called **Z-DNA**, has been discovered.

Both A-DNA and B-DNA have the familiar shape of a right-handed helical staircase (Figure 7.1). The rails are two antiparallel phosphate-sugar chains, and the rungs are purine-pyrimidine base pairs, which are hydrogen bonded to each other. In A-DNA there are an average of 10.9 base pairs per turn of the helix, which corresponds to an average helical-twist angle of 33.1° (10.9 × 33.1 = 360°) from one base pair to the next. The spacing along the helix axis from one base pair to the next is 2.9 Å. In B-DNA these values are

Figure 7.1 Schematic drawing of B-DNA. Each atom of the sugar-phosphate backbones of the double helix is represented as connected circles within ribbons. The two sugar-phosphate backbones are highlighted by orange ribbons. The base pairs that are connected to the backbone are represented as blue planks. Notice that in B-DNA the central axis of this double helix goes through the middle of the base pairs and that the base pairs are perpendicular to the axis.

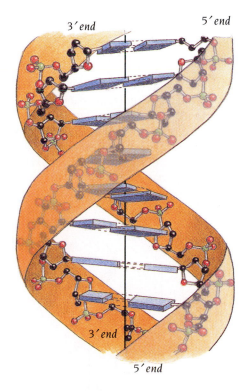

3′ end   5′ end

3′ end

5′ end

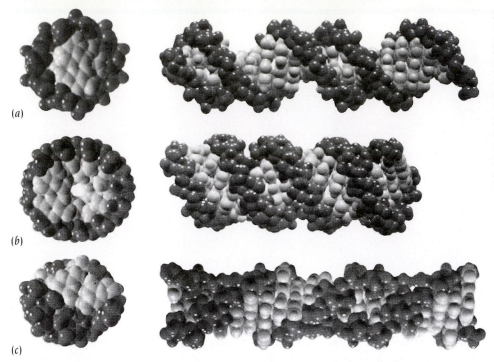

(a)

(b)

(c)

Figure 7.2 Three helical forms of DNA, each containing 22 nucleotide pairs, shown in both side and top views. The sugar-phosphate backbone is dark; the paired nucleotide bases are light. (a) B-DNA, which is the most common form in cells. (b) A-DNA, which is obtained under dehydrated nonphysiological conditions. Notice the hole along the helical axis in this form. (c) Z-DNA, which can be formed by certain DNA sequences under special circumstances. (Courtesy of Richard Feldmann.)

10.0 base pairs and 35.9° and 3.4 Å, respectively. There are, however, considerable variations in individual twist angles from the average values, and these variations are larger in A-DNA than in B-DNA. These variations are sequence dependent, and in B-DNA they might be important for the specificity of interactions with proteins.

The DNA helix has major and minor grooves

The sugar-phosphate backbones are bulky and form ridges on the edges of the helix, with grooves in between within which the bases are exposed (Figure 7.2). These grooves are of two different widths, reflecting the asymmetrical attachment of the base pairs to the sugar rings of the backbone. Whereas in a regular helix the distance between the attachment points for the rungs would be the same at the front and the back of each step (Figure 7.3), in the DNA molecule each base-pair "rung" is effectively wider at one edge than at the other (Figure 7.4) so that the helical molecule has one narrower groove, known as the **minor groove**, and one wider groove, known as the **major groove** (Figure 7.5).

In B-DNA because the helical axis runs through the center of each base pair and the base pairs are stacked nearly perpendicular to the helical axis (see Figures 7.1 and 7.5), the major and minor grooves are of similar depths.

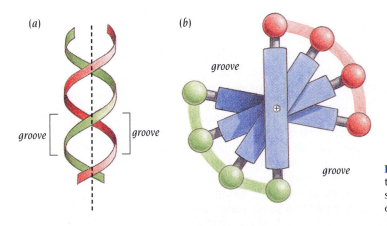

(a)

(b)

groove

groove

groove

groove

Figure 7.3 Schematic diagram illustrating that there are two similar grooves in a helical staircase. Four rungs are viewed from the top of the staircase in (b).

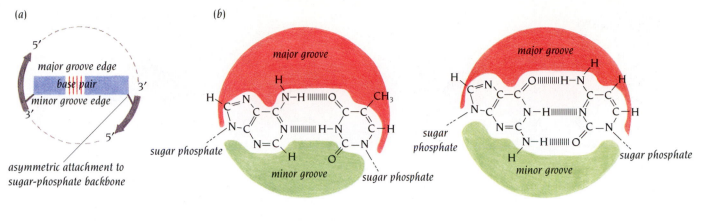

(a)

5′
major groove edge
base pair
minor groove edge
3′
3′
5′

asymmetric attachment to
sugar-phosphate backbone

(b)

major groove

sugar phosphate

minor groove

sugar phosphate

ADENINE : THYMINE

major groove

sugar
phosphate

minor groove

sugar phosphate

GUANINE : CYTOSINE

In A-DNA, on the other hand, the helical axis is shifted from the center of the bases into the major groove, bypassing the bases, and the base pairs are not perpendicular to this axis but are tilted between 13° and 19°. This arrangement makes the major groove very deep, extending from the surface all the way past the central axis and part of the way out toward the opposite side, while the minor groove is shallow, scarcely more than a helical depression spiraling around the outside of the cylinder (see Figure 7.5).

The edges of the base pairs form the floors of the two grooves. The edge of a base pair furthest from its attachment points to the sugar-phosphate backbones is the major groove edge; the one closest is the minor groove edge (see Figure 7.4a). These edges are accessible from the outside and form the basis for the sequence-specific recognition of DNA by proteins.

Figure 7.4 The edges of the base pairs in DNA that are in the major groove are wider than those in the minor groove, due to the asymmetric-attachment of the base pairs to the sugar-phosphate backbone (a). These edges contain different hydrogen bond donors and acceptors for potentially specific interactions with proteins (b).

Z-DNA forms a zigzag pattern

Z-DNA has a quite different structure (see Figure 7.2c). The helix is left-handed, and the sugar-phosphate backbone follows a zigzag path. The structure has been found for sequences with alternating G and C bases, such as CGCG and CGCGCG. Each cytosine has its sugar attached to the base in such a way that the pyrimidine ring swings away from the minor groove. This is the normal conformation for all four bases in A- and B-DNA. However, each guanine in Z-DNA has its sugar ring rotated 180° so that it bends inward toward the minor groove. The sugar-phosphate backbone of alternating C and G bases with their different conformations make a zigzag pattern around the helix. The shape of this helix is thin and elongated. It has a deep but quite narrow minor groove, whereas the major groove is pushed to the surface so that it is no longer a groove at all.

Figure 7.5 Schematic diagram illustrating the major and minor grooves in A- and B-DNA. The sugar-phosphate backbone is represented by connected circles in color and the base pairs as blue planks. Four base pairs are shown from the top of the helix to highlight how the grooves are formed due to the asymmetric connections. The position of the helix axis is marked by a cross.

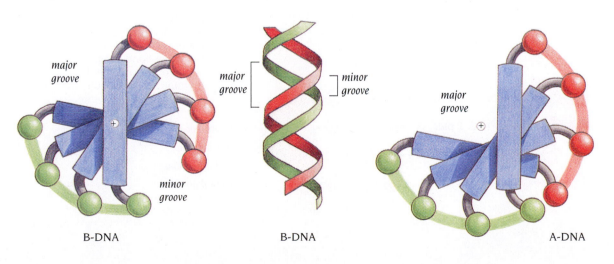

major groove

minor groove

B-DNA

major groove

minor groove

B-DNA

major groove

A-DNA

123

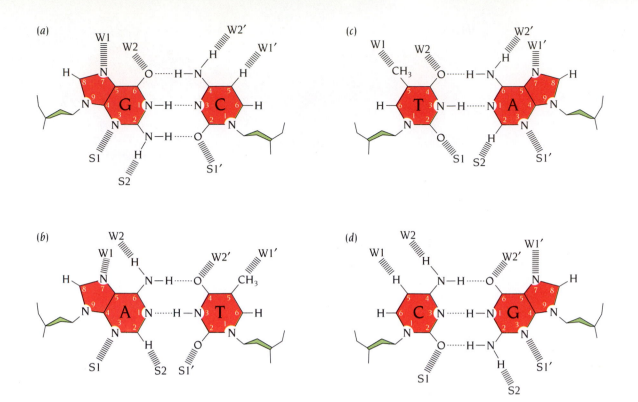

B-DNA is the preferred conformation *in vivo*

The specific protein–DNA interactions described in this book are all with DNA in its regular B-form, or, in some cases with distorted B-DNA. In biological systems DNA appears not to adopt the A conformation, although double-stranded RNA does preferentially adopt this conformation *in vivo*. Whether or not Z-DNA occurs in nature is a matter of controversy. However, the formation of A-DNA and Z-DNA *in vitro* does illustrate the large structural changes that DNA can be forced to undergo.

Specific base sequences can be recognized in B-DNA

Let us now look at how specific sequences might in principle be recognized in B-DNA. The only regions where the bases are available for interaction are at the floor of the grooves. These are paved with nitrogen and oxygen atoms that can make hydrogen bonds with the side chains of a protein. The methyl group of thymine and the corresponding hydrogen in cytosine provide additional discriminatory recognition groups (see Figure 7.4). These sites, which are represented in Figure 7.6 and schematically illustrated in Figure 7.7 in the form of a color code, form patterns that are different for the four possible Watson/Crick base pairs. If, for instance, we compare a GC and an AT base-pair reading from the purine to the pyrimidine ring, we can read the

Figure 7.6 The edges of the base pairs contain nitrogen and oxygen atoms that can make hydrogen bonds to protein side chains. An H atom in cytosine (C) and a methyl group in thymine (T) form additional sequence-specific recognition sites in DNA. W1, W2, W2′, and W1′ are the recognition sites at the edges of the base pairs in the major groove (W for wide) and S1, S2, and S1′ are those in the minor groove (S for small). The recognition sites are shown for all four base pairs: GC (a), AT (b), TA (c), and CG (d).

Figure 7.7 Color codes for the recognition patterns at the edges of the base pairs in the major (a) and minor (b) grooves of B-DNA. Hydrogen-bond acceptors are red; hydrogen-bond donors are blue. The methyl group of thymine is yellow, while the corresponding H atom of cytosine is white.

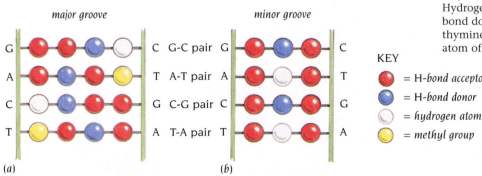

major groove   minor groove

G	🔴🔴🔵⚪	C	G-C pair	G	🔴🔵🔴	C
A	🔴🔵🔴🟡	T	A-T pair	A	🔴⚪🔴	T
C	⚪🔵🔴🔴	G	C-G pair	C	🔴🔵🔴	G
T	🟡🔴🔵🔴	A	T-A pair	T	🔴⚪🔴	A

KEY

🔴 = H-*bond acceptor*
🔵 = H-*bond donor*
⚪ = *hydrogen atom*
🟡 = *methyl group*

(a)   (b)

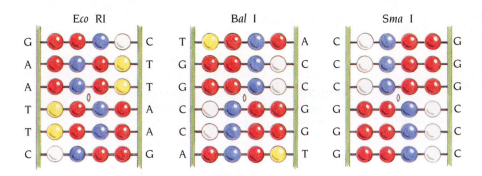

Figure 7.8 Sequence-specific recognition sites in the major groove of DNA for three restriction enzymes—Eco RI, Bal I, and Sma I. The DNA sequences that are recognized by these enzymes are represented by the color code defined in Figure 7.7.

recognition pattern in the following way. For the GC pair, the major groove exposes a hydrogen-bond acceptor, nitrogen 7 of guanine (G N7; W1), another acceptor, G O6 (W2), a hydrogen-bond donor, C NH4 (W2′), and, finally, a hydrogen atom at C5 (W1′) (see Figure 7.6a). For the AT pair, the corresponding groups are acceptor A N7, donor A NH6, acceptor T O4, and a methyl group at T5 (see Figure 7.6b).

These patterns of potential hydrogen-bond acceptors and donors are clearly quite different for the different base pairs in the major groove, and they could easily be recognized and distinguished by a protein molecule. This is not the case in the minor groove. If we look at the same two base pairs, we see that for the GC pair we have an acceptor G N3 (S1), a donor G NH2 (S2), and an acceptor C O2 (S1′). For the AT pair we find the following pattern: an acceptor A N3, a hydrogen atom at A2, and an acceptor T O2. In the minor groove there is only one distinguishing feature between these two base pairs: a hydrogen-bond donor group in GC compared with a neutral hydrogen atom in AT. We can represent these recognition patterns by a color code as shown in Figure 7.7. Hydrogen-bond donors and acceptors of the base pairs have different colors as well as the H and CH_3 groups of cytosine and thymine, respectively. In this way, each of the four different base pairs is represented by a unique color code, comprising four positions in the major groove and three in the minor groove. Clearly, the major groove is a much better candidate for sequence-specific recognition than the minor groove for two reasons. First, the major groove is wider than the minor, and the bases are thus more accessible to a protein molecule. Second, the pattern of possible hydrogen bonds from the edges of the base pairs to a protein are more specific and discriminatory in the major groove than in the minor (see Figure 7.7).

Only a rather limited number of base pairs is needed to provide unique and discriminatory recognition sites in the major groove. This is illustrated in Figure 7.8, which gives the color codes for the hexanucleotide recognition sites of three different restriction enzymes—Eco RI, Bal I, and Sma I. It is clear that these patterns are quite different, and each can be uniquely recognized by specific protein–DNA interactions.

Bacteriophage repressor proteins provide excellent examples of sequence-specific interactions between the side chains of a protein and bases lining the floor of the major groove of B-DNA. As we shall see, to fit the protein's recognition module into this groove it has to be made even wider; in other words, the B-DNA has to be distorted.

Conclusion

B-DNA is the conformation of DNA *in vivo*. It has a wide major groove and a narrow minor groove. The edges of the base pairs form the bottom of the grooves, where nitrogen and oxygen atoms are available to make hydrogen bonds with the side chains of a protein. Such hydrogen bonds form the basis for sequence-specific recognition of DNA.

Selected readings

General

Dickerson, R.E. The DNA helix and how it is read. *Sci. Am.* 249(6): 94–111, 1983.

Dickerson, R.E., et al. The anatomy of A-, B- and Z-DNA. *Science* 216: 475–485, 1982.

Felsenfeld, G. DNA. *Sci. Am.* 253(4): 58–66, 1985.

Kennard, O., Hunter, W.N. Oligonucleotide structure: a decade of results from single crystal x-ray diffraction studies. *Q. Rev. Biophys.* 22: 327–379, 1989.

Rich, A.A., Nordheim, A., Wang, A.H.-J. The chemistry and biology of left-handed Z-DNA. *Annu. Rev. Biochem.* 53: 791–846, 1984.

Saenger, W. *Principles of Nucleic Acid Structure.* Berlin: Springer, 1984.

Travers, A.A. DNA conformation and protein binding. *Annu. Rev. Biochem.* 58: 427–452, 1989.

Travers, A.A. *DNA–Protein Interactions.* London: Chapman and Hall, 1990.

Watson, J.D., Crick, F.H.C. Genetic implications of the structure of deoxyribonucleic acid. *Nature* 171: 964–967, 1953.

Specific structures

Arnott, S., Hukins, D.W.J. Optimised parameters for A-DNA and B-DNA. *Biochem. Biophys. Res. Commun.* 47: 1504–1509, 1972.

Franklin, R.E., Gosling, R.G. Molecular structure of nucleic acids. Molecular configuration in sodium thymonucleate. *Nature* 171: 740–741, 1953.

Seeman, N.C., Rosenberg, J.M., Rich, A. Sequence-specific recognition of double helical nucleic acids by proteins. *Proc. Natl. Acad. Sci. USA* 73: 804–809, 1976.

Wang, A.H.-J., et al. Molecular structure of a left-handed DNA fragment at atomic resolution. *Nature* 282: 680–686, 1979.

Watson, J.D., Crick, F.H.C. Molecular structure of nucleic acids. A structure for deoxyribose nucleic acid. *Nature* 171: 737–738, 1953.

Wilkins, M.H.F., Stokes, A.R., Wilson, H.R. Molecular structure of nucleic acids. Molecular structure of deoxypentose nucleic acids. *Nature* 171: 738–740, 1953.

Wing, R.M., et al. Crystal structure analysis of a complete turn of B-DNA. *Nature* 287: 755–758, 1980.

Structure, Function, and Engineering

Part 2

DNA Recognition in Procaryotes by Helix-Turn-Helix Motifs

8

Proteins that regulate transcription of DNA recognize specific DNA sequences through discrete DNA-binding domains within their polypeptide chains. These domains are in general relatively small, less than 100 amino acid residues. Many procaryotic DNA-binding domains contain a **helix-turn-helix motif** that recognizes and binds specific regulatory regions of DNA. The two α helices have the same orientation relative to each other, and they are connected by a loop region of similar structure in all DNA-binding helix-turn-helix motifs. In this chapter we will discuss the functional properties of the helix-turn-helix motif and the way this motif is integrated into structurally different DNA-binding domains of procaryotic repressors and activators. In the following two chapters we discuss the wider range of DNA binding motifs in eucaryotic DNA-binding proteins.

The mechanism of action of bacterial and bacteriophage repressors and activators is in principle very simple. Repressors bind tightly to the DNA at the promoter of a structural gene, preventing the RNA polymerase from gaining access and hence blocking the initiation of transcription. Activators, on the other hand, work by binding next to the promoter and helping the polymerase to bind to the adjacent promoter, thereby increasing the rate of transcription of the gene. However, as we will see, the subtle regulation of these binding interactions can be quite complex.

Some of the most thoroughly studied procaryotic regulator proteins belong to bacteriophage lambda and related phages. These phages produce two regulator proteins, namely, **repressor** and **Cro**. Repressor and Cro were given their names at an early date and subsequent studies have shown that, paradoxically, lambda Cro is a repressor protein, whereas lambda repressor has both repressor and activator functions; the names, however, have been kept for historical reasons. Repressor and Cro proteins operate a switch between two states of lambda phage replication. Both proteins contain the helix-turn-helix motif. Even though this motif was first observed in a different protein, CAP—which activates operons involved in sugar breakdown in *Escherichia coli*—we will focus our discussion on the two bacteriophage proteins since they are amongst the best-understood DNA regulatory proteins.

A molecular mechanism for gene control

Certain strains of *Escherichia coli* can be stimulated by irradiation with a moderate dose of ultraviolet (UV) light to stop normal growth and start producing bacteriophages that eventually lyse the bacterium. Bacteria of these so-called lysogenic strains carry the DNA of the phage integrated into their own

129

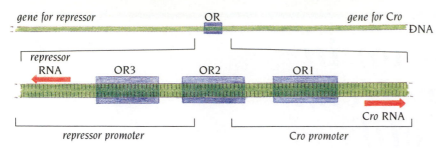

Figure 8.1 A region of DNA in the related bacteriophages lambda, 434, and P22 that controls the switch for synthesis of new phage particles. Two structural genes are involved in this switch; one coding for a repressor protein and one coding for the Cro protein. Between these genes there is an operator region (OR) that contains three protein binding sites—OR1, OR2, and OR3.

chromosomes, where it is dormant during normal cell growth; the phage DNA is replicated as an integral part of the bacterial chromosome, but the phage genes are not expressed. The phage in this state is called a temperate phage. Ultraviolet light switches on the phage genes which then produce new phages and the cell eventually dies. We are now able to explain in part the molecular mechanism of this genetic on-off switch from results obtained by a powerful combination of genetic and x-ray structural studies.

Repressor and Cro proteins operate a procaryotic genetic switch region

Three related species of temperate bacteriophages have been studied, lambda, 434, and P22. A relatively small region of the phage genome contains all the genetic components of the on-off switch (Figure 8.1). In each of the three species of phage this region comprises two structural genes coding for the two regulator proteins, Cro and repressor, that operate the switch; and the "right" **operator region** (OR) on which they act.

The two genes are transcribed in opposite directions from their two promoters, which occupy opposite ends of the operator region (Figure 8.1). When RNA polymerase is bound to the left-hand promoter, repressor is switched on, *Cro* and the lytic genes are repressed (Figure 8.2a), and the cell survives as a lysogenic strain. When the polymerase is bound to the right-hand promoter, *Cro* is switched on, along with the early lytic genes that lie to the right of *Cro*, and cell lysis results (Figure 8.2b).

The lysis–lysogeny decision depends upon which of the two promoters in the operator region is able to bind polymerase, and that, in turn, depends upon the binding of the Cro and repressor proteins to three binding sites—OR1, OR2, and OR3—in OR. These binding sites are situated in the middle of the operator in such a way that OR1 and OR2 overlap the promoter

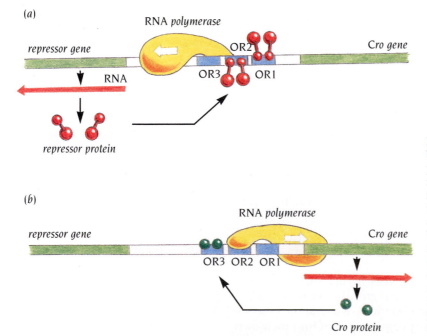

Figure 8.2 (a) The situation in the lysogenic bacterium. Repressor (red) and RNA polymerase (yellow) bound to the switch region in a lysogenic strain of *E. coli*. The repressor binds to OR1 and OR2, thereby turning off synthesis of Cro. The repressor also works as an activator for its own synthesis by facilitating RNA-polymerase binding to the repressor promoter through its binding to OR2. (b) The situation in the lytic phase. Synthesis of the Cro protein turns off synthesis of the repressor, since Cro binds to OR3 and blocks RNA-polymerase binding to the repressor promoter. Transcription of phage genes to the right can now occur.

of the *Cro* gene and OR2 and OR3 overlap the promoter for the repressor gene (see Figure 8.1). Both Cro and repressor bind as dimers to all three sites, but Cro binds with the highest affinity to OR3, and when it is bound, it blocks the access of RNA polymerase to the promoter for the repressor gene; conversely, repressor binds with the highest affinity to OR1, and when it is bound, it blocks the access of the polymerase to the *Cro* promoter.

Thus, repressor establishes the lysogenic state by binding to OR1 and switching off the lytic genes to the right of *Cro*. At the same time, it ensures the maintenance of this state by binding to OR2, where instead of blocking the access of the polymerase to its own promoter, it helps the polymerase to bind, thus producing more repressor. In the lysogen, repressor dimers are more or less continuously bound to OR1 and OR2 because once a molecule of repressor is bound at its high-affinity site at OR1, it helps a second molecule to bind at OR2 through a cooperative interaction between the two molecules. Thus repressor acts both to repress Cro synthesis and to activate its own synthesis.

Cro, by contrast, acts purely as a repressor. When it is bound to its high-affinity site at OR3, it prevents repressor synthesis by obstructing the access of polymerase to the left-hand promoter. In the absence of repressor, RNA polymerase can bind to the *Cro* promoter, and Cro can be synthesized along with the early phage genes to its right.

Thus, for a lysogen to switch over to the production of phage particles, repressor must be released from OR1 for long enough to allow Cro to be synthesized and bind to OR3. This is brought about by ultraviolet radiation, which activates a bacterial protease, Rec A, that cleaves the repressor molecules into two so that they can no longer form stable dimers. The protease activity of Rec A is activated when polymers of Rec A bind to single-stranded DNA formed as a result of DNA damage by UV light. The cleaved repressor fragments have a much lower affinity for OR1 and OR2 than the intact repressor molecule, and after a while the *Cro* promoter site becomes free to bind RNA polymerase, allowing transcription of the *Cro* gene along with the early genes for viral replication. The newly formed Cro protein molecules bind to OR3, blocking the access of RNA polymerase to the repressor promoter, and synthesis of more repressor molecules is prevented (Figure 8.2b). The switch has been flipped, and the machinery of the cell is converted to the synthesis of phage particles.

The x-ray structure of the complete lambda Cro protein is known

How do repressor and Cro recognize the specific operator regions and achieve this subtle differential binding to the switch regions? The sequences of OR1, OR2, and OR3 in lambda are similar but not identical (Table 8.1). They are

Table 8.1 The nucleotide sequences of the three protein-binding regions OR1, OR2, and OR3 of the operator of bacteriophage lambda

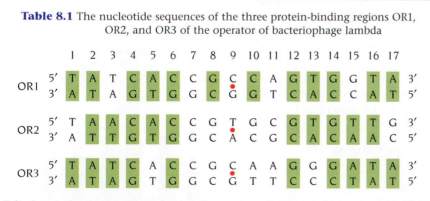

Palindromic base pairs that are most frequent at the two ends are green, and the pseudo-twofold symmetry axis is indicated by a red dot.

(a)

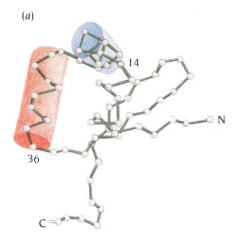

(b)

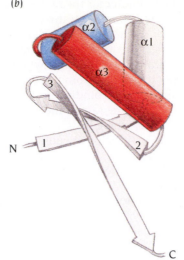

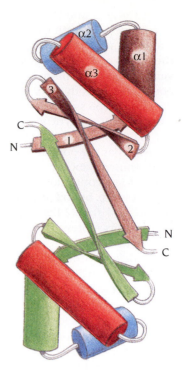

Figure 8.3 The DNA-binding protein Cro from bacteriophage lambda contains 66 amino acid residues that fold into three α helices and three β strands. (a) A plot of the C_α positions of the first 62 residues of the polypeptide chain. The four C-terminal residues are not visible in the electron density map. (b) A schematic diagram of the subunit structure. α helices 2 and 3 that form the helix-turn-helix motif are colored blue and red, respectively. The view is different from that in (a). [(a) Adapted from W.F. Anderson et al., *Nature* 290: 754–758, 1981. (b) Adapted from D. Ohlendorf et al., *J. Mol. Biol.* 169: 757–769, 1983.]

Figure 8.4 Cro molecules from bacteriophage lambda form dimers both in solution and in the crystal structure. The main dimer interactions are between β strands 3 from each subunit. In the diagram one subunit is green and the other is brown. Alpha helices 2 and 3, the helix-turn-helix motifs, are colored blue and red, respectively, in both subunits. (Adapted from D. Ohlendorf et al., *J. Mol. Biol.* 169: 757–769, 1983.)

also partly **palindromic**, especially at their ends. The palindromic nature of these sequences is important because it means that the halves of each binding site are related by an approximate twofold symmetry axis. Both Cro and repressor proteins are dimers, and the palindromic parts of OR1, OR2, and OR3 provide them with two almost but not quite identical recognition sites, one for each dimer subunit.

A first glimpse into the structural details of this system was obtained in 1981 when the group of Brian Matthews at Eugene, Oregon, determined the crystal structure of the Cro protein of phage lambda at 2.8 Å resolution. Cro is a small protein that forms stable dimers in solution. Each subunit is a single polypeptide chain of 66 amino acid residues with a very simple structure. It folds into three α helices and three strands of antiparallel β sheet and belongs therefore to the α + β class of structures, with the three α helices in a loop region between β strands 1 and 2 (Figure 8.3).

However, the α helices are not packed against each other in the usual way as described in Chapter 3. Instead, α helices 2 and 3, residues 15–36, form a unique helix-turn-helix arrangement that in 1981 had only been observed once, in a different bacterial DNA-binding protein, the catabolite gene-activating protein CAP.

Dimerization of pairs of Cro monomers depends primarily on interactions between β strand 3 from each subunit (Figure 8.4). These strands, which are at the carboxy end of the chains, are aligned in an antiparallel fashion and hydrogen bonded to each other so that the three-stranded β sheets of the monomers form a six-stranded antiparallel β sheet in the dimer (Figure 8.5).

The x-ray structure of the DNA-binding domain of the lambda repressor is known

Superficially, the lambda repressor protein is very different from lambda Cro. The polypeptide chain is much larger, 236 amino acids, and is composed of two domains that can be released as separate fragments by mild proteolysis. In repressor the domain responsible for dimerization is separate from the

Figure 8.5 The three β strands of each subunit in lambda Cro are aligned in the dimer so that a six-stranded antiparallel β sheet is formed as shown in this topology diagram. The β strands are colored as in Figure 8.4.

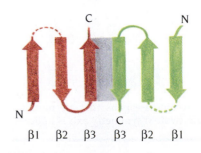

β1   β2   β3   β3   β2   β1

132

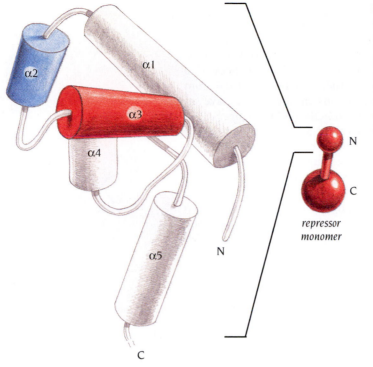

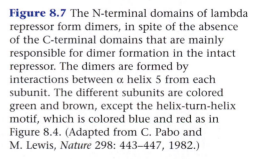

Figure 8.6 The N-terminal domain of lambda repressor, which binds DNA, contains 92 amino acid residues folded into five α helices. Two of these, α2 (blue) and α3 (red) form a helix-turn-helix motif with a very similar structure to that of lambda Cro shown in Figure 8.4. The complete repressor monomer contains in addition a larger C-terminal domain. (Adapted from C. Pabo and M. Lewis, *Nature* 298: 443–447, 1982.)

domain with DNA-binding functions; the C-terminal domains form the strong subunit interactions that hold the dimer together, while the N-terminal domains of 92 residues bind specifically to operator DNA. Although the N-terminal domains can form dimers on their own, they do so only weakly, and for this reason they also bind the operator more weakly than does the intact repressor molecule. This feature is crucial for the switch from lysogeny to the lytic cycle: the proteolytic cleavage that is activated by UV light separates the C-terminal domain from the N-terminal domain and this is how the affinity of repressor for OR1 is reduced and the switch from lysogeny to lysis initiated.

The x-ray structure of the N-terminal DNA-binding domain of the lambda repressor was determined to 3.2 Å resolution in 1982 by Carl Pabo at Harvard University and revealed a structure with striking similarities to that of Cro, although the β strands in Cro are replaced by α helices in repressor.

The polypeptide chain of the 92 N-terminal residues is folded into five α helices connected by loop regions (Figure 8.6). Again the helices are not packed against each other in the usual way for α-helical structures. Instead, α helices 2 and 3, residues 33–52, form a helix-turn-helix motif with a very similar structure to that found in Cro.

In spite of the absence of the C-terminal domains, the DNA-binding domains of lambda repressor form dimers in the crystals, as a result of interactions between the C-terminal helix number 5 of the two subunits that are somewhat analogous to the interactions of the C-terminal β strand 3 in the Cro protein (Figure 8.7). The two helices pack against each other in the normal way with an inclination of 20° between the helical axes. The structure of the C-terminal domain, which is responsible for the main subunit interactions in the intact repressor, remains unknown.

Both lambda Cro and repressor proteins have a specific DNA-binding motif

The specific arrangement of two α helices joined by a loop region in lambda Cro and repressor, as well as in CAP, constitute the helix-turn-helix DNA-binding motif (Figure 8.8), which also occurs in some eucaryotic transcription factors as discussed in Chapter 9. The orientation of the two helices and

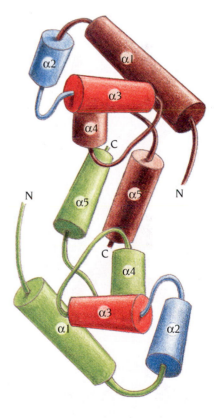

Figure 8.7 The N-terminal domains of lambda repressor form dimers, in spite of the absence of the C-terminal domains that are mainly responsible for dimer formation in the intact repressor. The dimers are formed by interactions between α helix 5 from each subunit. The different subunits are colored green and brown, except the helix-turn-helix motif, which is colored blue and red as in Figure 8.4. (Adapted from C. Pabo and M. Lewis, *Nature* 298: 443–447, 1982.)

the conformation of the loop regions are very similar in these three molecules, and this particular helix-turn-helix structure is unique to DNA-binding proteins.

Since these protein structures were determined in the absence of DNA, there was at that time no experimental evidence that these α helices were involved in DNA binding. However, Brian Matthews in 1981 realized from the structure of Cro that the subunit interactions in the dimer have an important consequence. They cause the second α helix (red in Figure 8.8) in the helix-turn-helix motifs of the two subunits to be at opposite ends of the elongated dimeric molecule, separated by a distance of 34 Å (see Figure 8.4). This distance corresponds almost exactly to one turn of a B-DNA double helix. As a result, if the second α helix of one subunit binds into the major groove of B-DNA, the corresponding α helix of the other subunit can also bind into the major groove one turn further along the DNA molecule. This α helix (the second in the helix-turn-helix motif, colored red in the illustrations) was, therefore, called the **recognition α helix**, and Matthews proceeded to build a model of a possible Cro–DNA complex.

Model building predicts Cro–DNA interactions

Matthews was able to show, by model building on a graphics display, that the two recognition helices of the Cro dimer indeed fitted very well into the major groove of a piece of regular B-DNA as seen in Figure 8.9. The orientation

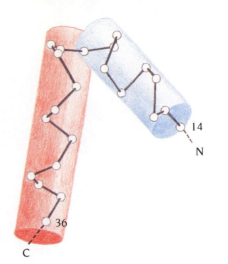

Figure 8.8 The DNA-binding helix-turn-helix motif in lambda Cro. C_α positions of the amino acids in this motif have been projected onto a plane and the two helices outlined. The second helix (red) is called the recognition helix because it is involved in sequence-specific recognition of DNA.

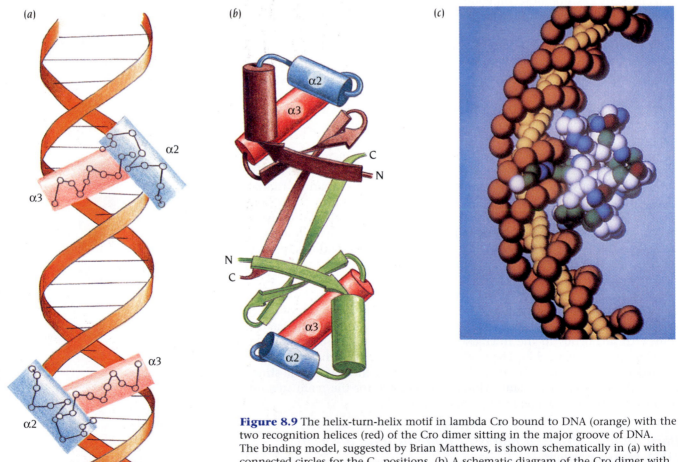

Figure 8.9 The helix-turn-helix motif in lambda Cro bound to DNA (orange) with the two recognition helices (red) of the Cro dimer sitting in the major groove of DNA. The binding model, suggested by Brian Matthews, is shown schematically in (a) with connected circles for the C_α positions. (b) A schematic diagram of the Cro dimer with different colors for the two subunits. (c) A schematic space-filling model of the dimer of Cro bound to a bent B-DNA molecule. The sugar-phosphate backbone of DNA is orange, and the bases are yellow. Protein atoms are colored red, blue, green, and white. [(a) Adapted from D. Ohlendorf et al., *J. Mol. Evol.* 19: 109–114, 1983. (c) Courtesy of Brian Matthews.]

of the two recognition α helices follows the orientation of the groove at both sites. Amino acid residues from these α helices can make contact with the edge of the base pairs in the major groove. These amino acids were thus assumed to be involved in recognizing specific operator regions.

Approximately 10 base pairs are required to make one turn in B-DNA. The centers of the palindromic sequences in the DNA-binding regions of the operator are also separated by about 10 base pairs (see Table 8.1). Thus if one of the recognition α helices binds to one of the palindromic DNA sequences, the second recognition α helix of the protein dimer is poised to bind to the second palindromic DNA sequence.

Both the relative orientation of the α helices and the distance between them are determined by the way the dimer is formed through interactions between the subunits. Three features of the Cro structure therefore were claimed to be important for specific binding to operator DNA: (1) the presence of a helix-turn-helix motif that provides a recognition α helix, which binds in the major groove of B-DNA; (2) the specific amino acid sequence of this α helix, which recognizes different operator regions; and (3) the subunit interactions that provide the correct distance and relative orientation between the two recognition α helices of the dimer, thereby increasing the affinity between Cro and operator DNA.

This model of Cro binding to DNA was arrived at by intuition and clever model building. Its validity was considerably strengthened when the same features were subsequently found in the DNA-binding domains of the lambda-repressor molecule. The helix-turn-helix motif with a recognition helix is present in the repressor, and moreover the repressor DNA-binding domains dimerize in the crystals in such a way that the recognition helices are separated by 34 Å as in Cro.

Genetic studies agree with the structural model

The presence of this common helix-turn-helix motif poised for DNA binding in lambda Cro and repressor provided considerable stimulus for further genetic and structural studies of these and other procaryotic DNA-binding proteins. All the results essentially supported the proposed mode of binding between these regulator proteins and DNA.

The most informative genetic experiments were made by the group of Mark Ptashne at Harvard University. In one experiment they redesigned the repressor from phage 434, replacing its recognition α helix with the corresponding α helix from Cro of phage 434 (a difference of five amino acids). The redesigned repressor thereby acquired the differential binding properties of the 434 Cro protein. In another experiment the group changed selected amino acid residues in the recognition α helix of the 434 repressor to those that occur in the repressor of phage P22 (Figure 8.10). The amino acids selected to be changed were those that in the structural model face the DNA and therefore presumably interact with it.

The consequences of these changes were impressive. The parental 434 repressor had no affinity for the P22 operator *in vivo*, but the redesigned 434 repressor controlled only P22 operators and not 434 operator regions *in vivo*. Purified redesigned 434 repressor bound specifically to P22 operator DNA *in vitro* and showed the same hierarchy of affinities for the three regions OR1, OR2, and OR3 as native P22 repressor.

These genetic experiments clearly demonstrated that the proposed structural model for the binding of these proteins to the phage operators was essentially correct. The second α helix in the helix-turn-helix motif is involved in recognizing operator sites as well as in the differential selection of operators by P22 Cro and repressor proteins. However, a note of caution is needed: many other early models of DNA–protein interactions proved to be misleading, if not wrong. Modeling techniques are more sophisticated today but are still not infallible and are certainly not replacements for experimental determinations of structure.

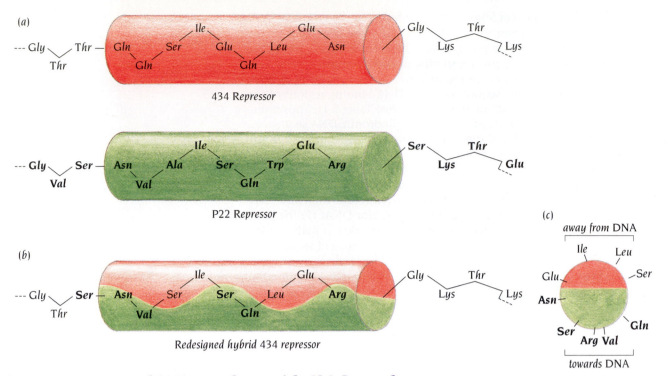

(a)

--- Gly — Thr — Gln — Ser — Glu — Leu — Asn — Gly — Thr
       Thr   Gln   Glu   Gln   Lys   Lys
             Ile   Glu

434 Repressor

--- Gly — Ser — Asn — Ala — Ser — Trp — Arg — Ser — Thr
       Val   Val   Gln   Lys   Glu
             Ile   Glu

P22 Repressor

(b)

--- Gly — Ser — Asn — Ser — Ser — Leu — Arg — Gly — Thr
       Thr   Val   Gln   Lys   Lys
             Ile   Glu

Redesigned hybrid 434 repressor

(c)

away from DNA

Ile   Leu
Glu   Ser
Asn
Ser   Gln
   Arg Val

towards DNA

The x-ray structure of DNA complexes with 434 Cro and repressor revealed novel features of protein–DNA interactions

The general features of the model for DNA binding were confirmed experimentally in 1987 when Stephen Harrison's group at Harvard University determined the structure of a complex of DNA and the DNA-binding domain of the 434 repressor to 3.2 Å resolution. However, it also became evident, both from the structure of this complex and from further site-directed mutagenesis studies, that the selective recognition of the different operator regions by the 434 repressor depends mostly on other factors than the amino acid residues of the recognition helix. The complexity of the fine tuning of DNA regulation has been clearly demonstrated by Harrison's subsequent studies of complexes between different operator DNA regions and both 434 Cro and the DNA-binding domain of 434 repressor.

For purely practical reasons, the complexes that Harrison first studied contained the N-terminal DNA-binding domain of the repressor from phage 434, which comprises 69 amino acids, complexed with a 14 base-pair piece of synthetic DNA ("14mer") having a completely palindromic sequence. In other words, the DNA in the complex had a strict twofold symmetry analogous to the twofold symmetry of the dimeric repressor molecule. This synthetic DNA thus contains identical halves, each of which, as we shall see, binds one subunit of Cro or one repressor fragment. The three 14 base-pair operator regions that the 434 repressor recognizes in the phage right-hand operator (OR) are not perfectly palindromic, however (Table 8.2). The 14mer is closest in sequence to a site (OL2) in a second operator region of the phage, the "left" operator (OL), as Table 8.2 shows. The only difference is an inversion of base pair 7 from A–T to T–A, and experiments had shown that this inversion did not alter the affinity for the DNA of either intact repressor or the N-terminal DNA-binding fragment.

The crystals of the complex with 14mer diffracted to only medium resolution, however, and by systematic variations of the length of the DNA fragment and its sequences at the ends, Harrison and coworkers were later able to find a piece of DNA that gave crystals that diffracted to high resolution, both with 434 Cro and the DNA-binding domain of 434 repressor. This DNA fragment contains 20 nucleotides in each chain, and the sequence of its middle region is identical to OR1 (see Table 8.2). The 5′ ends contain one nonpaired nucleotide that is involved in packing the fragments in the crystal.

Figure 8.10 The proposed DNA-binding surface of the recognition helix of bacteriophage 434 repressor was redesigned genetically to that of P22 repressor by changing six amino acid residues. The amino acid sequences of the recognition helices of the wild-type repressors are shown in (a) and that of the redesigned repressor in (b) and (c) viewed from the side and along the helix, respectively. The redesigned 434 repressor acquired all the DNA-binding properties of the P22 repressor. The six amino acid residues that were changed in the 434 repressor are shown in boldface in (b) and (c). (Adapted from R. Wharton and M. Ptashne, *Nature* 316: 601–605, 1985.)

Table 8.2 The six operator regions (OR1–OR3 and OL1–OL3) in bacteriophage 434

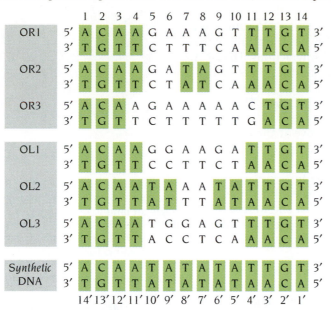

		1	2	3	4	5	6	7	8	9	10	11	12	13	14	
OR1	5′	A	C	A	A	G	A	A	A	G	T	T	T	G	T	3′
	3′	T	G	T	T	C	T	T	T	C	A	A	A	C	A	5′
OR2	5′	A	C	A	A	G	A	T	A	G	T	T	T	G	T	3′
	3′	T	G	T	T	C	T	A	T	C	A	A	A	C	A	5′
OR3	5′	A	C	A	A	G	A	A	A	A	A	C	T	G	T	3′
	3′	T	G	T	T	C	T	T	T	T	T	G	A	C	A	5′
OL1	5′	A	C	A	A	G	G	A	A	G	A	T	T	G	T	3′
	3′	T	G	T	T	C	C	T	T	C	T	A	A	C	A	5′
OL2	5′	A	C	A	A	T	A	A	A	T	A	T	T	G	T	3′
	3′	T	G	T	T	A	T	T	T	A	T	A	A	C	A	5′
OL3	5′	A	C	A	A	T	G	G	A	G	T	T	T	G	T	3′
	3′	T	G	T	T	A	C	C	T	C	A	A	A	C	A	5′
Synthetic DNA	5′	A	C	A	A	T	A	T	A	T	A	T	T	G	T	3′
	3′	T	G	T	T	A	T	A	T	A	T	A	A	C	A	5′
		14′	13′	12′	11′	10′	9′	8′	7′	6′	5′	4′	3′	2′	1′	

Each operator region is 14 base pairs long. The palindromic base pairs of these regions are marked in green. Crystal structures have been determined of complexes between both 434 Cro and the repressor fragment with synthetic DNA fragments—one 14 base pairs long (a 14mer), which is completely palindromic, and one 20 base pairs long (a 20mer), which contains the sequence of OR1 in its middle region.

Figure 8.11 The DNA-binding domain of 434 repressor. It is a dimer in its complexes with DNA fragments. Each subunit (green and brown) folds into a bundle of four α helices (1–4) that have a structure similar to the corresponding region of the lambda repressor (see Figure 8.7) including the helix-turn-helix motif (blue and red). A fifth α helix (5) is involved in the subunit interactions, details of which are different from those of the lambda repressor fragment. The structure of the 434 Cro dimer is very similar to the 434 repressor shown here.

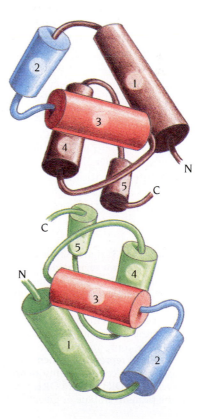

By comparing the crystal structures of these complexes with a further complex of the 434 repressor DNA-binding domain and a synthetic DNA containing the operator region OR3, Harrison has been able to resolve at least in part the structural basis for the differential binding affinity of 434 Cro and repressor to the different 434 operator regions.

The structures of 434 Cro and the 434 repressor DNA-binding domain are very similar

The 434 Cro molecule contains 71 amino acid residues that show 48% sequence identity to the 69 residues that form the N-terminal DNA-binding domain of 434 repressor. It is not surprising, therefore, that their three-dimensional structures are very similar (Figure 8.11). The main difference lies in two extra amino acids at the N-terminus of the Cro molecule. These are not involved in the function of Cro. By choosing the 434 Cro and repressor molecules for his studies, Harrison eliminated the possibility that any gross structural difference of these two molecules can account for their different DNA-binding properties.

The DNA-binding domain of 434 repressor also has significant sequence homology (26% identity) with the corresponding part of the lambda repressor and, consequently, a related three-dimensional structure (compare Figures 8.7 and 8.11). Like its lambda counterpart, the subunit structure of the DNA-binding domain of 434 repressor, as well as that of 434 Cro, contains a cluster of four α helices, with helices 2 and 3 forming the helix-turn-helix motif. The two helix-turn-helix motifs are at either end of the dimer and contribute the main protein–DNA interactions, while protein–protein interactions at the C-terminal part of the chains hold the two subunits together in the complexes. Both 434 Cro and repressor fragments are monomers in solution even at high protein concentrations, whereas they form dimers when they are bound to DNA. (It should be noted once again, however, that in the intact repressor the main dimerization interactions are believed to be formed by the C-terminal domain, which is not present in the crystal complex.)

The proteins impose precise distortions on the B-DNA in the complexes

The real significance of Harrison's work with these protein–DNA complexes lies not in the structure of the protein domains but rather in the details of the structures of the bound DNA and in the protein–DNA interactions. The DNA in all the complexes is in the B-form but with significant distortions. Examination of the local twist between base pairs showed that the DNA was overwound (larger twist) at its center and underwound at the ends. The helical axis was also somewhat bent toward the recognition helices at the ends of the protein dimer. These distortions narrow the minor groove at the center and widen it at the ends, as is shown in Figure 8.12, which compares the observed structure of DNA complexed with 434 repressor fragments to that of regular B-DNA. Since the same distortions have been observed in complexes of the 434 repressor fragment with various DNA sequences (the 14mer, the OR1-containing 20mer, an OR2-containing 20mer, and others) and in more than one crystal packing, it is reasonable to conclude that the conformational characteristics of the DNA fragment in these complexes result directly from the protein–DNA interactions.

In complexes with Cro, the overall bend and twist of the DNA are similar to those in the repressor complexes, but there is a significant difference in the local structure of two of the nucleotides in each half-site. Binding of 434 Cro or repressor fragment thus imposes a distinct local structure (Figure 8.13), as a result of differences in both the identity and conformations of various amino acid residues that interact with the DNA. The DNA conformational details are significant for the relative affinities of Cro and repressor for various sites, as we describe in a later section.

Sequence-specific protein–DNA interactions recognize operator regions

The protein–DNA interactions have been analyzed in detail at high resolution in the complex between the 434 repressor fragment and the OR1 containing 20mer DNA. A pseudo-twofold symmetry axis relates the halves of this complex. The symmetry is not exact since the nucleotide sequence of the DNA is slightly different in each half (see Table 8.2). However, the interactions between one protein subunit and one half of the DNA are very similar to those between the second subunit and the other half of the DNA since most of the bases that interact with the protein are identical in both halves. Details of the interaction are very similar to those in the complex with the palindromic synthetic 14mer of DNA shown in Figures 8.14 and 8.15. The base pairs at one end of the DNA, 1–14', 2–13', etc. are called base pairs 1, 2, etc.

The protein dimer binds so that the recognition α helices at opposite ends of the protein molecule are in the major groove of the DNA as predicted, where they interact with base pairs at the end of the DNA molecule. Since these binding sites are separated by one turn of the DNA helix, it follows that at the center of the DNA molecule the narrow groove faces the protein

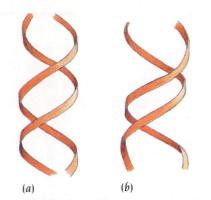

(a)   (b)

Figure 8.12 (above) The changes of DNA structure from regular B-DNA (a) to a distorted version (b) when 434 Cro and repressor fragment bind to operator regions. The distortions essentially involve bending of DNA and overwinding of the middle regions. The diagram shows the sugar-phosphate backbones of DNA as orange ribbons viewed in the narrow groove in the middle region of the operator.

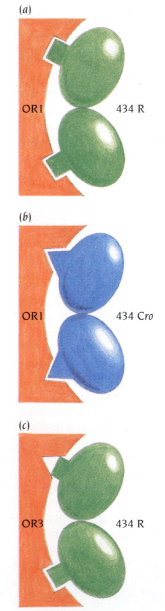

Figure 8.13 (right) Binding of 434 repressor fragment (a) and 434 Cro (b) to operator region OR1. This binding induces different structural changes in the region of the DNA that binds the proteins illustrated as different shapes of the binding regions of OR1. In the complex of operator OR3 with the 434 repressor fragment (c) the two half sites of OR3 are different: one is similar to OR1, with bound repressor, whereas the other has a different nucleotide sequence that adopts the Cro-type binding conformation on binding repressor. The binding surfaces of the DNA and repressor fragment do not complement each other as they do in the OR1 complex; consquently, the repressor fragment binds more weakly to OR3 than to OR1. DNA is schematically shown in orange and the proteins in blue or green.

(Figure 8.14). There are no interactions between the protein and the bases of the DNA in this middle region of the operator.

Residues of the recognition α helix project their side chains into the major groove and interact with the edges of the DNA base pairs on the floor of the groove (Figure 8.15a). Gln (Q) 28 forms two hydrogen bonds to N_6 and N_7 of A1 in base pair 1 (T14'–A1) (Figure 8.15b), and Gln 29 forms a hydrogen bond to O_6 of G13' in base pair 2 (G13'–C2). At base pair 3 (T12'–A3) no hydrogen bonding to the protein occurs and direct contacts are all hydrophobic. The methyl groups of the side chains of Thr (T) 27 and Gln (Q) 29 form a hydrophobic pocket to receive the methyl group of T12'.

The first three base pairs in all six operator regions recognized by phage 434 repressor are identical (see Table 8.2). This means that interactions between these three base pairs and the two glutamine residues (28 and 29) cannot contribute to the discrimination between the six binding sites in the DNA; rather, these interactions provide a general recognition site for operator regions. This simple pattern of hydrogen bonds and hydrophobic interactions therefore accounts for the specificity of phage 434 Cro and repressor proteins for 434 operator regions. The role of Gln 29 is particularly important since it interacts with both base pairs 2 and 3. Its hydrogen bond to guanine specifies C–G at position 2 and the hydrophobic pocket formed by the hydrophobic part of its side chain together with residue 27, specifies A–T at base pair 3.

This general recognition function is crucial for the bacteriophage. When glutamines 28 and 29 are replaced by any other amino acid, the mutant phages are no longer viable. Moreover, 434 Cro protein has glutamine residues at these two positions in its recognition helix as expected since it binds to the same set of operator regions.

Protein–DNA backbone interactions determine DNA conformation

It is apparent from the crystal structures of these protein–DNA complexes that the differential affinities of 434 repressor and Cro for the different operator regions are not determined by sequence-specific interactions between amino acid side chains of the recognition helix and base pairs in the major groove of DNA. Instead, they seem to be determined mainly by the ability of the DNA to undergo specific structural changes so that complementary surfaces are formed between the proteins and the DNA. Interactions between the DNA sugar-phosphate backbone and the proteins are important for establishing such structural changes and thus for positioning the recognition helix correctly in the major groove.

The interface between the repressor fragment and DNA covers an extended area. Most interactions between one monomer of the fragment and the DNA occur within a single half-site; in other words, each repressor subunit

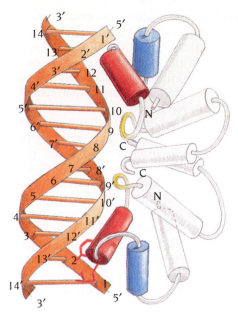

Figure 8.14 Overall view of the complex between 434 repressor fragment and a palindromic synthetic 14mer of DNA (see Table 8.2). The two binding sites of the repressor dimer to the DNA are identical. The recognition helices of the repressor are red, and the first helix of the helix-turn-helix motif is blue. (Adapted from J. Anderson et al., *Nature* 326: 846–852, 1987.)

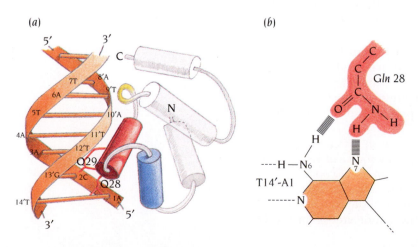

Figure 8.15 Sequence-specific protein–DNA interactions provide a general recognition signal for operator regions in 434 bacteriophage. (a) In this complex between 434 repressor fragment and a synthetic DNA there are two glutamine residues (28 and 29) at the beginning of the recognition helix in the helix-turn-helix motif that provide such interactions with the first three base pairs of the operator region. The side chain of Gln 28 forms two hydrogen bonds (b) to the edge of the adenine base of base pair T14'–A1 in the major groove of the DNA. (Adapted from J. Anderson et al., *Nature* 326: 846–852, 1987.)

"sees" only one half of the operator. In order for interactions to occur at both half-sites, giving tight binding, it is essential that the structural change in the DNA coincides with the dimer organization of the repressor. Thus, constraints on the conformation of the central base pairs come not only from local backbone interactions with the protein but also from the way in which the dimer interface in the protein fixes the relative position and orientation of the two recognition helices and their respective major groove contacts.

In all complexes studied the protein subunit is anchored across the major groove with extensive contacts along two segments of the sugar-phosphate backbone, one to either side of the groove. Hydrogen bonds between the DNA phosphate groups and peptide backbone NH groups are remarkably prevalent in these contacts (Figure 8.16).

One of these interaction regions involves the loop after the recognition helix, residues 40–44 of the repressor (yellow in Figures 8.14 and 8.15a), where three main-chain NH groups form hydrogen bonds with phosphates 9' and 10'. The side chain of Arg 43 in this loop projects into the minor groove and probably stabilizes its compression, by introducing a positive charge between the phosphate groups on opposite sides of the narrow groove. All residues in this loop, which are outside the helix-turn-helix motif, contribute to the surface complementarity between the protein and the sugar-phosphate surfaces of nucleotides 9' and 10'.

These and other interactions between groups on the protein and the DNA sugar-phosphate backbone stabilize the distorted DNA conformation. They involve a large number of residues that are distributed along most of the polypeptide chain. In addition, other residues that are involved in the subunit interactions in the protein dimer ensure that each subunit is properly poised for binding to its half site in the operator DNA. Thus, the "unit" that is responsible for the differential binding to different operator DNA regions is really an entire binding domain, appropriately dimerized, and nearly all the protein–DNA contacts contribute to this specificity.

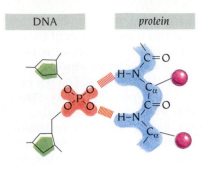

Figure 8.16 Nonspecific protein–DNA interactions are frequently formed by hydrogen bonds (red) between backbone phosphate oxygen atoms of DNA and main-chain NH groups of the protein.

Conformational changes of DNA are important for differential binding of repressor and Cro to different operator sites

Repressor and Cro distinguish the different operator sites in two distinct ways. The first involves base pair 4, which is A-T everywhere except in one half-site of OR3 (see Table 8.2). Cro, which binds most tightly to OR3, is indifferent to whether there is A-T or G-C at position 4'; repressor, which binds most tightly to OR1, strongly prefers A-T. The conformation of a synthetic DNA containing the OR3 operator site complexed with the repressor DNA-binding fragment, has provided an important clue to the molecular basis for this discrimination. Binding of the 434 repressor fragment to DNA induces different conformational changes in the halves of OR3. The left half, which contains the consensus sequence 5'-ACAA..., has the same conformation as both halves have in other complexes with the repressor fragment, whereas the right half has the conformation (around phosphate 11) that is found in complexes with Cro.

These results indicate that the repressor fragment is unable to impose upon the nonconsensus sequence on the right-hand side of OR3 the DNA conformation that is required for tight binding between repressor and DNA. It thus appears that the weaker binding of repressor to OR3 is a consequence of the resulting less perfect "fit" between protein and DNA backbone (see Figure 8.13).

The second way repressor and Cro distinguish different operator sites involves the central base pairs, which are the only source of variation among sites other than in OR3. The ability of the protein–DNA contacts to accomplish proper changes in the DNA structure, in particular the overwinding in the central region, can be modulated by the nucleotide sequence of the DNA. The overwinding causes a narrowing of the minor groove in the central region of the operator. It is generally believed that A-T base pairs can more readily be accommodated in narrow grooves than G-C base pairs. This means that DNA with A-T base pairs in the central region of operator sequences should be able to adopt the conformation necessary for proper DNA–protein interaction more readily than DNA with G-C base pairs in these positions. This is confirmed by mutation experiments using synthetic operator regions; these show that it is possible to change the affinity between repressor and DNA by changing the base pairs in the middle region. Thus, if base pairs 7 and 8 are changed from T-A and A-T to G-C and C-G, the affinity for the repressor fragment is decreased fifteenfold. In other words, the base sequence in this region, by influencing local conformational changes in the DNA, alters the affinity of repressor for DNA. Here we have a case of local DNA structure, rather than direct sequence-specific DNA–protein interactions, modulating repressor binding.

The essence of phage repressor and Cro

The DNA-binding proteins Cro and repressor of the three phages lambda, P22, and 434 all recognize operator regions in DNA through a helix-turn-helix motif. The second α helix of this motif, the recognition helix, binds in the major groove of DNA. These proteins form dimers such that two identical recognition helices, 34 Å apart at the two ends of the elongated dimer, can bind to the ends of the operator regions.

The first two amino acid residues of the recognition helix in these two proteins from phage 434 are glutamines, which form specific hydrogen bonds to the first two base pairs of the operator region. These interactions are identical in all operator regions of phage 434. A few residues on the recognition helix of the helix-turn-helix motif thus form the recognition signal of Cro and repressor proteins for the operator regions by specific interaction with the edge of the base pairs in the major groove of DNA, as predicted by Matthews.

Such sequence-specific interactions do not, however, account for the differential affinity of Cro and repressor for the operator regions. These affinities are instead determined by the ability of the DNA to undergo specific structural changes. At least two factors are important to the ability of the complex to achieve the proper DNA conformation (Figure 8.17). Interactions between the sugar-phosphate backbone of DNA and regions of the protein form a large interaction area of complementary surfaces that stabilizes the structural change in the DNA. These protein regions are not restricted to the

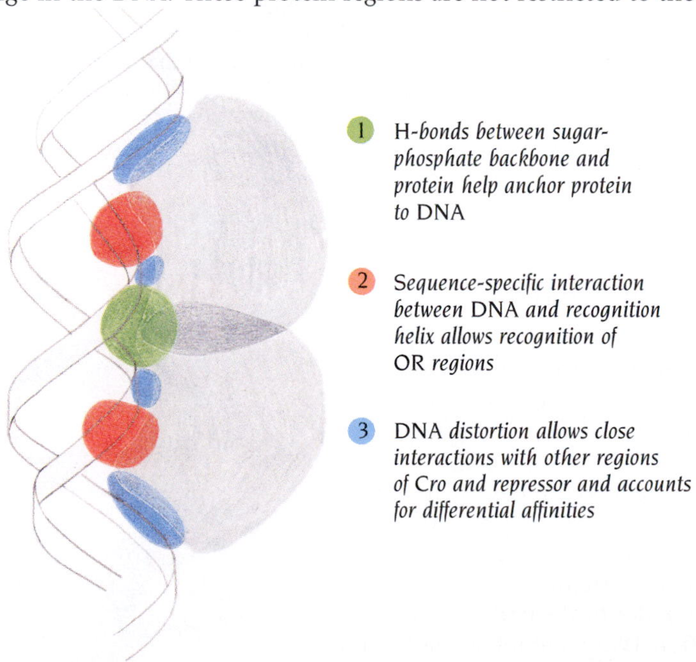

1 H-bonds between sugar-phosphate backbone and protein help anchor protein to DNA

2 Sequence-specific interaction between DNA and recognition helix allows recognition of OR regions

3 DNA distortion allows close interactions with other regions of Cro and repressor and accounts for differential affinities

Figure 8.17 Schematic diagram of the main features of the interactions between DNA and the helix-turn-helix motif in DNA-binding proteins.

helix-turn-helix motif but are spread over almost the entire polypeptide chain. The ability of these interactions to accomplish a structural change can be modulated by the sequence of DNA.

DNA binding is regulated by allosteric control

The DNA-binding capacity of most repressors and activators is regulated by small molecules, such as sugars, amino acids, or cyclic AMP, which bind to a distinct site on the protein. This binding causes a conformational change which alters the DNA binding site and therefore regulates its affinity for DNA. As described in Chapter 6, Jacques Monod, at the Pasteur Institute in Paris, introduced the term allosteric binding to emphasize that the small ligands, known as **allosteric effectors**, and the sites at which they bind are sterically quite distinct from the functional binding sites. We will discuss one example of allosteric control of DNA binding—the tryptophan (*trp*) repressor from *Escherichia coli*.

The trp repressor forms a helix-turn-helix motif

The ***trp* repressor** controls the operon for the synthesis of L-tryptophan in *Escherichia coli* by a simple **negative feedback loop**. In the absence of L-tryptophan, the repressor is inactive, the operon is switched on and the enzymes which synthesize L-tryptophan are produced. As the concentration of L-tryptophan increases, it binds to the repressor and converts it to an active form so that it can bind to the operator region and switch off the gene.

The structure of the *trp* repressor has been determined both with and without bound tryptophan to 1.8 Å resolution by the group of Paul Sigler, then at the University of Chicago. The *trp* repressor is a dimer, like the other proteins we have discussed in this chapter. The polypeptide chain has 107 amino acids and is folded into six α helices (Figure 8.18). It therefore has an α-type structure like the lambda repressor, but the arrangement of the helices in the subunit is quite different. The six helices do not pack to form a regular structure with a hydrophobic core, and it is doubtful if monomers with the arrangement of helices shown in Figure 8.18 would be stable alone. Stability is conferred by dimerization because the two subunits fit together to give a functional molecule that, in contrast to its subunits, has a compact globular form (Figure 8.19).

The helices at the N-terminal regions of the two polypeptide chains are intertwined and make extensive contacts in the central part of the molecule to form a stable core. This core supports two "heads", each comprising the last three helices from one polypeptide chain. Alpha helix 3 in the middle of the subunit chain is quite long and forms the main link between the core and the head.

Structural studies of a repressor–DNA complex have shown that helices 4 and 5 form a helix-turn-helix motif and that side chains from the recognition helix 5 form water-mediated interactions with bases in the major groove.

A conformational change operates a functional switch

What is the molecular mechanism by which the binding of tryptophan—the allosteric effector—allows the *trp* repressor to bind DNA? The structures of the *trp* repressor with and without bound tryptophan provided a simple answer. In the absence of tryptophan the heads, and in particular the recognition helices, are tilted inward toward the core (Figure 8.20a). This makes the distance between the two DNA binding sites of the repressor molecule too short by 5–6 Å to allow them to fit into the major groove. Specific DNA binding is not possible, and the repressor is inactive. Tryptophan binding causes a conformational change in the repressor that alters the orientations of the recognition helices. The heads tilt outward and two identical cavities are formed between the heads and the core (Figure 8.20b). Two tryptophan molecules

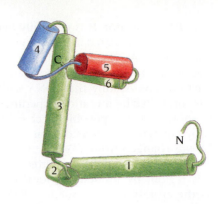

Figure 8.18 The subunit of the *trp* repressor. The subunit contains 107 amino acid residues that are folded into six α helices. Helices 4 (blue) and 5 (red) form the DNA-binding helix-turn-helix motif. (Adapted from R.W. Schevitz et al., *Nature* 317: 782–786, 1985.)

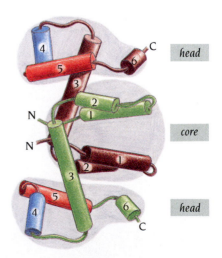

Figure 8.19 The α helices of the N-terminal region of the *trp* repressor are involved in subunit interactions and form a stable core in the middle of the dimer. Alpha helices 4–6, which include the helix-turn-helix motif, form two "head" regions at the two ends of the molecule. Alpha helix 3 connects the core to the head in both subunits. (Adapted from R.W. Schevitz et al., *Nature* 317: 782–786, 1985.)

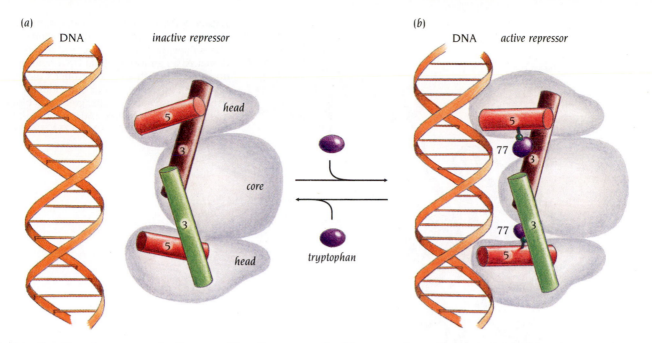

(a) DNA   inactive repressor   (b) DNA   active repressor

head

5

3

core   3

3   77   3

5

head   tryptophan   77   3

5

bind to the dimeric repressor in these cavities; they are wedged between the long connecting α helices (3) and the recognition helices of the heads. When tryptophan is bound in these cavities, the recognition helices are properly poised 34 Å apart for binding DNA in the major groove, as they are in the 434 repressor.

The elegant genetic studies by the group of Charles Yanofsky at Stanford University, conducted before the crystal structure was known, confirm this mechanism. The side chain of Ala 77, which is in the loop region of the helix-turn-helix motif, faces the cavity where tryptophan binds. When this side chain is replaced by the bulkier side chain of Val, the mutant repressor does not require tryptophan to be able to bind specifically to the operator DNA. The presence of a bulkier valine side chain at position 77 maintains the heads in an active conformation even in the absence of bound tryptophan. The crystal structure of this mutant repressor, in the absence of tryptophan, is basically the same as that of the wild-type repressor with tryptophan. This is an excellent example of how ligand-induced conformational changes can be mimicked by amino acid substitutions in the protein.

Figure 8.20 Schematic diagrams of docking the *trp* repressor to DNA in its inactive (a) and active (b) forms. When L-tryptophan, which is a corepressor, binds to the repressor, the "heads" change their positions relative to the core to produce the active form of the repressor, which binds to DNA. The structures of DNA and the *trp* repressor are outlined. The positions of the DNA recognition helices (5) and the helices (3) that connect the core with the heads are indicated. The approximate position of the side chain of residue 77 is marked as a purple ball (see text for the significance of this residue). (Adapted from R.-G. Zhang et al., *Nature* 327: 591–597, 1987.)

Lac repressor binds to both the major and the minor grooves inducing a sharp bend in the DNA

More than 30 years ago Jacob and Monod introduced the *Escherichia coli* **lac** operon as a model for gene regulation. The lac repressor molecule functions as a switch, regulated by inducer molecules, which controls the synthesis of enzymes necessary for *E. coli* to use lactose as an energy source. In the absence of lactose the repressor binds tightly to the operator DNA preventing the synthesis of these enzymes. Conversely when lactose is present, the repressor dissociates from the operator, allowing transcription of the operon.

The lac repressor monomer, a chain of 360 amino acids, associates into a functionally active homotetramer. It is the classic member of a large family of bacterial repressors with homologous amino acid sequences. PurR, which functions as the master regulator of purine biosynthesis, is another member of this family. In contrast to the lac repressor, the functional state of PurR is a dimer. The crystal structures of these two members of the Lac I family, in their complexes with DNA fragments, are known. The structure of the tetrameric lac repressor–DNA complex was determined by the group of Mitchell Lewis, University of Pennsylvania, Philadelphia, and the dimeric PurR–DNA complex by the group of Richard Brennan, Oregon Health Sciences University, Portland.

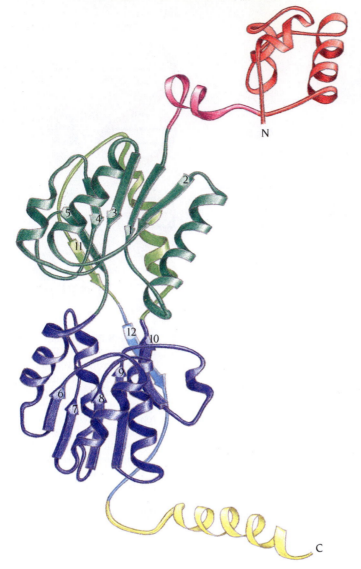

Figure 8.21 Richardson-type diagram of the structure of one subunit of the lac repressor. The polypeptide chain is arranged in four domains, an amino terminal DNA-binding domain (red) with a helix-turn-helix motif, a hinge helix (purple), a large core domain which has two subdomains (green and blue) and a C-terminal α helix. (Adapted from M. Lewis et al., *Science* 271: 1247–1254, 1996.)

The polypeptide chain of the lac repressor subunit is arranged in four domains (Figure 8.21): an N-terminal DNA-binding domain with a helix-turn-helix motif, a hinge helix which binds to the minor groove of DNA, a large core domain which binds the corepressor and has a structure very similar to the periplasmic arabinose-binding protein described in Chapter 4, and finally a C-terminal α helix which is involved in tetramerization. This α helix is absent in the PurR subunit structure; otherwise their structures are very similar.

The tetrameric structure of the lac repressor has a quite unusual V-shape (Figure 8.22). Each arm of the V-shaped molecule is a tight dimer, which is very similar in structure to the PurR dimer and which has the two N-terminal DNA binding domains close together at the tip of the arm. The two dimers of the lac repressor are held together at the other end by the four carboxy-terminal α helices, which form a four-helix bundle.

Each dimer arm binds a separate copy of a palindromic DNA sequence. The recognition helices of the two subunits insert into successive major grooves in a way similar to the binding of bacteriophage Cro and repressor molecules described above. In addition, the two hinge helices of the dimer, which are close together, interact with the minor groove between the two major-groove binding sites of the helix-turn-helix motifs (Figure 8.23). Both the recognition helix and the hinge helix form sequence-specific interactions between their amino acid side chains and the nucleotide bases of the DNA.

Figure 8.22 The lac repressor molecule is a V-shaped tetramer in which each arm is a dimer containing a DNA-binding site. The helix-turn-helix motifs (red) of each dimer bind in two successive major grooves and the hinge helices (purple) bind adjacent to each other in the minor groove between the two major groove binding sites. The four subunits of the tetramer are held together by the four C-terminal helices (yellow) which form a four helix bundle. The bound DNA fragments are bent. (Adapted from M. Lewis et al., *Science* 271: 1247–1254, 1996.)

This is an example of minor groove base recognition in the context of major groove recognition. Two bulky leucine residues from the hinge helices force the minor groove to open up and become wide and shallow. As a result the DNA is distorted from its B-form and sharply bent away from the protein (see Figure 8.22). Due to this bending, the distance between the two recognition helices in the major groove and their relative orientation are different from those in bacteriophage Cro and repressor complexes with DNA. The wild-type operator regions for the lac repressor are pseudo-palindromic with a central base pair between the pseudo-symmetric halves whereas the DNA in the crystals is strictly palindromic and lacks the central base pair. This difference would not affect binding of the recognition helices to DNA, but details of the DNA bending and binding of the hinge helices in the central part of the DNA could be somewhat different in the crystals compared with the situation *in vivo*. However, such differences are expected to be minor since the dimeric PurR repressor also bends DNA and essentially looks like one arm of the tetrameric lac repressor. The DNA molecule in the crystals of the PurR–DNA complex has the same number of base pairs as the PurR operator regions. The two arms of the V-shaped tetramer of the lac repressor each bind their DNA fragments in a similar way. Since the binding sites are far away from each other in the tetramer at the tips of the two arms of the V there is no interaction between the two DNA fragments (see Figure 8.22).

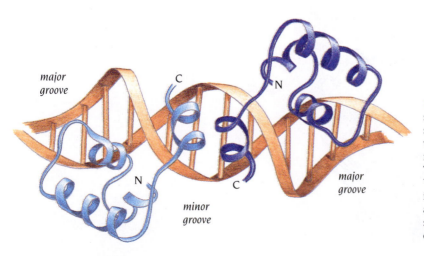

major groove

minor groove

major groove

Figure 8.23 The helix-turn-helix motifs of the subunits of both the PurR and the lac repressor subunits bind to the major groove of DNA with the N-terminus of the second helix, the recognition helix, pointing into the groove. The two hinge helices of each arm of the V-shaped tetramer bind adjacent to each other in the minor groove of DNA, which is wide and shallow due to distortion of the B-DNA structure. (Adapted from M.A. Schumacher et al., *Science* 266: 763–770, 1994.)

CAP-induced DNA bending could activate transcription

CAP, catabolite gene activating protein, is a DNA-binding protein that assists RNA polymerase in binding effectively to certain *Escherichia coli* promoters. In other words, it is a **positive control protein** that promotes more frequent initiation of RNA synthesis. Alone, CAP is a nonspecific DNA-binding protein but it is converted by binding cyclic AMP to a form that binds strongly to specific operator regions.

CAP controls a number of operons, all of which are involved in the breakdown of sugar molecules and one of which is the lac operon. When the level of the breakdown products of lactose is low, the concentration of cyclic AMP in the cell increases and CAP is switched on, binds to its specific operators, and increases the rate of transcription of adjacent operons.

Many biochemical and biophysical studies of CAP–DNA complexes in solution have demonstrated that CAP induces a sharp bend in DNA upon binding. This was confirmed when the group of Thomas Steitz at Yale University determined the crystal structure of cyclic AMP–DNA complex to 3 Å resolution. The CAP molecule comprises two identical polypeptide chains of 209 amino acid residues (Figure 8.24). Each chain is folded into two domains that have separate functions (Figure 8.24b). The larger N-terminal domain binds the allosteric effector molecule, cyclic AMP, and provides all the subunit interactions that form the dimer. The C-terminal domain contains the helix-turn-helix motif that binds DNA.

The CAP dimer interacts directly with 27 of the 30 base pairs of the DNA fragment present in the crystal. The amino ends of the recognition helices (red in Figure 8.24) of the helix-turn-helix motifs are bound in the major groove of DNA. These interactions provide sequence specificity between protein side chains and the exposed edges of base pairs. In addition 22 of the phosphate groups of DNA interact directly with protein side chains or main-chain amide groups providing a large number of contacts.

The central 10 base pairs of the palindromic DNA molecule have a regular B-DNA structure. Between base pairs 5 and 6 in each half of the fragment (base pairs are counted from the center) there is a 40° kink which causes these base pairs to be unstacked (Figure 8.24a). After this localized kink the two end regions have an essentially B-DNA structure. The kink occurs at a TG step in the sequence GTG. These TG steps at positions 5 and 6 are highly conserved in both halves of different CAP-binding sites, presumably in part because they facilitate kinking.

Figure 8.24 (a) Structure of the CAP–cyclic AMP–DNA complex. The two protein chains of the dimeric CAP molecule are colored green and brown. The helix-turn-helix motif is blue and red with the recognition helix and the C-terminus red. The DNA which is kinked where the recognition helices bind is orange and the cyclic AMP molecules are purple. (b) Schematic diagram of the structure of one subunit of CAP. The β strands (1–12) and α helices (A–F) are labeled from the N-terminus. The DNA-binding helix-turn-helix motif is colored blue and red with the recognition helix red. The binding site for cyclic AMP is purple. [(a) Adapted from S.C. Schultz et al., *Science* 253: 1001–1007, 1991. (b) Adapted from D. McKay and T. Steitz, *Nature* 290: 744–749, 1981.]

(*b*)

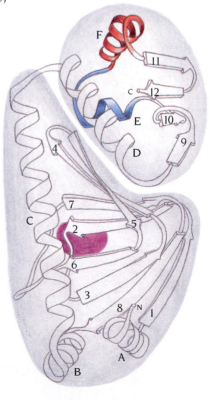

(*a*)

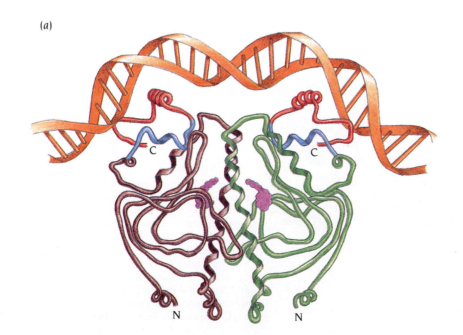

What contributes to the energy of bending in this complex? The central 10 base pairs are anchored to the protein by hydrogen bonds to the six DNA phosphates that face the protein in this region. The surface of the protein is complementary to B-DNA so that no bending of DNA is required to form these interactions. In contrast, interactions between CAP and the 10 base pairs at each end of DNA require that the DNA be bent. These base pairs interact extensively with the protein to provide in each half of the molecule five ionic interactions and four hydrogen bonds to DNA phosphates. These interactions could not occur with straight DNA and therefore can contribute to binding only if the DNA is bent. Thus, the energy required to bend the DNA in this complex is largely supplied by extensive hydrogen bonding and ionic interactions with the DNA phosphates of the two flanking segments.

Steitz has suggested that DNA bending by CAP could contribute to activation of transcription by looping the DNA around CAP to provide for contacts between RNA polymerase and DNA upstream of the CAP-binding site. Such a model could explain how CAP can activate transcription from a variety of distances from the RNA polymerase-binding site since the size of the loop could vary.

Conclusion

Many proteins that switch off or on gene expression in bacteria are dimeric molecules, and the DNA sequences that they specifically recognize are palindromic at their ends. The twofold symmetry of the protein is therefore matched by twofold symmetry at the ends of the recognition sequence.

The monomeric subunits have a helix-turn-helix motif that functions as the specific DNA-binding region. The second helix of this motif is the recognition helix, and its side chains form hydrogen bonds and hydrophobic contacts with nucleotide bases at the bottom of the major groove in B-DNA. These interactions form the recognition signal of Cro and repressor proteins for the operator regions. In the dimeric protein molecules recognition helices in each of the two helix-turn-helix motifs, at opposite ends of the elongated molecule, are separated by 34 Å, corresponding to one turn of B-DNA. When one recognition helix binds to the DNA major groove, the second, 34 Å away, is positioned to bind into the major groove one helical turn along the DNA.

Binding of 434 Cro or repressor to operator sites imposes specific structural changes on the DNA. The changes imposed by the two proteins are overall very similar, but they differ in one segment of the DNA backbone, that is important for the key specificity difference between Cro and repressor (their relative affinities for OR1 and OR3). The conformation of bound DNA is stabilized by interactions between its sugar-phosphate backbone and regions of the protein, which form a large interaction area with complementary surfaces. The ability of these interactions to accomplish a structural change can be modulated by the actual sequence of the DNA, accounting for affinity differences among the various sites.

Specific recognition of DNA targets by the helix-turn-helix motif involves not only interactions between side chains of the recognition helix and bases in the major groove of DNA, but is to a large extent governed by interactions within complementary surfaces between the protein and the DNA. These interactions frequently involve hydrogen bonds from protein main-chain atoms to the DNA backbone in both the major and minor grooves, and are dependent on sequence-specific deformability of the target DNA. Water-mediated interactions often occur even between bases in the major groove and side chains of the recognition helix.

Some of the procaryotic DNA-binding proteins are activated by the binding of an allosteric effector molecule. This event changes the conformation of the dimeric protein, causing the helix-turn-helix motifs to move so that they are 34 Å apart and able to bind to the major groove. The dimeric repressor for purine biosynthesis, PurR, induces a sharp bend in DNA upon binding caused by insertion of α helices in the minor groove between the two helix-turn-helix motif binding sites in the major groove. Each dimer of the V-shaped tetrameric lac repressor, which has sequence homology to PurR, binds independently to DNA fragments in a similar way to the PurR repressor dimer. Such induced bending of DNA may allow the lac repressor molecule to bind to two operator sites which are distant from each other in the operon.

Binding of the catabolite gene activator protein, CAP, to DNA bends the DNA by producing two localized kinks each of 40°. Such DNA bending could activate transcription by facilitating contacts between RNA polymerase and DNA upstream from the CAP binding site. Sequence specific binding of CAP derives both from sequence-dependent distortions of the DNA helix and from direct sequence-specific interactions between protein side chains and exposed edges of base pairs in the major groove.

Genetics and model building provide a framework for the interpretation of structural data, but they cannot reliably be used to predict the details of the stereochemistry of specific interactions within or between molecules.

Selected Readings

General

Anderson, W.F., et al. Proposed α-helical super-secondary structure associated with protein–DNA recognition. *J. Mol. Biol.* 159: 745–751, 1982.

Brennan, R.G., Matthews, B.W. Structural basis of DNA–protein recognition. *Trends Biochem. Sci.* 14: 286–290, 1989.

Brennan, R.G., Matthews, B.W. The helix-turn-helix DNA-binding motif. *J. Biol. Chem.* 264: 1903–1906, 1989.

Harrison, S.C., Aggarwal, A.K. DNA recognition by proteins with the helix-turn-helix motif. *Annu. Rev. Biochem.* 59: 933–969, 1990.

Ptashne, M. *A Genetic Switch: Gene Control And Phage Lambda.* Palo Alto: Blackwell, 1986.

Ptashne, M. Repressors. *Trends Biochem. Sci.* 9: 142–145, 1984.

Ptashne, M., Johnson, A.D. Pabo, C.O. A genetic switch in a bacterial virus. *Sci. Am.* 247: 128–140, 1982.

Sauer, R.T., et al. Homology among DNA-binding proteins suggests use of a conserved super-secondary structure. *Nature* 298: 447–451, 1982.

Steitz, T.A. Structural studies of protein–nucleic acid interaction: the sources of sequence-specific binding. *Q. Rev. Biophys.* 23: 205–280, 1990.

Steitz, T.A., et al. Structural similarity in the DNA-binding domains of catabolite gene activator and Cro repressor proteins. *Proc. Natl. Acad. Sci. USA* 79: 3097–3100, 1982.

Specific structures

Aggarwal, A.K., et al. Recognition of a DNA operator by the repressor of phage 434: a view at high resolution. *Science* 242: 899–907, 1988.

Anderson, J.E., Ptashne, M., Harrison, S.C. A phage repressor–operator complex at 7 Å resolution. *Nature* 316: 596–601, 1985.

Anderson, J.E., Ptashne, M., Harrison, S.C. Structure of the repressor–operator complex of bacteriophage 434. *Nature* 326: 846–852, 1987.

Anderson, W.F., et al. Structure of the Cro repressor from bacteriophage λ and its interaction with DNA. *Nature* 290: 754–758, 1981.

Boelens, R., et al. Complex of *lac* repressor headpiece with a 14 base-pair *lac* operator fragment studied by two-dimensional nuclear magnetic resonance. *J. Mol. Biol.* 193: 213–216, 1987.

Bushman, F.D., et al. Ethylation interference and x-ray crystallography identify similar interactions between 434 repressor and operator. *Nature* 316: 651–653, 1985.

Hochschild, A., Ptashne, M. Homologous interactions of λ repressor and λ Cro with the λ operator. *Cell* 44: 925–933, 1986.

Jordan, R.S., Pabo, C.O. Structure of the λ complex at 2.5 Å resolution: details of the repressor–operator interactions. *Science* 242: 893–899, 1988.

Koudelka, G.B., et al. DNA twisting and the affinity of bacteriophage 434 operator for bacteriophage 434 repressor. *Proc. Natl. Acad. Sci. USA* 85: 4633–4637, 1988.

Koudelka, G.B., Harrison, S.C., Ptashne, M. Effect of non-contacted bases on the affinity of 434 operator for 434 repressor and Cro. *Nature* 326: 886–891, 1987.

Lawson, C.L., Sigler, P.B. The structure of *trp* pseudorepressor at 1.65 Å shows why indole propionate acts as a *trp* "inducer." *Nature* 333: 869–871, 1988.

Lewis, M., et al. Crystal structure of the lactose operon repressor and its complexes with DNA and inducer. *Science* 271: 1247–1254, 1996.

Lim, W.A., Sauer, R.T. Alternative packing arrangements in the hydrophobic core of λ repressor. *Nature* 339: 31–36, 1989.

Mondragón, A., et al. Structure of the amino-terminal domain of phage 434 repressor at 2.0 Å resolution. *J. Mol. Biol.* 205: 189–200, 1989.

Mondragón, A., Wolberger, C., Harrison, S.C. Structure of phage 434 Cro protein at 2.35 Å resolution. *J. Mol. Biol.* 205: 179–188, 1989.

Ohlendorf, D.H., et al. Comparison of the structures of Cro and λ repressor proteins from bacteriophage λ. *J. Mol. Biol.* 169: 757–769, 1983.

Ohlendorf, D.H., et al. The molecular basis of DNA–protein recognition inferred from the structure of Cro repressor. *Nature* 298: 718–723, 1982.

Otwinowski, Z., et al. Crystal structure of *trp* repressor/operator complex at atomic resolution. *Nature* 335: 321–329, 1988.

Pabo, C.O., et al. Conserved residues make similar contacts in two repressor–operator complexes. *Science* 247: 1210–1213, 1990.

Pabo, C.O., Lewis, M. The operator-binding domain of λ repressor: structure and DNA recognition. *Nature* 298: 443–447, 1982.

Schevitz, R.W., et al. The three-dimensional structure of *trp* repressor. *Nature* 317: 782–786, 1985.

Schultz, S.C., Shields, G.C., Steitz, T.A. Crystal structure of a CAP–DNA complex: the DNA is bent by 90 degrees. *Science* 253: 1001–1007, 1991.

Schumacher, M.A., et al. Crystal structure of Lac I member, PurR, bound to DNA: minor groove binding by α helices. *Science* 266: 763–770, 1994.

Weber, I., Steitz, T.A. The structure of a complex of catabolite gene activator protein and cyclic AMP refined at 2.5 Å resolution. *J. Mol. Biol.* 198: 311–326, 1987.

Wharton, R.P., Brown, E.L., Ptashne, M. Substituting an α helix switches the sequence-specific DNA interactions of a repressor. *Cell* 38: 361–369, 1984.

Wharton, R.P., Ptashne, M. Changing the binding specificity of a repressor by redesigning an α helix. *Nature* 316: 601–605, 1985.

Wolberger, C., et al. Structure of phage 434 Cro/DNA complex. *Nature* 335: 789–795, 1988.

Zhang, R.-G., et al. The crystal structure of *trp* aporepressor at 1.8 Å shows how binding tryptophan enhances DNA affinity. *Nature* 327: 591–597, 1987.

DNA Recognition by Eucaryotic Transcription Factors

9

The regulation of transcription in eucaryotes is in general much more complex and currently less well understood than the rather simple switch mechanisms that regulate procaryotic gene expression, examples of which we discussed in Chapter 8. There are three classes of RNA polymerases in eucaryotes. Two of these polymerases, RNA polymerase I and RNA polymerase III, transcribe the genes encoding ribosomal RNAs and transfer RNAs, respectively; genes that code for the messenger RNAs of proteins are transcribed by RNA polymerase II, and it is these genes that will be our principal focus. Complex sets of regulatory elements control the initiation of transcription of these structural genes. Distal to the RNA polymerase II initiation site there are different combinations of specific DNA sequences, each of which is recognized by a corresponding site-specific DNA-binding protein. These proteins are called **transcription factors**, and each combination of DNA sequence and cognate transcription factor constitutes a **control module**. The essence of transcriptional regulation in eucaryotes is to use different combinations of a large set of control modules to regulate the expression of each gene. Given the large number of modules that is now being discovered in the human genome, the number of possible combinations is almost unlimited.

The DNA part of each control module can be divided into three main regions, the **core** or **basal promoter elements**, the **promoter proximal elements** and the **distal enhancer elements** (Figure 9.1). The best characterized core promoter element is the **TATA box**, a DNA sequence that is rich in A-T base pairs and located 25 base pairs upstream of the transcription start site. The TATA box is recognized by one of the basal transcription factors, the **TATA box-binding protein, TBP**, which is part of a multisubunit complex called **TFIID**. This complex in combination with RNA polymerase II and other basal transcription factors such as TFIIA and TFIIB form a preinitiation complex for transcription.

The promoter proximal elements are usually 100 to 200 base pairs long and relatively close to the site of initiation of transcription. Within each of these elements there are DNA sequences specifically recognized by several different transcription factors which either interact directly with the preinitiation complex or indirectly through other proteins.

Figure 9.1 The transcriptional elements of a eucaryotic structural gene extend over a large region of DNA. The regulatory sequences can be divided into three main regions: (1) the basal promoter elements such as the TATA box, (2) the promoter proximal elements close to the initiation site, and (3) distal enhancer elements far from the initiation site.

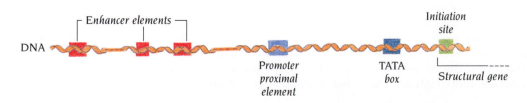

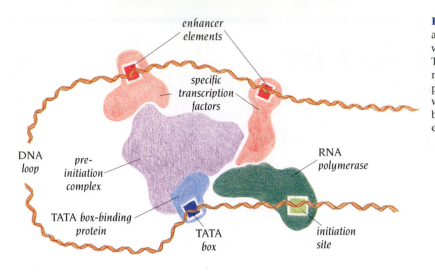

Figure 9.2 Schematic model for transcriptional activation. The TATA box-binding protein, which bends the DNA upon binding to the TATA box, binds to RNA polymerase and a number of associated proteins to form the preinitiation complex. This complex interacts with different specific transcription factors that bind to promoter proximal elements and enhancer elements.

Enhancer elements by contrast are short DNA sequences that occur further upstream or downstream from the initiator site than the promoter proximal elements. Enhancers contain specific sequences recognized by cognate transcription factors. The remarkable feature of enhancers is their distance from the promoters they control. They are often a few thousand base pairs from the promoter but may be 20,000 base pairs or more distant. This means that in eucaryotes transcription is regulated by a series of DNA sequences, each specifically recognized by a DNA-binding protein, dispersed over very long stretches of DNA (Figure 9.2). Efficient gene expression, therefore, can take place only in cells in which the appropriate DNA-binding proteins are present. Every cell of an organism has the same DNA control sequences, but not the complete or the same set of DNA-binding proteins. The regulation of transcription, including cell-specific gene expression, depends on the complement of transcription factors present at any one time in each cell.

Transcription factors can be divided into two broad classes. The general factors of the preinitiation complex are required for the expression of all structural genes transcribed by RNA polymerase II and are, therefore, ubiquitous. Each specific transcription factor that binds either to promoter proximal DNA sequences or to distal enhancer elements is present in only one or a few types of cells, or in cells under particular physiological or environmental conditions. The presence or absence of such transcription factors and their activation or inactivation therefore determine most of the functions of eucaryotic cells, from the changes they undergo during development and differentiation in a multicellular organism to their responses to environmental changes, such as heat shock.

Transcription is activated by protein–protein interactions

The general transcription factor TFIID is believed to be the key link between specific transcription factors and the general preinitiation complex. However, the purification and molecular characterization of TFIID from higher eucaryotes have been hampered by its instability and heterogeneity. All preparations of TFIID contain the TATA box-binding protein in combination with a variety of different proteins called **TBP-associated factors, TAFs**. When the preinitiation complex has been assembled, strand separation of the DNA duplex occurs at the transcription start site, and RNA polymerase II is released from the promoter to initiate transcription. However, TFIID can remain bound to the core promoter and support rapid reinitiation of transcription by recruiting another molecule of RNA polymerase.

The polypeptide chains of the specific transcription factors usually have two different functions: one is to bind to a specific DNA sequence and another is to activate transcription. These two functions are often embodied in separate domains of the protein. How the activating domains function to increase the initiation of transcription is not well understood, but in higher eucaryotes it is believed that they often act through interactions with TAFs to enhance the binding of RNA polymerase to the promoter by contributing to the formation of a stable initiation complex at the TATA box (see Figure 9.2). When the specific transcription factor is part of a DNA control module many hundreds of base pairs away from the promoter site, the intervening DNA is thought to loop out to enable the activating region of the protein to participate in the formation of the transcriptional initiation complex.

A complete transcriptional initiation complex comprising many different protein molecules as well as a large DNA molecule is very difficult to reconstitute *in vitro* and has not yet been crystallized. We therefore know little about the interactions in such complexes of the transcription factors with each other, with RNA polymerase, or with TAFs. Biochemical evidence from cooperative binding studies indicates that there are defined spatial relationships between the different proteins, but as yet we do not understand them at atomic resolution. However, we have made considerable progress toward understanding the structural basis for the recognition of target DNA sequences by specific transcription factors. The DNA-binding domains of the transcription factors are usually short polypeptide chains of around 100 amino acids and they bind to short DNA regions of less than 20 nucleotides. These protein and DNA molecules are easily produced by recombinant DNA technology and chemical synthesis, respectively. The structures of a large number of such protein:DNA complexes have recently been determined by NMR and x-ray techniques. Although the protein molecules used in these studies are only fragments of complete transcription factors, we know from biochemical and genetic studies that their DNA-binding properties in general are the same as those of the intact transcription factors.

Genes that encode specific transcription factors are now being cloned and sequenced at a rapid pace. By comparing the deduced amino acid sequences of these proteins it has become apparent that their DNA-binding regions are built up from a very limited number of structural motifs. Both x-ray and NMR studies have given detailed structural information on complexes of DNA with members of the four most important families of specific transcription factors: those with helix-turn-helix motifs as in procaryotes, and the leucine zipper, helix-loop-helix and zinc-containing motifs that we shall discuss in Chapter 10. About 80% of the known transcription factors have DNA-binding domains that belong to one or another of these four families, the first of which we shall discuss later in this chapter, while the next chapter is devoted to the other three classes.

The TATA box-binding protein is ubiquitous

TBP, the TATA box-binding protein, was first isolated and purified from yeast in 1988 and found to be composed of a single polypeptide chain with a molecular weight of 27 kDa. Yeast TBP is capable of binding specifically to the TATA box of many promoters. Isolation of the corresponding cDNA clone led quickly to the identification of genes encoding homologous proteins from a wide variety of species including those of plants, insects and mammals. Comparison of the deduced amino acid sequences of TBPs reveals that they are composed of a highly conserved C-terminal domain of 180 amino acids and an N-terminal domain that varies in length and shows little or no sequence conservation among different classes of organisms.

TBP mutants lacking the N-terminal region are fully functional in promoter binding and stimulation of basal transcription and therefore these two functions must be provided by the C-terminal domain. Furthermore, the C-terminal domain of yeast TBP contains all the functions essential for normal yeast cell growth and for responses to specific transcriptional activators with a net negative charge. This C-terminal domain contains two homologous

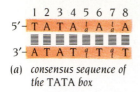

(a) consensus sequence of the TATA box

(b) the DNA fragment used in crystals of the complex with yeast TBP

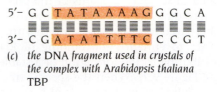

(c) the DNA fragment used in crystals of the complex with Arabidopsis thaliana TBP

Figure 9.3 Comparison of the consensus nucleotide sequence of the TATA box (a) and the sequences of the DNA fragments used in the crystal structure determinations of the TATA box-binding proteins from yeast (b) and the plant *Arabidopsis thaliana* (c).

repeats of 88 amino acids each and mutation studies have indicated that these repeats bind to the TATA consensus sequence (Figure 9.3a) with a dissociation constant in the nanomolar range.

The three-dimensional structures of TBP–TATA box complexes are known

The three-dimensional structures of two complexes between DNA fragments containing a TATA box and the C-terminal domain of TBP were determined in 1993. The group of Paul Sigler at Yale University, worked with yeast TBP whilst the group of Stephen Burley at Rockefeller University, used TBP from the plant *Arabidopsis thaliana*. Both groups used recombinant proteins that were produced in *Escherichia coli*. Sigler's group produced complete yeast TBP then enzymatically cleaved off the N-terminal domain. The C-terminal domain was cocrystallized with a 29 nucleotide hairpin DNA sequence bearing one of the TATA boxes present in the yeast genome (Figure 9.3b). The plant protein studied by Burley had the N-terminal region of 20 amino acids preceding the DNA-binding C-terminal domain, and was cocrystallized with a 14 base pair construction containing the TATA element of the adenovirus late promoter (see Figure 9.3c). Much of the short N-terminal segment was not visible in the crystal structure, presumably because it is disordered. The two structures are surprisingly similar bearing in mind that one protein is from yeast, the other from a plant, and that the two TATA boxes are from such diverse sources as yeast and an animal virus. The results are therefore in all probability valid for all TBP–TATA box interactions, and provide a framework for detailed functional studies of this fundamental part of gene regulation in all species.

A β sheet in TBP forms the DNA-binding site

The two homologous repeats, each of 88 amino acids, at both ends of the TBP DNA-binding domain form two structurally very similar motifs. The two motifs each comprise an antiparallel β sheet of five strands and two helices (Figure 9.4). These two motifs are joined together by a short loop to make a 10-stranded β sheet which forms a saddle-shaped molecule. The loops that connect β strands 2 and 3 of each motif can be visualized as the stirrups of this molecular saddle. The underside of the saddle forms a concave surface built up by the central eight strands of the β sheet (see Figure 9.4a). Side chains from this side of the β sheet, as well as residues from the stirrups, form the DNA-binding site. No α helices are involved in the interaction area, in contrast to the situation in most other eucaryotic transcription factors (see below).

The side of the β sheet that faces away from DNA is covered by two long α helices. One of these helices contains a number of basic residues from the middle segment of the polypeptide chain while the second helix is formed by the C-terminal residues. Residues from these two helices and from the short loop that joins the two motifs (red in Figure 9.4) are likely candidates for interactions with other subunits of the TFIID complex, and with specific transcription factors.

(a)

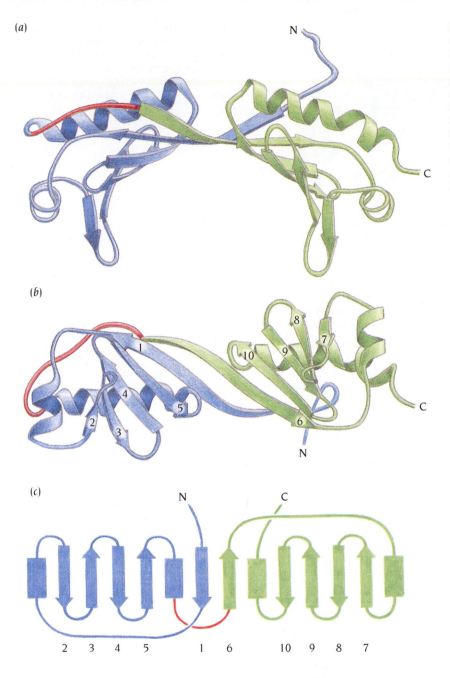

N

C

(b)

8

7

10 9

1

4

5

2

3

6

C

N

(c)

N   C

2   3   4   5   1   6   10 9   8   7

Figure 9.4 Schematic diagrams of the structure of the TATA box-binding protein. (a and b) Richardson-type diagrams illustrating the structural similarity of the two domains (blue and green), which together form a saddle-shaped molecule with stirrups, with the saddle formed from an eight-stranded β sheet. A loop region connecting the two motifs is shown in red. The view in (b) is rotated 90° from that in (a). (c) Topological diagram of the secondary structure elements along the polypeptide chain. (Adapted from D.B. Nikolov et al., *Nature* 360: 40–46, 1992.)

TBP binds in the minor groove and induces large structural changes in DNA

In the two complexes studied by x-ray crystallography, the interactions between TBP and the DNA, as well as the deformation of the B-DNA structure, are very similar, and we will illustrate some of these details for the yeast structure. Minor details of the two complexes vary due to differences in some of the side chains and nucleotides that are present in the interaction areas.

All eight base pairs of the TATA box are in contact with the protein and their structure deviates greatly from the canonical B-form DNA (Figure 9.5). When looking at the TBP structure alone, in the orientation shown in Figure 9.4a, one could imagine that the saddle would straddle normal B-DNA with the helical axis of the DNA perpendicular to a line connecting the two stirrups (perpendicular to the plane of the paper). But this is not the case. Instead, the DNA is sharply bent in the TATA box region so that the local helical axis is almost parallel to the line from stirrup to stirrup. Remarkably, the normal B-DNA structure abruptly returns outside the TATA box. The helical axes of the DNA at each end of the TATA box form an angle of about 100°

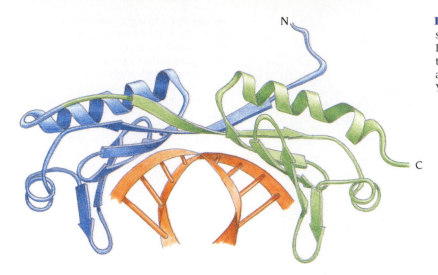

N

C

Figure 9.5 Schematic diagram illustrating the structure of the complex between TBP and a DNA fragment containing the TATA box. Both the stirrups and the underside of the saddle are in contact with the DNA. (Adapted from Y. Kim et al., *Nature* 365: 514–520, 1993.)

to each other, instead of the expected 180° if the DNA was not bent (Figure 9.6). Thus the deformation is restricted to and persists over the entire TATA box, producing severe bends that radically alter the direction taken by the flanking B-DNA duplexes. In procaryotes, the catabolite-activating protein, CAP, induces DNA bending of a similar magnitude when it binds to its specific binding site, as was described in Chapter 8.

Although all the nucleotides within the TATA box deviate in their conformations from regular B-DNA, there are two positions that are especially distorted. Between the first two and the last two base pairs of the TATA box there are very sharp kinks in the DNA. Between these kinks the DNA is smoothly curved and partially unwound. Two pairs of phenylalanine residues are partially inserted between the first two and the last two base pairs of the TATA box. These phenylalanine wedges prevent the stacking of the base pairs, keeping them apart, which causes a significant increase in the rise of the DNA at this base pair step, hence a severe kink. The kinks at each end of the TATA box, in combination with the partial unwinding of the DNA in between, produce a wide and shallow minor groove. This groove is exposed at the convex outer surface of the TATA box region and interacts intimately with the entire concave undersurface of the TBP saddle between the stirrups (see Figure 9.5). All three distortions of the TATA box—the bending of the DNA, the widening of the minor groove and the unwinding of the DNA—are mechanically coupled and overall there is no topological strain in the DNA.

Most sequence-specific regulatory proteins bind to their DNA targets by presenting an α helix or a pair of antiparallel β strands to the major groove of DNA. Recognition of the TATA box by TBP is therefore exceptional; it utilizes a concave pleated sheet protein surface that interacts with the minor groove of DNA. Since the minor groove has very few sequence-specific

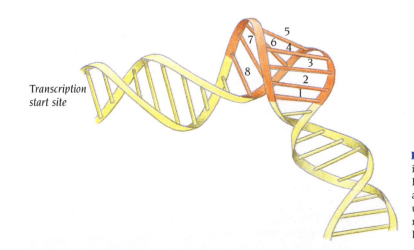

Transcription start site

5
7
6
4
8
3
2
1

Figure 9.6 The DNA fragment is sharply bent in the TATA box region (orange) so that the DNA helices on both sides form an angle of about 110° to each other instead of 180° for undistorted B-DNA. The view of the DNA is rotated about 90° compared with that of Figure 9.5.

functional groups (see Chapter 7) it is appropriate to ask what factors account for the 100,000-fold preference of TBP for a TATA-box sequence over random DNA.

The interaction area between TBP and the TATA box is mainly hydrophobic

The interaction area between the underside of the TBP saddle and the minor groove of DNA is formed by two large, complementary surfaces with no water molecules between them. A surprisingly large amount of this area is hydrophobic, in contrast to protein–DNA interactions that involve the major groove, which are mainly hydrophilic and sometimes mediated by water molecules (see Chapter 8). In TBP, side chains from the eight central β strands interact with both the phosphate sugar backbone and the minor groove of the eight nucleotides of the TATA box. Fifteen side chains projecting from the β strands make hydrophobic contacts with the sugars and bases of DNA. The phosphate groups are hydrogen bonded to arginine and lysine side chains at the edges of the interaction area.

The only DNA sequence-specific aspect of these hydrophobic contacts is that they exclude G-C base pairs that are, of course, absent from the TATA box. The amino group of guanine (see Chapter 7) would, if present, sterically disrupt or modify the interface. Surprisingly, 12 of the 16 hydrogen bond acceptors of the minor groove (N3 of adenine and O2 of thymine) are buried in this interaction area without forming hydrogen bonds either to protein side chains or to water molecules. Studies of site-directed mutations have shown that burying polar side chains in a hydrophobic environment without satisfying their hydrogen-bonding requirements is energetically unfavorable. Clearly, the large number of energetically favorable hydrophobic interactions between TBP and TATA-box DNA must compensate for these unfavorable interactions in order to achieve binding constants in the nanomolar range.

The only sequence-specific hydrogen bonds between TBP side chains and the bases in the minor groove occur at the very center of the TATA box (Figure 9.7). The amide groups of two asparagine side chains donate four hydrogen bonds, two each to adjacent bases on the same DNA strand (Asn 69

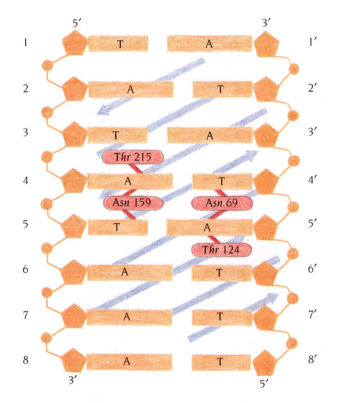

Figure 9.7 Sequence specific interactions between TBP and the TATA box. Asn 69 and Thr 124 from one domain and the equivalent residues Asn 159 and Thr 215 from the second domain interact with the palindromic TATA sequence of the central region of the TATA box.

157

to O2 of T4′ and to N3 of A5′, and Asn 159 to O2 of T5 and to N3 of A4). In addition, N3 of A5′ and A4 accept hydrogen bonds from Thr 124 and Thr 215 respectively. There are also two conserved valine residues on each side of base pairs 4 and 5, Val 71 and 122 on one side and Val 161 and 213 on the other side. The side chains of the valine residues would cause steric interference with the NH_2 substituent from G-C or C-G base pair. The flanking valine residues in combination with the six hydrogen bonds specify A-T or T-A at positions 4 and 5 of the TATA box.

Like Thr 124 and Thr 215, the Asn 69 and Asn 159 residues occupy equivalent positions in the two homologous motifs of TBP. By analogy with the symmetric binding of a dimeric repressor molecule to a palindromic sequence described in Chapter 8, the two motifs of TBP form symmetric sequence-specific hydrogen bonds to the quasi-palindromic DNA sequence at the center of the TATA box. The consensus TATA-box sequence has an A-T base pair at position 4, but either a T-A or an A-T base pair at the symmetry-related position 5, and the sequence is, therefore, not strictly palindromic. However, the hydrogen bonds in the minor groove can be formed equally well to an A-T base pair or to a T-A base pair, because O2 of thymine and N3 of adenine occupy nearly stereochemically equivalent positions, and it is sufficient, therefore, for the consensus sequence of the TATA box to be quasi-palindromic.

In conclusion, one important factor that contributes to the strong affinity of TBP proteins to TATA boxes is the large hydrophobic interaction area between them. Major distortions of the B-DNA structure cause the DNA to present a wide and shallow minor groove surface that is sterically complementary to the underside of the saddle structure of the TBP protein. The complementarity of these surfaces, and in addition the six specific hydrogen bonds between four side chains from TBP and four hydrogen bond acceptors from bases in the minor groove, are the main factors responsible for causing TBP to bind to TATA boxes 100,000-fold more readily than to a random DNA sequence.

Functional implications of the distortion of DNA by TBP

The untwisting and strand separation of DNA are distortions necessary for transcription. These distortions occur at specific DNA sequences that contain a T-A step in the double helix, either in isolation, as described in Chapter 8 for CAP promoter regions in bacteria, or as the tandem repeat T-A-T-A in the TATA box of eucaryotic DNA. The B-DNA structure of such sequences is less stable than those containing G-C base pairs because the stacking interactions between A-T and T-A base pairs are relatively weak. TBP exploits this property of the TATA box to distort and bend the DNA and initiate transcription. The TATA box thus carries information for transcriptional initiation primarily at a physical level: its DNA is easily deformed. Chemical information, in the form of sequence-specific interactions between nucleotides and the protein, plays only a minor role, in contrast to the binding of specific transcription factors to the major groove of DNA. This is also reflected in the sequences of different TATA boxes, where the actual combinations of T-A and A-T base pairs are quite diverse. The minor groove recognition and the bendability of DNA works equally well for different combinations provided there are no G-C or C-G base pairs involved.

The sharp bend of DNA at the TATA box induced by TBP binding is favorable for the formation of the complete DNA control module; in particular, for the interaction of specific transcription factors with TFIID. Since these factors may bind to DNA several hundred base pairs away from the TATA box, and at the same time may interact with TBP through one or several TAFs, there must be several protein–DNA interactions within this module that distort the regular B-DNA structure (see Figure 9.2). The DNA bend caused by the binding of TBP to the TATA box is one important step to bring activators near to the site of action of RNA polymerase.

TFIIA and TFIIB bind to both TBP and DNA

TFIIA and **TFIIB** are two basal transcription factors that are involved in the nucleation stages of the preinitiation complex by binding to the TBP–TATA box complex. Crystal structures of the ternary complex TFIIA–TBP–TATA box have been determined by the groups of Paul Sigler, Yale University, and Timothy Richmond, ETH, Zurich, and that of the TFIIB–TBP–TATA box by Stephen Burley and collaborators. The TBP–DNA interactions and the distortions of the DNA structure are essentially the same in these ternary complexes as in the binary TBP–TATA complex.

TFIIB is arranged in two domains, both of which have the cyclin fold described in Chapter 6. Both domains bind to the TBP–TATA box complex at the C-terminal stirrup and helix of TBP. The phosphate and sugar moieties of DNA form extensive non-sequence-specific contacts with TFIIB both upstream and downstream of the middle of the TATA box.

TFIIA also has two domains, one of which is a four-helix bundle and the other an antiparallel β sandwich. The β sandwich interacts with the N-terminal half of TBP and thus positions TFIIA on the other side of the complex compared with TFIIB. This domain also interacts with phosphates and sugars of DNA upstream of the TATA box. The four-helix bundle domain makes no contact with DNA or TBP and is far removed from the position of TFIIB.

These structures have given us a first glimpse of the preinitiation complex for transcription by RNA polymerase II. Only TBP provides sequence-specific recognition of DNA. The transcription factors TFIIA and TFIIB bind to both TBP and DNA in such a way that they are separated from each other and provide large surfaces available for binding to other components of the preinitiation complex. The components of the preinitiation complex assemble on the promoters with high cooperativity. TBP exhibits no detectable affinity toward either TFIIA or TFIIB in the absence of DNA. Similarly, the promoter segment binds TFIIA and TFIIB poorly or not at all in the absence of TBP. However, the bend imposed on the promoter by TBP repositions the charges and functional groups on the DNA so that TFIIA and TFIIB bind with high affinity, anchored properly in the complex by specific interactions with TBP. In essence, the TBP–promoter complex is a nucleoprotein unit that binds the basal factors TFIIA and TFIIB with high affinity and specificity.

Homeodomain proteins are involved in the development of many eucaryotic organisms

Eucaryotes have many more genes and a broader range of specific transcription factors than procaryotes and gene expression is regulated by using sets of these factors in a combinatorial way. Eucaryotes have found several different solutions to the problem of producing a three-dimensional scaffold that allows a protein to interact specifically with DNA. In the next chapter we shall discuss some of the solutions that have no counterpart in procaryotes. However, the procaryotic helix-turn-helix solution to this problem (see Chapter 8) is also exploited in eucaryotes, in **homeodomain** proteins and some other families of transcription factors.

The **homeobox**, a DNA sequence of about 180 base pairs within the coding region of certain genes, was first discovered in the fruitfly *Drosophila* during studies of mutations that cause bizarre disturbances of the fly's body plan, so-called homeotic transformations. In the mutation *Antennapedia*, for example, legs grow from the head in place of antennae. Such homeotic mutations cause a whole set of cells to be misinformed as to their location in the organism and consequently to form a structure appropriate to another place. Homeoboxes have since been found in several hundred different genes from both vertebrates and invertebrates, and there are varying degrees of DNA sequence homology between different members of this superfamily. A number of these homeobox genes have retained both their precisely ordered tandem arrangement in the genome, and their developmental roles in axial patterning across more than 500 million years of evolution.

Homeoboxes code for homeodomains, sequences of 60 amino acids that function as the DNA-binding regions of transcription factors. Each homeobox gene in *Drosophila* is expressed only in its own characteristic subset of embryonic cells, and almost every embryonic cell contains a unique combination of homeodomain proteins.

Monomers of homeodomain proteins bind to DNA through a helix-turn-helix motif

Comparisons of the amino acid sequences of homeodomains with the sequences of procaryotic DNA-binding proteins suggested that part of the homeodomain forms a helix-turn-helix motif similar to the DNA-binding motif of Cro and repressor, described in Chapter 8. This suggestion was verified by the NMR determination of the solution structures of the *Antennapedia* homeodomain and its complex with a DNA fragment in the laboratory of Kurt Wüthrich, ETH, Zurich, and by an x-ray structure determination of a DNA complex with a different homeodomain from the *Drosophila* gene *engrailed*, in the laboratory of Carl Pabo at Johns Hopkins University, Baltimore. Subsequently, the x-ray structure of a DNA complex of the yeast Mat α2 repressor, which regulates the expression of cell-type specific genes in yeast, was determined by the group of Carl Pabo. The overall structures of these three complexes are almost identical, differing only in the details of their protein–DNA interactions.

The homeodomain is built up from three α helices, connected by rather short loop regions (Figure 9.8). The three helices are formed by residues 10 to 22, 28 to 38, and 42 to 59 (see Figure 9.11). Alpha helices 2 and 3 and the loop region between them form the helix-turn-helix motif, and the structure of the region between residues 30 to 50 is very similar to the helix-turn-helix motif of Cro and repressor. Indeed the three-dimensional structures of this region of these three homeodomains are as similar to the procaryotic helix-turn-helix motifs as they are to each other. The remaining regions of the homeodomain structure, however, differ from the structures of procaryotic DNA-binding proteins. Homeodomain helix 1 is differently arranged compared with the additional helices in λ or trp repressor, and homeodomain helix 3 is considerable longer than the recognition helices of the procaryotic helix-turn-helix motifs (Figure 9.9). Interestingly, the additional helical turns of helix 3 in the *Antennapedia* homeodomain are apparently induced by DNA binding because in the absence of DNA these residues are disordered.

The homeodomain frequently binds to DNA as a monomer, in contrast to procaryotic DNA-binding proteins containing the helix-turn-helix motif, which usually bind as dimers. *In vitro* the homeodomain binds specifically to

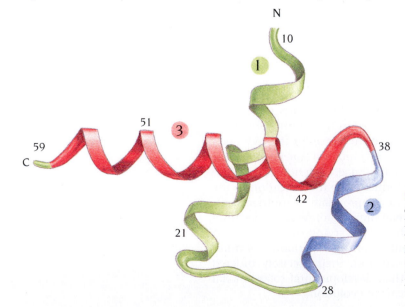

Figure 9.8 Schematic diagram of the three-dimensional structure of the Antennapedia homeodomain. The structure is built up from three α helices connected by short loops. Helices 2 and 3 form a helix-turn-helix motif (blue and red) similar to those in procaryotic DNA-binding proteins. (Adapted from Y.Q. Qian et al., *Cell* 59: 573–580, 1989.)

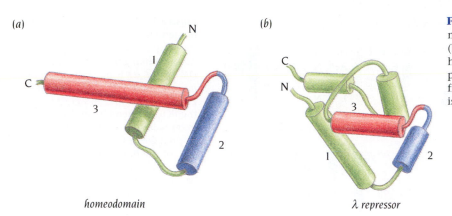

(a)

N

1

C

3

2

homeodomain

(b)

C

N

3

1

2

λ repressor

Figure 9.9 Comparison of the helix-turn-helix motifs in homeodomains (a) and λ repressor (b). The recognition helix (red) of the homeodomain is longer than in the procaryotic repressor motif. In addition the first helix of the homeodomain [(green in (a)] is oriented differently.

DNA fragments containing the sequence 5′-A-T-T-A-3′ with a dissociation constant of about 1 nanomolar. It also binds nonspecifically to DNA fragments with different sequences, but with about 100 times lower affinity. In all three homeodomain structures the DNA fragment is essentially regular B-DNA. As shown in Figure 9.10, the recognition helix of the homeodomain is positioned in the major groove of the B-DNA and the extended polypeptide chain at the N-terminus, before helix 1, is positioned on the opposite side of the DNA, in the minor groove. The homeodomain–DNA complex is stabilized by a number of salt bridges and hydrogen bonds between the DNA backbone and protein side chains, involving mainly lysine and arginines, from four regions of the protein: the N-terminus, the loop between helices 1 and 2, helix 2 and the recognition helix 3 (Figure 9.11). The positions of the residues involved in these nonspecific DNA contacts vary slightly in these homeodomain complexes but the overall arrangement of the homeodomains bound to DNA is virtually identical. The amino acid sequence identity between the three is only 20%, and such remarkable conformational identity between such distantly related members of the homeodomain family strongly suggests that their conserved structure–function relationships have played important roles in the evolution of eucaryotic organisms.

Residues 3, 5, 6, and 8 in the N-terminal arm lie in the minor groove and form contacts with either the edge of the bases or with the DNA backbone. Almost all homeodomains contain four conserved residues, Asn 51, Arg 53, Trp 48 and Phe 49, in the middle of the long recognition helix. The first two conserved polar residues interact with DNA. The second two are part of the hydrophobic core of the homeodomain, and are important for the accurate positioning of the recognition helix and the N-terminal arm with respect to

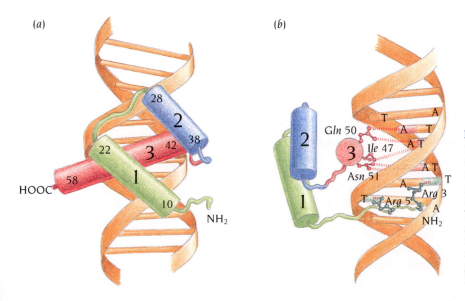

(a)

28

2

22   3   42   38

1

58

HOOC

10

NH₂

(b)

2

Gln 50

3   Ile 47

Asn 51

Arg 3

1   Arg 5

T   A

A   T

A·T

A·T

A   T

T   A

NH₂

Figure 9.10 Schematic diagrams illustrating the complex between DNA (orange) and one monomer of the homeodomain. The recognition helix (red) binds in the major groove of DNA and provides the sequence-specific interactions with bases in the DNA. The N-terminus (green) binds in the minor groove on the opposite side of the DNA molecule and arginine side chains make nonspecific interactions with the phosphate groups of the DNA. (Adapted from C.R. Kissinger et al., *Cell* 63: 579–590, 1990.)

161

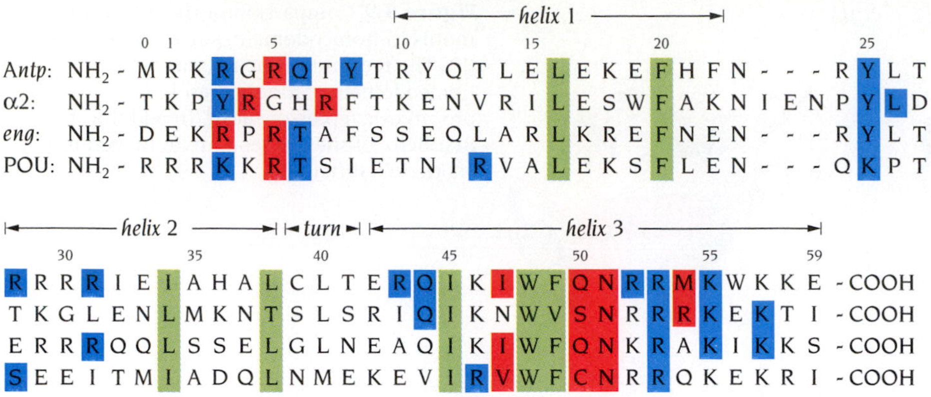

Figure 9.11 Amino acid sequences of homeodomains from four different transcription factors: Antp is from the *Antennapedia* gene in the fruitfly *Drosophila*, α2 is from the yeast *Mat α2* gene, eng is from the engrailed gene in *Drosophila* and POU is from the POU homeodomain in the mammalian gene *Oct-1*. Residues colored green form the hydrophobic core of the homeodomain, blue form nonspecific interactions with the DNA backbone and red form contacts with the edges of the DNA bases.

each other and to the remaining parts of the homeodomain structure. The side chains of Asn 51 and residues 47, 50, and 54 point into the major groove and make either direct or water-mediated contacts with the base edges of the DNA recognition site (see Figure 9.10). These contacts allow homeodomains to bind to operator sites containing the consensus sequence 5'-A-T-T-A-3', and also to some extent differentiate between different operator sites by interacting with flanking bases.

The NMR study by Wüthrich and coworkers has shown that there is a cavity between the protein and the DNA in the major groove of the *Antennapedia* complex. There are several water molecules in this cavity with a residence time with respect to exchange with bulk water in the millisecond to nanosecond range. These observations indicate that at least some of the specific protein–DNA interactions are short-lived and mediated by water molecules. In particular, the interactions between DNA and the highly conserved Gln 50 and the invariant Asn 51 are best rationalized as a fluctuating network of weak-bonding interactions involving interfacial hydration water molecules.

In vivo *specificity of homeodomain transcription factors depends on interactions with other proteins*

Homeodomains that have closely related amino acid sequences bind with comparable affinities to the same DNA sequence *in vitro*. But *in vivo* transcription factors with similar homeodomains must activate the expression of their different target genes in a highly selective and precise way to avoid, for example, catastrophic deformations of the developing body plan of an organism. How is this selective activity achieved? How do functionally different homeodomain proteins achieve the necessary selectivity and specificity when their DNA binding regions are virtually identical? The answer lies in the fact that homeodomain proteins, like other classes of transcription factors, do not operate alone but in concert with other transcription factors and associated proteins.

These additional proteins cooperate with the homeodomain to provide the required specificity for the target genes. For example combinations of the **Mat α2** homeodomain transcription factor with either of two different transcription factors **Mcm1** or **Mat a1** specify two different cell types in yeast. Mat a1 has a homeodomain whereas Mcm1 is a different type of transcription factor. Mat α2 contains an N-terminal dimerization domain that binds to Mcm1 and a C-terminal region that binds to Mat a1 in each case to form a DNA-binding heterodimer. The crystal structure of the Mat α2–Mat a1 heterodimer in a complex with DNA has been determined by the group of Cynthia Wolberger at Johns Hopkins University, and compared to the Mat α2–DNA complex.

(a) Mat *a*1

(b) Mat α2

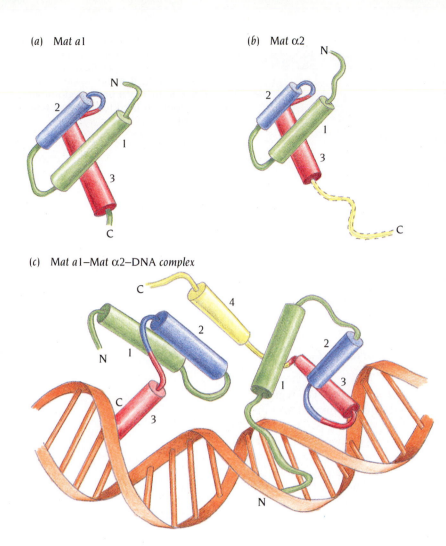

(c) Mat *a*1–Mat α2–DNA *complex*

Figure 9.12 Schematic diagram of the structure of the heterodimeric yeast transcription factor Mat α2–Mat *a*1 bound to DNA. Both Mat α2 and Mat *a*1 are homeodomains containing the helix-turn-helix motif. The first helix in this motif is colored blue and the second, the recognition helix, is red. (a) The assumed structure of the Mat *a*1 homeodomain in the absence of DNA, based on its sequence similarity to other homeodomains of known structure. (b) The structure of the Mat α2 homeodomain. The C-terminal tail (dotted) is flexible in the monomer and has no defined structure. (c) The structure of the Mat *a*1–Mat α2–DNA complex. The C-terminal domain of Mat α2 (yellow) folds into an α helix (4) in the complex and interacts with the first two helices of Mat α2, to form a heterodimer that binds to DNA. (Adapted from B.J. Andrews and M.S. Donoviel, *Science* 270: 251–253, 1995.)

Mat *a*1 and Mat α2 bind in tandem to DNA with their recognition helices in the major groove (Figure 9.12). The C-terminal tail of Mat α2 provides the only contacts with Mat *a*1 in the heterodimer. A portion of the tail forms a short α helix with three leucine residues which project into a hydrophobic pocket between helices 1 and 2 of the Mat *a*1 homeodomain. In the absence of Mat *a*1 this tail is flexible with no ordered structure. The 21-base pair DNA in this complex has a significant overall bend of 60°, in contrast to the B-DNA in the Mat α2–DNA complex. The DNA is smoothly bent without dramatic local distortion or kinks. Without the observed bend in the DNA the C-terminal tail of Mat α2 would not be able to reach its binding site on Mat *a*1. Proper dimerization and positioning of the two recognition helices consequently require the DNA to be bent and provide the necessary energy for the bending.

How is the binding specificity of the heterodimer achieved compared with the specificity of Mat α2 alone? The crystal structure rules out the simple model that the contacts made between the Mat α2 homeodomain and DNA are altered as a result of heterodimerization. The contacts between the Mat α2 homeodomain and DNA in the heterodimer complex are virtually indistinguishable from those seen in the structure of the Mat α2 monomer bound to DNA. However, there are at least two significant factors that may account for the increased specificity of the heterodimer. First, the Mat *a*1 homeodomain makes significant contacts with the DNA, and the heterodimeric complex will therefore bind more tightly to sites that provide the contacts required by both partners. Second, site-directed mutagenesis experiments have shown that the protein–protein interactions involving the

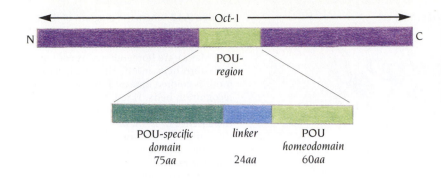

Figure 9.13 The DNA-binding region of the protein Oct-1, the POU region (green), comprises two domains, the POU-specific domain (dark green) and the POU homeodomain (light green) joined by a linker region (blue). These two domains bind to DNA in a tandem arrangement.

C-terminal tail of Mat α2 and the Mat a1 homeodomain are crucial for the specificity of DNA binding. The heterodimer interface may increase specificity by dictating the precise spacing of the two binding sites for the recognition helices, analogous to the 434 Cro and repressor proteins discussed in Chapter 8. This could be one way in which regions of transcription factors outside the DNA-binding domains, the so-called activating regions, provide specificity of transcriptional activation *in vivo*.

POU regions bind to DNA by two tandemly oriented helix-turn-helix motifs

A tandem arrangement of two DNA-binding domains, called the POU region, is present in a class of transcription factors involved in development and in the expression of growth factors, histones and immunoglobulins. The POU region is a 150–160 amino acid sequence consisting of a homeodomain and a POU-specific domain joined by a short variable linker region (Figure 9.13). NMR studies by the groups of Robert Kaptein in Utrecht University and Peter Wright at The Scripps Research Institute, La Jolla, established that the POU-specific domain has a structure very similar to the λ repressor, in spite of having no apparent sequence homology; as described in Chapter 8, this structure is very simple, consisting of four α helices where helices 2 and 3 form a helix-turn-helix motif. The POU region has, therefore, two helix-turn-helix motifs that are similar while the remaining structures of its two domains are quite different.

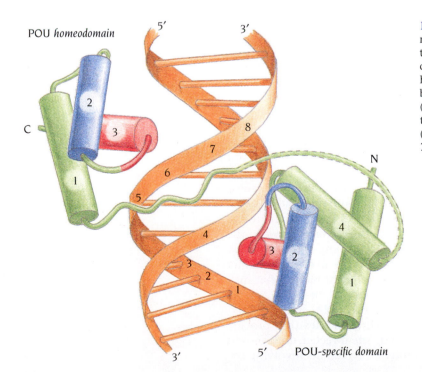

Figure 9.14 The two domains of the POU region bind in tandem on opposite sides of the DNA double helix. Both the POU-specific domain and the POU homeodomain have a helix-turn-helix motif (blue and red) which binds to DNA with their recognition helices (red) in the major groove. The linker region that joins these domains is partly disordered. (Adapted from J.D. Klemm et al., *Cell* 77: 21–32, 1994.)

The x-ray structure of a DNA complex with the complete POU region from the transcription factor Oct-1, which regulates transcription of small nuclear RNA genes and the histone H2B gene, was subsequently determined by the group of Carl Pabo. The Oct-1 POU region recognizes the octamer nucleotide sequence 5'-A-T-G-C-A-A-A-T-3' *in vivo*, hence the name Oct-1. Pabo found that the two helix-turn-helix domains bind in tandem along the octamer DNA sequence (Figure 9.14). The λ repressor-like POU-specific domain makes its essential contacts with the 5' half of the octamer site, and the POU homeodomain makes its contacts with the AAAT sequence in the 3' half. These two domains bind on opposite sides of the DNA double helix with their recognition helices in the major groove. There are no protein–protein contacts between the POU-specific domain and the POU homeodomain, and the 24-residue long linker region between the domains is disordered. This tandem arrangement of the two domains increases the binding specificity of the POU region to the octamer site above the specificities of the individual domains. Such a tandem arrangement of DNA-binding domains is also utilized by the "classic" zinc fingers to enhance their specificity, as will be described in Chapter 10.

Both domains of the POU region bind to DNA by the usual combination of non specific binding to the DNA backbone and specific binding to the bases. The contacts between the homeodomain and DNA are similar to those of the engrailed homeodomain (compare Figures 9.10b and 9.15a) and the

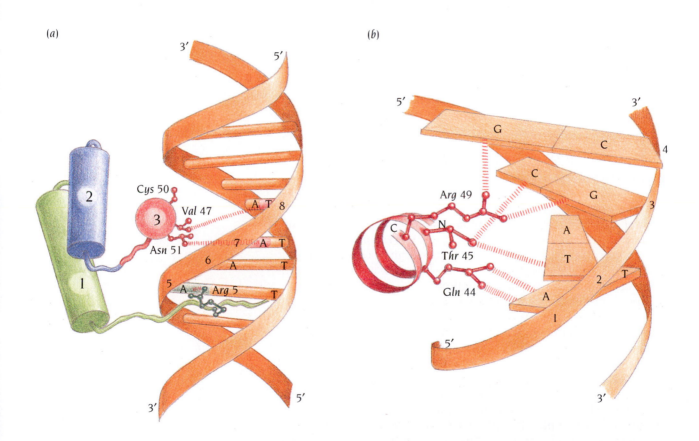

(a)

(b)

Figure 9.15 Diagram illustrating the sequence-specific contacts between DNA and the POU region. (a) Contacts from the POU homeodomain. Three residues from the recognition helix, Val 47, Cys 50 and Asn 51, interact with base pairs 7 to 8 in the major groove of the second half of the POU-binding site. In addition, Arg 5 in the N-terminal arm interacts with base pair 5 of the DNA in the minor groove. (b) Contacts from the POU-specific region. Three residues from the recognition helix, Gln 44, Thr 45 and Arg 49, interact with bases in the major groove of the first half of the POU-binding site, base pairs 1 to 4. These contacts are from the opposite side of the DNA molecule compared with those in (a). The numbering system starts from the first residue of the homeodomain. (Adapted from J.D. Klemm et al., *Cell* 77: 21–32, 1994.)

contacts between DNA and the POU-specific domain are almost identical to those observed in the λ repressor–DNA complex (Figure 9.15b). Helix 3 of the POU-specific domain is positioned in the major groove and makes all the base contacts with the ATGC subsite. The first residue of this helix, Gln 44, makes two hydrogen bonds to the adenine at base pair 1 and is also involved in a network of hydrogen bonds involving Arg 20, Gln 27, Glu 51 and the DNA backbone. These four residues (Arg 20, Gln 27, Gln 44 and Glu 51) are conserved in all of the over 20 different POU-specific domain sequences known, and identical residues are found at corresponding positions of the 434 repressor– and 434 Cro–DNA complexes. Such similarities raise the possibility that the eucaryotic POU-specific domains are evolutionarily related to bacteriophage repressors.

Much remains to be learnt about the function of homeodomains in vivo

Although homeobox genes are easy to identify from their nucleotide sequences, the target genes for homeodomain proteins have been difficult to locate because we lack knowledge about the additional factors required for the sequence specificity of these transcription factors *in vivo*. Which gene products could be involved? When tissues are formed in multicellular organisms, different types of cells are brought into contact with each other through complex cellular interactions, and signals are exchanged between cells leading to organized programs of cell division and cell differentiation. Prime candidate proteins involved in these complex interactions are cell adhesion molecules such as tenacin and cadherin, as well as growth factors and associated proteins involved in signal transduction, some of which we shall discuss in Chapter 13. How homeodomain proteins regulate the expression of such molecules is a key element of one of the central questions in molecular cell biology: how do cells form tissues during embryonic development, producing the characteristic body plan of a species?

Understanding tumorigenic mutations

A protein with the innocuous name **p53** is one of the most frequently cited biological molecules in the Science Citation Index. The "p" in p53 stands for protein and "53" indicates a molecular mass of 53 kDa. The p53 protein plays a fundamental role in human cell growth and mutations in this protein are frequently associated with the formation of tumors. It is estimated that of the 6.5 million people diagnosed with one or another form of cancer each year about half have p53 mutations in their tumor cells and that the vast majority of these mutations are single point mutations.

Wild-type p53 inhibits tumor formation, and its gene is therefore called a tumor suppressor gene. Details of exactly how this suppression is accomplished are currently unknown, but one important role of p53 is to maintain the integrity of the genome during cell division by controlling a critical step in the cell cycle. A complex series of molecular interactions regulates the cell cycle; the chief players are proteins called cyclins and their associated enzymes, cyclin-dependent kinases, which we discussed in Chapter 6. These kinases are inhibited by another protein, p21, whose expression is promoted by p53. In the presence of p21 the cell cycle is halted before the cell is committed to divide. Presumably this gives the cell time either to repair damaged DNA or, if it is beyond repair, to initiate programmed cell death (apoptosis).

One of the most important molecular functions of p53 is therefore to act as an activator of p21 transcription. The wild-type protein binds to specific DNA sequences, whereas tumor-derived p53 mutants are defective in sequence-specific DNA binding and consequently cannot activate the transcription of p53-controlled genes. As we will see more than half of the over one thousand different mutations found in p53 involve amino acids which are directly or indirectly associated with DNA binding.

The monomeric p53 polypeptide chain is divided in three domains

The polypeptide chain of p53 is divided in three domains, each with its own function (Figure 9.16). Like many other transcription factors, p53 has an N-terminal activation domain followed by a DNA-binding domain, while the C-terminal 100 residues form an oligomerization domain involved in the formation of the p53 tetramers. Mutants lacking the C-terminal domain do not form tetramers, but the monomeric mutant molecules retain their sequence-specific DNA-binding properties *in vitro*.

Intact p53 has so far resisted crystallization, and it is too large for current NMR technology, so the three-dimensional structure of the complete molecule is not known. However, the group of Nikola Pavletich at the Memorial Sloan-Kettering Cancer Center in New York has determined the crystal structure of the DNA-binding domain (residues 102–292) bound to a 21-base pair DNA fragment containing a specific p53-binding sequence. Furthermore, the same group has determined the x-ray structure of a region of the oligomerization domain, residues 325–356, involved in the formation of tetramers. These structural studies have enhanced our understanding of tumorigenic mutants of p53, and they give a solid basis both for mutational studies of p53 and for the design of cancer drugs.

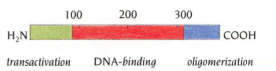

Figure 9.16 The polypeptide chain of the protein p53 is divided into three domains with different functions: transactivation, DNA binding and oligomerization.

The oligomerization domain forms tetramers

The x-ray studies have shown that the 32-amino acid peptide comprising residues 325–356 of the oligomerization domain of p53 is sufficient for tetramer formation. The structure of each unit is very simple; a β-strand-turn-α-helix motif (Figure 9.17). Two such units form a dimer through interactions between the β strands and the α helices. The β strands form an antiparallel two-stranded β sheet and the two α helices are arranged in an antiparallel fashion. The interactions that hold the subunits together as dimers are mainly provided through eight backbone hydrogen bonds in the two-stranded β sheet. In addition there is a small interior core of hydrophobic side chains from both the β strands and the α helices.

Two such dimers form the tetramer through mainly hydrophobic interactions between the α helices. The β strands are on the outside of the tetramer and are not involved in the dimer–dimer interactions. The arrangement of the four α helices is unusual and provides a rare example of four α helices packed against each other in a way different from the four-helix bundle motif.

Although most of the p53 mutations detected in tumors are located in the DNA-binding domain, a few have also been observed in the oligomerization domain. The structure provides a clear explanation of the effects of two such mutations. One is a mutation of the β-strand residue Leu 330 to His. The side chain of this Leu residue in the wild-type protein is in the center of the hydrophobic core of the dimer; replacing it with the hydrophilic histidine side chain of the mutant destabilizes the core, prevents formation of dimers, and therefore inhibits p53 function. The second example is the mutation of a glycine residue in the turn between the β strand and the α helix. The turn comprises only this glycine residue, which adopts an unusual conformation in a region of the Ramachandran plot (see Figure 1.7) that is energetically

Figure 9.17 Schematic diagram illustrating the tetrameric structure of the p53 oligomerization domain. The four subunits have different colors. Each subunit has a simple structure comprising a β strand and an α helix joined by a one-residue turn. The tetramer is built up from a pair of dimers (yellow-blue and red-green). Within each dimer the β strands form a two-stranded antiparallel β sheet which provides most of the subunit interactions. The two dimers are held together by interactions between the four α helices, which are packed in a different way from a four-helix bundle. (Adapted from P.D. Jeffrey et al., *Science* 267: 1498–1502, 1995.)

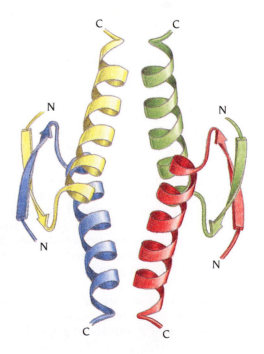

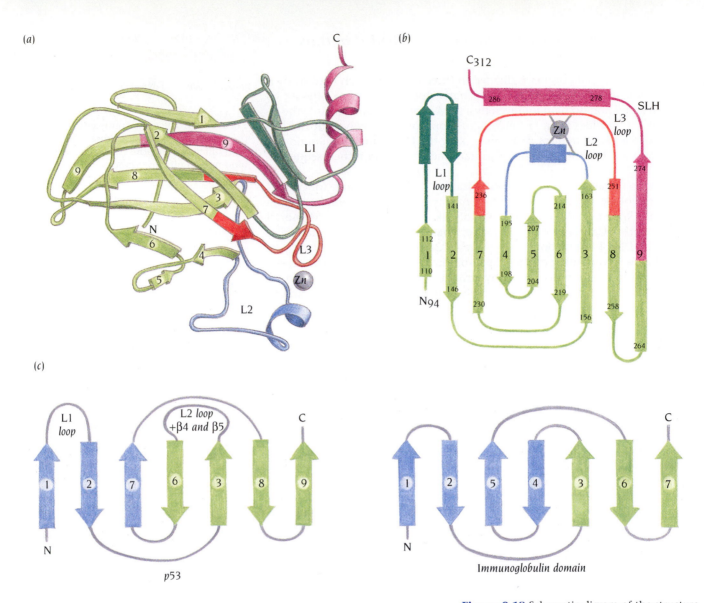

(a) (b)

(c)

p53

Immunoglobulin domain

unfavorable for any other residue than glycine. Mutation of this glycine residue prevents the subunit from folding into the correct structure for dimerization and hence abolishes p53 function.

The DNA-binding domain of p53 is an antiparallel β barrel

Transcription factors frequently utilize a few standard motifs for DNA binding, for example zinc-fingers or helix-turn-helix motifs, but the DNA-binding domain of p53 has none of these. Instead it binds to DNA using protruding loops that are anchored to an antiparallel β-barrel scaffold. The fold of this barrel is similar to the immunoglobulin fold which is the basic domain structure of immunoglobulins, as will be discussed in Chapter 15. The β barrel of p53 comprises nine β strands, seven of which form the immunoglobulin fold (Figure 9.18). From the ribbon diagram in Figure 9.18a it can be seen that β strands 4 and 5 are short and at the edges of the β sheets that build up the compressed barrel. The remaining seven strands of the barrel have not only the topology of an immunoglobulin fold (Figure 9.18c) but also an overall structure that is similar to members of the immunoglobulin superfamily, such as one domain of the MHC-binding coreceptor CD4, which also acts as a receptor for HIV.

p53 is not the only transcription factor having the immunoglobulin fold; it also occurs in the mammalian transcription factor NF-κB, which regulates the expression of a number of genes involved in the response to infection

Figure 9.18 Schematic digram of the structure of the DNA-binding domain of p53. (a) The DNA binding domain of p53 folds into an antiparallel β barrel with long loop regions—L1 (dark green), L2 (blue) and L3 (red)—at one end of the molecule. The conformations of loops L2 and L3 are stabilized by a zinc atom. The C-terminus of the chain (purple) is at the same end of the molecule as these loops. (b) Topological diagram of the DNA-binding domain of p53. Nine antiparallel β strands form a β barrel, the central part of which has a fold similar to that of immunoglobulin domains. A region called SLH (strand-loop-helix) spans residues 271–286. (c) Simplified topological diagram of p53 compared with a topological diagram of the constant immunoglobulin domain (see Figure 15.7). Strands of the same color belong to the same β sheet. Strand numbers 4 and 5 are considered to be part of the loop between strands 3 and 6. The middle strand in these diagrams belongs to different sheets but in the actual structures these strands are at the edges of the two sheets, in only slightly different positions relative to the common structure. [(b) Adapted from Y. Cho et al., *Science* 265: 346–355, 1994.]

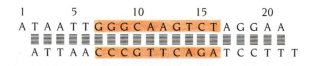

```
 1        5        10       15       20
A T A A T T G G G C A A G T C T A G G A A
■■■■■■■■■■■■■■■■■■■■■■■■■■■■■■■■■■■■■■■■■■■
A T T A A C C C G T T C A G A T C C T T T
```

Figure 9.19 Nucleotide sequence of the 21-base pair DNA fragment cocrystallized with the DNA-binding domain of p53. The p53 binds in a sequence-specific manner to the shaded region.

and stress, including immunoglobulin genes. NF-κB is a member of a family of transcription factors that is characterized by a homologous sequence of about 280 amino acid residues, called the REL-homology region. The groups of Steve Harrison and Paul Sigler determined the structure of the REL-homology region from NF-κB and showed that it has two immunoglobulin domains that both use loop regions to bind DNA. Even though p53 does not have a REL-homology region in its amino acid sequence, it is a member of the family of transcription factors that use loop regions of the immunoglobulin fold for DNA binding.

At one end of the barrel the β strands are close together, forming a compact structure with short loop regions connecting the strands. The β strands are highly twisted and their directions diverge as they reach the other end of the barrel. Consequently the barrel is more open at this end, the loops are longer and they extend outside the barrel; this is the end of the molecule to which DNA binds (see Figure 9.20). The conformations of two of these loops are stabilized by a zinc atom, which is bound to two cysteine side chains from one loop and to one cysteine and one histidine side chain from the other. There is no structural resemblance to the zinc finger domains discussed in the next chapter, in spite of the similar zinc ligands: these ligands only reflect the preference of zinc to form strong bonds to nitrogen- and sulfur-containing side chains.

Two loop regions and one α helix of p53 bind to DNA

Figure 9.19 shows the sequence of the DNA that was used for the structure determination of the p53–DNA complex; the bases involved in sequence-specific binding to the protein are shaded. One molecule of the DNA-binding domain of p53 binds to the minor and the major grooves of the DNA making sequence-specific interactions with both strands (Figure 9.20).

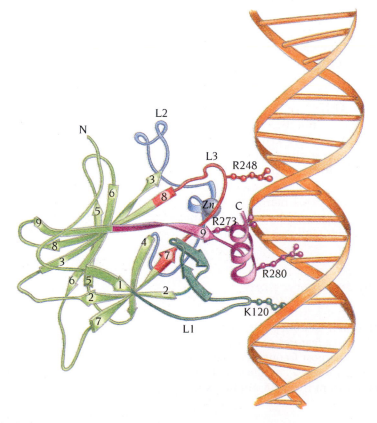

Figure 9.20 Diagram illustrating the sequence-specific interactions between DNA and p53. The C-terminal α helix and loop L1 of p53 bind in the major groove of the DNA. Arg 280 from the α helix and Lys 120 from L1 form important specific interactions with bases of the DNA. In addition, Arg 248 from loop L3 binds to the DNA in the minor groove. (Adapted from Y. Cho et al., *Science* 265: 346–355, 1994.)

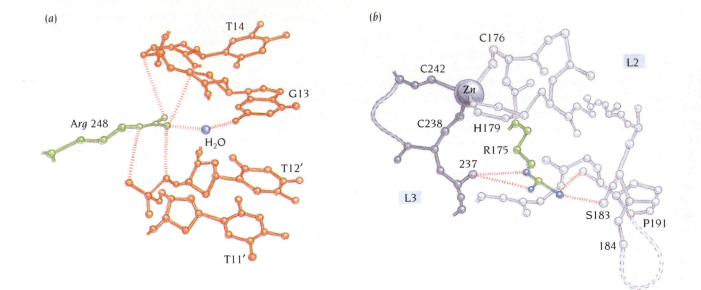

(a)

T14

G13

Arg 248

H_2O

T12'

T11'

(b)

C176

L2

C242

Zn

C238

H179

R175

237

L3

S183

P191

184

Interactions that are not sequence specific are also an important part of the binding and occur between the sugar and phosphate residues of the DNA and the side-chain and main-chain atoms of the protein. In the crystals the DNA fragment retains the B-DNA structure with only minor distortions.

Two of the four loop regions (L1 and L3) at the open end of the barrel are directly involved in DNA binding. In addition, the COOH terminus of the polypeptide chain extends from the barrel and forms a protruding α helix that is positioned in the major groove of the DNA. Residues from this α helix as well as from the preceding loop (from β strand 9) participate in both specific and nonspecific binding. The most critical of these sequence-specific contacts in the major groove of DNA is made by Arg 280 from the α helix, which donates hydrogen bonds to the base of G10' (see Figure 9.20). Because the C-G 10 base pair (see Figure 9.19) is invariant in all p53-specific binding sites and several point mutations of Arg 280 have been observed in tumors, we can infer that this contact is important for the proper function of p53. Loop L1, which connects the first two strands of the barrel, interacts with a second area of the major groove. The side chain of Lys 120 forms hydrogen bonds with the base of G8 (see Figure 9.20), and its main-chain NH group interacts with one phosphate oxygen atom.

The minor groove interactions occur in the AT-rich region (base pairs 11 and 12) and involve Arg 248 from loop L3, which connects β strands 7 and 8 (Figures 9.20 and 9.21a). The side chain of Arg 248 is wedged into the minor groove and makes contact with the sugar and phosphate groups of T12' and T14, as well as with a water molecule that in turn is hydrogen bonded to the base G13. These interactions introduce slight distortions of the B-DNA structure, by compressing the minor groove. This narrowing of the groove allows Arg 248 to pack tightly against the sugar phosphate groups inside the minor groove. It is likely that an AT-rich sequence is required for this compression to occur.

Tumorigenic mutations occur mainly in three regions involved in DNA binding

By examining some of the over one thousand tumor-causing point mutations of p53 in the light of its structure, we can identify features of p53 that are necessary for tumor suppression. The amino acids most frequently changed in cancer cells are at or near the protein–DNA interface; residues that are infrequently mutated, if at all, are in general far from the DNA-binding site.

Figure 9.21 Detailed views of two arginine residues of p53 that are frequently mutated in tumors. (a) Interactions between Arg 248 (green) and the DNA (orange). The side chain of Arg 248 is wedged into the minor groove and makes contacts with the sugar and phosphate groups of T12' and T14. In addition, a water molecule (blue) mediates a hydrogen bond between Arg 248 and the base of G13. (b) Interactions between L2 and L3 in p53. The zinc atom is bound to two cysteine residues in L3 and to one cysteine and one histidine in L2. In addition, residue Arg 175 from L2 forms two hydrogen bonds to the main chain C=O group of residue 237 in L3 and one hydrogen bond to the side chain of Ser 183 in L2. Mutations in residues 175, 183 or 237 would distort loop L3 and prevent proper binding of p53 to DNA.

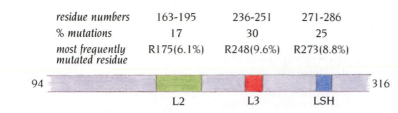

residue numbers	163-195	236-251	271-286
% mutations	17	30	25
most frequently mutated residue	R175(6.1%)	R248(9.6%)	R273(8.8%)

94 — L2 — L3 — LSH — 316

Figure 9.22 Most tumorigenic mutations of p53 are found in the regions of the polypeptide chain that are involved in protein–DNA interactions. These regions are loops L2 (green) and L3 (red) and a region called LSH (blue) which comprises part of β strand 9 as well as the C-terminal α helix.

Mutations are particularly common in three regions of the structure (Figure 9.22). One of these regions is at the C-terminus (residues 271–286), and includes part of β strand 9, the C-terminal α helix and the loop between them. This region is called SLH for strand-loop-helix (see Figure 9.18). The other two regions are loop L3 (residues 236–251), which interacts with the minor groove of DNA, and loop L2 (residues 163–195), which is wedged between L3 and the β barrel and stabilizes the position of L3.

Thirty percent of the tumor-derived mutations are in L3, which contains the single most frequently mutated residue, Arg 248. Clearly the interaction between DNA and the specific side chain of an arginine residue inside the minor groove is of crucial importance for the proper function of p53. It is an open question whether this interaction is needed for the recognition of specific DNA sequences, or is required for the proper distortion of the DNA structure, or a combination of both. Other residues that are frequently mutated in this region participate in interactions with loop L2 and stabilize the structures of loops L2 and L3. Mutations of these residues presumably destabilize the structure so that efficient DNA binding can no longer take place.

Although the L2 loop does not interact directly with DNA, the numerous interactions between both main-chain atoms and the side chains of loops L2 and L3 are necessary to keep L3 in the correct position for DNA binding. The importance of the interactions between loops L2 and L3 is illustrated by Arg 175 in L2 (Figure 9.21b). The side chain of Arg 175 forms two hydrogen bonds to the main-chain C=O group of residue 237 in L3 and one hydrogen bond to the side-chain oxygen atom of Ser 183 in L2. It is therefore an essential residue for maintaining the structural stability of both L2 and L3, and in fact it is the most tumorigenic residue in L2, being mutated in 6.1% of p53-induced tumors. Another example of the importance of loop L2 for function of p53 is provided by mutations that affect the binding of the zinc atom. The zinc atom that stabilizes the structure of loops L2 and L3 is bound to two side chains from L2 (Cys 176 and His 179) and to two side chains from L3 (Cys 238 and Cys 242) (Figure 9.21b). The crystal structure shows that mutations in these four important residues would distort loop L3 and prevent efficient binding to DNA, and such mutations are indeed frequently found in tumors.

Some of the residues from the SLH region bind directly to DNA in the major groove as described earlier (see Figure 9.20). The most frequently mutated residue in this region, Arg 273, binds to the phosphate group of T11'. In addition it participates in an extended network of hydrogen bonds with other side chains of the SLH region, including Arg 280, which makes the critical sequence-specific DNA interaction with the base of G10'. Mutations of Arg 280 are found in 2.1% of p53-induced tumors.

In conclusion it is obvious from a comparison of the structure of the DNA-binding domain of p53 and the pattern of p53 mutations observed in tumors that mutations that disturb or abolish specific p53 DNA-binding also abolish the ability of p53 to suppress tumors. The mutated residues are of two different types; those that participate directly in DNA binding and those in the two loop regions that are important for the overall structural integrity of the DNA-binding regions. The function of p53 may be particularly sensitive to the second class of mutations. The challenge for designers of anticancer drugs is to try to synthesize molecules that can stabilize these loop structures in mutant p53 molecules, thereby restoring the protein's normal DNA-binding function.

Conclusions

The general transcription factor TBP sharply bends and unwinds TATA box DNA, causing the minor groove of DNA to open up and become wide and shallow enough to accomodate a β sheet. Hydrophobic interactions occur between the side chains of the β sheet residues and sugars and bases of the DNA. The mechanism of DNA binding by TBP is therefore very different from that of the helix-turn-helix homeobox transcription factors that, like Cro and repressor in bacteriophages, form hydrophilic interactions with the major groove of DNA, in a base-specific manner. However, unlike Cro and repressor, which bind as homodimers, homeobox factors bind as monomers. In addition to binding in the major groove, the homeodomain proteins also exhibit minor groove base recognition. TBP and the helix-turn-helix transcription factors, like all other transcription factors, form salt bridges between their lysine and arginine side chains and phosphate groups of the DNA to stabilize the DNA:protein complex. The monomeric POU transcription factors have two tandemly oriented helix-turn-helix motifs in the same polypeptide chain, which bind to two sequential binding sites in the major groove of the DNA.

DNA binding by TBP is strongly dependent on the presence of T-A base pairs in the TATA box. Bending allows remote sites on the DNA, with their bound cognate specific transcription factors, to come close together such that the proteins can interact to form the transcriptional preinitiation complex.

The DNA-binding domains of the tetrameric tumor suppressor protein, p53, has an antiparallel β barrel scaffold with an immunoglobulin fold, from one end of which projects two loop regions that are directly involved in binding DNA in the major groove. In addition, the carboxy terminus of the DNA-binding domain extends from the barrel as an α helix and is also involved in binding to the major groove of DNA. Most of the tumorigenic point mutations of p53 are in these DNA-binding regions. Transcription factors such as NF-κB that contain REL-homology regions also use the immunoglobulin fold as scaffold for loops that bind DNA.

Selected readings

General

Burley, S.K. Picking up the TAB. *Nature* 381: 112–113, 1996.

Burley, S.K., Roeder, R.G. Biochemistry and structural biology of transcription factor IID. *Annu. Rev. Biochem.* 65: 769–799, 1996.

Conaway, R.C., Conaway, J.W. General initiation factors for RNA polymerase II. *Annu. Rev. Biochem.* 62: 161–190, 1993.

Gehring, W.J., et al. Homeodomain–DNA recognition. *Cell* 78: 211–223, 1994.

Goodrich, J.A., Tjian, R. TBP:TAF complexes: selectivity factors for eucaryotic transcription. *Curr. Opin. Cell Biol.* 6: 403–409, 1994.

Harris, C., Hollstein, M. Clinical implications of the p53 tumor-suppressor gene. *N. Engl. J. Med.* 329: 1318–1327, 1993.

Harrison, S.C., Sauer, R.T. (eds.) Protein–nucleic acid interactions. *Curr. Opin. Struc. Biol.* 4: 1–35, 1994.

Hori, R., Carey, M. The role of activators in assembly of RNA polymerase II transcription complexes. *Curr. Opin. Genet. Dev.* 4: 236–244, 1994.

Kornberg, T.B. Understanding the homeodomain. *J. Biol. Chem.* 268: 26813–26816, 1993.

Pabo, C.O., Sauer, R.T. Transcription factors: structural families and principles for DNA recognition. *Annu. Rev. Biochem.* 61: 1053–1098, 1992.

Philips, S.E.V., Moras, D. Protein–nucleic acid interactions. *Curr. Opin. Struct. Biol.* 3: 1–16, 1993.

Phillips, S.E.V. Built by association: structure and function of helix-loop-helix DNA-binding proteins. *Structure* 2: 1–4, 1994.

Rosenfeld, M.G. POU-domain transcription factors; pou-erful developmental regulators. *Genes Dev.* 5: 897–907, 1991.

Struhl, K. Duality of TBP, the universal transcription factor. *Science* 263: 1103–1104, 1994.

Tjian, R., Maniatis, T. Transcriptional activation: a complex puzzle with few easy pieces. *Cell* 77: 5–8, 1994.

Specific structures

Billeter, M., et al. Determination of the nuclear magnetic resonance solution structure of an *Antennapedia* homeodomain:DNA complex. *J. Mol. Biol.* 234: 1084–1097, 1993.

Cho, Y., et al. Crystal structure of a p53 tumor suppressor:DNA complex: understanding tumorigenic mutations. *Science* 265: 346–355, 1994.

Clore, G.M., et al. High-resolution structure of the oligomerization domain of p53 by multidimensional NMR. *Science* 265: 386–391, 1994.

Decker, N., et al. Solution structure of the POU-specific DNA-binding domain of Oct-1. *Nature* 362: 852–855, 1993.

Fairman, R., et al. Multiple oligomeric states regulate the DNA binding of helix-loop-helix peptides. *Proc. Natl. Acad. Sci. USA* 90: 10429–10433, 1993.

Geiger, J.H., et al. Crystal structure of the yeast TFIIA/TBP/DNA complex. *Science* 272: 830–836, 1996.

Ghosh, G., et al. Structure of NF-κB p50 homodimer bound to a κB site. *Nature* 373: 303–310, 1995.

Jeffrey, P.D., Gorina, S., Pavletich, N.P. Crystal structure of the tetramerization domain of the p53 tumor suppressor at 1.7 Ångstroms. *Science* 267: 1498–1502, 1995.

Kim, J.L., Nikolov, D.B., Burley, S.K. Crystal structure of TBP recognizing the minor groove of a TATA element. *Nature* 365: 520–527, 1993.

Kim, Y., et al. Crystal structure of a yeast TBP/TATA-box complex. *Nature* 365: 512–520, 1993.

Kissinger, C.R., et al. Crystal structure of an engrailed homeodomain:DNA complex at 2.8 Å resolution: a framework for understanding homeodomain–DNA interactions. *Cell* 63: 579–590, 1990.

Klemm, J.D., et al. Crystal structure of the Oct-1 POU domain bound to an octamer site: DNA recognition with tethered DNA-binding modules. *Cell* 77: 21–32, 1994.

Li, T., et al. Crystal structure of the MATa1/MATα2 homeodomain heterodimer bound to DNA. *Science* 270: 262–269, 1995.

Muller, C.W., et al. Structure of the NF-κB p50 homodimer bound to DNA. *Nature* 373: 311–317, 1995.

Nikolov, D.B., et al. Crystal structure of a TFIIB–TBP–TATA-element ternary complex. *Nature* 377: 119–128, 1995.

Nikilov, D.B., et al. Crystal structure of TFIID TATA-box binding protein. *Nature* 360: 40–46, 1992.

Tan, S., et al. Crystal structure of a yeast TFIIA/TBP/DNA complex. *Nature* 381: 127–134, 1996.

Wilson, D.S., et al. Crystal structure of a paired (PAX) class cooperative homeodomain dimer on DNA. *Cell* 82: 709–719, 1995.

Wolberger, C., et al. Crystal structure of a Mat α2 homeodomain:operator complex suggests a general model for homeodomain–DNA interactions. *Cell* 67: 517–528, 1991.

Specific Transcription Factors Belong to a Few Families

Many procaryotic transcription factors are dimers that recognize palindromic DNA sequences in such a way that each subunit binds one half-site. Dimerization and DNA-binding activities usually reside in different domains or regions of the protein, as was discussed in Chapter 8. The POU transcription factors exhibit a variation of this theme by having two similar domains in one polypeptide chain (see Chapter 9). A similar arrangement is found in another important class of specific eucaryotic transcription factors—those that have zinc atoms in their DNA-binding motifs. As we will see, the polypeptide chains of these proteins may contain several homologous domains, each capable of specific DNA binding and each containing zinc as an integral part of the DNA-binding domain. These proteins belong to several different groups with different structures and different modes of DNA binding. We will discuss three of them that illustrate three different ways by which zinc-containing motifs recognize specific binding sites arranged contiguously along DNA segments.

As we have already mentioned in Chapters 8 and 9, one unifying concept to emerge from the study of different DNA-binding motifs is that these motifs provide three-dimensional scaffolds that match the contours of DNA. These scaffolds ensure proper positioning of the interacting protein surfaces against the DNA, allowing both sequence-specific and nonspecific interactions to occur. In some cases it is the amino acid sequences of the scaffolds, rather than the sequences of amino acids involved in DNA interactions, that make it possible to identify a DNA-binding motif in a protein sequence. This principle applies in particular to two important families of eucaryotic transcription factors that contain leucine zippers. Later in this chapter we will discuss one example each from these two families: GCN4, which belongs to the basic region/leucine zipper (b/zip) family, and Max, which is a member of the basic region/helix-loop-helix/leucine zipper (b/HLH/zip) family. The HLH motif, which is also present in many transcription factors that do not have the leucine zipper region, is very different from the helix-turn-helix motif that is so frequent in procaryotic DNA-binding proteins (see Chapter 8) and in homeodomains (see Chapter 9).

The DNA-binding motifs discussed in this and the preceding two chapters are those most frequently found in procaryotes and eucaryotes. However, other motifs are known, for example the β sheet motif of the met repressor in *Escherichia coli* which binds to the major groove of DNA. No doubt others remain to be discovered.

Several different groups of zinc-containing motifs have been observed

More than a thousand different transcription factors contain zinc as an essential element of their DNA-binding domains. The polypeptide chains of such zinc-containing motifs are usually short, about 50 amino acids or less, with regular patterns of cysteine and/or histidine residues along the chains. These residues bind to the zinc atoms, thereby providing a scaffold for the folding of the motif into a small compact domain. Several different subgroups of such motifs have so far been discovered, each with a different pattern of zinc-binding residues and with different three-dimensional structures.

The classic zinc finger motif was first described in 1985 in the laboratory of Aaron Klug at the MRC Laboratory of Molecular Biology in Cambridge, UK, where it was inferred from an analysis of the amino acid sequence of the transcription factor TFIIIA from *Xenopus laevis*. This factor, which regulates transcription of ribosomal 5S RNA, can be isolated in large quantities from frog oocytes. Each oocyte contains about 20,000 molecules of TFIIIA complexed with 5S RNA to form 7S ribonucleoprotein particles. The amino acid sequence of the 344-residue polypeptide chain of TFIIIA contains nine repeated sequences of about 30 residues each. The repeats are not identical in sequence but each contains two Cys residues at the amino end and two His residues at the carboxy end. Since the protein contains intrinsic zinc atoms and its transcriptional activity is dependent on the presence of these zinc atoms, Klug suggested that the Cys and His atoms are ligands to a zinc atom and that the loop between these residues forms the DNA-binding region. Each of the nine repeats was therefore called a zinc finger. As we shall see, these deductions were subsequently confirmed by structural studies. Since 1985, thousands of different genes coding for similar motifs have been identified and sequenced from a variety of eucaryotic organisms. They all contain two cysteine residues separated by two or four amino acids and two histidine residues separated by three to five amino acids (Figure 10.1a). The linker region between the last cysteine and the first histidine is 12 residues long.

Figure 10.1 (a) The classic zinc finger motif comprises about 30 amino acid residues, with two cysteine and two histidine residues which bind to a zinc atom. The linker region between the last cysteine and the first histidine is 12 residues long and is called the finger region. (b) Schematic diagram of the three-dimensional structure of a chemically synthesized 25-residue peptide with an amino acid sequence corresponding to one of the zinc fingers in an embryonic protein, Xfin, from *Xenopus laevis*. The structure is built up from an antiparallel β hairpin motif (residues 1 to 10) followed by an α helix (residues 12 to 24). The four zinc ligands Cys 3, Cys 6, His 19, and His 23 anchor one end of the helix to one end of the β sheet. Models, quite similar to the observed structure, were predicted from amino acid sequences of members of this zinc finger family. (Adapted from M.S. Lee et al., *Science* 245: 635–637, 1989.)

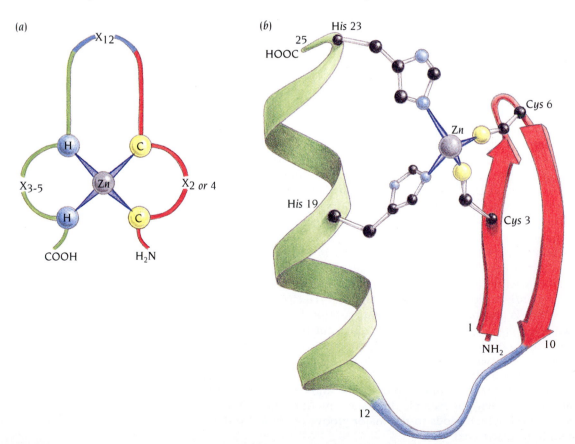

TFIIIA has nine zinc fingers, but other proteins have anything from one to more than 60 zinc finger motifs within their sequence.

The classic zinc finger is a motif that is repeated in tandem to recognize DNA sequences of different lengths, with each finger interacting with a small number of base pairs. Other zinc finger motifs bind as monomers to discrete sites or as dimers to palindromic DNA sequences, and frequently several binding sites in the DNA are separated by hundreds of base pairs in the transcriptional control region. The strength of the interaction can be varied by changes in the sequence of both the protein and the DNA and by varying the length of the spacing between the fingers. These changes allow a high level of specificity in recognition, and the modular design offers a large number of combinatorial possibilities for the specific recognition of DNA.

The classic zinc fingers bind to DNA in tandem along the major groove

Many attempts have been made to crystallize the transcription factor TFIIIA, but so far no crystals have been obtained that diffract to high resolution. However, the structures of a number of other individual classic zinc fingers have been determined. The first were obtained in the laboratories of Rachel Klevit, University of Washington, Seattle, and Peter Wright at the Scripps Clinic, La Jolla, from chemically synthesized peptides. Wright used a peptide corresponding to one zinc finger of a protein from *Xenopus laevis*, encoded by an embryonic gene called *Xfin* (*X* for *Xenopus*, *fin* for fingers). This protein has no less than 37 tandemly repeated classic zinc fingers.

The structure of the Xfin synthetic peptide confirms that a classic zinc finger is an independent folding-unit; it can be described as a "miniglobular protein" with a hydrophobic core and with polar side chains on the surface (Figure 10.1b). Residues 1 to 10 form an antiparallel hairpin motif with the zinc ligand Cys 3 in the first β strand and the second zinc ligand, Cys 6, in the tight turn between the β strands. The hairpin is followed by a helix, residues 12 to 24, of about three and a half turns. The remaining two zinc ligands, His 19 and His 23, are in the C-terminal half of this helix. The helix is distorted between these zinc ligands to form a so-called 3_{10} helix, in which hydrogen bonds occur between every third residue of the helix instead of every fourth residue as in a normal α helix. The zinc atom is buried in the interior of the protein and is necessary for the formation of a stable finger structure, since a single folded conformation only occurs in the presence of zinc ions in aqueous solution. Obviously the two ends of the molecule are held together by the binding of side chains to the zinc atom. However, as will be described in Chapter 17 it has been possible to design a peptide sequence that has a stable zinc finger fold without binding zinc.

It has also been possible to determine the x-ray structures of classic zinc finger motifs from several proteins bound to specific DNA fragments. We will here describe one such structure containing three zinc fingers from a mouse protein, Zif 268, which is expressed at an early developmental stage of the mouse. Nikola Pavletich and Carl Pabo at the Johns Hopkins University School of Medicine, Baltimore, determined the x-ray structure to 2.1 Å resolution of a recombinant polypeptide derived from Zif 268 bound to a 10-base

Figure 10.2 (a) Amino acid sequence of a fragment of the Zif 268 protein that contains three zinc fingers. Residues forming the β strands and α helices are red and green, respectively, and those involved in the turn between the last β strand and the α helix are blue. (b) The nucleotide sequence of the DNA fragment that was used in the x-ray structure determination of the Zif 268 fragment complexed with DNA.

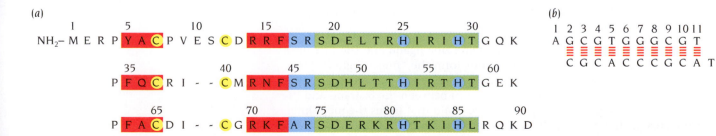

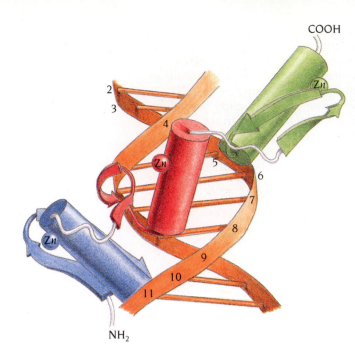

COOH

NH₂

Figure 10.3 Schematic diagram of the structure of three zinc fingers of Zif 268 bound to DNA. The three zinc fingers, which bind in tandem to the major groove of DNA, are colored blue, red and green from the N-terminus. The zinc fingers have the same structure and bind in a similar way with the N-terminus of the α helix pointing into the major groove. (Adapted from N.P. Pavletich et al., *Science* 261: 1701–1707, 1993.)

pair DNA fragment, with a sequence corresponding to the consensus sequence of Zif 268 binding sites (Figure 10.2b). The 90-residue Zif 268 polypeptide comprises three zinc finger motifs joined by linker regions each of four residues (Figure 10.2a). The structures of the three zinc finger motifs in the Zif 268–DNA complex are very similar to each other and are essentially the same as that of the Xfin motif shown in Figure 10.1b, while the DNA fragment in the complex is essentially undistorted B-DNA. Complex formation therefore does not appreciably change the structure of either the zinc finger motifs or the DNA.

The DNA fragment binds to the polypeptide with a dissociation constant of 6 nanomoles, indicating tight binding, and the overall structure of the complex reveals why tandemly repeated zinc fingers are such efficient motifs for protein–DNA recognition. The three zinc fingers are arranged in a semi-helical structure that follows the major groove of the B-DNA (Figure 10.3). The helix of each finger fits directly into the major groove, with residues from the N-terminal region of each helix in contact with base pairs in the major groove. Each of the three zinc fingers uses its helix in a similar fashion and each finger makes its sequence-specific contacts with a subsite of three base pairs. The three tandemly oriented zinc fingers therefore bind to nine consecutive base pairs in the major groove. Although the α helix fits into the major groove, it is tilted with respect to the groove such that the C-terminal region of the helix is pointing away from the bases and does not make contacts with DNA. The β sheet is on the other side of the helix, away from the base pairs, and is shifted toward one side of the major groove. The two strands of the β sheet have very different roles in the complex. The first does not make any contacts with the DNA whereas the second β strand contacts the sugar-phosphate backbone along one strand of the DNA.

The finger region of the classic zinc finger motif interacts with DNA

The 12 residues between the second cysteine zinc ligand and the first histidine ligand of the classic zinc finger motif form the "finger region". Structurally, this region comprises the second β strand, the N-terminal half of the helix and the two residues that form the turn between the β strand and the helix. This is the region of the polypeptide chain that forms the main interaction area with DNA and these interactions are both sequence specific,

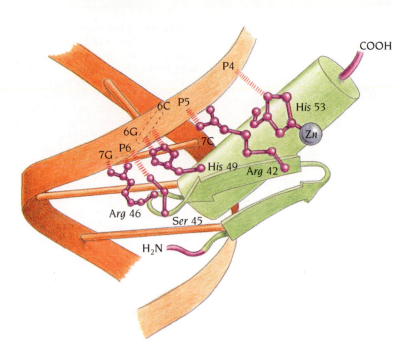

Figure 10.4 Detailed view of the binding of the second zinc finger of Zif 268 to DNA. Two side chains, Arg 46 and His 49, form sequence-specific interactions with DNA. There are also three nonspecific interactions between phosphate groups of the DNA and the side chains of Arg 42, Ser 45, and His 53.

between side chains of the protein and the bases of DNA, and nonspecific, mainly between the phosphate oxygen atoms of DNA and side chains of the protein.

A typical example of these interactions, for the second zinc finger of the Zif 268–DNA complex, is shown in Figure 10.4. Five side chains from the polypeptide bind to DNA, two of them in a sequence-specific manner. Arg 46, which is the last residue in the loop region and immediately precedes the helix, forms two hydrogen bonds with the guanine base of the G-C base pair 7. His 49, which is the third residue of the helix, forms hydrogen bonds with the guanine base of the G-C base pair 6. In addition to these sequence-specific hydrogen bonds, there are three nonspecific interactions between phosphate oxygen atoms of base pairs 4, 5, and 6 and the side chains of His 53, Arg 42, and Ser 45, respectively. The participation of His 53 in interactions with DNA is noteworthy, since this side chain is also one of the zinc ligands. The role of this side chain is therefore twofold: to stabilize the finger structure by providing a zinc ligand from one of the nitrogen atoms of the imidazole ring, and to participate in DNA binding by forming a hydrogen bond from the second nitrogen atom of the ring.

The same general picture of zinc finger binding to DNA observed in the Zif 268–DNA complex has also been found in two other DNA complexes formed by classic zinc fingers from the proteins GLI and TTK. GLI is the product of a human oncogene and TKK is a transcriptional control regulator of the *Drosophila* developmental gene *fushi-tarazu*. The GLI–DNA structure contains five zinc finger motifs, of which fingers 4 and 5 are specifically bound to DNA. The TTK–DNA structure contains two zinc fingers, both of which are specifically bound to DNA. The arrangement of these fingers in the major groove of DNA is very similar to the arrangement of two consecutive zinc fingers in the Zif 268–DNA complex. In all cases, the specific DNA contacts are made from the amino terminal half of the helix. This helix is oriented around its axis in a similar way in all zinc fingers, such that the side chain immediately preceding the first His ligand, residue 6 of the helix, is facing the edge of the bases in DNA. Consequently helix residues 2 and 3 are also facing the bases, as is the last residue of the loop region. Due to the geometry of the α helix, these four residues are all on the same side of the helix. Almost all sequence-specific contacts observed in these structures involve different combinations of these four residues (Figure 10.5). The few exceptions involve the remaining three residues of this N-terminal region of the helix.

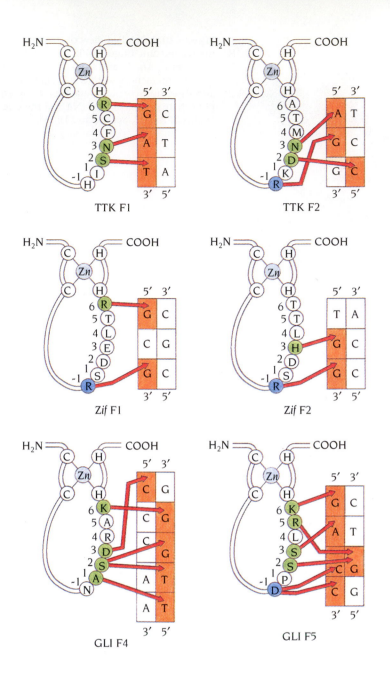

Figure 10.5 Comparison of the sequence-specific binding to DNA of six different zinc fingers. Residues in the N-terminus of the α helix in the finger regions are numbered 1 to 6. The residue immediately preceding the α helix is numbered -1. Amino acid residues and nucleotides that make sequence-specific contacts are colored. In spite of the structural similarities between the zinc fingers and their overall mode of binding, there is no simple rule that governs which bases the fingers contact.

The comparison of sequence-specific contacts shown in Figure 10.5 reveals that there is no specific correspondence between an amino acid residue in one of the four critical positions of the α helix and the base that it contacts. Although contiguous zinc fingers contact contiguous sets of base pairs, there is no simple rule that governs which bases the fingers contact. Furthermore, the number of bases in each set varies from three to five, and sometimes the sets of bases are strictly contiguous while in other cases they overlap or are spaced by one base pair that is not in contact with any finger. In addition, contacts are made in a seemingly random fashion to bases of both DNA strands. It is therefore not possible at present to predict from the amino acid sequence of a zinc finger the DNA sequence to which it will bind, or vice versa. However, one sequence-specific interaction occurs more

Figure 10.6 One sequence-specific interaction occurs more frequently than others in protein–DNA complexes: two hydrogen bonds form between an arginine side chain of the protein and a guanine base of the DNA, as shown in this diagram.

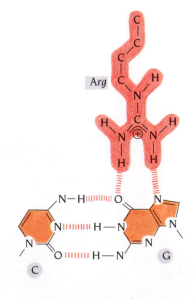

180

frequently than the others: that between an arginine side chain and a guanine base (Figure 10.6). The two strong hydrogen bonds that can be formed between these two groups make an essential contribution to sequence-specific recognition, as in the other families of DNA-binding proteins discussed in Chapters 8 and 9. The nonspecific interactions anchor the zinc fingers into the major groove of the DNA and are important for the tandem arrangement of the fingers along the groove. The interactions occur mainly from the linker regions between the zinc finger motifs and from side chains of the second β strand, which is aligned along one of the backbone chains of the DNA. Furthermore, the hydrogen bond from the histidine zinc ligand, exemplified above in the zinc finger 2 of the Zif 268–DNA complex, is found in most zinc finger–DNA interactions.

In conclusion, the structures of the classic zinc finger–DNA complexes reveal a simple and efficient mechanism for recognizing specific sites on double-stranded DNA. The α helix of each zinc finger fits directly into the major groove of B-DNA and side chains from the N-terminal region of this helix contact the edges of the base pairs. Most sequence-specific interactions involve side chains from the same four positions of the helix and arginine-guanine contacts occur frequently. The zinc fingers are tandemly arranged in the major groove and successive zinc fingers bind to successive subsites of three to five base pairs along the DNA.

Two zinc-containing motifs in the glucocorticoid receptor form one DNA-binding domain

The glucocorticoid receptor is a member of a family of transcription factors that includes the thyroid hormone receptor, the retinoic acid receptors, the vitamin D3 receptor and different steroid hormone receptors (Figure 10.7). All members of this family have a highly conserved DNA-binding domain containing zinc that consists of about 70 residues and that binds to activating elements of DNA called hormone-response elements. In addition, they all contain a variable C-terminal ligand-binding domain, and some of them also have a large N-terminal domain, involved in transcriptional activation.

Protein fragments, produced from recombinant DNA and containing the glucocorticoid receptor DNA-binding domain, exhibit sequence-specific DNA binding to glucocorticoid response elements. These protein fragments contain two zinc atoms that are required for proper folding and DNA binding. The three-dimensional solution structures of several such receptor fragments have been determined by NMR methods. The first, from the glucocorticoid receptor and from the estrogen receptor, were determined in the laboratories of Robert Kaptein, University of Utrecht, and of Aaron Klug, University of Cambridge, UK, respectively.

The amino acid sequence of the glucocorticoid fragment indicated that there are two zinc-containing regions, each binding a zinc atom through four cysteine residues instead of two Cys and two His residues (Figure 10.8a). In the three-dimensional structure these two zinc-binding motifs (red and green

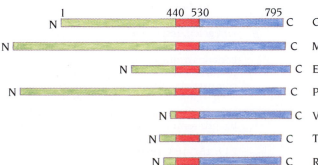

Figure 10.7 Evolutionarily related members of the receptor family of transcription factors that have four cysteine residues bound to zinc in each of two regions of the sequence. The DNA-binding domains (red) have highly homologous amino acid sequences, whereas the ligand-binding domains (blue) are more variable. Residue numbers of the domain boundaries are given for the glucocorticoid receptor. Exchange of individual domains between different members of the family suggests that they can function as interchangeable modules.

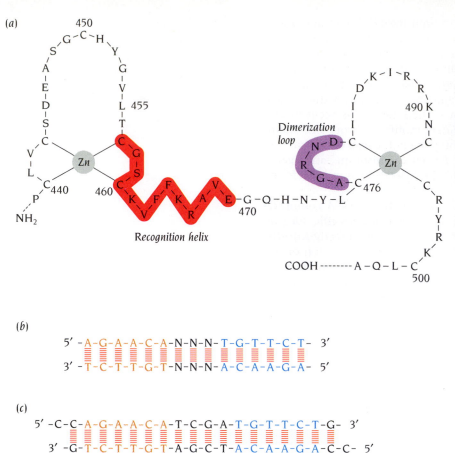

(a)

Recognition helix

Dimerization loop

Figure 10.8 (a) Amino acid sequence of the zinc-containing DNA-binding domain of the glucocorticoid receptor. There are two zinc atoms within this domain, each bound to four cysteine residues. One of these stabilizes the recognition helix (red) that provides sequence-specific binding to DNA and the other zinc region contains a loop (purple) that is involved in formation of the dimeric molecule. (b) Nucleotide sequence of the region of DNA that binds to the glucocorticoid receptor, the glucocorticoid response element, GRE. The two palindromic half-sites (blue and orange) are separated by a spacer region of three base pairs. (c) The DNA fragment that was used in the crystal structure determination of the complex with the DNA-binding domain of the glucocorticoid receptor.

(b)

5′ -A-G-A-A-C-A-N-N-N-T-G-T-T-C-T- 3′
3′ -T-C-T-T-G-T-N-N-N-A-C-A-A-G-A- 5′

(c)

5′ -C-C-A-G-A-A-C-A-T-C-G-A-T-G-T-T-C-T-G- 3′
3′ -G-T-C-T-T-G-T-A-G-C-T-A-C-A-A-G-A-C-C- 5′

in Figure 10.9) are not separated into discrete units but are interwoven into a single globular domain with extensive interactions between them. In each of the two zinc motifs the second pair of cysteine zinc ligands initiates an amphipathic α helix. The hydrophobic sides of these two α helices pack against each other to form a compact core with a hydrophobic interior. Hydrophilic side chains on the other side of the first α helix are exposed to solvent and form the DNA interaction area. The two zinc atoms and the

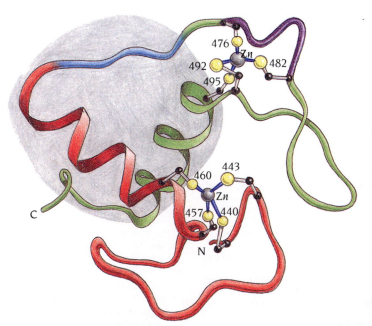

Figure 10.9 Schematic diagram of the three-dimensional structure of the DNA-binding domain of the glucocorticoid receptor. The two zinc-binding regions, defined from the amino acid sequence (see Figure 10.8a) are colored red and green, respectively, and the region that joins them is blue. Each has one α helix and contains a zinc atom bound to four cysteine residues. The two α helices are part of a compact globular core, shown in gray. The dimerization loop is shown in purple. The zinc atoms and the residues between the zinc ligands form protrusions from this globular core. (Adapted from T. Härd et al., *Science* 249: 157–160, 1990.)

protein regions between the zinc ligands form protrusions from this globular core. A comparison of a classic zinc finger (see Figure 10.1b) with the zinc-binding motifs in the glucocorticoid receptor (see Figure 10.9) shows that their three-dimensional structures are quite different. This difference is reflected in their different modes of binding to DNA.

A dimer of the glucocorticoid receptor binds to DNA

The crystal structure of the DNA-binding domain of the glucocorticoid receptor bound to a cognate DNA fragment was determined to 2.9 Å resolution in 1991 by the group of Paul Sigler at Yale University. The regions of DNA that bind to this receptor, the glucocorticoid response elements, GRE, comprise two identical palindromic half-sites, each of six base pairs with the sequence shown in Figure 10.8b. These half-sites are separated by a spacer region of three base pairs whose sequence is essentially unimportant but whose length is crucial for proper binding of the receptor to the two half-sites.

Solution studies have shown that the complete receptor molecule is a dimer that is stabilized by the ligand-binding C-terminal domain. The DNA-binding domains alone in solution are monomers but they dimerize upon binding to DNA. Sigler's group found by serendipity that this dimer interaction is very specific and crucial for binding to response elements with the proper length of the spacer region. They used a DNA fragment (see Figure 10.8c) that contained a spacer of four instead of three base pairs because they wanted an exactly symmetrical DNA fragment. Nevertheless, the monomeric domains dimerized strongly, with the domains arranged such that they could interact correctly only with DNA half-sites separated by the normal three base pairs (Figure 10.10). In the crystal, using a spacer of four base pairs, only one domain interacts properly with one half-site of the DNA. The extra base pair in the spacer displaces the second half-site by one base pair, and as a result the second domain is presented with a noncognate sequence with which it forms a "nonspecific" interface. When presented with a DNA target with the correct three-base pair spacer, both DNA-binding domains bound symmetrically in a specific manner. Aaron Klug's group has since shown that both domains of the estrogen receptor also bind to their cognate DNA half-sites when the spacer has only three base pairs.

The two domains in the dimer bound to DNA are very similar (see Figure 10.10). There are, however, significant differences between the structure of the monomer in solution and that of the subunits in the dimer. Although their overall folds are similar, binding to DNA induces substantial conformational changes especially in the dimer interface area. Most of the interdomain contacts are made by residues in the loop between the first two cysteine zinc ligands of the second zinc region, Cys 476 and Cys 482 (see Figure 10.9). This

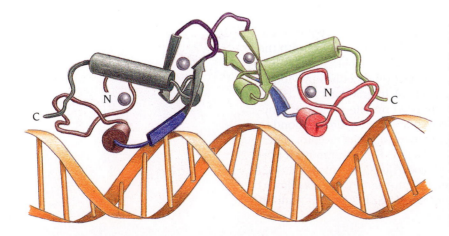

Figure 10.10 Structure of the complex between the dimeric glucocorticoid receptor molecule and a DNA fragment (orange). The two zinc-binding regions of each subunit have different colors; brown and dark green in one subunit and red and light green in the second. The linker region is blue. The recognition helices of the dimer (red and brown) are positioned in the major groove. The distance between them, which corresponds to one turn of the DNA helix, is fixed by the dimerization loop (purple). This region undergoes a significant conformational change when the dimer binds to DNA. (Adapted from B.F. Luisi et al., *Nature* 352: 497–505, 1991.)

loop, called the "D (for dimerization) Box" is not well defined in the monomer structure but is a fixed, well defined β turn in the dimer (see Figure 10.10). Furthermore, the structure around the second zinc is not well defined in the free monomer, but becomes a well defined β-strand-loop-helix in the DNA-bound dimer that forms the intersubunit interface of the dimer. In short, DNA acts as an allosteric effector of its own recognition by supporting tertiary structural changes that lead to a new quaternary structure uniquely suited to bind cooperatively to the correct DNA target.

An α helix in the first zinc motif provides the specific protein–DNA interactions

The two DNA-binding domains in the dimer bind to one face of the DNA, with the interactive surfaces of each subunit in successive major grooves (see Figure 10.10). The amphipatic α helix in the first zinc region, residues 457 to 469 in the glucocorticoid receptor, forms sequence-specific interactions with the edge of the bases in the major groove, and is therefore called the recognition helix. There are no protein contacts to the bases in the intervening minor groove. The two recognition helices of the dimer have the correct orientation and distance between them to fit into two successive major grooves on one side of the DNA double helix. This mode of DNA contact, which is very sensitive to the number of nucleotides separating successive binding sites, bears a striking resemblance to the binding of Cro and repressor to procaryotic DNA (compare Figures 8.14 and 10.10) but is quite different from the way that two successive classic zinc fingers bind (see Figure 10.3).

The recognition helix is positioned in the major groove by a number of nonspecific contacts between phosphate groups and protein side chains, as in many other protein–DNA interactions. These contacts are made mainly by residues from the two loop regions between the second and third cysteine zinc ligands in both zinc motifs (see Figure 10.8a). One of the nitrogen atoms of His 451 from the first zinc motif in the glucocorticoid receptor forms a hydrogen bond to one of the phosphates of base pair T-A 8. The orientation of the His 451 side chain is defined by a hydrogen bond network at the tip of this loop involving Ser 448. The His 451 residue is conserved in this receptor family, and Ser or Thr always occur at position 448, indicating that this network of hydrogen bonds is of general importance for hormone receptor recognition of response elements. A further important nonspecific interaction is provided by Arg 489 in the second zinc region, which also hydrogen bonds to a phosphate in the DNA backbone. Arg 489 is another conserved residue in this receptor family and its functional importance is shown by a spontaneous mutation that occurs in humans: Arg 489 is replaced with Gln in the vitamin D receptor of patients with hereditary vitamin D-resistant rickets.

Three residues in the recognition helix provide the sequence-specific interactions with DNA

Three residues in the NH$_2$ terminal half of the glucocorticoid receptor recognition helix—Lys 461, Val 462, and Arg 466—make specific contacts with the edges of the bases in the major groove (Figure 10.11). Arg 466 makes two hydrogen bonds to G4, similar to the Arg–G interaction described previously as a frequent feature of the classic zinc finger–DNA interactions (see Figure 10.6). If Arg is replaced with Lys or Gly, the mutated protein no longer functions. Lys 461 provides one direct and one water-mediated hydrogen bond to G7'. Both Lys 461 and Arg 466 are absolutely conserved in the receptor superfamily, and both their targeted base pairs G-C 4 and C-G 7 occur in most hormone-response elements. Finally, in the glucocorticoid receptor Val 462, which is not conserved within the hormone receptor family, makes a favorable van der Waals contact with the methyl group of T5. Thymine at position

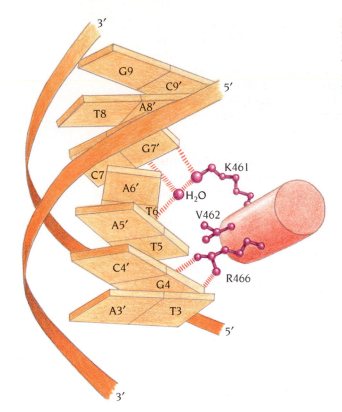

Figure 10.11 Sequence-specific interactions between DNA (orange) and the recognition helix (red) of the glucocorticoid receptor. Three residues, Lys 461, Val 462 and Arg 466 make specific contacts with the edges of the bases in the major groove.

5 is a unique and consistent feature of all glucocorticoid response elements but is not conserved in the response elements of other members of the hormone receptor family.

The two zinc ions fulfill important but different functions in the DNA-binding domains. The first zinc ion is important for DNA-binding because it properly positions the recognition helix; the last two cysteine zinc ligands are part of this helix. The second zinc ion is important for dimerization since the five-residue loop between the first two cysteine zinc ligands is the main component of the dimer interaction area.

The retinoid X receptor forms heterodimers that recognize tandem repeats with variable spacings

Generally, the family of nuclear receptors have low target affinity, high amino acid sequence homology and similar DNA response elements. How then are they targeted to their appropriate genes? The steroid hormone receptors bind exclusively as homodimers to response elements where the half-sites are organized in a palindromic orientation. Other nuclear receptors, such as those for vitamin D (VDR), thyroid hormone (TR) or *trans*-retinoic acid (RAR), form heterodimers with the *cis*-retinoic acid receptor (RXR) that bind to response elements with half-sites organized as direct repeats. Different heterodimers recognize response elements with different lengths of the spacer region. Paul Sigler's group has determined the x-ray structure of the heterodimer of the DNA-binding domains of RXR and TR bound to DNA, and has suggested a mechanism by which different heterodimers involving the common partner RXR can recognize response elements of identical target sequences but with different lengths of the spacer regions.

The individual domains of the two receptors both have structures similar to that of the glucocorticoid receptor, and they bind to DNA in a similar way, with their recognition helices in the major groove. The dimer contacts are, however, totally different. In the glucocorticoid receptor, which binds to a palindromic DNA sequence like the 434 repressor described in Chapter 8, the domains interact symmetrically in a head-to-head fashion: equivalent residues in the two domains interact across a twofold symmetry axis. By contrast, in the RXR–TR heterodimer, which binds to direct DNA repeats like the classic zinc finger transcription factors, the domains bind to each other in a head-to-tail fashion: residues in one domain interact with totally different residues in the other domain. As a consequence the dimer is polar, as is the response element, and in the complex the RXR domain binds exclusively to the 5′ half-site of the response element even though the two direct repeats have the same nucleotide sequence.

The two domains are oriented in the dimer in such a way that each recognition helix binds to its half-site in the major groove of B-DNA with a spacer of four base pairs between the half-sites. The dimer interface region straddles the minor groove of the spacer region. Residues from the second zinc motif of RXR interact with residues in the first zinc motif of TR. These residues are also involved in nonsequence-specific interactions with the DNA backbone. Thus the DNA contacts support the dimer interface, which in turn supports both the specific and nonspecific contacts with DNA. These mutually supporting interactions between the protein domains and the DNA backbone reinforce a lesson learned from the glucocorticoid receptor–DNA complex concerning the importance of spacing beween the half-sites: the DNA-binding domains of the nuclear receptors, which are monomers in the absence of DNA and which generally bind weakly as monomers to a single half-site, bind with high affinity and strong cooperativity to response elements in which the appropriate half-sites are spaced correctly.

RXR forms heterodimers with VDR, TR and RAR that bind to response elements that differ only in their inter half-site spacing (Figure 10.12), of three, four and five base pairs, respectively. The DNA-binding domains of these four nuclear receptors have highly similar amino acid sequences and hence similar three-dimensional structures. How can RXR form dimers with these similar partners such that their two recognition helices can bind to the corresponding two specific B-DNA target sequences in the major groove whose relative positions are different both in orientation and length of separation due to the different lengths of the spacer between them? Using the structure of the RXR–TR–DNA complex, Sigler's group has tried to answer this question by modeling possible structures for the other two dimers, assuming that RXR and its binding mode is the same in these complexes. They found that by rotating the second RXR domain about 35° together with a translation of 3.4 Å, plausible new interface regions occur that allow dimers to form with the appropriate DNA-binding properties for the different complexes. It is significant that the amino acid residues that participate in these interface regions occupy nonconserved positions within the receptor family and are relatively specific to a particular receptor, thereby ensuring a unique mode of dimerization on the appropriate response element.

In summary, a DNA-supported asymmetric interface located within the DNA-binding domains of these nuclear receptors provides the molecular basis for receptor heterodimers to distinguish between closely related response elements. RXR can provide a repertoire of different dimerization surfaces, each one unique for a specific partner, allowing dimers to form that are adapted to the length of the spacer region in their corresponding response elements.

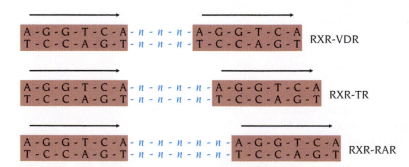

Figure 10.12 Response elements for heterodimers of the nuclear receptor for *cis*-retinoic acid (RXR) with the receptors for vitamin D (VDR), thyroid hormone (TR) and *trans*-retinoic acid (RAR). The half-sites of these response elements have identical nucleotide sequences and are organized as direct repeats. They differ in the number of base pairs in the spacer region between the half-sites. This difference forms the basis for the ability of the heterodimers to discriminate between the different response elements.

Yeast transcription factor GAL4 contains a binuclear zinc cluster in its DNA-binding domain

A third subfamily of zinc-containing motifs has been found in several transcription factors from yeast and fungi. The most thoroughly studied member of this family, GAL4, activates transcription of genes required for the breakdown of the sugars galactose and melibiose. GAL4 binds as a dimer to DNA recognition sequences that are 17 base pairs in length, and some of the genes that are activated by GAL4 have four of these upstream-activating sequences (UAS).

The GAL4 protein is large, having 881 amino acids, and different functions have been assigned to several regions of the polypeptide chain. In addition to the DNA-binding region, residues 1–49, and three acidic activating regions, residues 94–106, 148–196 and 768–881, there is a dimerization region formed by residues 50–94. The structure of a GAL4 fragment comprising residues 1–65 complexed with a DNA sequence of 19 base pairs has been determined by the group of Stephen Harrison, Harvard University. This protein fragment, which binds as a dimer to DNA, can be divided into three structurally and functionally distinct regions: a zinc cluster domain at the N-terminus, a dimerization region at the C-terminus and a 9-residue linker region in between (Figure 10.13).

Residues 50–64 of the GAL4 fragment fold into an amphipathic α helix and the dimer interface is formed by the packing of these helices into a coiled coil, like those found in fibrous proteins (Chapters 3 and 14) and also in the leucine zipper families of transcription factors to be described later. The fragment of GAL4 comprising only residues 1–65 does not dimerize in the absence of DNA, but the intact GAL4 molecule does, because in the complete molecule residues between 65 and 100 also contribute to dimer interactions.

In contrast to the zinc-containing DNA-binding domains described so far, the GAL4 family contains a cluster of two zinc atoms liganded to six cysteine residues, and two of these cysteines (residues 11 and 28) are bound to

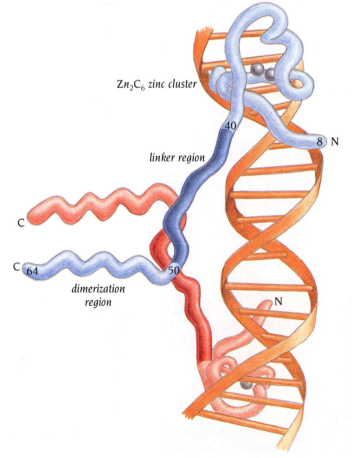

Figure 10.13 Schematic diagram of the dimeric GAL4 transcription factor bound to DNA. The structure of each subunit is divided into three distinct regions: a dimerization region joined by a linker region to a DNA-binding Zn_2Cys_6 zinc cluster region. The two dimerization α helices are packed into a coiled coil like a leucine zipper, and the length of the linker regions determines the distance between the DNA-binding sites. (Adapted from R. Marmorstein et al., *Nature* 356: 408–414, 1992.)

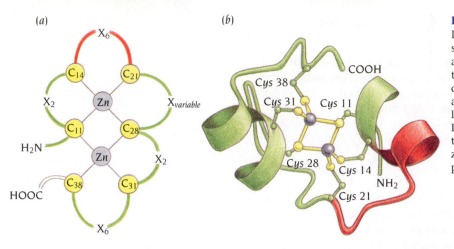

(a)

(b)

Figure 10.14 Schematic diagrams of the DNA-binding zinc cluster region of the GAL4 subunit. (a) The zinc cluster contains two zinc atoms each bound to four cysteine residues, two of which bridge the zinc atoms. The diagram illustrates the number of amino acids in the loop regions between the cysteine ligands. (b) Richardson-type diagram of the DNA-binding region. The red region provides the sequence-specific DNA interactions. The zinc cluster stabilizes the structure to give the proper fold for DNA binding.

both zinc atoms, forming two bridges between them (Figure 10.14). The amino terminal polypeptide chain of GAL4 wraps around the zinc cluster and contains two short α helices, one of which provides the specific interactions with DNA (red in Figure 10.14b). It is clear from the structure that the function of the zinc cluster is to ensure the proper fold for DNA-binding by this region of the polypeptide chain.

The zinc cluster regions of GAL4 bind at the two ends of the enhancer element

The 19-base pair DNA fragment used in the structure determination contains a GAL4 upstream-activating sequence (UAS) of 17 base pairs that is semi-palindromic, with a highly conserved CCG sequence at either end, reading outwards 5′ to 3′ from the center of the sequence. These two conserved, three-base pair recognition modules are always separated by 11 base pairs in UAS that bind GAL4. In the UAS of each member of this subfamily of enhancer elements the length of the spacer between the two CCG sequences is constant and characteristic, but other subfamilies have spacers of different lengths.

The zinc cluster modules of the GAL4 dimer bind in the major groove of DNA, in contact with the CCG base pairs at the extreme ends of the consensus region (see Figure 10.13). Since the DNA in the complex is essentially B-DNA, these binding sites are separated by more than one full turn of DNA. The two zinc cluster modules bind in an almost identical manner to the two sites, and they are anchored to the sugar-phosphate backbone by hydrogen bonds to phosphates 5 and 6 (Figure 10.15). These interactions position the module so that the C-terminus of the first α helix (red in Figure 10.14b) points into the groove. At the C-terminus of an α helix the C=O groups of the main chain are not involved in helical hydrogen bonds (see Figure 2.2) but point away from the helix. Consequently in the GAL4–DNA complex these main-chain carbonyl groups are directed towards the edge of the bases in the major groove. The carbonyl group of residue 17 accepts a hydrogen bond from cytosine 8′ and the carbonyl of residue 18 accepts hydrogen bonds from both cytosines 7 and 8′. In addition, the side chain of Lys 18 donates hydrogen bonds to both guanines 6′ and 7′. This pattern of contacts uniquely specifies the CCG sequence at base pairs 6 to 8 (see Figure 10.15). There are no other direct interactions between the protein and DNA base pairs.

The GAL4 recognition module therefore contains only one protein side chain, Lys 18, that provides specific interactions with the DNA. The remaining specific interactions with DNA are from main-chain atoms and depend critically on the correct conformation of the protein. The correct positioning of the C-terminus of the α helix is particularly important for recognition. This is to date the only example of a protein–DNA interaction in which

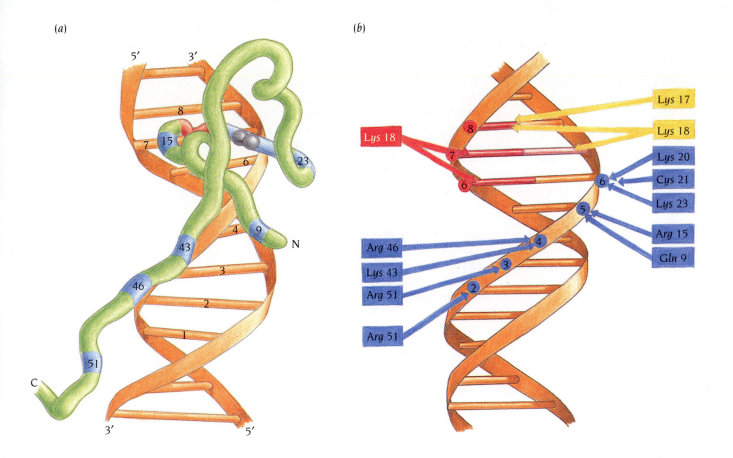

(a)

(b)

the main-chain carbonyl groups at the C-terminus of an α helix participate in specific DNA contacts. Since Lys 18 is conserved in most members of this family of transcription factors, it is quite clear that all their recognition modules bind in the same way to DNA, and that discrimination between different upstream-activating sequences depends on factors other than the zinc-containing modules.

The linker region also contributes to DNA binding

The linker region connecting the recognition zinc module with the dimerization region of the GAL4 fragment is an extended chain (see Figure 10.13). There are several nonspecific contacts from this linker region and the beginning of the dimerization region to phosphate groups of the DNA (see Figure 10.15). The path of each linker follows one strand of the DNA backbone from the zinc module towards the minor groove, thus positioning the dimerization domain over the minor groove. In addition, the structure formed by residues 47–51 of both subunits is sterically complementary to the phosphate backbones of the minor groove, and this contributes to the binding affinity. It is obvious from the structure of this GAL4 fragment that the linker regions define the distance between the two sites at which the recognition modules of the dimer bind to the DNA. A large region of the DNA major groove, between the binding sites at the ends of the 17-base pair DNA fragment, is not covered by the protein and is accessible to solvent (see Figure 10.13).

In conclusion, recognition of DNA enhancer elements by a GAL4 dimer involves three components: (1) a specificity for the conserved CCG triplets, which is provided by direct contacts between the major groove bases and the C-terminus of an α helix in the zinc-containing recognition modules; (2) a requirement for symmetrical sites, provided by a coiled-coil dimerization region that lies over the minor groove; and (3) a preference for a spacer between the CCG triplets of 11 base pairs, which is imposed by the structure of the linker region.

Figure 10.15 Schematic diagrams of the binding of one subunit of GAL4 to DNA. (a) Shows the structure of the DNA-binding domain in complex with the DNA; and (b) shows the interactions between amino acid residues and the DNA. The zinc cluster region binds in the major groove of DNA and is anchored to the sugar-phosphate backbone by nonsequence-specific interactions (blue). The C-terminus of the first α helix points into the major groove, and two main-chain C=O groups from residues 17 and 18 (yellow) form hydrogen bonds to the edge of the bases. There is only one sequence-specific interaction with an amino acid side chain, which is provided by Lys 18 (red). The linker region is an extended chain that follows one strand of the DNA and provides several nonspecific contacts (blue) to the DNA. The numbering of the base pairs starts from the center of the DNA fragment. (Adapted from R. Marmorstein et al., *Nature* 356: 408–414, 1992.)

189

DNA-binding site specificity among the C₆-zinc cluster family of transcription factors is achieved by the linker regions

At least 11 other fungal DNA-binding proteins are known to contain repeated Cys sequences like those found in the GAL4 recognition module. Residues 15–20 of GAL4, which are in closest proximity to DNA in the complex, are highly conserved in these homologous proteins. Lys 18, which makes the only direct sequence-specific contact with DNA, is conserved in all but three cases. In addition, these proteins all recognize CCG triplets in their upstream-activating sequences. Presumably, therefore, all the individual zinc-containing DNA-binding motifs of these proteins bind to CCG triplets in the same way. In which case how do the proteins recognize their proper upstream-activating sequences?

The group of Mark Ptashne at Harvard University, has tried to answer this question by domain-swapping experiments between GAL4 and the related yeast transcription factor, PPR1. Both Lys 18 and Lys 19 in GAL4 are conserved in PPR1. Both proteins recognize a DNA site containing two palindromic CCG triplets, but the triplets are separated by 6 base pairs in the enhancer elements controlled by PPR1 instead of the 11 in those regulated by GAL4. The PPR1 protein binds *in vitro* only to DNA fragments with a spacer of 6 base pairs and GAL4 only to those with 11 base pairs.

The Ptashne group first constructed a chimeric protein comprising the DNA-binding zinc module of PPR1 (residues 6–40 in the GAL4 numbering system) and the linker and dimerization regions of GAL4 (residues 41–100) (Figure 10.16). In spite of having the DNA-binding module of PPR1, this protein had the binding specificity of GAL4; in other words, it bound much more strongly to DNA fragments in which the CCG triplets were separated by 11 base pairs than by 6 base pairs. To determine if the linker or the dimerization region is responsible for this specificity, they constructed a second chimeric protein containing the zinc module, the linker region and the first residues of the dimerization module (residues 6–57) from PPR1 and the main part of the dimerization module (residues 58–100) from GAL4. In contrast to the first chimera, this protein has a binding specificity for PPR1 binding sites that is 200 times stronger than for GAL4 sites. Similar results have been obtained using a third member of this family which has a spacer of 10 base pairs between the CCG triplets.

Subsequently Stephen Harrison's group determined the x-ray structure of a PPR1–DNA complex and showed that the zinc cluster domain of PPR1 and its mode of binding to DNA was very similar to that of GAL4, and that PPR1 also dimerized through a coiled-coil region. However, the linker region

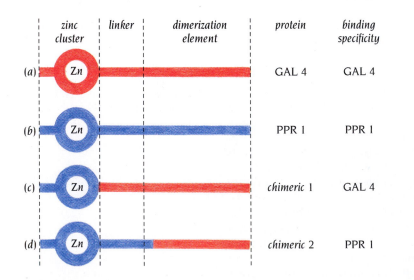

Figure 10.16 Domain-swapping experiments with GAL4 (a) and a related transcription factor PPR1 (b) have shown that the specificity of DNA binding depends on the size of the linker region. A chimeric protein with the PPR1 zinc cluster combined with GAL4 linker and dimerization regions (c) has GAL4 DNA-binding specificity, whereas a chimeric protein with PPR1 zinc cluster and linker regions combined with a GAL4 dimerization region (d) has PPR1 specificity.

was very different, comprising a β hairpin instead of an extended chain. This gives a more compact asymmetric dimer, and since the DNA-binding zinc cluster regions are closer together they are adapted to bind a DNA region with a spacer of six base pairs between the two CCG recognition sequences.

These results prove that the zinc cluster domains of this family of transcription factors work interchangeably to recognize the conserved CCG triplets, while the region of about 20 amino acids comprising the linker and the beginning of the dimerization domain directs the protein to its proper upstream-activating sequences, with its appropriately sized spacer. Zinc and its ligands fix the protein chain in a conformation that allows specific DNA binding.

Families of zinc-containing transcription factors bind to DNA in several different ways

We have now seen examples of three different ways in which zinc-containing transcription factors recognize their DNA-binding sites, and how the specificity of the individual members within these three families is achieved. In all three cases the DNA recognition module consists of a small polypeptide chain wrapped around a zinc atom, such that an α helix is exposed to bind into the major groove of DNA. But this is achieved in three different ways with three different conformations of the polypeptide chains. The common role of zinc and its ligands is to fix firmly the conformation of the polypeptide chain, so that the crucial α helix is held in a position to make specific contacts to exposed bases of DNA, while the rest of the polypeptide chain anchors the recognition module to the major groove of DNA by nonspecific contacts to the DNA backbone.

The classic zinc fingers bind in tandem to successive recognition sites in the major groove of DNA. Variability within the family is achieved by varying the number of zinc fingers as well as the side chains of their recognition helices. Members of the nuclear receptor family bind to DNA response elements with very similar target sequences. Specificity is achieved in several different ways; the target sequences are organized as palindromes or direct repeats with linker regions of different lengths, and the DNA-binding domains can form homo- or heterodimers. Finally, the GAL4 family of transcription factors all recognize conserved CCG triplets at the ends of the enhancer elements in the same way, using interchangeable zinc cluster recognition modules. The specificity of recognition is achieved by varying the linker regions between the dimerization domains and the recognition modules, so that they are compatible with the different lengths of the spacer DNA between the CCG triplets in the different enhancer elements.

The world of zinc-containing DNA-binding proteins is by no means exhausted by these three subfamilies. Several other subfamilies are already known with different three-dimensional structures and different sequence patterns of cysteine and histidine residues that form the zinc ligands. Further subfamilies may well be discovered as the genomes of different species are sequenced; whether or not any fundamentally new principles for DNA–protein recognition will be discovered amongst these new subfamilies remains to be seen.

Leucine zippers provide dimerization interactions for some eucaryotic transcription factors

The leucine zipper motif (see Chapter 3) was first recognized in the amino acid sequences of a yeast transcription factor GCN4, the mammalian transcription factor C/EBP, and three oncogene products, Fos, Jun and Myc, which also act as transcription factors. When the sequences of these proteins are plotted on a helical wheel, a remarkable pattern of leucine residues

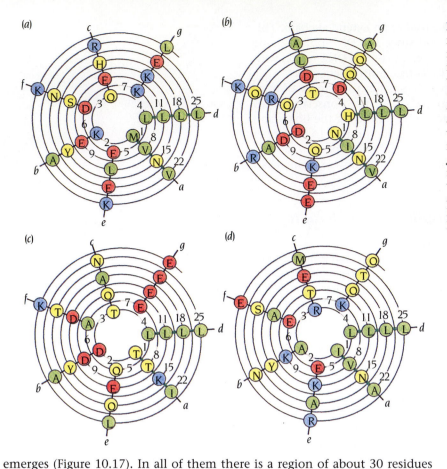

Figure 10.17 Amino acid sequences, represented as α helical wheels with 3.5 residues per turn, of a region of 28 residues from the DNA-binding domains of the transcription factors (a) GCN4, (b) Max, (c) Fos and (d) Jun. Positively charged residues are blue, negatively charged red; polar residues are yellow and hydrophobic, green. Residues with a spacing of seven residues are connected into spikes of the wheel which are labeled a to g. Side chains from each residue in a spike form a ridge on the outside of the α helix which is parallel to the helical axis. Spike d in all four sequences forms a heptad pattern of leucine residues that is characteristic of a leucine zipper coiled-coil α-helical structure.

emerges (Figure 10.17). In all of them there is a region of about 30 residues where the sequence can be arranged in modules of seven residues, and in almost all of these modules the fourth residue is leucine (position d in Figure 10.17), hence the name leucine zipper. In addition, the first residue of each module is frequently hydrophobic (position a in Figure 10.17). As was described in Chapter 3, the first structure determination of a synthetic peptide containing such modules showed that the peptide dimerizes and forms two parallel coiled-coil α helices with a helical repeat of 3.5 residues per turn, so that the interaction pattern of side chains between the helices repeats integrally every seven residues (see Figure 3.3) . The hydrophobic side chains in positions a and d of the heptad repeats form a hydrophobic core between the helices in this coiled-coil with the leucine residues facing each other (Figure 10.18a). This hydrophobic core is one major determinant of the stability of the dimer. The side chains that are immediately outside this core (positions e and g in Figure 10.17) are frequently charged and can either promote dimer formation by forming complementary charge interactions between the monomers, or prevent dimer formation by the repulsion of like charges (Figure 10.18b). Leucine zippers can form either homodimers or heterodimers in which the two monomers are either the same or different transcription factors, respectively.

The proto-oncogenic transcription factors Fos and Jun provide an illustrative example of differential dimer formation. These proteins are found in the transcription factor complex AP1 (active gene regulatory protein), which promotes transcription of many different genes including some that induce cell proliferation. The mammalian genome contains a family of related genes for Jun proteins, including c-Jun which is the cellular counterpart of the viral oncogene v-Jun. Members of this family can form both homodimers and heterodimers. In contrast, Fos cannot form homodimers and does not by itself bind to DNA. From the sequence of Fos in Figure 10.17c it is obvious that dimer formation is prevented by the strong charge repulsion of the five glutamic acid residues in the e and g positions with no compensating positive charge. However, Fos can form a heterodimer with Jun due to the

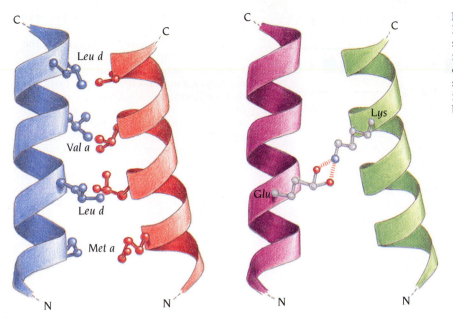

Figure 10.18 Side-chain interactions in the leucine zipper structure. (a) The hydrophobic side chains in spikes *a* and *d* (see Figure 10.17) form a hydrophobic core between the two coiled α helices. (b) Charged side chains in spikes *e* and *g* can promote dimer formation by forming complementary charge interactions between the two α helices.

complementary positive charges in the *e* and *g* positions of Jun (Figure 10.17d). An x-ray structure determination of the Fos-Jun heterodimer bound to DNA by the group of Stephen Harrison has confirmed the existence of such *g/e* salt bridges (see Figure 10.18b). The heterodimer binds to DNA with the same target specificity as the Jun homodimer and with an affinity to the AP1 binding site that is 10 times higher.

The ability of the leucine zipper proteins to form heterodimers greatly expands the repertoire of DNA-binding specificities that these proteins can display. As illustrated in Figure 10.19, for example, three distinct DNA-binding specificities could, in principle, be generated from two types of monomer, while six could be created from three types of monomer and so on. This is an example of combinatorial control, in which combinations of proteins, rather than individual proteins, control a cellular process. It is one of the most important mechanisms used by eucaryotic cells to control gene expression.

The GCN4 basic region leucine zipper binds DNA as a dimer of two uninterrupted α helices

The yeast protein GCN4 is a typical member of the basic region-leucine zipper (b/Zip) family of transcription factors, which has more than 50 known members from yeast, mammalian and plant cells. Monomeric GCN4, which is 281 amino acids long, binds specifically to the promoter regions of more than 30 genes that specify enzymes involved in amino acid biosynthesis, inducing their transcription in response to amino acid starvation. The dimerization and DNA-binding properties reside in a C-terminal region comprising about 55 amino acids. The amino acid sequence shows that this region of GCN4 is divided into a basic region of about 20 residues followed

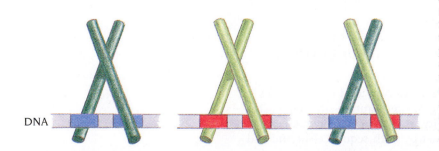

Figure 10.19 Heterodimerization of leucine zipper proteins can alter their DNA-binding specificity. Leucine zipper homodimers bind to symmetric DNA sequences, as shown in the left-hand and center drawings. These two proteins recognize different DNA sequences, as indicated by the red and blue regions in the DNA. The two different monomers can combine to form a heterodimer that recognizes a hybrid DNA sequence, composed of one red and one blue region. (Adapted from B. Alberts et al, *Molecular Biology of the Cell*, 3rd ed. New York: Garland Publishing, 1994.)

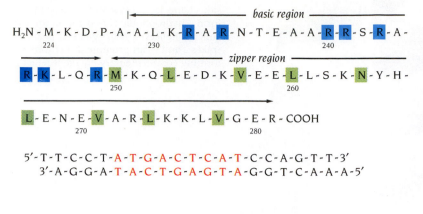

(a) Amino acid and DNA sequences used in the structure determination

$\leftarrow$ ——— basic region ———

H₂N - M - K - D - P - A - A - L - K - R - A - R - N - T - E - A - A - R - R - S - R - A -
224   230   240

———— zipper region ————

R - K - L - Q - R - M - K - Q - L - E - D - K - V - E - E - L - L - S - K - N - Y - H -
                    250   260

L - E - N - E - V - A - R - L - K - K - L - V - G - E - R - COOH
        270   280

5'- T - T - C - C - T - A - T - G - A - C - T - C - A - T - C - C - A - G - T - T -3'
3'- A - G - G - A - T - A - C - T - G - A - G - T - A - G - G - T - C - A - A - A -5'

(b) AP1-binding site

5'- A - T - G - A - C - T - C - A - T -3'
3'- T - A - C - T - G - A - G - T - A -5'

(c) Symmetric GCN4-binding site

5'- A - T - G - A - C - G - T - C - A - T -3'
3'- T - A - C - T - G - C - A - G - T - A -5'

Figure 10.20 (a) Amino acid sequence of the DNA-binding domain of the transcription factor GCN4 and nucleotide sequence of the DNA used for the x-ray structure determination of the complex. The positively charged residues arginine and lysine in the basic region that are involved in DNA binding are colored blue. The residues in the *a* and *d* positions that form the hydrophobic core of the leucine zipper region are green. The pseudosymmetric nucleotide sequence to which GCN4 binds is colored red. (b) The consensus nucleotide sequence of the *in vivo* AP1 pseudo-symmetric DNA recognition sites of GCN4. (c) A symmetric nucleotide sequence to which GCN4 can also bind.

by a C-terminal leucine zipper region (Figure 10.20a). The basic region contains eight charged residues, mainly arginines, which are involved in DNA binding and which are disordered when the protein is in solution in the absence of DNA. The transcriptional activation function is localized to a short acidic region of about 40 residues around residue 100. Surprisingly, large portions of the protein-coding sequence outside the DNA-binding and activation regions can be deleted from the gene without significantly affecting transcriptional activation.

The DNA recognition sites of GCN4 *in vivo* are similar to those for the Fos/Jun heterodimer, the AP1 site, the consensus sequence of which is shown in Figure 10.20b. This sequence is pseudo-symmetric, but GCN4 also binds to a perfectly symmetric sequence, which has one base pair inserted in the middle of the recognition site (Figure 10.20c). The dimeric GCN4 molecule is therefore able to bind to the two half-sites in both these DNA recognition sequences even though they have spacer regions of different lengths. The group of Stephen Harrison at Harvard University has determined the crystal structure of the GCN4 b/zip region (the amino acid sequence shown in Figure 10.20a) complexed with a DNA fragment of 20 base pairs containing the pseudo-palindromic nucleotide sequence shown in Figure 10.20b.

The structure is in principle very simple (Figure 10.21). Each monomer of the GCN4 fragment forms a smoothly curved, continous α helix (see Figure 10.21a). The leucine zipper region of the monomers pack into a coiled coil, essentially identical to the isolated leucine zipper (see Figure 3.3). The two α helices diverge from the dimer axis in a segment comprising the junction between the leucine zipper and the basic regions. This fork creates a smooth bend in each α helix which displaces the basic regions away from the dimer interface so that they can pass through the major groove of DNA, with one α helix on each side of the DNA (see Figure 10.21b and c). The DNA has essentially the B-DNA structure. Each basic region binds to one half-site with numerous contacts to the DNA, and the structure looks like α-helical forceps gripping the major groove of DNA.

GCN4 binds to DNA with both specific and nonspecific contacts

The consensus nucleotide sequence (see Figure 10.20b) used in the crystals is a symmetrized version of naturally occuring AP1 recognition sites, but GCN4 binds to this sequence with a high affinity. Each half-site in this DNA is bound to one monomer of the GCN4 dimer by both sequence-specific and

194

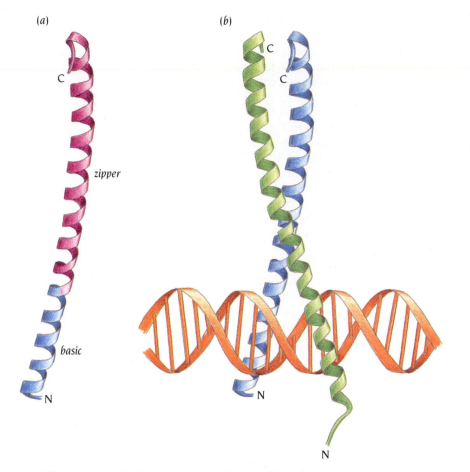

(a) C

zipper

basic

N

(b) C C

N

N

(c)

nonspecific contacts. In the same way in both half-sites, four amino acid side chains form sequence-specific contacts with bases (Figure 10.22). Asn 235, which is strictly conserved in all members of the b/zip family, is at the center of the interaction area. Its side chain forms two hydrogen bonds; the oxygen atom accepts a hydrogen bond from a nitrogen atom of the base C2, and the nitrogen atom of Asn 235 donates a hydrogen bond to an oxygen atom of T3. These contacts require that the α helix of the GCN4 basic region

Figure 10.21 The structure of a complex between the DNA-binding domain of GCN4 and a fragment of DNA. (a) Each monomer of the GCN4 domain forms a smoothly curved continuous α helix comprising both the basic and the leucine zipper regions. (b) The monomers are held together in a dimer in the zipper region. They diverge from each other in the basic regions, which are bound to DNA in the major groove on opposite sides of the B-DNA fragment (c), like helical forceps gripping the DNA. (Adapted from T.E. Ellenberger et al., *Cell*, 71: 1223–1237, 1992.)

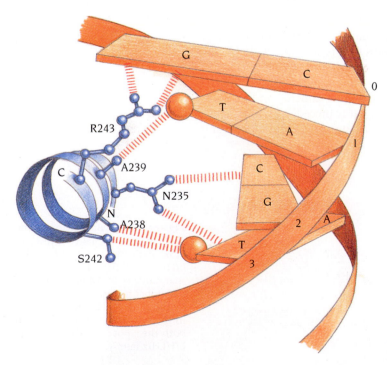

G

C

T

A

0

R243

A239

C

C

N235

G

N

A238

1

2

A

T

S242

3

Figure 10.22 Sequence-specific interactions between one of the α-helical basic regions of GCN4 (blue) and bases in the DNA fragment (orange). The methyl groups of the thymine bases are shown as spheres.

lies deep in the major groove, and they specify two of the four base pairs in each half-site. Asn 235 lies in a hydrophobic pocket formed by the methyl side chains of alanines 238 and 239, which form hydrophobic contacts with the methyl groups of T3 and T1, respectively. In addition, the methylene group of Ser 242 contacts the methyl group of T3. These hydrophobic contacts specify the thymines of base pairs 1 and 3 of each half-site.

The asymmetry of the AP1 site causes GCN4 to contact the central base pair in an asymmetric manner. Arg 243 in one monomer (shown in Figure 10.22) donates hydrogen bonds to the guanine of the central G-C base pair in the bidentate manner described earlier for zinc fingers (see Figure 10.6). Arg 243 in the second monomer can not of course bind in the same way. Instead it forms nonspecific hydrogen bonds to phosphate oxygen atoms in the central region of the site. Despite this local asymmetry in the center of the binding site, the positions of the two α helices in the major groove of the DNA and the side chain contacts to bases T1, C2, and T3 are the same in the two half-sites.

The position of the basic region of the protein on the DNA is also achieved by a number of basic and polar residues donating hydrogen bonds to phosphate oxygens of the DNA backbone. All the eight positively charged residues of the basic region are involved in these contacts, most of which are conserved between the half-sites, with the exception of those near the center of the binding site.

How does the GCN4 dimer bind to half-sites comprising T1, C2, and T3 when these half sites are separated by two nucleotides instead of one? The group of Tim Richmond at ETH in Zurich has answered this question by studying the crystal structure of GCN4 complexed with a DNA fragment containing the sequence shown in Figure 10.20c. They found that flexibility in both the DNA and the protein allows the GCN4 basic regions to interact with the major groove of each half-site in essentially equivalent manners in the two complexes. The basic region helices spread apart about 10 degrees, and the DNA bends towards the coiled-coil region and is inserted a bit further up into the fork. As a consequence, the actual specific protein–base pair contacts are all conserved. The combination of protein and DNA flexibility apparently allows GCN4 and other b/zip proteins to tolerate different spacings between half-sites.

The HLH motif is involved in homodimer and heterodimer associations

The coiled-coil structure of the leucine zipper motif is not the only way that homodimers and heterodimers of transcription factors are formed. As we saw in Chapter 3 when discussing the RNA-binding protein ROP, the formation of a four-helix bundle structure is also a way to achieve dimerization, and the helix-loop-helix (HLH) family of transcription factors dimerize in this manner. In these proteins, the helix-loop-helix region is preceded by a sequence of basic amino acids that provide the DNA-binding site (Figure 10.23), and

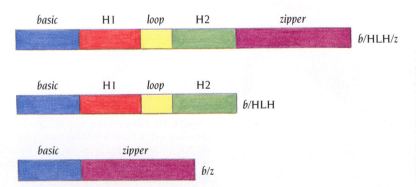

Figure 10.23 Domain arrangement along the polypeptide chains of three families of transcription factors; b/z , b/HLH and b/HLH/z. All three have a basic region (blue) that binds DNA. Dimerization is achieved by the zipper region (purple) in the b/z family, by the H1-loop-H2 region (red-yellow-green) in the b/HLH family and by a combination of both the zipper and the HLH regions in the b/HLH/z family.

therefore these proteins are called the b/HLH transcription factors, just as members of the leucine zipper family are called the b/zip factors. Members of the b/HLH family have substantial amino acid sequence identity and they bind to the consensus DNA sequence 5′-CANNTG-3′, where N is any nucleotide.

The myogenic proteins are an important class of b/HLH transcription factors that are crucially involved in the development of muscle cells. A mature muscle cell is distinguished from other cell types by a large number of characteristic proteins, including specific types of actin and myosin (see Chapter 14). The entire program of muscle differentiation can be triggered *in vitro* in cultured skin fibroblast cells by introducing any one of the myogenic proteins. Introduction of one myogenic protein activates the production of all the others, which in combination with already present but inactive HLH proteins then activate the muscle-specific genes. Thus, muscle differentiation is determined by specific combinations of these transcription factors, including heterodimers.

The crystal structure of the b/HLH domain of one myogenic protein, MyoD, complexed with a DNA fragment containing the consensus recognition sequence has been determined by the group of Carl Pabo at Massachusetts Institute of Technology, Boston. The DNA fragment, having two symmetric binding sites, binds to one dimer of the b/HLH domain. Each monomer of the protein has a very simple structure comprising two α helices joined by a loop region (Figure 10.24). The first α helix contains the basic region followed by the H1 region; the second helix, joined to the first by a loop, contains the H2 region of the HLH motif. The helical H1 and H2 regions of each monomer are involved in the dimer interface and participate in forming the four-helix bundle of the dimer (Figure 10.25). Conserved hydrophobic residues in both H1 and H2 form the hydrophobic core of the four-helix bundle. This core stabilizes the dimer and keeps the monomers together with the precise geometry that allows the two basic regions to interact properly with the two recognition sites of the DNA fragment, which are separated by two base pairs. These rather stringent stereochemical requirements impose constraints on the amino acid sequence of the HLH motif; this is reflected in the high degree of sequence homology that allows new proteins to be identified from their sequences as members of this family.

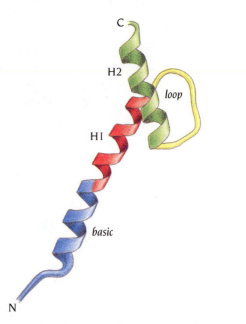
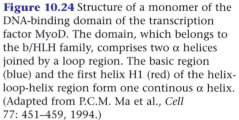

Figure 10.24 Structure of a monomer of the DNA-binding domain of the transcription factor MyoD. The domain, which belongs to the b/HLH family, comprises two α helices joined by a loop region. The basic region (blue) and the first helix H1 (red) of the helix-loop-helix region form one continous α helix. (Adapted from P.C.M. Ma et al., *Cell* 77: 451–459, 1994.)

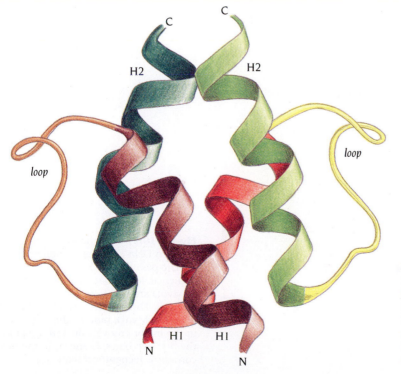

Figure 10.25 Structure of the dimerization region of MyoD. The α helices H1 (red and brown) and H2 (light and dark green) of the two monomers form a four-helix bundle that keeps the dimer together. The loops (yellow and orange) are on the outside of the four-helix bundle. (Adapted from P.C.M. Ma et al., *Cell* 77: 451–459, 1994.)

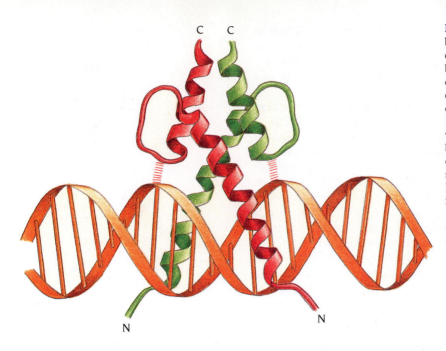

Figure 10.26 Schematic diagram of the binding of MyoD to DNA. MyoD binds as a dimer (red and green) with the N-terminal basic regions interacting with the major groove of DNA (orange). The monomers diverge from each other after the H1 region and bind on opposite sides of DNA, as for GCN4 (see Figure 10.21). The four-helix bundle, together with contacts with phosphates of the DNA backbone, rigidify the fork and allow MyoD only to bind to enhancer elements with a specific spacing between the half-sites. The DNA structure is essentially B-DNA. (Adapted from P.C.M. Ma et al., *Cell* 77: 451–459, 1994.)

The α-helical basic region of the b/HLH motif binds in the major groove of DNA

The helical basic regions of the MyoD dimer bind in the major groove of DNA in a very similar fashion to the basic regions of the leucine zipper protein GCN4 (Figure 10.26). The DNA structure is essentially that of B-DNA with a slightly narrower major groove and a wider minor groove. Three residues from each of the two α helices make sequence specific contacts to the edges of the bases in the major groove (Figure 10.27). These contacts are

(*a*)

5′ C A C G T G 3′
3′ G T G C A C 5′

(*b*)

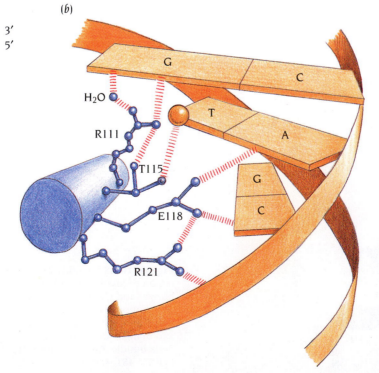

Figure 10.27 Sequence-specific contacts between DNA and one monomer of MyoD. (a) The palindromic MyoD recognition sequence. Bases that contact one monomer of MyoD are shown in red. (b) Three residues, Glu 118, Thr 115 and Arg 111 from the basic region of MyoD, bind to the edges of the bases in the major groove. Glu 118 recognizes the first two bases, C and A, in the MyoD consensus recognition sequence.

198

identical in the two half-sites. Glu 118 plays an especially important role in the recognition of the specific nucleotides

5'-CAC-3'
3'-GTG-5'

in each half site of the recognition site for the MyoD dimer (see Figure 10.27a). The side chain of Glu 118 accepts a hydrogen bond from the nitrogen atom of N4 of the first cytosine and donates a hydrogen bond to N6 of the following adenine. In addition, this side chain makes water-mediated contacts to two other bases further away in the recognition site. The side chain conformation of Glu 118 is stabilized by forming a salt bridge to Arg 121 which is located in the next turn of the α helix. Thr 115 and Arg 111, whose side chains also face the major groove, form hydrogen bonds with T and G in the complementary strand (see Figure 10.27). The methyl group of Thr 115 forms a hydrophobic contact with the methyl group of thymine and Arg 111 donates a hydrogen bond to a nitrogen atom of guanine. Thus four bases of the three-base pair recognition site form sequence-specific contacts with side chains of the helical basic region of MyoD (see Figure 10.27a).

In addition to these sequence-specific contacts there are a large number of nonspecific contacts between phosphate groups of the DNA and polar or positively charged groups on the protein, notably in the basic region (where five of the eight positively charged residues are involved in such contacts), but also in the HLH domain of MyoD. The half-site spacing is invariant in the b/HLH family, in contrast to the b/zip family. This is because the small four-helix bundle in b/HLH locks the basic regions in position and prevents them spreading apart, as occurs in GCN4, for example.

The b/HLH/zip family of transcription factors have both HLH and leucine zipper dimerization motifs

In Chapter 13 we discuss growth factors that activate phosphorylation cascades leading to changes in gene expression. Several of the genes that are first activated by growth factors, the early-response genes, code for transcription factors that in turn activate other genes, the delayed-response genes, which are more specific for the different cell types. The best-studied early-response genes are the *myc*, *fos* and *jun* oncogenes. Myc has a critical role in the normal control of cell proliferation. Cells in which Myc expression is specifically prevented will not divide even in the presence of growth factors. When Myc is overexpressed or hyperactivated by mutations, these cells will start uncontrolled proliferation leading to cancerous growth.

As we have seen, the Fos and Jun proteins belong to the b/zip family of transcription factors. The *myc* gene product also contains a leucine zipper motif but in addition has an HLH motif; it therefore belongs to the b/HLH/zip family of transcription factors (see Figure 10.23). Members of this family occur in diverse eucaryotes ranging from mammals to yeast and are important in regulating metabolism, cell differentiation and development. Some members of this family have no basic region and therefore cannot bind DNA. They can, however, form heterodimers with other members that do contain the basic region, preventing them from binding DNA and thereby acting as negative regulators of gene expression. The Myc protein does not normally form homodimers and therefore does not by itself bind DNA. However, Myc can form a heterodimer with a different member of the same family, Max, that binds to DNA under physiological conditions. The association of Max and Myc is required for malignant transformation by the *myc* oncogene. Max probably also plays an essential role in orchestrating the biological activities of many other b/HLH/zip transcription factors.

In contrast to Myc, Max can form homodimers that bind tightly to DNA. These homodimers recognize the same consensus sequence as members of the b/HLH family, 5'-CANNTG-3'. The three-dimensional structure of the b/HLH/zip domain of Max complexed with a DNA fragment containing the sequence 5'-CACGTG-3' has been determined by the group of Stephen

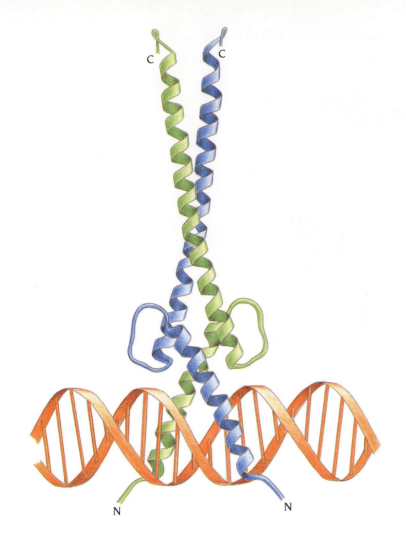

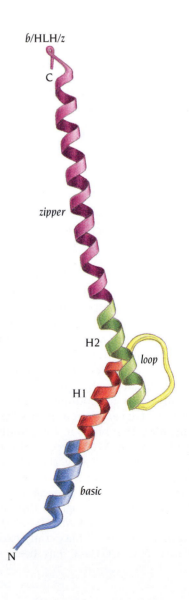

Figure 10.28 Schematic diagram of the binding of the transcription factor Max to DNA. The two monomers of Max (blue and green) form a dimer through both the helix-loop-helix regions which form a four-helix bundle like MyoD, and the zipper regions, which are arranged in a coiled coil. The N-terminal basic regions bind to DNA in a way similar to GCN4 and MyoD. (Adapted from A.R. Ferre-D'Amare et al., *Nature* 363: 38–45, 1993.)

Burley at Rockefeller University, New York. The overall structure of the b/HLH part of the domain is very similar to that in MyoD (see Figures 10.26 and 10.28), while the leucine zippers of the dimer form a helical coiled coil very similar to that of GCN4 described earlier (see Figures 10.21b and 10.28). In the Max monomer the zipper helix is a continuation of the helical H2 region, so that H2 and zip form one long α helix (Figure 10.29). The overall structure of the monomer is therefore two α helices joined by a loop, where the first helix comprises the basic region followed by the H1 region of the HLH motif, and the second helix is formed by the H2 region of HLH followed by the zipper region.

The loop regions of both the b/HLH and the b/HLH/zip families are highly variable both in sequence and in length (Figure 10.30). Homologous protein sequences have loops ranging from 5 to 23 residues. Since the two helices in the HLH motif are parallel, the loop must be sufficiently long to connect one end of the four-helix bundle with the other end (see Figure 10.25). Given that the distance between successive α carbon atoms in an extended polypeptide chain is 3.8 Å and that the distance between the end of H1 and the beginning of H2 in the bundle is 15 Å in these structures, it follows that the loop must be at least four or five residues long. This has been confirmed by shortening the loop of MyoD to four residues by deletion mutations, whereupon the protein loses its DNA-binding activity.

Figure 10.29 The structure of the Max monomer is essentially built up from two long α helices joined by a loop region (yellow). The basic region (blue) and H1 (red) of the helix-loop-helix region form one continous α helix, and H2 (green) and the zipper region (purple) form a second continous α helix.

		H1		loop		H2	

```
Max   NH₂ - D H I K D S F H S L R D S V P - - - - - - S L Q G E K A S - R A Q I L D K A T E Y I Q Y M - COOH
C-Myc     - N E L K R S F F A L R D Q I P - - - - - E L E N N E K A P - K V V I L K K A T A Y I L S V -
MyoD      - S K V N E A F E T L K R C T S - - - - - - S N P N Q R L P - K V E I L R N A I R Y I E G L -
CBF1      - E N I N T A I N V L S D L L P - - - - - - - V R E S S - K A A I L A R A A E Y I Q K L -
Pho4      - N R L A V A L H E L A S L I P - A E W K Q Q N V S A A P S - K A T T V E A A C R Y I R H L -
```

Max and MyoD recognize the DNA HLH consensus sequence by different specific protein–DNA interactions

The helical basic region of Max binds to the major groove of DNA in the same manner as MyoD (see Figures 10.26 and 10.28). The DNA structure is essentially that of B-DNA with minor distortions as in the MyoD complex. Both Max and MyoD recognize the same consensus half-site DNA sequence

    5'-CAC-3'
    3'-GTG-5'

However, specific contacts with the bases of DNA are formed by different protein side chains in the two complexes. In both complexes a central glutamic acid residue, Glu 32 in Max and Glu 118 in MyoD, recognizes the 5' C and A bases by similar hydrogen bonds (compare Figures 10.27 and 10.31). However, in Max, His 28, which is located one turn of the α helix away from Glu 32, recognizes the 5' G in the complementary strand (Figure 10.31), which is not recognized by MyoD. Conversely, the methyl groups of T recognized by Thr 115 of MyoD is not recognized by Max. Finally, in both Max and MyoD, an arginine residue, Arg 36 in the former and Arg 111 in the latter, recognizes a G residue in the half-site (Figures 10.27 and 10.31).

Conclusion

A large group of eucaryotic transcription factors use liganded zinc atoms to stabilize a DNA-binding motif, and most classic zinc finger transcription factors contain several and sometimes over 30 concatenated zinc finger motifs. These motifs bind to a sequence of binding sites in the target DNA. Although each motif only recognizes three or four bases via a loop region at the tip of each finger, a zinc finger transcription factor containing multiple fingers, each with its own short recognition site, becomes a highly discriminating DNA-binding protein.

Figure 10.30 Amino acid sequences of the helix-loop-helix region of some members of the b/HLH and b/HLH/zip families of transcription factors. Residues that form the hydrophobic core of the four-helix bundle are colored green and a conserved lysine residue is blue. The loop region between H1 and H2 is highly variable in length but must be at least four or five residues long.

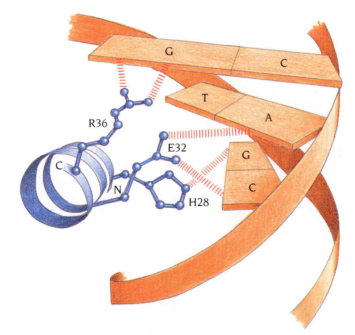

Figure 10.31 Sequence-specific interactions between DNA and one monomer of Max. Three residues, His 28, Glu 32 and Arg 36, form specific interactions with the edges of the bases in the major groove of DNA. Like MyoD, a Glu residue recognizes the first two bases, C and A, of the recognition sequence.

The nuclear receptors utilize pairs of zinc atoms to stabilize a DNA-binding domain. The steroid hormone receptors are homodimers with two DNA-binding sites that recognize an almost palindromic sequence in the DNA. The monomeric DNA-binding domains dimerize upon binding to DNA in such a way that the two recognition helices of the dimer can bind to target sequences of the response element with the appropriate spacer region. These dimers are symmetrical, like the dimers of the 434 repressor. Other nuclear receptors form heterodimers that are nonsymmetric and bind to response elements in which the target sequences are organized as direct repeats. These heterodimers discriminate response elements on the basis of the different lengths of their spacer region. A DNA-supported asymmetric dimerization interface located within the DNA-binding domains allows these heterodimers to distinguish between closely related response elements.

The C_6-zinc cluster transcription factors utilize two zinc atoms bound by six cysteine residues to stabilize the scaffold of the monomeric subunit. Again, the active factor is a dimer with a distinct dimerization region. The dimer resembles the letter T: the dimerization region is in the upright stem and the two DNA binding motifs are at the ends of the crossbar. All members of this family recognize the same base sequence, the triplet CCG; specificity is achieved by the length of the crossbar, in other words, the separation of the CCG triplet along the DNA.

Dimerization of the C_6-zinc cluster transcription factors involves an α-helical coiled coil in the dimerization region. Coiled coils, often called leucine zippers, are also found in a large group of transcription factors that do not contain zinc. The leucine zipper is made up of two α helices in a coiled coil with every seventh residue leucine or some other large hydrophobic residue, such as isoleucine or valine. Leucine zipper transcription factors (b/zip) include factors characterized by heterodimerization, for example Fos and Jun. The α-helical DNA-binding motifs of the heterodimers recognize quite different base sequences and are continous with the α helices of the zipper.

Helix-loop-helix (b/HLH) transcription factors are either heterodimers or homodimers with basic α-helical DNA-binding regions that lie across the major groove, rather than along it, and these helices extend into the four-helix bundle that forms the dimerization region. A modification of the b/HLH structure is seen in some transcription factors (b/HLH/zip) in which the four-helix bundle extends into a classic leucine zipper.

Homodimerization, the common feature of procaryotic transcription factors, is in eucaryotes extended by heterodimerization (as in the leucine zippers) and by concatenation (as in zinc fingers and POU). These devices increase the range of DNA sequences that can be recognized and the degree of binding specificity. In evolutionary terms this increase in range and specificity became necessary as genomes became larger and more complex. The real complexity of regulation of gene expression in eucaryotes has so far barely been touched upon by structural biologists, because it lies in the assembly of aggregates of general and specific transcription factors at the promoter region. The biochemical data are strong enough to conclude that the various transcription factors bound to distinct sites at a complex promoter have specific three-dimensional relationships, and so form a macromolecular assembly analogous to a ribosome, proteosome or spliceosome.

Selected readings

General

Alber, T. How GCN4 binds DNA. *Curr. Biol.* 3: 182–184, 1993.

Berg, J.M., Shi, Y. The galvanization of biology: a growing appreciation for the roles of zinc. *Science* 271: 1081–1085, 1996.

Buratowski, S. The basics of basal transcription by RNA Polymerase II. *Cell* 77: 1–3, 1994.

Fairman, R., et al. Multiple oligomeric states regulate the DNA binding of helix-loop-helix peptides. *Proc. Natl. Acad. Sci. USA* 90: 10429–10433, 1993.

Freedman, L.P., Luisi, B.F. On the mechanism of DNA binding by nuclear hormone receptor: a structural and functional perspective. *J. Cell. Biochem.* 51: 140–150, 1993.

Harrison, S .C., Sauer, R.T. (eds.) Protein–nucleic acid interactions. *Curr. Opin. Struct. Biol.* 4: 1–35, 1994.

Kerppola, T., Currant, T. Zen and the art of Fos and Jun. *Nature* 373: 199–200, 1995.

Pabo, C.O., Sauer, R.T. Transcription factors: structural families and principles for DNA recognition. *Annu. Rev. Biochem.* 61: 1053–1098, 1992.

Phillips, S.E.V. Built by association: structure and function of helix-loop-helix DNA-binding proteins. *Structure* 2: 1–4, 1994.

Philips, S.E.V., Moras, D. (eds.) Protein–nucleic acid interactions. *Curr. Opin. Struct. Biol.* 3: 1–16, 1993.

Schwabe, J.W.R., Klug, A. Zinc mining for protein domains. *Nature Struct. Biol.* 1: 345–350, 1994.

Tjian, R., Maniatis, T. Transcriptional activation: a complex puzzle with few easy pieces. *Cell* 77: 5–8, 1994.

Specific structures

Ellenberger, T.E., et al. The GCN4 basic region leucine zipper binds DNA as a dimer of uninterrupted α helices: crystal structure of the protein–DNA complex. *Cell* 71: 1223–1237, 1992.

Fairall, L., et al. The crystal structure of a two zinc finger peptide reveals an extension to the rules for zinc finger/DNA recognition. *Nature* 366: 483–487, 1993.

Ferré-D'Amaré, A.R., et al. Recognition by Max of its cognate DNA through a dimeric b/HLH/Z domain. *Nature* 363: 38–45, 1993.

Ferré-D'Amaré, A.R., et al. Structure and function of the b/HLH/Z domain of USF. *EMBO J.* 13: 180–189, 1994.

Glover, J.N.M., Harrison, S.C. Crystal structure of the heterodimeric bZIP transcription factor c-Fos-c-Jun bound to DNA. *Nature* 373: 257–261, 1995.

Hard, T., et al. Solution structure of the glucocorticoid receptor DNA-binding domain. *Science* 249: 157–160, 1990.

Hegde, R.S., et al. Crystal structure at 1.7 Å of the bovine papillomavirus-1 E2 DNA-binding domain bound to its DNA target. *Nature* 359: 505–512, 1992.

König, P., Richmond, T.J. The x-ray structure of the GCN4-bZIP bound to ATF/CREB site DNA shows the complex depends on DNA flexibility. *J. Mol. Biol.* 233: 139–154, 1993.

Lee, M.S., et al. Three-dimensional solution structure of a single zinc finger DNA-binding domain. *Science* 245: 635–637, 1989.

Luisi, B.F., et al. Crystallographic analysis of the interaction of the glucocorticoid receptor with DNA. *Nature* 352: 497–505, 1991.

Ma, P.C.M., et al. Crystal structure of MyoD bHLH domain–DNA complex: perspectives on DNA recognition and implications for transcriptional activation. *Cell* 77: 451–459, 1994.

Marmorstein, R., et al. DNA recognition by GAL4: structure of a protein–DNA complex. *Nature* 356: 408–414, 1992.

Marmorstein, R., Harrison, S.C. Crystal structure of a PPR1–DNA complex: DNA recognition by proteins containing a Zn2Cys6 binuclear cluster. *Genes Dev.* 8: 2504–2512, 1994.

O'Shea, E.K., et al. Mechanism of specificity in the fos-jun oncoprotein heterodimer. *Cell* 68: 699–708, 1992.

O'Shea, E.K., et al. X-ray structure of the GCN4 leucine zipper, a two-stranded, parallel coiled coil. *Science* 254: 539–544, 1991.

Parraga, G., et al. Zinc-dependent structure of a single-finger domain of yeast ADR1. *Science* 241: 1489–1492, 1988.

Pavletich, N.P., Pabo, C.O. Crystal structure of a five-finger GLI–DNA complex: new perspectives on zinc fingers. *Science* 261: 1701–1707, 1993.

Pavletich, N.P., Pabo, C.O. Zinc finger-DNA recognition: crystal structure of a Zif268–DNA complex at 2.1 Å. *Science* 252: 809–817, 1991.

Rastinejad, F., et al. Structural determinants of nuclear receptor assembly on DNA direct repeats. *Nature* 375: 203–211, 1995.

Reece, R.J., Ptashne, M. Determinants of binding-site specificity among yeast C_6 zinc cluster proteins. *Science* 261: 909–911, 1993.

Schwabe, J.W.R., et al. The crystal structure of the estrogen receptor DNA-binding domain bound to DNA: how receptors discriminate between their response elements. *Cell* 75: 567–578, 1993.

An Example of Enzyme Catalysis: Serine Proteinases

In 1946 Linus Pauling first formulated the basic principle underlying enzyme catalysis, namely, that an enzyme increases the rate of a chemical reaction by binding and stabilizing the transition state of its specific substrate tighter than the ground state. However, for many years it was not generally appreciated that the high affinity of an enzyme for the transition state of a substrate plays a major role in determining substrate specificity as well as the rate of catalysis. In the past few years, kinetic studies of site-directed mutants, combined with x-ray structures, have made it possible to identify unambiguously the role of particular amino acids in both the substrate specificity and the catalytic reaction of an enzyme as well as providing information about the energetic basis of catalysis itself. The full consequences of Pauling's principle emerged only when it was found that mutants designed to change an enzyme's catalytic rate also changed its substrate specificity and vice versa.

In this chapter we shall illustrate some fundamental aspects of enzyme catalysis using as an example the serine proteinases, a group of enzymes that hydrolyze peptide bonds in proteins. We also examine how the transition state is stabilized in this particular case.

Proteinases form four functional families

Proteinases are widely distributed in nature, where they perform a variety of different functions. Viral genes code for proteinases that cleave the precursor molecules of their coat proteins, bacteria produce many different extracellular proteinases to degrade proteins in their surroundings, and higher organisms use proteinases for such different functions as food digestion, cleavage of signal peptides, and control of blood pressure and blood clotting. Many proteinases occur as domains in large multifunctional proteins, but others are independent smaller polypeptide chains. *In vivo* the activity of many proteinases is controlled by endogenous protein inhibitors that complex with the enzymes and block them. The three-dimensional structures of a large number of the smaller proteinases and of their complexes with protein inhibitors have been determined, and this wealth of data allows some general conclusions to be drawn.

All the well-characterized proteinases belong to one or other of four families: serine, cysteine, aspartic, or metallo proteinases. This classification is based on a functional criterion, namely, the nature of the most prominent functional group in the active site. Members of the same functional family are usually evolutionarily related, but there are exceptions to this rule. We have chosen two **serine proteinases**, mammalian **chymotrypsin** and bacterial **subtilisin**, as representative examples to illustrate one of the catalytic mechanisms leading to proteolysis. Before describing the structures, mechanism, and engineering of these two enzymes, however, we shall define some basic enzymological concepts.

The catalytic properties of enzymes are reflected in K_m and k_{cat} values

Leonor Michaelis and Maud Menten laid the foundation for enzyme kinetics as early as 1913 by proposing the following scheme:

$$E + S \underset{}{\overset{K_d}{\rightleftharpoons}} ES \underset{}{\overset{k_{cat}}{\rightleftharpoons}} E + P$$

Enzyme and substrate first reversibly combine to give an **enzyme–substrate (ES) complex**. Chemical processes then occur in a second step with a rate constant called k_{cat}, or the **turnover number**, which is the maximum number of substrate molecules converted to product per active site of the enzyme per unit time. The k_{cat} is, therefore, a rate constant that refers to the properties and reactions of the ES complex. For simple reactions k_{cat} is the rate constant for the chemical conversion of the ES complex to free enzyme and products.

These definitions are valid only when the concentration of the enzyme is very small compared with that of the substrate. Moreover, they apply only to the initial rate of formation of products: in other words, the rate of formation of the first few percent of the product, before the substrate has been depleted and products that can interfere with the catalytic reaction have accumulated.

The Michaelis–Menten scheme nicely explains why a maximum rate, V_{max}, is always observed when the substrate concentration is much higher than the enzyme concentration (Figure 11.1). V_{max} is obtained when the enzyme is saturated with substrate. There are then no free enzyme molecules available to turn over additional substrate. Hence, the rate is constant, V_{max}, and is independent of further increase in the substrate concentration.

The substrate concentration when the half maximal rate, $(V_{max}/2)$, is achieved is called the K_m. For many simple reactions it can easily be shown that the K_m is equal to the dissociation constant, K_d, of the ES complex. The K_m, therefore, describes the affinity of the enzyme for the substrate. For more complex reactions, K_m may be regarded as the overall dissociation constant of all enzyme-bound species.

The quantity k_{cat}/K_m is a rate constant that refers to the overall conversion of substrate into product. The ultimate limit to the value of k_{cat}/K_m is therefore set by the rate constant for the initial formation of the ES complex. This rate cannot be faster than the diffusion-controlled encounter of an enzyme and its substrate, which is between 10^8 to 10^9 per mole per second. The quantity k_{cat}/K_m is sometimes called the **specificity constant** because it describes the specificity of an enzyme for competing substrates. As we shall see, it is a useful quantity for kinetic comparison of mutant proteins.

Enzymes decrease the activation energy of chemical reactions

The Michaelis complex, ES, undergoes rearrangement to one or several **transition states** before product is formed. Energy is required for these rearrangements. The input energy required to bring free enzyme and substrate to the highest transition state of the ES complex is called the **activation energy** of the reaction (Figure 11.2). In the absence of enzyme, spontaneous conversion of substrate to product also proceeds through transition states that require activation energy. The rate of a chemical reaction is strictly dependent on its

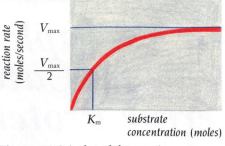

Figure 11.1 A plot of the reaction rate as a function of the substrate concentration for an enzyme catalyzed reaction. V_{max} is the maximal velocity. The Michaelis constant, K_m, is the substrate concentration at half V_{max}. The rate v is related to the substrate concentration, [S], by the Michaelis–Menten equation:

$$v = \frac{V_{max} \times [S]}{K_m + [S]}$$

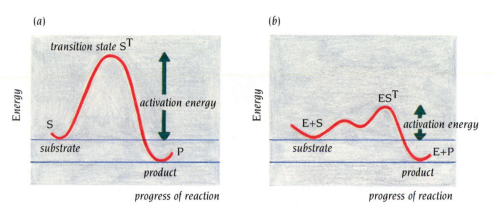

(a)

transition state S^T

Energy

activation energy

S

substrate

P

product

progress of reaction

(b)

Energy

ES^T

E+S

activation energy

substrate

E+P

product

progress of reaction

Figure 11.2 Enzymes accelerate chemical reactions by decreasing the activation energy. The activation energy is higher for a noncatalyzed reaction (a) than for the same reaction catalyzed by an enzyme (b). Both reactions proceed through one or several transition states, S^T. Only one transition state is shown in (a), whereas the two bumps in (b) represent two different transition states.

activation energy, and the more than 1 millionfold enhancement of rate achieved by enzyme catalysis results from the ability of the enzyme to decrease the activation energy of the reaction (Figure 11.2).

This decrease in activation energy is achieved by enzymes in several different ways: for example, by providing catalytically active groups for a specific reaction mechanism, by binding several substrates in an orientation appropriate to the reaction catalyzed, and, most importantly, by using the differential binding energy of the substrate in its transition state compared with its normal state. The activation energy for the conversion of ES to E + P is lower if the enzyme binds more tightly to the transition state of S than to its normal structure (Figure 11.3). The higher affinity of the enzyme for the transition state makes the transition energetically favorable and thus decreases the activation energy. If, on the other hand, the enzyme were to bind the unaltered substrate more strongly than the transition state, the decrease in binding energy on the formation of the transition state would increase the activation energy and catalysis would not be achieved (see Figure 11.3). It is therefore catalytically advantageous for the enzyme's active site to be complementary to the transition state of the substrate rather than to the normal structure of the substrate.

Since this differential binding energy relates to the complete substrate molecule, including groups that determine the substrate specificity, it is obvious that specificity and catalytic rate are interdependent. The importance of the differential binding energy for catalysis is nicely illustrated by the recent production of antibodies with catalytic activity. Such antibodies were raised against small hapten molecules that simulate a transition state structure for a specific chemical reaction, such as ester hydrolysis. These antibodies not only bound the transition state more tightly than the normal structure of the ester, but they also exhibited significant catalytic activity even though they had not been selected to have any catalytically competent residues in the binding site.

Figure 11.3 One of the most important factors in enzyme catalysis is the ability of an enzyme to bind substrate more tightly in its transition state than in its ground state. The difference in binding energy between these states lowers the activation energy of the reaction. This is illustrated by energy profiles for an enzyme in its wild-type form (a), for a mutant that stabilizes the substrate in its transition state and therefore decreases the activation energy from ES to the transition state ES^T giving higher rates (b), and for a mutant that stabilizes the substrate in its ground state giving lower rates (c). (Adapted from A. Fersht, *Enzyme Structure and Mechanism*, 2nd ed. pp. 314–315. New York: W.H. Freeman, 1984.)

(a)

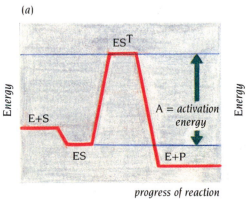

Energy

ES^T

E+S

A = activation energy

ES

E+P

progress of reaction

wild type enzyme

(b)

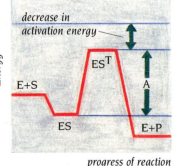

Energy

decrease in activation energy

E+S

ES^T

A

ES

E+P

progress of reaction

mutant that stabilizes transition state

(c)

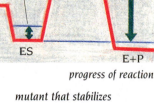

Energy

increase in activation energy

ES^T

E+S

A

ES

E+P

progress of reaction

mutant that stabilizes the ground state

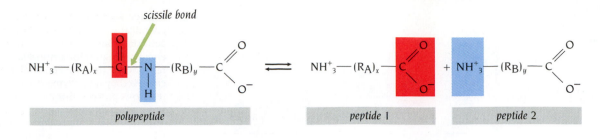

scissile bond

polypeptide ⇌ peptide 1 + peptide 2

Serine proteinases cleave peptide bonds by forming tetrahedral transition states

The serine proteinases have been very extensively studied, both by kinetic methods in solution and by x-ray structural studies to high resolution. From all these studies the following reaction mechanism has emerged.

A serine proteinase cleaves peptide bonds within a polypeptide to produce two new smaller peptides (Figure 11.4). The reaction proceeds in two steps. The first step produces a covalent bond between C_1 of the substrate and the hydroxyl group of a reactive Ser residue of the enzyme (Figure 11.5a). Production of this acyl-enzyme intermediate proceeds through a negatively charged transition state intermediate where the bonds of C_1 have tetrahedral geometry in contrast to the planar triangular geometry in the peptide group. During this step the peptide bond is cleaved, one peptide product is attached to the enzyme in the acyl-enzyme intermediate, and the other peptide product rapidly diffuses away. In the second step of the reaction, deacylation, the acyl-enzyme intermediate is hydrolyzed by a water molecule to release the second peptide product with a complete carboxy terminus and to restore the Ser-hydroxyl of the enzyme (Figure 11.5b). This step also proceeds through a negatively charged tetrahedral transition state intermediate (Figure 11.5b). What are the structural requirements for the enzyme to perform these reactions?

Figure 11.4 Serine proteinases catalyze the hydrolysis of peptide bonds within a polypeptide chain. The bond that is cleaved is called the scissile bond. $(R_A)_x$ and $(R_B)_y$ represent polypeptide chains of varying lengths.

Figure 11.5 (a) Formation of an acyl-enzyme intermediate involving a reactive Ser residue of the enzyme is the first step in hydrolysis of peptide bonds by serine proteinases. First, a transition state is formed where the peptide bond is cleaved in which the C_1 carbon has a tetrahedral geometry with bonds to four groups, including the reactive Ser residue of the enzyme and a negatively charged oxygen atom. (b) Deacylation of the acyl-enzyme intermediate is the second step in hydrolysis. This is essentially the reverse of the acylation step, with water in the role of the NH_2 group of the polypeptide substrate. The base shown in the figure is a His residue of the protein that can accept a proton during the formation of the tetrahedral transition state.

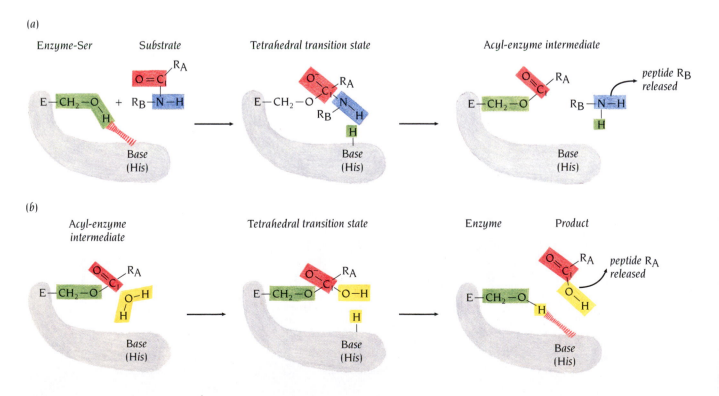

208

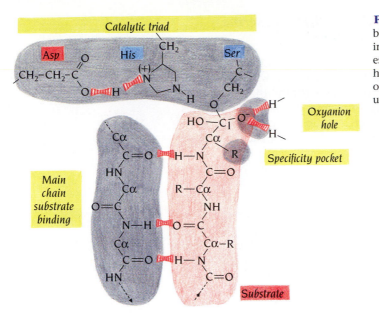

Figure 11.6 A schematic view of the presumed binding mode of the tetrahedral transition state intermediate for the deacylation step. The four essential features of the serine proteinases are highlighted in yellow: the catalytic triad, the oxyanion hole, the specificity pocket, and the unspecific main-chain substrate binding.

Four important structural features are required for the catalytic action of serine proteinases

The serine proteinases have four important structural features that facilitate this mechanism of catalysis (Figure 11.6).

1. The enzyme provides a general base, a His residue, that can accept the proton from the hydroxyl group of the reactive Ser thus facilitating formation of the covalent tetrahedral transition state. This His residue is part of a **catalytic triad** consisting of three side chains from Asp, His, and Ser, which are close to each other in the active site, although they are far apart in the amino acid sequence of the polypeptide chain (Figure 11.6).

2. Tight binding and stabilization of the tetrahedral transition state intermediate is accomplished by providing groups that can form hydrogen bonds to the negatively charged oxygen atom attached to C_1. These groups are in a pocket of the enzyme called the **oxyanion hole** (see Figure 11.6). The positive charge that develops on the His residue after it has accepted a proton also stabilizes the negatively charged transition state. These features also presumably destabilize binding of substrate in the normal state.

3. Most serine proteinases have no absolute substrate specificity. They can cleave peptide bonds with a variety of side chains adjacent to the peptide bond to be cleaved (the scissile bond). This is because polypeptide substrates exhibit a nonspecific binding to the enzyme through their main-chain atoms, which form hydrogen bonds in a short antiparallel β sheet with main-chain atoms of a loop region in the enzyme (see Figure 11.6). One of these hydrogen bonds is long (3.6 Å) in enzyme-substrate complexes but short in complexes that simulate the transition state. This nonspecific binding therefore also contributes to stabilization of the transition state.

4. Even though these enzymes have no absolute specificity, many of them show a preference for a particular side chain before the scissile bond as seen from the amino end of the polypeptide chain. The preference of chymotrypsin to cleave after large aromatic side chains and of trypsin to cleave after Lys or Arg side chains is exploited when these enzymes are used to produce peptides suitable for amino acid sequence determination and fingerprinting. In each case, the preferred side chain is oriented so as to fit into a pocket of the enzyme called the **specificity pocket**.

Convergent evolution has produced two different serine proteinases with similar catalytic mechanisms

These four features all occur in an almost identical fashion in all members of the chymotrypsin superfamily of homologous enzymes, which includes among others chymotrypsin, trypsin, elastase, and thrombin. Reasonably, one might imagine that such a combination of four characteristic features had arisen only once during evolution to give an ancestral molecule from which all serine proteinases diverged. However, subtilisin, a bacterial serine proteinase with an amino acid sequence and, as we will see, a three-dimensional structure quite different from the mammalian serine proteinases, exhibits these same four characteristic features. Subtilisin is not evolutionarily related to the chymotrypsin family of enzymes; nevertheless, the atoms in subtilisin that participate in the catalytic triad, in the oxyanion hole, and in substrate binding are in almost identical positions relative to one another in the three-dimensional structure as they are in chymotrypsin and its relatives. Starting from unrelated ancestral proteins, convergent evolution has resulted in the same structural solution to achieve a particular catalytic mechanism. The serine proteinases, in other words, provide a spectacular example of convergent evolution at the molecular level, which we can best appreciate by explaining in detail the structures of chymotrypsin and subtilisin.

The chymotrypsin structure has two antiparallel β-barrel domains

In 1967 the group of David Blow at the MRC Laboratory of Molecular Biology, Cambridge, UK, determined the three-dimensional structure of chymotrypsin. This was one of the very first enzyme structures known at high resolution. Since then a large number of serine proteinase structures, complexed both with small peptide inhibitors and large endogenous polypeptide inhibitors, have been determined to high resolution mainly by the groups of Michael James, Edmonton, and Robert Huber, Munich.

The polypeptide chain of chymotrypsinogen, the inactive precursor of chymotrypsin, comprises 245 amino acids. During activation of chymotrypsinogen residues 14–15 and 147–148 are excised. The remaining three polypeptide chains are held together by disulfide bridges to form the active chymotrypsin molecule.

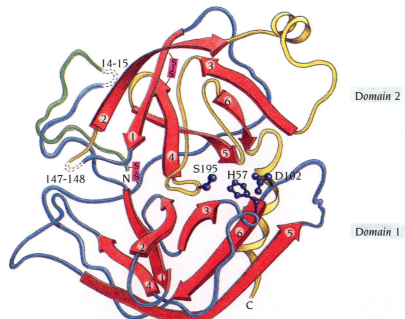

Figure 11.7 Schematic diagram of the structure of chymotrypsin, which is folded into two antiparallel β domains. The six β strands of each domain are red, the side chains of the catalytic triad are dark blue, and the disulfide bridges that join the three polypeptide chains are marked in violet. Chain A (green, residues 1–13) is linked to chain B (blue, residues 16–146) by a disulfide bridge between Cys 1 and Cys 122. Chain B is in turn linked to chain C (yellow, residues 149–245) by a disulfide bridge between Cys 136 and Cys 201. Dotted lines indicate residues 14–15 and 147–148 in the inactive precursor, chymotrypsinogen. These residues are excised during the conversion of chymotrypsinogen to the active enzyme chymotrypsin.

The polypeptide chain is folded into two domains (Figure 11.7), each of which contains about 120 amino acids. The two domains are both of the antiparallel β-barrel type, each containing six β strands with the same topology (Figure 11.8). Even though the actual structure looks complicated, the topology is very simple, a Greek key motif (strands 1–4) followed by an antiparallel hairpin motif (strands 5 and 6).

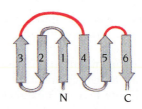

The active site is formed by two loop regions from each domain

The active site is situated in a crevice between the two domains. Domain 1 contributes two of the residues in the catalytic triad, His 57 and Asp 102, whereas the reactive Ser 195 is part of the second domain (see Figure 11.7).

Inhibitors as well as substrates bind in this crevice between the domains. From the numerous studies of different inhibitors bound to serine proteinases we have chosen as an illustration the binding of a small peptide inhibitor, Ac-Pro-Ala-Pro-Tyr-COOH to a bacterial chymotrypsin (Figure 11.9). The enzyme–peptide complex was formed by adding a large excess of the substrate Ac-Pro-Ala-Pro-Tyr-CO-NH$_2$ to crystals of the enzyme. The enzyme molecules within the crystals catalyze cleavage of the terminal amide group to produce the products Ac-Pro-Ala-Pro-Tyr-COOH and NH$_3^+$. The ammonium ions diffuse away, but the peptide product remains bound as an inhibitor to the active site of the enzyme.

This inhibitor does not form a covalent bond to Ser 195 but one of its carboxy oxygen atoms is in the oxyanion hole forming hydrogen bonds to the main-chain NH groups of residues 193 and 195. The tyrosyl side chain is positioned in the specificity pocket, which derives its specificity mainly from three residues, 216, 226, and 189, as we shall see later. The main chain of

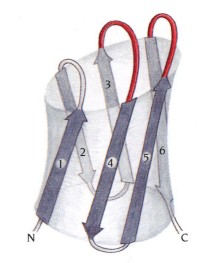

Figure 11.8 Topology diagrams of the domain structure of chymotrypsin. The chain is folded into a six-stranded antiparallel β barrel arranged as a Greek key motif followed by a hairpin motif.

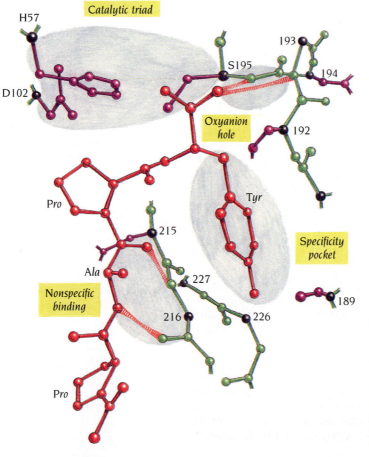

Figure 11.9 A diagram of the active site of chymotrypsin with a bound inhibitor, Ac-Pro-Ala-Pro-Tyr-COOH. The diagram illustrates how this inhibitor binds in relation to the catalytic triad, the substrate specificity pocket, the oxyanion hole and the nonspecific substrate binding region. The inhibitor is red. Hydrogen bonds between inhibitor and enzyme are striped. (Adapted from M.N.G. James et al., *J. Mol. Biol.* 144: 43–88, 1980.)

the inhibitor forms a short stretch of antiparallel β sheet with residues 215–216 of the enzyme forming hydrogen bonds to the NH and CO groups of residue 216.

A closer examination of these essential residues, including the catalytic triad, reveals that they are all part of the same two loop regions in the two domains (Figure 11.10). The domains are oriented so that the ends of the two barrels that contain the Greek key crossover connection (described in Chapter 5) between β strands 3 and 4 face each other along the active site. The essential residues in the active site are in these two crossover connections and in the adjacent hairpin loops between β strands 5 and 6. Most of these essential residues are conserved between different members of the chymotrypsin superfamily. They are, of course, surrounded by other parts of the polypeptide chains, which provide minor modifications of the active site, specific for each particular serine proteinase.

His 57 and Ser 195 are within loop 3–4 of domains 1 and 2, respectively. The third residue in the catalytic triad, Asp 102, is within loop 5–6 of domain 1. The rest of the active site is formed by two loop regions (3–4 and 5–6) of domain 2. As in so many other protein structures described previously, the barrels apparently provide a stable scaffold to position a few loop regions that constitute the essential features of the active site.

Did the chymotrypsin molecule evolve by gene duplication?

Although the two domains of chymotrypsin have similar three-dimensional structures there is no amino acid sequence identity between them. Nevertheless, based on the argument that three-dimensional structure is more conserved than amino acid sequence, it has been suggested that the members of the chymotrypsin superfamily evolved by gene duplication of a single ancestral proteinase domain. The putative ancestral domain, obviously, could not have had the catalytic triad in present-day serine proteinases since the contemporary triad is derived from both domains. However, this is less of an obstacle to the gene-duplication hypothesis than it seems at first sight. The ancestral domain could have been a barrel structure similar to the second domain of chymotrypsin, which contains most of the essential features of the active site, including the reactive serine residue. We also now know from experiments with genetically engineered mutants in which the triad has been abolished that the catalytic triad is not absolutely essential for catalytic activity. As we will see later, these mutants retain some proteinase activity. It is, therefore, quite possible that there was a single ancestral gene specifying a single domain with some catalytic activity. This activity could then have been enhanced by a gene-duplication event followed by mutation and evolution leading to the catalytic triad of today. The fact that the active-site residues that comprise the catalytic triad of chymotrypsin and its relatives are clustered in the same two loop regions of domains 1 and 2 supports such an evolutionary history.

Different side chains in the substrate specificity pocket confer preferential cleavage

The serine proteinases all have the same substrate, namely, polypeptide chains of proteins. However, different members of the family preferentially cleave polypeptide chains at sites adjacent to different amino acid residues. The structural basis for this preference lies in the side chains that line the substrate specificity pocket in the different enzymes.

This is nicely illustrated by members of the chymotrypsin superfamily: the enzymes chymotrypsin, trypsin, and elastase have very similar three-dimensional structures but different specificity. They preferentially cleave adjacent to bulky aromatic side chains, positively charged side chains, and small uncharged side chains, respectively. Three residues, numbers 189, 216, and 226, are responsible for these preferences (Figure 11.11). Residues 216

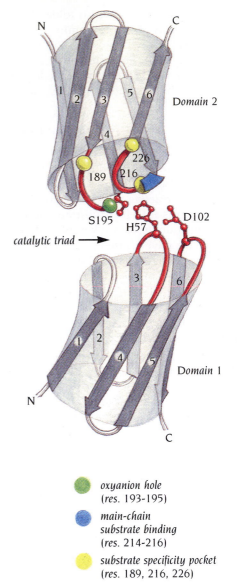

● oxyanion hole
(res. 193-195)

● main-chain
substrate binding
(res. 214-216)

● substrate specificity pocket
(res. 189, 216, 226)

Figure 11.10 Topological diagram of the two domains of chymotrypsin, illustrating that the essential active-site residues are part of the same two loop regions (3–4 and 5–6, red) of the two domains. These residues form the catalytic triad, the oxyanion hole (green), and the substrate binding regions (yellow and blue) including essential residues in the specificity pocket.

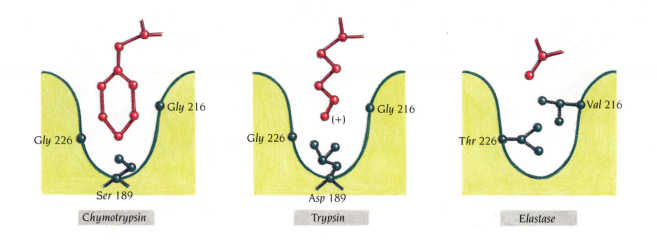

| Chymotrypsin | Trypsin | Elastase |

and 226 line the sides of the pocket. In trypsin and chymotrypsin these are both glycine residues that allow side chains of the substrate to penetrate into the interior of the specificity pocket. In elastase they are Val and Thr, respectively, that fill up most of the pocket with hydrophobic groups (Figure 11.11). Consequently, elastase does not cleave adjacent to large or charged side chains but adjacent to small uncharged side chains instead.

Residue 189 is at the bottom of the specificity pocket. In trypsin the Asp residue at this position interacts with the positively charged side chains Lys or Arg of a substrate. This accounts for the preference of trypsin to cleave adjacent to these residues. In chymotrypsin there is a Ser residue at position 189, which does not interfere with the binding of the substrate. Bulky aromatic groups are therefore preferred by chymotrypsin since such side chains fill up the mainly hydrophobic specificity pocket. It has now become clear, however, from site-directed mutagenesis experiments that this simple picture does not tell the whole story.

Engineered mutations in the substrate specificity pocket change the rate of catalysis

How would substrate preference be changed if the glycine residues in trypsin at positions 216 and 226 were changed to alanine rather than to the more bulky valine and threonine groups that are present in elastase? This question was addressed by the groups of Charles Craik, William Rutter, and Robert Fletterick in San Francisco, who have made and studied three such trypsin mutants: one in which Ala is substituted for Gly at 216, one in which the same substitution is made at Gly 226, and a third containing both substitutions.

Model building shows that both Arg- and Lys-containing substrates should be accommodated by the substrate specificity pocket after these Gly to Ala changes but that some details of the binding mode at the bottom of the pocket would be altered. The Ala 226 substitution would introduce a methyl group in the region where the end of the substrate's side chain binds (Figure 11.12) and would therefore be expected to accommodate Lys better than Arg, since the latter has a longer and more bulky side chain. Based on these steric arguments alone, one would therefore predict that the K_m for an Arg-containing substrate would be larger (less favorable binding) and that the K_m for Lys would be essentially unaltered. The specificity constant, k_{cat}/K_m, would decrease more for an Arg-containing substrate than for one with Lys.

Model building also predicts that the Ala 216 mutant would displace a water molecule at the bottom of the specificity pocket that in the wild type enzyme binds to the NH_3^+ group of the substrate Lys side chain (Figure 11.12). The extra CH_3 group of this mutant is not expected to disturb the binding of the Arg side chain. One would therefore expect that the K_m for Lys

Figure 11.11 Schematic diagrams of the specificity pockets of chymotrypsin, trypsin and elastase, illustrating the preference for a side chain adjacent to the scissile bond in polypeptide substrates. Chymotrypsin prefers aromatic side chains and trypsin prefers positively charged side chains that can interact with Asp 189 at the bottom of the specificity pocket. The pocket is blocked in elastase, which therefore prefers small uncharged side chains.

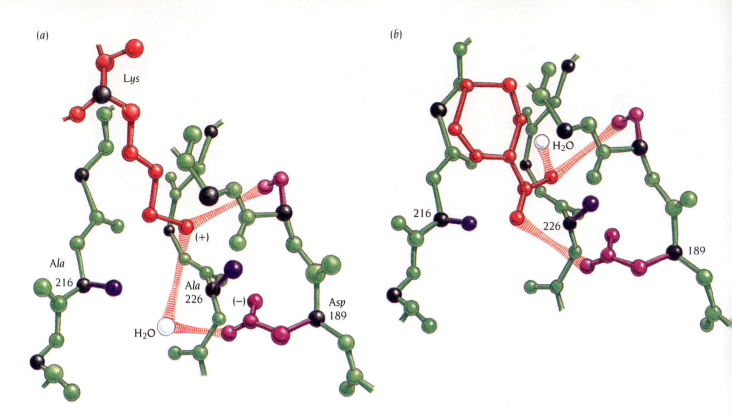

(a)

Lys

Ala
216

Ala
226

(+)

(−)

Asp
189

H_2O

(b)

H_2O

216

226

189

Figure 11.12 Schematic diagram of the specificity pocket of mutant trypsin with Ala (purple) at positions 216 and 226. (a) The position of a bound Lys side chain (red) in this pocket as observed in the structure of a complex between trypsin (green) and a peptide inhibitor. The NH_3^+ group of the Lys side chain interacts with the COO^- group of Asp 189 through a water molecule. (b) No structure is available for an Arg side chain in the substrate specificity pocket of trypsin. It is assumed that the complex of trypsin (green) with benzamide (red) is a good model for arginine binding in this pocket. One NH_2 group of benzamide interacts directly with the COO^- group of Asp 189 and the second NH_2 group interacts with a water molecule and the OH group of Ser 190. (Adapted from C.S. Craik et al., *Science* 228: 291–297, 1985.)

substrates would increase and therefore k_{cat}/K_m would decrease more for Lys- than for Arg-containing substrates. For the double mutant where both Gly 216 and Gly 226 are changed to Ala, one would predict an increase in the K_m values for both Lys- and Arg-containing substrates.

The experimentally determined k_{cat} and K_m values for the wild-type enzyme and the mutants are shown in Table 11.1. The dramatic kinetic effects of these mutations are best illustrated with the Arg substrate. The three mutants have roughly similar K_m values 15–35 times higher than for the wild type, but the k_{cat} values decrease by factors of 10 to about 1000. The mutants were designed to change the specificity, but by far the largest changes occur in the catalytic rates. Apparently, these mutations affect the structure of the enzyme in additional ways, possibly by causing conformational changes outside the specificity pocket, so that the stabilization of the transition state is reduced and consequently the activation energies for the reactions are different.

The changes in the specificity constants, on the other hand, were as expected from the predictions. The ratio of the k_{cat}/K_m values for the Arg and Lys substrates (last column in Table 11.1) gives a measure of the relative specificities. This ratio decreases for the Ala 226 mutant and increases for the Ala 216 mutant as predicted. However, the changes in these values depend not

Table 11.1 Kinetic data for wild-type and mutant trypsins

Enzyme	Arg			Lys			(k_{cat}/K_m) Arg / (k_{cat}/K_m) Lys
	k_{cat}	K_m	k_{cat}/K_m	k_{cat}	K_m	k_{cat}/K_m	
Wild type	1	1	1	0.9	10	0.1	10
Gly 216, Gly 226→Ala	0.001	15	0.0001	0.0005	25	0.00002	25
Gly 226→Ala	0.01	35	0.0003	0.1	250	0.0005	0.5
Gly 216→Ala	0.7	30	0.02	0.2	280	0.001	20

The substrates used were D-Val-Leu-Arg-amino fluorocoumarin (Arg) and D-Val-Leu-Lys-amino fluorocoumarin (Lys). For clarity the K_m and k_{cat} values have been normalized to those of the wild-type enzyme for the Arg substrate.

only on changes in the K_m values, which reflect binding of substrate, but even more on changes in the k_{cat} values, which reflect catalytic rate. It can, therefore, be argued that the agreement with prediction is fortuitous.

The simple lesson to be learnt from these experiments is that critical amino acid residues can have pleiotropic roles in determining a protein's structure and therefore its function.

The Asp 189-Lys mutation in trypsin causes unexpected changes in substrate specificity

Asp 189 at the bottom of the substrate specificity pocket interacts with Lys and Arg side chains of the substrate, and this is the basis for the preferred cleavage sites of trypsin (see Figures 11.11 and 11.12). It is almost trivial to infer, from these observations, that a replacement of Asp 189 with Lys would produce a mutant that would prefer to cleave substrates adjacent to negatively charged residues, especially Asp. On a computer display, similar Asp–Lys interactions between enzyme and substrate can be modeled within the substrate specificity pocket but reversed compared with the wild-type enzyme.

The results of experiments in which the mutation was made were, however, a complete surprise. The Asp 189-Lys mutant was totally inactive with both Asp and Glu substrates. It was, as expected, also inactive toward Lys and Arg substrates. The mutant was, however, catalytically active with Phe and Tyr substrates, with the same low turnover number as wild-type trypsin. On the other hand, it showed a more than 5000-fold increase in k_{cat}/K_m with Leu substrates over wild type. The three-dimensional structure of this interesting mutant has not yet been determined, but the structure of a related mutant Asp 189-His shows the histidine side chain in an unexpected position, buried inside the protein.

As these experiments with engineered mutants of trypsin prove, we still have far too little knowledge of the functional effects of single point mutations to be able to make accurate and comprehensive predictions of the properties of a point-mutant enzyme, even in the case of such well-characterized enzymes as the serine proteinases. Predictions of the properties of mutations using computer modeling are not infallible. Once produced, the mutant enzymes often exhibit properties that are entirely surprising, but they may be correspondingly informative.

The structure of the serine proteinase subtilisin is of the α/β type

Subtilisins are a group of serine proteinases that are produced by different species of bacilli. These enzymes are of considerable commercial interest because they are added to the detergents in washing powder to facilitate removal of proteinaceous stains. Numerous attempts have therefore recently been made to change by protein engineering such properties of the subtilisin molecule as its thermal stability, pH optimum, and specificity. In fact, in 1988 subtilisin mutants were the subject of the first US patent granted for an engineered protein.

The subtilisin molecule is a single polypeptide chain of 275 amino acids with no similarities in the amino acid sequence to chymotrypsin. The three-dimensional structure of subtilisin BPN' from *Bacillus amyloliquefaciens* was determined in 1969 by the group of Joseph Kraut in San Diego, California, and that of subtilisin Novo in 1971 by the group of Jan Drenth in Groningen, The Netherlands. The main feature of the subtilisin structure is a region of five parallel β strands (blue in Figure 11.13) surrounded by four helices, two on each side of the parallel β sheet. This α/β structure is thus quite different from the double antiparallel β-barrel structure of chymotrypsin (compare with Figure 11.7).

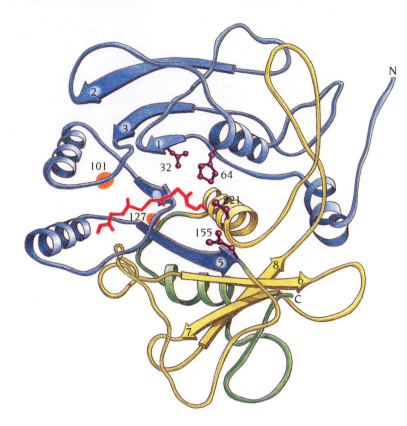

Figure 11.13 Schematic diagram of the three-dimensional structure of subtilisin viewed down the central parallel β sheet. The N-terminal region that contains the α/β structure is blue. It is followed by a yellow region, which ends with the fourth α helix of the α/β structure. The C-terminal part is green. The catalytic triad Asp 32, His 64, and Ser 221 as well as Asn 155, which forms part of the oxyanion hole are shown in purple. The main chain of part of a polypeptide inhibitor is shown in red. Main-chain residues around 101 and 127 (orange circles) form the nonspecific binding regions of peptide substrates.

The active sites of subtilisin and chymotrypsin are similar

The active site of subtilisin is outside the carboxy ends of the central β strands analogous to the position of the binding sites in other α/β proteins as discussed in Chapter 4. Details of this active site are surprisingly similar to those of chymotrypsin, in spite of the completely different folds of the two enzymes (Figures 11.14 and 11.9). A catalytic triad is present that comprises residues Asp 32, His 64 and the reactive Ser 221. The negatively charged oxygen atom of the tetrahedral transition state binds in an oxyanion hole,

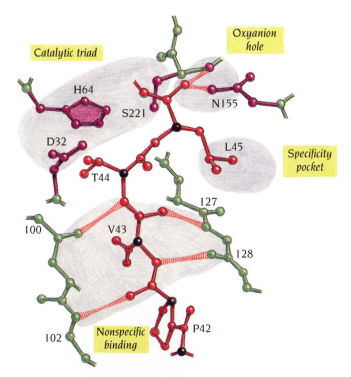

Figure 11.14 Schematic diagram of the active site of subtilisin. A region (residues 42–45) of a bound polypeptide inhibitor, eglin, is shown in red. The four essential features of the active site—the catalytic triad, the oxyanion hole, the specificity pocket, and the region for non-specific binding of substrate—are highlighted in yellow. Important hydrogen bonds between enzyme and inhibitor are striped. This figure should be compared to Figure 11.9, which shows the same features for chymotrypsin. (Adapted from W. Bode et al., *EMBO J.* 5: 813–818, 1986.)

forming hydrogen bonds with the side-chain amide group of Asn 155 and the main-chain nitrogen atom of Ser 221. Peptide substrates and inhibitors bind nonspecifically by forming a small antiparallel pleated sheet, which in subtilisin comprises three β strands (Figure 11.14). There is also a hydrophobic specificity pocket adjacent to the scissile bond.

All the four essential features of the active site of chymotrypsin are thus also present in subtilisin. Furthermore, these features are spatially arranged in the same way in the two enzymes, even though different framework structures bring different loop regions into position in the active site. This is a classical example of convergent evolution at the molecular level.

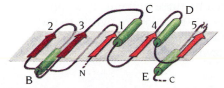

Figure 11.15 Topology diagram of the α/β region of subtilisin illustrating that $β_2$-$α_B$-$β_3$ has a different hand than the other β-α-β units.

A structural anomaly in subtilisin has functional consequences

There is one anomalous and puzzling feature of the subtilisin structure. We mentioned in Chapter 4 that virtually all β-α-β motifs were of the same hand, they were right-handed. Subtilisin contains the one exception to this general rule, which is illustrated in the topology diagram Figure 11.15. There are three β-α-β motifs in subtilisin, $β_2$-$α_B$-$β_3$, $β_3$-$α_C$-$β_4$, and $β_4$-$α_D$-$β_5$. If these motifs were of the same hand, the three α helices $α_B$, $α_C$, and $α_D$ would be on the same side of the β sheet. However, $α_B$ is beneath the sheet in the topology diagram in contrast to the other two helices because $β_2$-$α_B$-$β_3$ is left-handed. Why has this exception to the general rule of right-handed β-α-β motifs evolved?

The answer is quite clear. His 64, which is part of the catalytic triad, is in the first turn of helix $α_B$ (Figure 11.13). This helix would be on the other side of the β sheet, far removed from the active site if the motif $β_2$-$α_B$-$β_3$ were right-handed. Therefore, to produce a proper catalytic triad of Asp 32, His 64, and Ser 221, helix $α_B$ must be on the same side of the β sheet as Ser 221; consequently, the motif has evolved to be left-handed.

Transition-state stabilization in subtilisin is dissected by protein engineering

Two essential features are required to stabilize the covalent tetrahedral transition state in serine proteinases—the oxyanion hole, which provides hydrogen bonds to the negatively charged oxygen atom in the transition state, and the histidine residue of the catalytic triad, which provides a positive charge. The charge on this histidine is, in turn, stabilized by the aspartic acid side chain of the catalytic triad (Figure 11.6). The histidine residue also plays a second role in the catalytic mechanism by accepting a proton from the reactive serine residue and then donating that proton to the nitrogen atom of the leaving group. The effects on the catalytic rate of the different side chains involved in the catalytic triad and the oxyanion hole have been examined by P. Carter, J.A. Wells, and D. Estell at Genentech, USA, by analyses of mutants of subtilisin with one or several of these side chains have been changed.

Catalysis occurs without a catalytic triad

By changing Ser 221 in subtilisin to Ala the reaction rate (both k_{cat} and k_{cat}/K_m) is reduced by a factor of about 10^6 compared with the wild-type enzyme. The K_m value and, by inference, the initial binding of substrate are essentially unchanged. This mutation prevents formation of the covalent bond with the substrate and therefore abolishes the reaction mechanism outlined in Figure 11.5. When the Ser 221 to Ala mutant is further mutated by changes of His 64 to Ala or Asp 32 to Ala or both, as expected there is no effect on the catalytic reaction rate, since the reaction mechanism that involves the catalytic triad is no longer in operation. However, the enzyme still has an appreciable catalytic effect; peptide hydrolysis is still about 10^3–10^4 times the nonenzymatic rate. Whatever the reaction mechanism used by these mutants, it is apparent that the remaining parts of the active site must bind more tightly to the substrate in its transition state than in its initial state, thereby giving a higher reaction rate than in the absence of enzyme.

Substrate molecules provide catalytic groups in substrate-assisted catalysis

The single mutation His 64–Ala decreases the reaction rate of subtilisin for most substrates by the same factor (approx. 10^6) as the mutation of Ser 221. This histidine (His 64), therefore, seems to be as important as Ser 221 for the formation of a covalent tetrahedral intermediate. However, model building suggested that it might be possible at least partly to compensate for the loss of this histidine in the catalytic triad of the mutant protein with a histidine side chain from a peptide substrate (Figure 11.16). Experiments confirmed this prediction and showed that the mutant His 64–Ala catalyzes hydrolysis of a peptide substrate about 400 times faster when the peptide has histidine at the appropriate position in its sequence. However, the rate is still several orders of magnitude below the rate of the wild-type enzyme, presumably because of the slightly different position and orientation of the histidine side chain. Nevertheless, the principle of **substrate-assisted catalysis** has been demonstrated: an essential group that is lacked by a mutant enzyme can be replaced by a similar group from the substrate. One consequence of substrate-assisted catalysis is that the mutant enzyme is highly specific for substrates containing the essential group. The His 64–Ala mutant of subtilisin, for example, has a specificity factor (ratios of k_{cat}/K_m) of about 200 for substrates containing histidine.

The single mutation Asp 32–Ala reduces the catalytic reaction rate by a factor of about 10^4 compared with wild type. This rate reduction reflects the role of Asp 32 in stabilizing the positive charge that His 64 acquires in the transition state. A similar reduction of k_{cat} and k_{cat}/K_m (2.5×10^3) is obtained for the single mutant Asn 155–Thr. Asn 155 provides one of the two hydrogen bonds to the substrate transition state in the oxyanion hole of subtilisin.

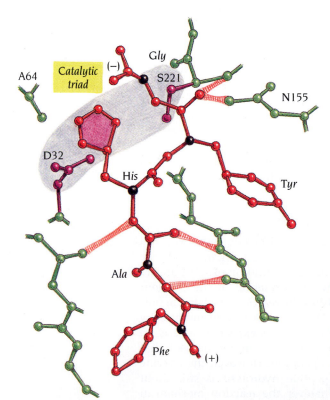

Figure 11.16 Substrate-assisted catalysis. Schematic diagram from model building of a substrate, NH$_2$-Phe-Ala-His-Tyr-Gly-COOH (red), bound to the subtilisin mutant His 64–Ala. The diagram illustrates that the His residue of the substrate can occupy roughly the same position in this mutant as His 64 in wild-type subtilisin (see Figure 11.14) and thereby partly restore the catalytic triad.

Model building shows that the OH group of Thr in the mutant is too far away to provide such a hydrogen bond. The loss of this feature of the stabilization of the transition state thus reduces the rate by more than a thousandfold.

The subtilisin mutants described here illustrate the power of protein engineering as a tool to allow us to identify the specific roles of side chains in the catalytic mechanisms of enzymes. In Chapter 17 we shall discuss the utility of protein engineering in other contexts, such as design of novel proteins and the elucidation of the energetics of ligand binding to proteins.

Conclusion

Enzymes increase the rate of chemical reactions by decreasing the activation energy of the reactions. This is achieved primarily by the enzyme preferentially binding to the transition state of the substrate. Catalytic groups of the enzyme are required to achieve a specific reaction path for the conversion of substrate to product.

Serine proteinases such as chymotrypsin and subtilisin catalyze the cleavage of peptide bonds. Four features essential for catalysis are present in the three-dimensional structures of all serine proteinases: a catalytic triad, an oxyanion binding site, a substrate specificity pocket, and a nonspecific binding site for polypeptide substrates. These four features, in a very similar arrangement, are present in both chymotrypsin and subtilisin even though they are achieved in the two enzymes in completely different ways by quite different three-dimensional structures. Chymotrypsin is built up from two β-barrel domains, whereas the subtilisin structure is of the α/β type. These two enzymes provide an example of convergent evolution where completely different loop regions, attached to different framework structures, form similar active sites.

The catalytic triad consists of the side chains of Asp, His, and Ser close to each other. The Ser residue is reactive and forms a covalent bond with the substrate, thereby providing a specific pathway for the reaction. His has a dual role: first, it accepts a proton from Ser to facilitate formation of the covalent bond; and, second, it stabilizes the negatively charged transition state. The proton is subsequently transferred to the N atom of the leaving group. Mutations of either of these two residues decrease the catalytic rate by a factor of 10^6 because they abolish the specific reaction pathway. Asp, by stabilizing the positive charge of His, contributes a rate enhancement of 10^4.

The oxyanion binding site stabilizes the transition state by forming two hydrogen bonds to a negatively charged oxygen atom of the substrate. Mutations that prevent formation of one of these bonds in subtilisin decrease the rate by a factor of about 10^3.

Mutations in the specificity pocket of trypsin, designed to change the substrate preference of the enzyme, also have drastic effects on the catalytic rate. These mutants demonstrate that the substrate specificity of an enzyme and its catalytic rate enhancement are tightly linked to each other because both are affected by the difference in binding strength between the transition state of the substrate and its normal state.

Selected readings

General

Blow, D.M. Structure and mechanism of chymotrypsin. *Acc. Chem. Res.* 9: 145–152, 1976.

Fersht, A. *Enzyme Structure and Mechanism*, 2nd ed. New York: W.H. Freeman, 1984.

Huber, R., Bode, W. Structural basis of the activation and action of trypsin. *Acc. Chem. Res.* 11: 114–122, 1978.

James, M.N.G. An x-ray crystallographic approach to enzyme structure and function. *Can. J. Biochem.* 58: 251–270, 1980.

Jencks, W.P. Binding energy, specificity, and enzymatic catalysis: the Circe effect. *Adv. Enzymol.* 43: 219–410, 1975.

Knowles, J.R. Tinkering with enzymes: what are we learning? *Science* 236: 1252–1258, 1987.

Kraut, J. How do enzymes work? *Science* 242: 533–540, 1988.

Kraut, J. Serine proteases: structure and mechanism of catalysis. *Annu. Rev. Biochem.* 46: 331–358, 1977.

Neurath, H. Evolution of proteolytic enzymes. *Science* 224: 350–357, 1984.

Pauling, L. Nature of forces between large molecules of biological interest. *Nature* 161: 707–709, 1948.

Steitz, T.A., Shulman, R.G. Crystallographic and NMR studies of the serine proteases. *Annu. Rev. Biochem. Biophys.* 11: 419–444, 1982.

Stroud, R.M. A family of protein-cutting proteins. *Sci. Am.* 231(1): 74–88, 1974.

Walsh, C. *Enzymatic Reaction Mechanisms*. New York: W.H. Freeman, 1979.

Warshel, A., et al. How do serine proteases really work? *Biochemistry* 28:3629–3637, 1989.

Wells, J.A., et al. On the evolution of specificity and catalysis in subtilisin. *Cold Spring Harbor Symp. Quant. Biol.* 52: 647–652, 1987.

Specific structures

Bode, W., et al. Refined 1.2 Å crystal structure of the complex formed between subtilisin Carlsberg and the inhibitor eglin c. Molecular structure of eglin and its detailed interaction with subtilisin. *EMBO J.* 5: 813–818, 1986.

Bode, W., et al. The refined 1.9 Å crystal structure of human α-thrombin: interaction with D-Phe-Pro-Arg chloromethylketone and significance of the Tyr-Pro-Pro-Trp insertion segment. *EMBO J.* 8: 3467–3475, 1989.

Bryan, P., et al. Site-directed mutagenesis and the role of the oxyanion hole in subtilisin. *Proc. Natl. Acad. Sci. USA* 83: 3743–3745, 1986.

Carter, P., Wells, J.A. Dissecting the catalytic triad of a serine protease. *Nature* 332: 564–568, 1988.

Carter, P., Wells, J.A. Engineering enzyme specificity by "substrate-assisted catalysis." *Science* 237: 394–399, 1987.

Craik, C.S., et al. Redesigning trypsin: alteration of substrate specificity. *Science* 228: 291–297, 1985.

Craik, C.S., et al. The catalytic role of the active site aspartic acid in serine proteases. *Science* 237: 909–913, 1987.

Cunningham, B.C., Wells, J.A. Improvement in the alkaline stability of subtilisin using an efficient random mutagenesis and screening procedure. *Prot. Eng.* 1: 319–325, 1987.

Drenth, J., et al. Subtilisin novo. The three-dimensional structure and its comparison with subtilisin BPN. *Eur. J. Biochem.* 26: 177–181, 1972.

Estell, D.A., Graycar, T.P., Wells, J.A. Engineering an enzyme by site-directed mutagenesis to be resistant to chemical oxidation. *J. Biol. Chem.* 260: 6518–6521, 1985.

Fehlhammer, H., Bode, W., Huber, R. Crystal structure of bovine trypsinogen at 1.8 Å resolution. II. Crystallographic refinement, refined crystal structure and comparison with bovine trypsin. *J. Mol. Biol.* 111: 415–438, 1977.

Fujinaga, M., et al. Crystal and molecular structures of the complex of α-chymotrypsin with its inhibitor turkey ovomucoid third domain at 1.8 Å resolution. *J. Mol. Biol.* 195: 397–418, 1987.

Graf, L., et al. Selective alteration of substrate specificity by replacement of aspartic acid 189 with lysine in the binding pocket of trypsin. *Biochemistry* 26: 2616–2623, 1987.

Grütter, M.G., et al. Crystal structure of the thrombin-hirudin complex: a novel mode of serine protease inhibition. *EMBO J.* 9: 2361–2365, 1990.

James, M.N.G., et al. Structures of product and inhibitor complexes of Streptomyces griseus protease A at 1.8 Å resolution. A model for serine protease catalysis. *J. Mol. Biol.* 144: 43–88, 1980.

Krieger, M., Kay, L.M., Stroud, R.M. Structure and specific binding of trypsin: comparison of inhibited derivatives and a model for substrate binding. *J. Mol. Biol.* 83: 209–230, 1974.

Matthews, B.W., Sigler, P.B., Henderson, R., Blow, D.M. Three-dimensional structure of tosyl-α-chymotrypsin. *Nature* 214: 652–656, 1967.

McLachlan, A.D. Gene duplications in the structural evolution of chymotrypsin. *J. Mol. Biol.* 128: 49–79, 1979.

Poulos, T.L., et al. Polypeptide halomethyl ketones bind to serine proteases as analogs of the tetrahedral intermediate. *J. Biol. Chem.* 251: 1097–1103, 1976.

Read, R.J., James, M.N.G. Refined crystal structure of *Streptomyces griseus* trypsin at 1.7 Å resolution. *J. Mol. Biol.* 200: 523–551, 1988.

Rühlman, A., et al. Structure of the complex formed by bovine trypsin and bovine pancreatic trypsin inhibitor. *J. Mol. Biol.* 77: 417–436, 1973.

Shotton, D.M., Watson, H.C. Three-dimensional structure of tosyl-elastase. *Nature* 225: 811–816, 1970.

Sigler, P.B., et al. Structure of crystalline α-chymotrypsin II. A preliminary report including a hypothesis for the activation mechanism. *J. Mol. Biol.* 35: 143–164, 1968.

Smith, S.O., et al. Crystal versus solution structures of enzymes: NMR spectroscopy of a crystalline serine protease. *Science* 244: 961–964, 1989.

Sprang, S., et al. The three-dimensional structure of Asn[102] mutant of trypsin: role of Asp[102] in serine protease catalysis. *Science* 237: 905–909, 1987.

Thomas, P.G., Russel, A.J., Fersht, A. Tailoring the pH dependence of enzyme catalysis using protein engineering. *Nature* 318: 375–376, 1985.

Tsukada, H., Blow, D.M. Structure of α-chymotrypsin refined at 1.68 Å resolution. *J. Mol. Biol.* 184: 703–711, 1985.

Wang, D., Bode, W., Huber, R. Bovine chymotrypsinogen A. X-ray crystal structure analysis and refinement of a new crystal form at 1.8 Å resolution. *J. Mol. Biol.* 185: 595–624, 1985.

Wells, J.A., et al. Designing substrate specificity by protein engineering of electrostatic interactions. *Proc. Natl. Acad. Sci. USA* 84: 1219–1223, 1987.

Wells, J.A., et al. Recruitment of substrate-specificity properties from one enzyme into a related one by protein engineering. *Proc. Natl. Acad. Sci. USA* 84: 5167–5171, 1987.

Wright, C.S., Alden, R.A., Kraut, J. Structure of subtilisin BPN' at 2.5 Å resolution. *Nature* 221: 235–242, 1969.

Membrane Proteins

12

Cells and organelles within them are bounded by membranes, which are extremely thin (4.5 nm) films of lipids and protein molecules. The lipids form a bilayered sheet structure that is hydrophilic on its two outer surfaces and hydrophobic in between. Protein molecules are embedded in this layer, and in the simplest case they are arranged with three distinct regions: one hydrophobic transmembrane segment and two hydrophilic regions, one on each side of the membrane. Those proteins whose polypeptide chain traverses the membrane only once usually form functional globular domains on at least one side of the membrane (Figure 12.1a). Often these can be cleaved off by proteolytic enzymes. The hemagglutinin and neuraminidase of influenza virus (discussed in Chapter 5), G-proteins and receptors (discussed in Chapter 13), and HLA proteins (discussed in Chapter 15) are examples of such cleavage products that can be handled as functional soluble globular domains. The polypeptide chain of other transmembrane proteins passes through the membrane several times, usually as α helices but in some cases as β strands (Figure 12.1b,c). In these cases the hydrophilic regions on either side of the membrane are the termini of the chain and the loops between the membrane-spanning parts. Proteolytic cleavage of these hydrophilic regions produces a number of fragments, and function is not preserved. Some proteins do not traverse the membrane but are instead attached to one side either through α helices that lie parallel to the membrane surface (Figure 12.1d) or by fatty acids, covalently linked to the protein, that intercalate in the lipid bilayer of the membrane.

Figure 12.1 Four different ways in which protein molecules may be bound to a membrane. Membrane-bound regions are green and regions outside the membrane are red. Alpha-helices are drawn as cylinders and β strands as arrows. From left to right are (a) a protein whose polypeptide chain traverses the membrane once as an α helix, (b) a protein that forms several transmembrane α helices connected by hydrophilic loop regions, (c) a protein with several β strands that form a channel through the membrane, and (d) a protein that is anchored to the membrane by one α helix parallel to the plane of the membrane.

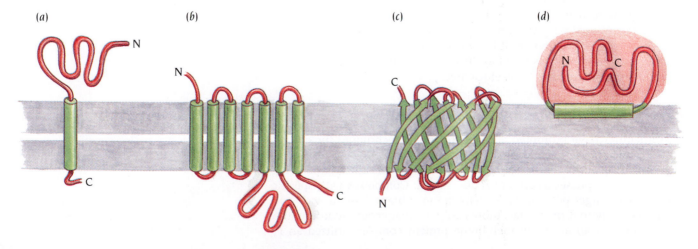

A biological membrane functions basically as a permeability barrier that establishes discrete compartments and prevents the random mixing of the contents of one compartment with those of another. However, biological membranes are more than passive containers. The embedded proteins serve as highly active mediators between the cell and its environment or the interior of an organelle and the cytosol. They catalyze specific transport of metabolites and ions across the membrane barriers. They convert the energy of sunlight into chemical and electrical energy, and they couple the flow of electrons to the synthesis of ATP. Furthermore, they act as signal receptors and transduce signals across the membrane. The signals can be, for example, neurotransmitters, growth factors, hormones, light or chemotactic stimuli. The transmembrane proteins of the plasma membrane are also involved in cell–cell recognition.

In this chapter we describe some examples of structures of membrane-bound proteins known to high resolution, and outline how the elucidation of these structures has contributed to understanding the specific function of these proteins, as well as some general principles for the construction of membrane-bound proteins. In Chapter 13 we describe some examples of the domain organization of receptor families and their associated proteins involved in signal transduction through the membrane.

Membrane proteins are difficult to crystallize

Membrane proteins, which have both hydrophobic and hydrophilic regions on their surfaces, are not soluble in aqueous buffer solutions and denature in organic solvents. However, if detergents, such as octylglucoside, are added to an aqueous solution, these proteins can be solubilized and purified in their native conformation. The hydrophobic part of the detergent molecules binds to the protein's hydrophobic surfaces, while the detergents' polar head-groups face the solution (Figure 12.2a). In this way the protein–detergent complex acquires an essentially hydrophilic surface with the hydrophobic parts buried inside the complex.

Such solubilized protein–detergent complexes are the starting material for purification and crystallization. For some proteins, the addition of small amphipathic molecules to the detergent-solubilized protein promotes crystallization, probably by facilitating proper packing interactions between the molecules in all three dimensions in a crystal (Figure 12.2b). Therefore, many different amphipathic molecules are added in separate crystallization experiments until, by trial and error, the correct one is found.

Despite considerable efforts very few membrane proteins have yielded crystals that diffract x-rays to high resolution. In fact, only about a dozen such proteins are currently known, among which are porins (which are outer membrane proteins from bacteria), the enzymes cytochrome c oxidase and prostaglandin synthase, and the light-harvesting complexes and photosynthetic reaction centers involved in photosynthesis. In contrast, many other membrane proteins have yielded small crystals that diffract poorly, or not at all, using conventional x-ray sources. However, using the most advanced synchrotron sources (see Chapter 18) it is now possible to determine x-ray structures from protein crystals as small as 20 μm wide which will permit more membrane protein structures to be elucidated.

Novel crystallization methods are being developed

These difficulties have prompted a search for novel techniques for crystallization of membrane proteins. Two approaches have given promising results; one using antibodies to solubilize the proteins and the second using continuous lipidic phases as crystallization media. Complexes with specific antibodies have larger polar surfaces than the membrane protein itself and are therefore likely to form crystals more easily in an aqueous enviroment. A recent example of an antibody–membrane protein complex utilized an F_v

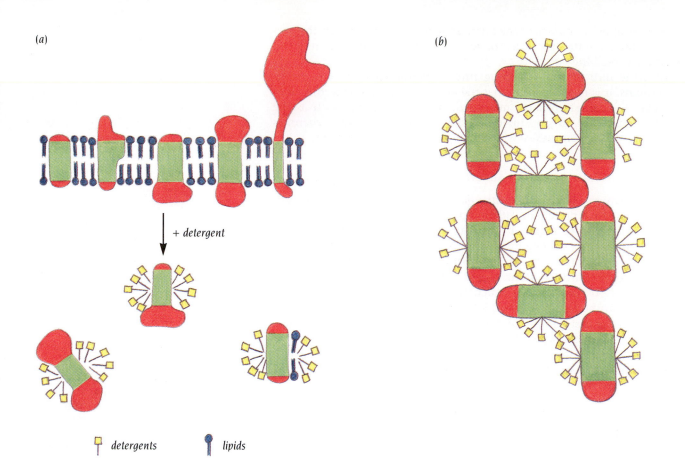

(a)   (b)

+ detergent

� detergents   ⦙ lipids

fragment (see Chapter 15) to crystallize a bacterial cytochrome c oxidase. In these crystals the major packing interactions are formed by the polar surfaces of the complex.

A continuous lipidic cubic phase is obtained by mixing a long-chain lipid such as monoolein with a small amount of water. The result is a highly viscous state where the lipids are packed in curved continuous bilayers extending in three dimensions and which are interpenetrated by communicating aqueous channels. Crystallization of incorporated proteins starts inside the lipid phase and growth is achieved by lateral diffusion of the protein molecules to the nucleation sites. This system has recently been used to obtain three-dimensional crystals 20 x 20 x 8 µm in size of the membrane protein bacteriorhodopsin, which diffracted to 2 Å resolution using a microfocus beam at the European Synchrotron Radiation Facility.

Figure 12.2 (a) Schematic drawing of membrane proteins in a typical membrane and their solubilization by detergents. The hydrophilic surfaces of the membrane proteins are indicated by red. (b) A membrane protein crystallized with detergents bound to its hydrophobic protein surface. The hydrophilic surfaces of the proteins and the symbols for detergents are as in (a). (Adapted from H. Michel, *Trends Biochem. Sci.* 8: 56–59, 1983.)

Two-dimensional crystals of membrane proteins can be studied by electron microscopy

The first really useful information about the structure of membrane proteins came not from x-ray crystallography but from high-resolution electron microscopy of **two-dimensional crystals**. Two-dimensional crystals can be thought of as crystalline membranes in which the membrane protein is arranged on a two-dimensional lattice. Naturally abundant membrane proteins sometimes form two-dimensional crystals *in vivo* or in isolated native membranes, in particular when some components such as lipids or other proteins are selectively extracted. A different way of making two-dimensional crystals is by reconstitution of detergent-solubilized membrane proteins into bilayers, which provide a natural, membrane-like environment for the protein. When the detergent is removed from a lipid–protein detergent mixture by dialysis or absorption, the hydrophobic effect causes the hydrophobic fatty acid tails of the lipids to associate with each other, and with the hydrophobic surface of the membrane protein. In this way, the protein is incorporated into lipid sheets or vesicles. In favorable conditions it can then form a crystalline lattice.

Given the difficulty of obtaining three-dimensional crystals of membrane proteins, it is not surprising that the electron microscope technique is now widely used to study large membrane-bound complexes such as the acetylcholine receptor, rhodopsin, ion pumps, gap junctions, water channels and light-harvesting complexes, which crystallize in two dimensions.

Bacteriorhodopsin contains seven transmembrane α helices

The purple membrane of *Halobacterium halobium* contains ordered sheets of **bacteriorhodopsin**, a protein of 248 amino acid residues which binds retinal, the same photosensitive pigment that is used to capture light in our eyes. Bacteriorhodopsin uses the energy of light to pump protons across the membrane. Richard Henderson and Nigel Unwin at LMB, Cambridge, UK, pioneered high-resolution three-dimensional reconstruction of tilted low-dose electron microscopy images using such two-dimensional crystals. The 7-Å model of bacteriorhodopsin (Figure 12.3a) that they obtained in this way in 1975 provided the first glimpse of how membrane proteins are constructed. The fundamental observation that this protein has a number of **transmembrane α helices** (Figure 12.3b) has had a great impact on subsequent theories and experiments on membrane proteins; it also provided the first

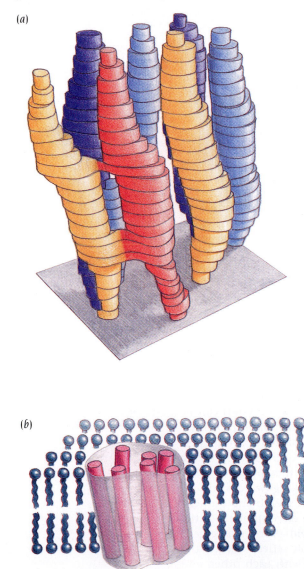

(a)

(b)

Figure 12.3 Two-dimensional crystals of the protein bacteriorhodopsin were used to pioneer three-dimensional high-resolution structure determination from electron micrographs. An electron density map to 7 Å resolution (a) was obtained and interpreted in terms of seven transmembrane helices (b). In 1990 the resolution was extended to 3 Å, which confirmed the presence of the seven α helices (c). This structure also showed how these helices were connected by loop regions and where the retinal molecule was bound to bacteriorhodopsin. (Courtesy of R. Henderson.)

(c)

experimental evidence behind the now extensively used methods to predict transmembrane helices from amino acid sequences.

This electron microscopy reconstruction has since been extended to high resolution (3 Å) where the connections between the helices and the bound retinal molecule are visible together with the seven helices (Figure 12.3c). The helices are tilted by about 20° with respect to the plane of the membrane. This is the first example of a high-resolution three-dimensional protein structure determination using electron microscopy. The structure has subsequently been confirmed by x-ray crystallographic studies to 2 Å resolution.

Bacteriorhodopsin is a light-driven proton pump

Halobacteria have the simplest biological system for the conversion of light to chemical energy. Under conditions of low oxygen tension and intense illumination the cells synthesize bacteriorhodopsin. When the bound retinal absorbs a photon it undergoes an isomerization from *trans* to *cis* (Figure 12.4) and, as a consequence, protons are pumped from the cytosol to the extracellular space, creating a proton gradient. This gradient is used to generate ATP and to transport ions and molecules across the membrane. The mechanism by which bacteriorhodopsin acts as a **proton pump** has been studied by many biophysical methods over several decades and the results, in conjunction with Henderson's structural studies, have given the following simplified scheme for proton pumping.

Retinal is bound in a pocket of bacteriorhodopsin about equidistant from the two sides of the membrane (Figure 12.5). The pigment forms a Schiff base with a lysine residue, Lys 216; in other words, it is covalently linked to the nitrogen atom of the lysine side chain that is protonated and therefore has a positive charge (see Figure 12.5). This positive charge is positioned in a channel of the protein that extends across the membrane and through which protons are pumped from the cytosolic to the extracellular side. The channel is narrow on the cytosolic side and lined with hydrophobic residues with the exception of Asp 96, which has been shown by studies of mutant proteins to be essential for proton pumping. In contrast, the channel is wide and hydrophilic on the extracellular side and includes a second essential acidic residue, Asp 85.

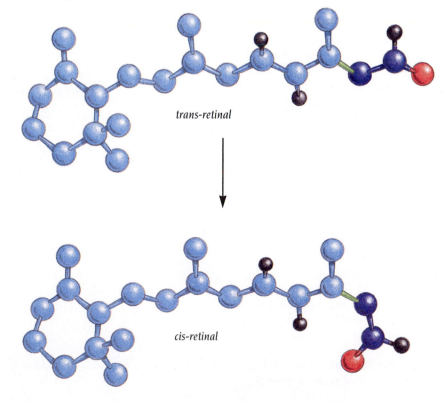

trans-retinal

cis-retinal

Figure 12.4 The light-absorbing pigment retinal undergoes a conformational change called isomerization, when it absorbs light. One part of the molecule (dark blue and red) rotates 180° around a double bond between two carbon atoms (green). The geometry of the molecule is changed by this rotation from a *trans* form to a *cis* form. Carbon atoms are blue, hydrogen atoms gray and the oxygen atom red.

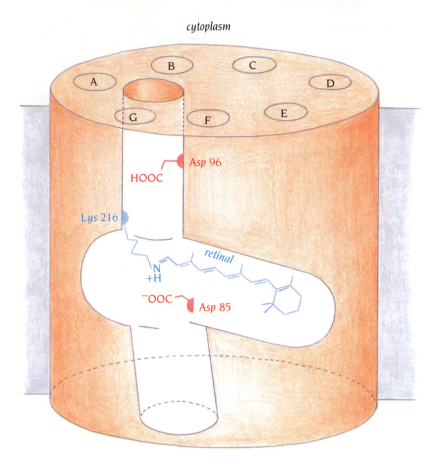

cytoplasm

Figure 12.5 Schematic diagram of the bacteriorhodopsin molecule illustrating the relation between the proton channel and bound retinal in its *trans* form. A to E are the seven transmembrane helices. Retinal is covalently bound to a lysine residue. The relative positions of two Asp residues, which are important for proton transfer, are also shown. (Adapted from R. Henderson et al., *J. Mol. Biol.* 213: 899–929, 1990.)

In the unisomerized *trans* state of retinal, Asp 85 is close to the positive charge of the Schiff base (Figure 12.6a). The structural change of the retinal molecule due to the *trans* to *cis* photoisomerization causes the Schiff base to change its position relative to Asp 85, which induces transfer of the Schiff base proton to the aspartate group (Figure 12.6b). Once the Schiff base–Asp 85 ion pair is converted to a neutral pair by this proton transfer the protein undergoes a conformational change from the T state to the relaxed R state (see Chapter 6). X-ray diffraction and electron microscopy studies have shown that this conformational change involves a reorganization of some of the transmembrane helices that bind retinal, with the consequence that the Schiff base is moved from the extracellular part of the channel to the cyto-plasmic part, away from Asp 85 towards Asp 96. Asp 85 then delivers a pro-ton through the hydrophilic part of the channel to the extracellular space and Asp 96 reprotonates the Schiff base, which subsequently reverts to the *trans* state and the protein changes its conformation back to the T-state ready for another cycle of photoisomerization-induced proton transfer (Figure 12.6c,d). In short, light causes a chemical change at the active site that alters the conformation of the protein, which in turn drives protons from the cytosolic side of the membrane to the extracellular side.

Porins form transmembrane channels by β strands

Gram-negative bacteria are surrounded by two membranes, an inner plasma membrane and an outer membrane. These are separated by a periplasmic space. Most plasma membrane proteins contain long, continuous sequences of about 20 hydrophobic residues that are typical of transmembrane α helices such as those found in bacteriorhodopsin. In contrast, most outer membrane proteins do not show such sequence patterns.

This enigma was resolved in 1990 when the x-ray structure of an outer membrane protein, **porin**, showed that the transmembrane regions were β

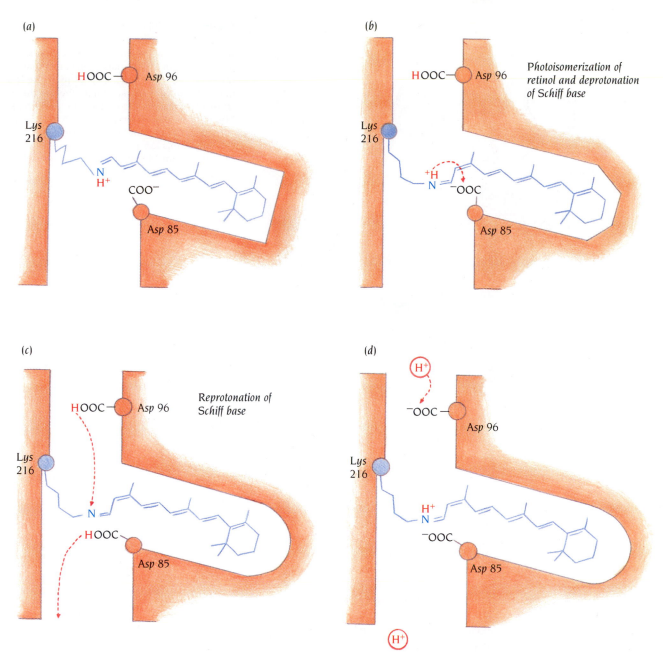

(a)

HOOC— Asp 96

Lys 216

N
H⁺

COO⁻

Asp 85

(b)

HOOC— Asp 96

Photoisomerization of retinol and deprotonation of Schiff base

Lys 216

⁺H
N

⁻OOC

Asp 85

(c)

HOOC— Asp 96

Reprotonation of Schiff base

Lys 216

N

HOOC

Asp 85

(d)

H⁺

⁻OOC— Asp 96

Lys 216

H⁺
N

⁻OOC

Asp 85

H⁺

strands and not α helices. Sequence comparisons have since shown that most if not all bacterial outer membrane porins have transmembrane β strands. Porins are among the most abundant proteins in bacteria. Each *Escherichia coli* cell contains about 100,000 copies of porin molecules in its outer membrane. Each porin forms an open water-filled channel that allows passive diffusion of nutrients and waste products across the outer membrane. These channels are restricted in size, and this excludes larger, potentially toxic compounds from entering the cell.

Porin channels are made by up and down β barrels

The first x-ray structure of a porin was determined by the group of Georg Schulz and Wolfram Welte at Freiburg University, Germany, who succeeded in growing crystals of a porin from *Rhodobacter capsulatus* that diffracted to 1.8 Å resolution. Since then the x-ray structures of several other porin molecules have been determined and found to be very similar to the *R. capsulatus* porin despite having no significant sequence identity.

Each subunit of the trimeric porin molecule from *R. capsulatus* folds into a 16-stranded up and down antiparallel β barrel in which all β strands form

Figure 12.6 Schematic diagram illustrating the proton movements in the photocycle of bacteriorhodopsin. The protein adopts two main conformational states, tense (T) and relaxed (R). The T state binds *trans*-retinal tightly and the R state binds *cis*-retinal. (a) Structure of bacteriorhodopsin in the T state with *trans*-retinal bound to Lys 216 via a Schiff base. (b) A proton is transferred from the Schiff base to Asp 85 following isomerization of retinal and a conformational change of the protein. (c) Structure of bacteriorphodopsin in the R state with *cis*-retinal bound. A proton is transferred from Asp 96 to the Schiff base and from Asp 85 to the extracellular space. (d) A proton is transferred from the cytoplasm to Asp 96. (Adapted from R. Henderson et al., *J. Mol. Biol.* 213: 899–929, 1990.)

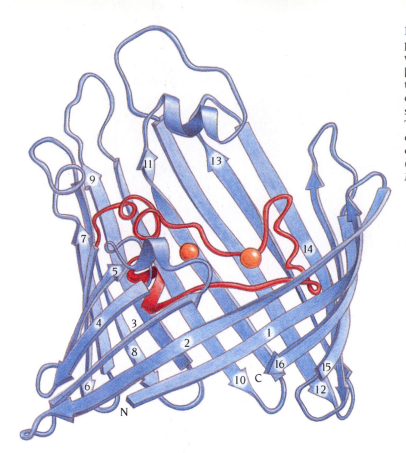

Figure 12.7 Ribbon diagram of one subunit of porin from *Rhodobacter capsulatus* viewed from within the plane of the membrane. Sixteen β strands form an antiparallel β barrel that traverses the membrane. The loops at the top of the picture are extracellular whereas the short turns at the bottom face the periplasm. The long loop between β strands 5 and 6 (red) constricts the channel of the barrel. Two calcium atoms are shown as orange circles. (Adapted from M.S. Weiss and G.E. Schulz, *J. Mol. Biol.* 227: 493-509, 1992.)

hydrogen bonds to their neighbors (Figure 12.7). All other known porin barrels also contain 16 β strands except a maltoporin from *E. coli*, which contains 18 β strands. In contrast to the β barrels we have previously discussed, which have a tightly packed hydrophobic core, the porin barrels contain a central channel because of the large number of β strands. The channel is, however, partially blocked by a long loop region between β strands 5 and 6 that projects into the channel. This arrangement creates a narrow region in the middle of the channel, called the eyelet, about 9 Å long and 8 Å in diameter, which defines the size of solute molecules that can traverse the channel (see Figure 12.8).

The eyelet is lined almost exclusively with positively and negatively charged groups that are arranged on opposite sides of the channel, causing a transversal electric field across the pore. One His, two Lys and three Arg residues form the positive side and four Glu and seven Asp residues are on the negative side. The large local surplus of negative charges is partially compensated by two bound calcium atoms. This asymmetric arrangement of charges no doubt contributes to the selection of molecules that can pass through the channel.

Since the outside of the barrel faces hydrophobic lipids of the membrane and the inside forms the solvent-exposed channel, one would expect the β strands to contain alternating hydrophobic and hydrophilic side chains. This requirement is not strict, however, because internal residues can be hydrophobic if they are in contact with hydrophobic residues from loop regions. The prediction of transmembrane β strands from amino acid sequences is therefore more difficult and less reliable than the prediction of transmembrane α helices.

Each porin molecule has three channels

The complete porin molecule is a stable trimer of three identical subunits, three each with a functional channel (Figure 12.8). About one-third of the

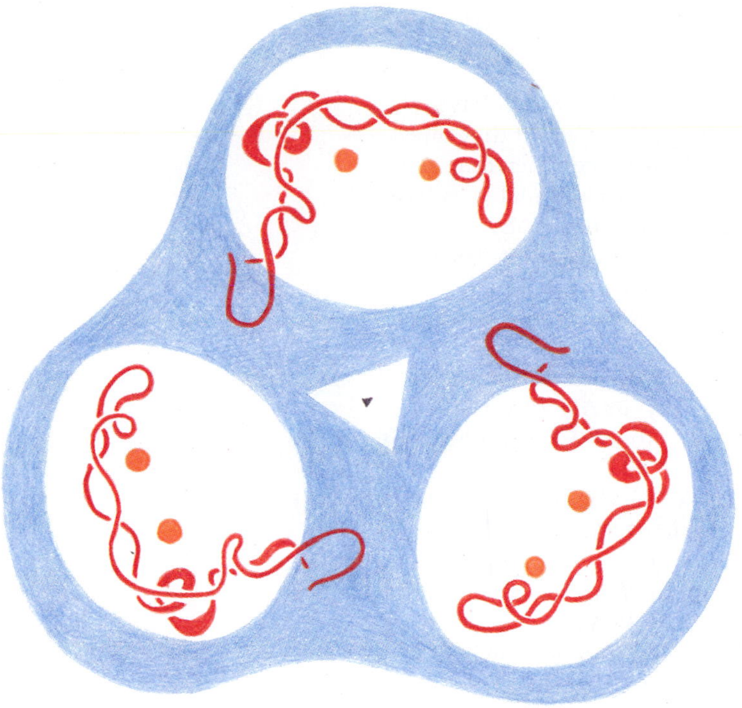

Figure 12.8 Schematic diagram of the trimeric porin molecule viewed from the extracellular space. Blue regions illustrate the walls of the three porin barrels, the loop regions that constrict the channel are red and the calcium atoms are orange.

barrel's outer surface is involved in subunit interactions with the other two subunits, comprising polar interactions from loop regions and hydrophobic interactions from side chains of the β strands. The bottom part of the trimer facing the periplasmic space has a flat and smooth surface made up of the short loop regions at this end of the three β barrels (as shown in Figure 12.7). In contrast the upper part has long loop regions and is funnel-shaped, with the channels of the three barrels providing three outlets from the single funnel. The inner sides of the funnel are lined with hydrophilic residues that are in contact with solvent from the extracellular space.

The journey of an extracellular solute molecule through the channel may now be depicted in the following way. Large molecules are prevented from entering by the size of the funnel of the trimer. Further screening occurs at the entrance of the channel in the individual barrels. Finally, a molecule small enough to enter the central channel then encounters the eyelet, where the charged side chains determine the size limitations and the ion selectivity of the pore. After this narrow passage the molecule is effectively released into the periplasmic space. It should, however, be borne in mind that this picture describes only one state of the channel. Triggers such as an electric potential or a change in osmotic pressure can modify the structure of the channel and therefore its permeability, but as yet we have no structural information on such changes.

The outer surface of the trimeric porin molecule shows a pronounced partitioning with respect to hydrophobicity. Polar side chains of the loop regions are abundant at the top of the trimer, followed by a hydrophobic band with a width of 25 Å that encircles the molecule. This band presumably forms the region that is embedded in the core of the outer membrane, which has a thickness of about 25 Å. The top and the bottom of this hydrophobic band are largely composed of aromatic residues, Phe and Tyr, whereas the central region is composed of small to medium-sized aliphatic residues. This suggests that aromatic rings are energetically favored at the interface between the inner lipid part of the membrane and the hydrophilic regions facing the extracellular and periplasmic spaces. A similar distribution of aromatic and aliphatic residues is also present in other membrane proteins such as the photosynthetic reaction center and bacteriorhodopsin.

Ion channels combine ion selectivity with high levels of ion conductance

Outer membrane channel-forming proteins such as porins, which have relatively large and permissive pores, would have disastrous effects if they directly connected the inside of a cell to the extracellular space. Therefore, most channel proteins in the plasma membranes of plant and animal cells have narrow and highly selective pores that are concerned specifically with the transport of inorganic ions, and so are referred to as **ion channels**. The function of such channels is to allow specific inorganic ions, mainly K^+, Na^+, Ca^{2+} and Cl^- to diffuse rapidly across the lipid bilayer and therefore balance differences in electric charge between the two sides of the membrane, the membrane potential. The membrane potential of resting cells is determined largely by K^+, which is actively pumped into the cell by an ATP-driven Na^+, K^+ pump, but which can also move freely in or out through **K^+ leak channels** in the plasma membrane.

The K^+ leak channels are highly selective for K^+ ions by a factor of 10,000 over Na^+. Two aspects of ion conduction by these K^+ channels have intrigued biophysicists. First, what is the chemical basis for this selectivity since both K^+ and Na^+ are featureless spheres? Steric occlusion cannot account for the selectivity since Na^+ is smaller than K^+ (0.95 Å and 1.35 Å radii respectively). Second, how can K^+ channels be so selective and at the same time exhibit a throughput of 10^8 ions per second, which approaches the diffusion-limited rate? The selectivity implies strong interactions between K^+ and the pore and intuitively one would assume that the off velocity of K^+ binding would be low, and consequently the rate of release would be low. A recent x-ray structure determination of a K^+ channel from the bacterium *Streptomyces lividans* by the group of Roderick MacKinnon at Rockefeller University, New York, has revealed the structural basis for combining ion selectivity with a high rate of ion conduction in these and other ion channels.

The K⁺ channel is a tetrameric molecule with one ion pore in the interface between the four subunits

The polypeptide chain of the bacterial K^+ channel comprises 158 residues folded into two transmembrane helices, a pore helix and a cytoplasmic tail of 33 residues that was removed before crystallization. Four subunits

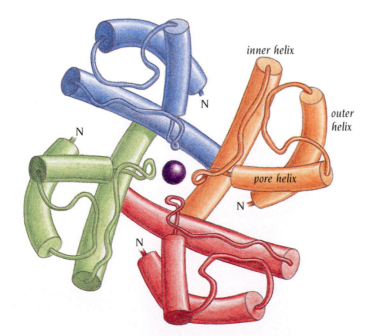

Figure 12.9 Schematic diagram of the structure of a potassium channel viewed perpendicular to the plane of the membrane. The molecule is tetrameric with a hole in the middle that forms the ion pore (purple). Each subunit forms two transmembrane helices, the inner and the outer helix. The pore helix and loop regions build up the ion pore in combination with the inner helix. (Adapted from S.A. Doyle et al., *Science* 280: 69–77, 1998.)

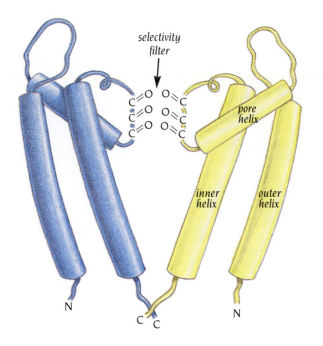

selectivity
filter

pore
helix

inner
helix

outer
helix

N

C C

N

Figure 12.10 Diagram showing two subunits of the K⁺ channel, illustrating the way the selectivity filter is formed. Main-chain atoms line the walls of this narrow passage with carbonyl oxygen atoms pointing into the pore, forming binding sites for K⁺ ions. (Adapted from D.A. Doyle et al., *Science* 280: 69–77, 1998.)

arranged around a central fourfold symmetry axis form the K⁺ channel molecule (Figure 12.9). The subunits pack together in such a way that there is a hole in the center which forms the ion pore through the membrane.

The C-terminal transmembrane helix, the inner helix, faces the central pore while the N-terminal helix, the outer helix, faces the lipid membrane. The four inner helices of the molecule are tilted and kinked so that the subunits open like petals of a flower towards the outside of the cell (Figure 12.10). The open petals house the region of the polypeptide chain between the two transmembrane helices. This segment of about 30 residues contains an additional helix, the pore helix, and loop regions which form the outer part of the ion channel. One of these loop regions with its counterparts from the three other subunits forms the narrow selectivity filter that is responsible for ion selectivity. The central and inner parts of the ion channel are lined by residues from the four inner helices.

The ion pore has a narrow ion selectivity filter

The overall length of the ion pore is 45 Å and its diameter varies along its length (Figure 12.11). As expected for a K⁺ channel, there is a surplus of

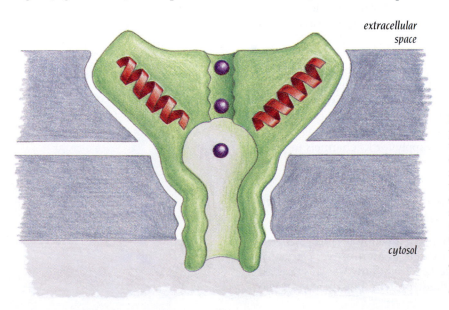

extracellular
space

cytosol

Figure 12.11 Schematic diagram of the ion pore of the K⁺ channel. From the cytosolic side the pore begins as a water-filled channel that opens up into a water-filled cavity near the middle of the membrane. A narrow passage, the selectivity filter, links this cavity to the external solution. Three potassium ions (purple spheres) bind in the pore. The pore helices (red) are oriented such that their carboxyl end (with a negative dipole moment) is oriented towards the center of the cavity to provide a compensating dipole charge to the K⁺ ions. (Adapted from D.A. Doyle et al., *Science* 280: 69–77, 1998.)

negative charges at both ends of the pore, which attract positively charged ions. From the cytosolic side, the pore begins as a channel 18 Å long, which opens into a wider cavity of about 10 Å diameter near the middle of the membrane. A narrow passage, the selectivity filter, links this cavity to the external solution. Main-chain atoms from all four subunits line the walls of this passage with carbonyl oxygen atoms pointing into the channel (see Figure 12.10). Three metal-binding sites have been identified in the ion pore (green in Figure 12.11), two within the selectivity filter and one in the cavity.

The structure of the selectivity filter has two essential features. First, the main-chain atoms create a stack of sequential oxygen rings along the passage, providing several closely spaced binding sites of the required dimensions for coordinating naked, dehydrated K^+ ions. The K^+ thus have only a small distance to diffuse from one site to the next within the selectivity filter. Second, the side chains of the residues that provide these binding sites point away from the channel and pack against the side chains from the pore helices. This packing firmly fixes the positions of the main-chain atoms, including the oxygen atoms that bind K^+. Since the side chains involved in these packing interactions are invariant in all known K^+ channels it is reasonable to assume that the carbonyl oxygen atoms are fixed in positions with the correct dimensions to provide strong binding sites for K^+. The resolution of the structure determination is, however, too low to establish details of these binding sites.

On the basis of the structure, MacKinnon has suggested a plausible mechanism for the ion selectivity and conductivity of the channel. When an ion, which in solution has a water hydration shell, enters the selectivity filter it dehydrates. Binding to the carbonyl oxygen atoms in the filter compensates the energetic cost of dehydration. The dimensions of the binding sites are such that a K^+ ion fits in the filter precisely so that the energetic costs and gains are well balanced, but the firm packing of the side chains prevents the carbonyl oxygen atoms from approaching close enough to compensate for the cost of dehydration of a Na^+ ion.

In the crystal, the selectivity filter has two K^+ ions, one bound at each end of the filter, separated by a distance of 7.5 Å. This is the same distance as the average distance between K^+ ions in 4 M KCl. There is thus a high local concentration of K^+ ions in the filter, implying that the filter attracts and concentrates K^+ ions. However, since there are no negative ions within the filter to balance these positive charges there is also a repulsive force between the two K^+ ions. Since in the crystal there is no concentration gradient of K^+ ions across the channel there is no net flow of the ions. When, however, channel molecules are embedded in cell membranes across which there is a K^+ ion concentration gradient, the ions flow through the channel. They are forced through by a combination of the mutual repulsion of the ions in the channel and the membrane potential. Within the filter the ions cascade from one carbonyl atom to the next.

All K^+ channels are tetrameric molecules. There are two closely related varieties of subunits for K^+ channels, those containing two membrane-spanning helices and those containing six. However, residues that build up the ion channel, including the pore helix and the inner helix, show a strong sequence similarity among all K^+ channels. Consequently, the structural features and the mechanism for ion selectivity and conductance described for the bacterial K^+ channel in all probability also apply for K^+ channels in plant and animal cells.

The bacterial photosynthetic reaction center is built up from four different polypeptide chains and many pigments

The crystallographic world was stunned when at a meeting in Erice, Sicily, in 1982, Hartmut Michel of the Max-Planck Institute in Martinsried, Germany, displayed the x-ray diagram shown in Figure 12.12. Not only was this the first x-ray picture to high resolution of a membrane protein, but the crystal was formed not from a small protein of trivial function but from a large complex of polypeptide chains that represents a class of proteins having a function of central importance for life on earth. The protein complex was a **photo-synthetic reaction center** from the photosynthetic purple bacterium *Rhodopseudomonas viridis*, which converts the energy of captured sunlight into electrical and chemical energy in the first steps of photosynthesis by pumping protons from one side of a membrane to the other. The structure has subsequently been resolved to 2.5 Å by H. Michel in collaboration with Hans Deisenhofer and Robert Huber at the same institute.

The interiors of rhodopseudomonad bacteria are filled with photosynthetic vesicles, which are hollow, membrane-enveloped spheres. The photosynthetic reaction centers are embedded in the membrane of these vesicles. One end of the protein complex faces the inside of the vesicle, which is known as the periplasmic side; the other end faces the cytoplasm of the cell. Around each reaction center there are about 100 small membrane proteins, the antenna pigment protein molecules, which will be described later in this chapter. Each of these contains several bound chlorophyll molecules that catch photons over a wide area and funnel them to the reaction center. By this arrangement the reaction center can utilize about 300 times more photons than those that directly strike the special pair of chlorophyll molecules at the heart of the reaction center.

The reaction center is built up from four polypeptide chains, three of which are called L, M, and H because they were thought to have light, medium, and heavy molecular masses as deduced from their electrophoretic mobility on SDS-PAGE. Subsequent amino acid sequence determinations showed, however, that the H chain is in fact the smallest with 258 amino acids, followed by the L chain with 273 amino acids. The M chain is the largest polypeptide with 323 amino acids. This discrepancy between apparent relative masses and real molecular weights illustrates the uncertainty in deducing molecular masses of membrane-bound proteins from their mobility in electrophoretic gels.

The L and M subunits show about 25% sequence identity and are therefore homologous and evolutionarily related proteins. The H subunit, on the other hand, has a completely different sequence. The fourth subunit of the reaction center is a cytochrome that has 336 amino acids with a sequence that is not similar to any other known cytochrome sequence.

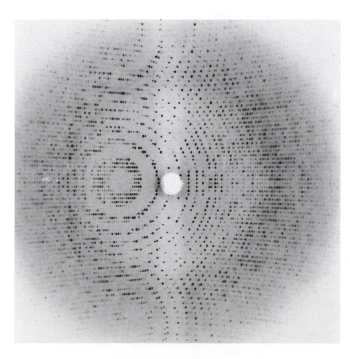

Figure 12.12 X-ray diffraction pattern from crystals of a membrane-bound protein, the bacterial photosynthetic reaction center. (Courtesy of H. Michel.)

235

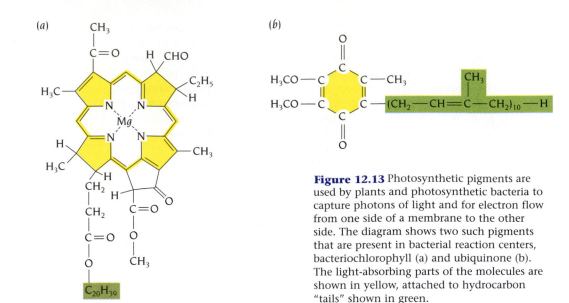

Figure 12.13 Photosynthetic pigments are used by plants and photosynthetic bacteria to capture photons of light and for electron flow from one side of a membrane to the other side. The diagram shows two such pigments that are present in bacterial reaction centers, bacteriochlorophyll (a) and ubiquinone (b). The light-absorbing parts of the molecules are shown in yellow, attached to hydrocarbon "tails" shown in green.

In addition to these polypeptide chains, the reaction center contains a number of pigments. There are four bacteriochlorophyll molecules (Figure 12.13a), two of which form the strongly interacting dimer called "the special pair." Furthermore, there is one Fe atom, a carotenoid, two quinone molecules (Figure 12.13b) and two bacteriopheophytin molecules, which are chlorophyll molecules without the central Mg^{2+} atom. Finally, the cytochrome subunit has four bound heme groups. The crystal structure shows how the polypeptide chains bind these pigments into a functional unit allowing electrons to flow from one side of the membrane to the other.

The L, M, and H subunits have transmembrane α helices

The L and the M subunits are firmly anchored in the membrane, each by five hydrophobic transmembrane α helices (yellow and red, respectively, in Figure 12.14). The structures of the L and M subunits are quite similar as expected from their sequence similarity; they differ only in some of the loop regions. These loops, which connect the membrane-spanning helices, form rather flat hydrophilic regions on either side of the membrane to provide interaction areas with the H subunit (green in Figure 12.14) on the cytoplasmic side and with the cytochrome (blue in Figure 12.14) on the periplasmic side. The H subunit, in addition, has one transmembrane α helix at the carboxy terminus of its polypeptide chain. The carboxy end of this chain is therefore on the same side of the membrane as the cytochrome. In total, eleven transmembrane α helices attach the L, M, and H subunits to the membrane.

No region of the cytochrome penetrates the membrane; nevertheless, the cytochrome subunit is an integral part of this reaction center complex, held through protein–protein interactions similar to those in soluble globular multisubunit proteins. The protein–protein interactions that bind cytochrome in the reaction center of *Rhodopseudomonas viridis* are strong enough to survive the purification procedure. However, when the reaction center of *Rhodobacter sphaeroides* is isolated, the cytochrome is lost, even though the structures of the L, M, and H subunits are very similar in the two species.

Alpha helices D and E from the L and M subunits (Figure 12.14) form the core of the membrane-spanning part of the complex. These four helices are tightly packed against each other in a way quite similar to the four-helix bundle motif in water-soluble proteins. Each of these four helices provides a histidine side chain as ligand to the Fe atom, which is located between the helices close to the cytoplasm. The role of the Fe atom is probably to

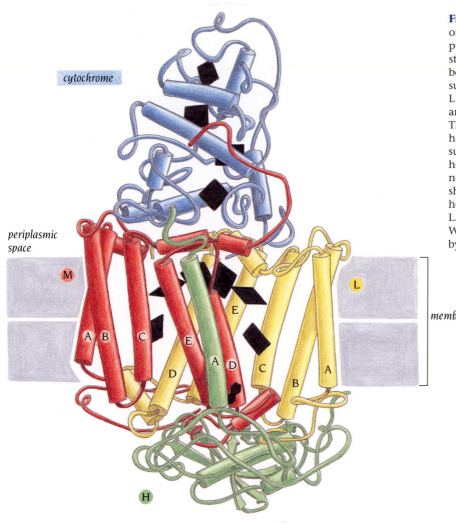

cytochrome

periplasmic space

M

L

membrane

A B C

E

E

D

A D

C

B

A

H

Figure 12.14 The three-dimensional structure of a photosynthetic reaction center of a purple bacterium was the first high-resolution structure to be obtained from a membrane-bound protein. The molecule contains four subunits: L, M, H, and a cytochrome. Subunits L and M bind the photosynthetic pigments, and the cytochrome binds four heme groups. The L (yellow) and the M (red) subunits each have five transmembrane α helices A–E. The H subunit (green) has one such transmembrane helix, AH, and the cytochrome (blue) has none. Approximate membrane boundaries are shown. The photosynthetic pigments and the heme groups appear in black. (Adapted from L. Stryer, *Biochemistry*, 3rd ed. New York: W.H. Freeman, 1988, after a drawing provided by Jane Richardson.)

stabilize the structure of the four-helix bundle. Since its removal does not change the rate of electron flow through the system, the Fe atom cannot have a functional role in electron transfer. The remaining three helices of each subunit are arranged around the core in a manner that is not found in water-soluble proteins. Presumably, their positions are at least partly determined by the positions of the loop regions outside the membrane and not by close packing of the α helices inside the membrane. It is interesting that none of the α helices are perpendicular to the assumed plane of the membrane; instead, they are all tilted at angles of about 20° to 25°, similar to the tilt of the transmembrane helices in bacteriorhodopsin (see Figure 12.3) and other transmembrane proteins.

The photosynthetic pigments are bound to the L and M subunits

The structurally similar L and M subunits are related by a pseudo-twofold symmetry axis through the core, between the helices of the four-helix bundle motif. The photosynthetic pigments are bound to these subunits, most of them to the transmembrane helices, and they are also related by the same twofold symmetry axis (Figure 12.15). The pigments are arranged so that they form two possible pathways for electron transfer across the membrane, one on each side of the symmetry axis.

This symmetry is important in bringing the two chlorophyll molecules of the "special pair" into close contact, giving them their unique function in initiating electron transfer. They are bound in a hydrophobic pocket close to the symmetry axis between the D and E transmembrane α helices of both

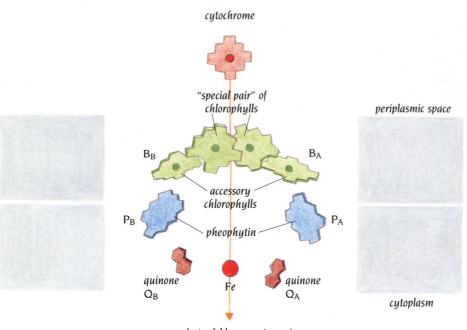

cytochrome

"special pair" of
chlorophylls

B_B

B_A

accessory
chlorophylls

P_B

pheophytin

P_A

quinone
Q_B

Fe

quinone
Q_A

periplasmic space

cytoplasm

pseudo-twofold symmetry axis

Figure 12.15 Schematic arrangement of the photosynthetic pigments in the reaction center of *Rhodopseudomonas viridis*. The twofold symmetry axis that relates the L and the M subunits is aligned vertically in the plane of the paper. Electron transfer proceeds preferentially along the branch to the right. The periplasmic side of the membrane is near the top, and the cytoplasmic side is near the bottom of the structure. (From B. Furugren, courtesy of the Royal Swedish Academy of Science.)

subunits. The pyrrol ring systems of these chlorophyll molecules are parallel and interact closely with each other; in particular, two pyrrol rings, one from each molecule, are stacked on top of each other, 3 Å apart (Figure 12.15). The twofold symmetry axis passes between these stacked pyrrol rings.

This pair of chlorophyll molecules, which as we shall see accepts photons and thereby excites electrons, is close to the membrane surface on the periplasmic side. At the other side of the membrane the symmetry axis passes through the Fe atom. The remaining pigments are symmetrically arranged on each side of the symmetry axis (Figure 12.15). Two bacteriochlorophyll molecules, the accessory chlorophylls, make hydrophobic contacts with the special pair of chlorophylls on one side and with the pheophytin molecules on the other side. Both the accessory chlorophyll molecules and the pheophytin molecules are bound between transmembrane helices from both subunits in pockets lined by hydrophobic residues from the transmembrane helices (Figure 12.16).

The functional reaction center contains two quinone molecules. One of these, Q_B (Figure 12.15), is loosely bound and can be lost during purification. The reason for the difference in the strength of binding between Q_A and Q_B is unknown, but as we will see later, it probably reflects a functional asymmetry in the molecule as a whole. Q_A is positioned between the Fe atom and one of the pheophytin molecules (Figure 12.15). The polar-head group is outside the membrane, bound to a loop region, whereas the hydrophobic tail is

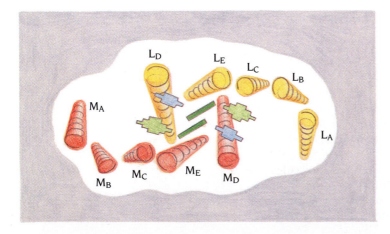

L_D L_E L_C L_B

M_A

L_A

M_B M_C M_E M_D

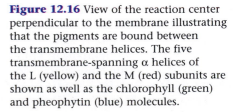

Figure 12.16 View of the reaction center perpendicular to the membrane illustrating that the pigments are bound between the transmembrane helices. The five transmembrane-spanning α helices of the L (yellow) and the M (red) subunits are shown as well as the chlorophyll (green) and pheophytin (blue) molecules.

bound to the pheophytin molecule and to hydrophobic side chains of trans-membrane helices of the L subunit. If Q_B is lost during purification prior to crystallization, the corresponding binding site on the other side of the pseudo-twofold axis is empty. Certain weedkillers that are inhibitors of photosynthesis in plants can bind in the crystal to this empty binding site.

Reaction centers convert light energy into electrical energy by electron flow through the membrane

In photosynthesis light energy is converted to electrical energy by an electron flow that causes the separation of negatively and positively charged molecules. Many molecules can absorb photons and use the energy of this process to donate an electron to a nearby electron acceptor. The electron donor then becomes positively charged and the electron acceptor negatively charged. In most cases, however, the transfer of electrons back from the acceptor to the donor is as fast as the forward reaction and the absorbed energy is lost, usually as fluorescent radiation. The arrangement of photoreaction centers in both bacteria and green plants results in a very fast forward reaction and a slow back reaction; therefore, the electric charges induced by the absorbed light energy stay separated. This separation of charge represents a storage of energy because energy would be released if the charges were able to come together. This is the basic primary process of photosynthesis, the detailed mechanism of which we do not yet understand. However, by interpreting spectroscopic and genetic experiments in terms of the structure of the bacterial reaction center we may come to understand this process and be able to re-create this primary biological function—that of a solar-powered electrolytic battery.

In the bacterial reaction center the photons are absorbed by the special pair of chlorophyll molecules on the periplasmic side of the membrane (see Figure 12.14). Spectroscopic measurements have shown that when a photon is absorbed by the special pair of chlorophylls, an electron is moved from the special pair to one of the pheophytin molecules. The close association and the parallel orientation of the chlorophyll ring systems in the special pair facilitates the excitation of an electron so that it is easily released. This process is very fast; it occurs within 2 picoseconds. From the pheophytin the electron moves to a molecule of quinone, Q_A, in a slower process that takes about 200 picoseconds. The electron then passes through the protein, to the second quinone molecule, Q_B. This is a comparatively slow process, taking about 100 microseconds.

There are two pheophytin molecules, one on each side of the twofold axis, that in principle could accept the electrons (see Figure 12.15). However, only one pathway, on the right side of the symmetry axis that is shown in Figure 12.15, is used for electron transfer. Electrons do not pass through the chain of pigments on the left side, which appear to have no role in the charge separation. The best guess as to why these pigments are present is that the L and M chains have evolved from an ancestral chain that formed symmetrical homodimers in which both pigment chains were utilized. Presumably, the present-day reaction centers are more efficient for charge separation than the ancestral homodimers. Q_A is most stable when excited by two electrons, and if each electron arrived randomly at Q_A and Q_B the energy stored in Q_A after absorption of one electron might be dissipated before a second electron was absorbed.

One apparent discrepancy between the spectroscopic data and the crystal structure is that no spectroscopic signal has been measured for participation of the accessory chlorophyll molecule B_A in the electron transfer process. However, as seen in Figure 12.15, this chlorophyll molecule is between the special pair and the pheophytin molecule and provides an obvious link for electron transfer in two steps from the special pair through B_A to the pheophytin. This discrepancy has prompted recent, very rapid measurements of the electron transfer steps, still without any signal from B_A. This means either that the electron bypasses B_A and is transferred directly over the very long distance of about 25 Å from the special pair to pheophytin, or that the transfer through B is too rapid to detect with current technology, less than 0.01 picosecond. Neither of these conclusions is compatible with current theories for electron transfer. However, the components for electron transfer are embedded in the protein environment of subunits L and M with special properties that are not taken into account in the theories.

While this electron flow takes place, the cytochrome on the periplasmic side donates an electron to the special pair and thereby neutralizes it. Then the entire process occurs again: another photon strikes the special pair, and another electron travels the same route from the special pair on the periplasmic side of the membrane to the quinone, Q_B, on the cytosolic side, which now carries two extra electrons. This quinone is then released from the reaction center to participate in later stages of photosynthesis. The special pair is again neutralized by an electron from the cytochrome.

The charge separation that stores the energy of the photons is now complete: two positive charges have been left on the cytochrome side of the membrane, and two electrons have traveled through the membrane to a quinone molecule on the cytosolic side. In the photosynthetic reaction centers charge separation is remarkably efficient for capturing light energy. The forward electron transfer from the reaction center to Q_A is more than eight orders of magnitude faster than the back reaction. This large difference allows the reaction center to capture the energy of between 98 and 100% of the photons it absorbs. As a solar cell, it is extraordinarily efficient: the energy stored in separated charges is about half of the energy inherent in the photons. The rest of the energy is lost in other ways, some of which are the reactions that drive the electrons along the pathway of photosynthetic pigment molecules.

Antenna pigment proteins assemble into multimeric light-harvesting particles

If the special pair of chlorophyll molecules of the reaction center was the only photon acceptor in the bacterial membrane, only a tiny fraction of the incoming sunlight would be captured and converted to chemical energy. All photosynthetic organisms have therefore evolved a system of **light-harvesting complexes**, which surrounds the reaction centers and increases the photon capturing area. The reaction centers receive practically all their light energy from such complexes. Detailed structural information is now available for the arrangement of the light-absorbing pigment molecules around the reaction center in photosynthetic bacteria. The pigments are firmly bound to small

Figure 12.17 Computer-generated diagram of the structure of light-harvesting complex LH2 from *Rhodopseudomonas acidophila*. Nine α chains (gray) and nine β chains (light blue) form two rings of transmembrane helices between which are bound nine carotenoids (yellow) and 27 bacteriochlorophyll molecules (red, green and dark blue). (Courtesy of M.Z. Papiz.)

hydrophobic protein molecules that are embedded in the membrane and which assemble into two types of multimeric complexes, called LH1 and LH2. The crystal structures of two LH2 complexes have been determined to high resolution, one from the purple bacterium *Rhodopseudomonas acidophila* by the group of Richard Cogdell and Neil Isaacs, Glasgow University, UK, and the other from *Rhodospirillum molischianum* by the group of Hartmut Michel at the Max-Planck Institute in Frankfurt, Germany. In addition, an 8.5-Å electron crystallography map of LH1 using two-dimensional crystals has given a broad outline of the structure of that complex. Strangely, plants seem to have evolved a totally different system of light-harvesting complexes. Werner Kühlbrandt at EMBL, Heidelberg, has determined the structure of a plant light-harvesting complex by electron crystallography and shown that the structure is quite different from those of the purple bacteria. We will here discuss only the bacterial system.

Chlorophyll molecules form circular rings in the light-harvesting complex LH2

The structure of the LH2 complex of *R. acidophila* is both simple and elegant (Figure 12.17). It is a ring of nine identical units, each containing an α and a β polypeptide of 53 and 41 residues, respectively, which both span the membrane once as α helices (Figure 12.18). The two polypeptides bind a total of three chlorophyll molecules and two carotenoids. The nine heterodimeric units form a hollow cylinder with the α chains forming the inner wall and the β chains the outer wall. The hole in the middle of the cylinder is empty, except for lipid molecules from the membrane.

Two of the chlorophyll molecules from each unit are in the space between the two walls and form a ring of 18 chlorophyll molecules near the periplasmic membrane surface (Figure 12.19a). The planar chlorophyll rings are oriented almost perpendicular to the plane of the membrane. Each magnesium atom of these chlorophyll molecules is bound to a histidine residue of either the α or the β chain. The third chlorophyll molecule is bound between the β chains forming part of the outer wall and oriented parallel to the plane of the membrane (Figure 12.19b). The magnesium atom is bound to an oxygen atom of the formylated N-terminus of the α chain and has a much more polar environment than the other two chlorophyll molecules. The third chlorophyll molecule therefore absorbs light at a considerably shorter wavelength than the other two molecules, with an absorption maximum at 800 nm compared with 850 nm. This arrangement allows the complex efficiently to capture a much broader energy band of the sunlight than would be possible if all the chlorophyll molecules had similar environments.

In contrast to bacteriorhodopsin or the reaction center, there is no direct contact within the membrane between the α helices in this complex. The helices are held together through contacts mediated by the pigments and by contacts at the ends of the polypeptide chains outside the membrane.

The LH2 complex of *R. molischianum* is very similar to that of *R. acidophila* except that the complex is built up into a ring of eight identical units instead of nine. In addition the third chlorophyll is coordinated to the O atom of an Asp residue of the α chain. Since the α and β chains of all bacterial light-harvesting complexes show some sequence similarity one can safely predict that they are all arranged in essentially the same way, except that the number of units forming the cylinder may differ. This prediction has been used to build a model of LH1 using the electron crystallographic map to 8.5 Å.

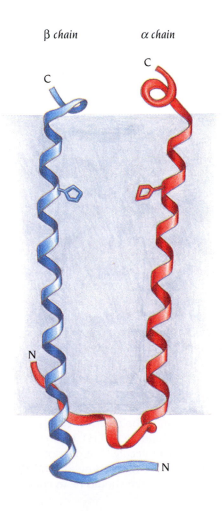

β chain   *α chain*

Figure 12.18 Ribbon diagram showing the α (red) and the β (blue) chains of the light-harvesting complex LH2. Each chain forms one transmembrane α helix, which contains a histidine residue that binds to the Mg atom of one bacteriochlorophyll molecule. (Adapted from G. McDermott et al., *Nature* 374: 517–521, 1995.)

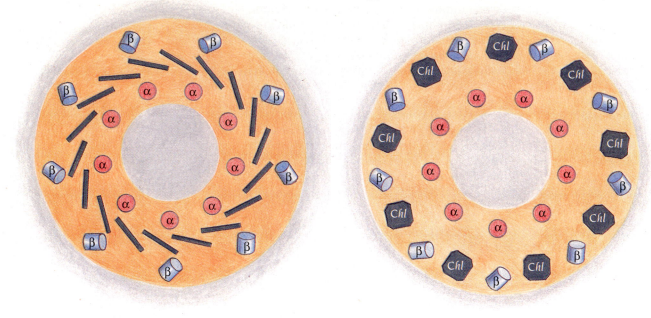

The reaction center is surrounded by a ring of 16 antenna proteins of the light-harvesting complex LH1

The light-harvesting complex LH1 is directly associated with the reaction center in purple bacteria and is therefore referred to as the core or inner antenna, whereas LH2 is known as the peripheral antenna. Both are built up from hydrophobic α and β polypeptides of similar size and with low but significant sequence similarity. The two histidines that bind to chlorophyll with absorption maxima at 850 nm in the periplasmic ring of LH2 are also present in LH1, but the sequence involved in binding the third chlorophyll in LH2 is quite different in LH1. Not surprisingly, the chlorophyll molecules of the periplasmic ring are present in LH1 but the chlorophyll molecules with the 800 nm absorption maximum are absent.

The electron crystallographic map of LH1 at 8.5 Å resolution (Figure 12.20) shows a striking similarity in molecular design to LH2. The LH1 ring clearly consists of the same basic units as LH2. The α and β polypeptides form an inner and an outer wall with the pigment molecules in between. However, the LH1 ring contains 16 rather than 9 units and has a hole 68 Å in diameter in the middle. By analogy with the structure of LH2 it can be assumed that LH1 has a ring of 32 chlorophyll molecules in the region between these two walls at the same distance from the periplasmic side of the membrane as in LH2. These chlorophyll molecules presumably have an even more hydrophobic environment than those in LH2 since they have an absorption maximum at 875 nm.

On the basis of these structural results it is now possible to derive the following schematic picture (Figure 12.21) of the photosynthetic apparatus in purple bacteria, and to begin to understand the design of this highly efficient apparatus for capturing photons from the sun and funneling them to the reaction center with minimum loss of energy.

Spectroscopic measurements show that the reaction center and LH1 are tightly associated and therefore it is assumed that the ring of pigments in LH1 surrounds the reaction center. Careful model building indicates that the hole in the middle of LH1 is large enough to accommodate the whole reaction center molecule. We do not know exactly how the LH2 complexes are arranged in the membrane around the LH1–reaction center complex, but at least some of them should be in contact with the outer rim of LH1 for efficient

Figure 12.19 Schematic diagrams illustrating the arrangement of bacteriochlorophyll molecules in the light-harvesting complex LH2, viewed from the periplasmic space. (a) Eighteen bacteriochlorophyll molecules (green) are bound between the two rings of α (red) and β (blue) chains. The planes of these molecules are oriented perpendicular to the plane of the membrane and the molecules are bound close to the periplasmic space. (b) Nine bacteriochlorophyll molecules (green) are bound between the β chains (blue) with their planes oriented parallel to the plane of the membrane. These molecules are bound in the middle of the membrane.

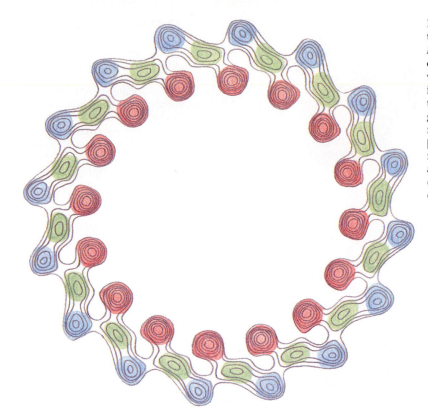

Figure 12.20 Electron density projection of the light-harvesting complex LH1 from *Rhodospirillum rubrum* determined by electron crystallography. On the basis of comparison with the LH2 complex the red regions can be interpreted as corresponding to the α chains, the blue regions to the β chains and each green region to two bacteriochlorophyll molecules bound between the α and the β chains. The ring of 16 units has a hole in the middle that is large enough to accommodate a complete reaction center molecule. (Adapted from Karrasch et al., *EMBO J.* 14: 631–638, 1995.)

energy transfer. There are 8 to 10 LH2 complexes for each reaction center and consequently around 300 energy-capturing chlorophyll molecules per reaction center.

In the photosynthetic membrane there is a downhill flow of energy from the light-harvesting proteins to the reaction centers. In purple bacteria, LH2 absorbs radiation at a shorter wavelength (higher energy) than LH1 and therefore delivers it to LH1, which in turn passes it on to the reaction center. A photon that is absorbed by the 800-nm chlorophylls in LH2 is rapidly transmitted to the energetically lower periplasmic ring of 850-nm chlorophyll in the same complex. Spectroscopic measurements have shown that the energy absorbed by a chlorophyll in the periplasmic rings spreads to the others, within 0.2 to 0.3 picoseconds. When a photon is absorbed by any of these chlorophylls it becomes in effect delocalized between the chlorophyll molecules of the ring. It can then easily jump to another chlorophyll in an adjacent complex, where it again becomes delocalized until it ends up at the reaction center, as schematically illustrated in Figure 12.22.

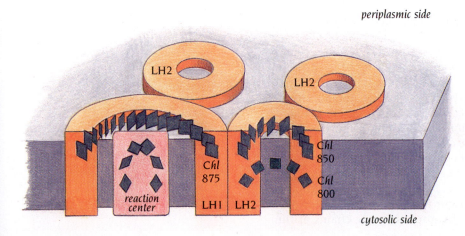

Figure 12.21 Schematic diagram of the relative positions of bacteriochlorophylls (green) in the photosynthetic membrane complexes LH1, LH2, and the reaction center. The special pair of bacteriochlorophyll molecules in the reaction center is located at the same level within the membrane as the periplasmic bacteriochlorophyll molecules Chl 875 in LH1 and the Chl 850 in LH2. (Adapted from W. Kühlbrandt, *Structure* 3: 521–525, 1995.)

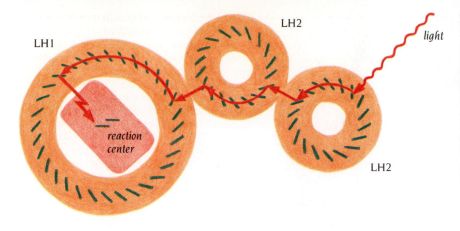

Figure 12.22 Schematic diagram showing the flow of excitation energy in the bacterial photosynthetic apparatus. The energy of a photon absorbed by LH2 spreads rapidly through the periplasmic ring of bacterio-chlorophyll molecules (green). Where two complexes touch in the membrane , the energy can be transmitted to an adjacent LH2 ring. From there it passes by the same mechanism to LH1 and is finally transmitted to the special chlorophyll pair in the reaction center. (Adapted from W. Kühlbrandt, *Structure* 3: 521–525, 1995.)

Modeling of the reaction center inside the hole of LH1 shows that the primary photon acceptor—the special pair of chlorophyll molecules—is located at the same level in the membrane, about 10 Å from the periplasmic side, as the 850-nm chlorophyll molecules in LH2, and by analogy the 875-nm chlorophyll molecules of LH1. Furthermore, the orientation of these chlorophyll molecules is such that very rapid energy transfer can take place within a plane parallel to the membrane surface. The position and orientation of the chlorophyll molecules in these rings are thus optimal for efficient energy transfer to the reaction center.

Energy transfer and energy conversion in photosynthetic systems occur virtually without energy loss, whereas solid-state solar cells operate at efficiencies of around 20%. By learning some of the lessons implicit in the arrangement and the reactions of the pigment molecules in the bacterial photosynthetic membrane, and applying these principles to the design of new types of solar cells, it may one day become possible to devise a system of clean, environmentally compatible energy production that operates efficiently at low light intensities.

Transmembrane α helices can be predicted from amino acid sequences

We have seen in previous chapters that only short continuous regions of the polypeptide chains contribute to the hydrophobic interior of water-soluble globular proteins. In such proteins α helices are generally arranged so that one side of the helix is hydrophobic and faces the interior while the other side is hydrophilic and at the surface of the protein as discussed in Chapter 3. Beta strands are usually short, with the residues alternating between the hydrophobic interior and hydrophilic surface. Loop regions between these secondary structure elements are usually very hydrophilic. Therefore, in soluble globular proteins, regions of more than 10 consecutive hydrophobic amino acids in the sequence are rarely encountered.

In contrast, the transmembrane helices observed in the reaction center are embedded in a hydrophobic surrounding and are built up from continuous regions of predominantly hydrophobic amino acids. To span the lipid bilayer, a minimum of about 20 amino acids are required. In the photosynthetic reaction center these α helices each comprise about 25 to 30 residues, some of which extend outside the hydrophobic part of the membrane. From the amino acid sequences of the polypeptide chains, the regions that comprise the transmembrane helices can be predicted with reasonable confidence.

Naively, one might assume that it should be possible to scan the sequence and pick out regions with about 20 consecutive hydrophobic amino acids. However, no such regions occur in the reaction center proteins. Just as in soluble proteins there are hydrophobic side chains at the

Table 12.1 Hydrophobicity scales

Amino acid	Phe	Met	Ile	Leu	Val	Cys	Trp	Ala	Thr	Gly	Ser	Pro	Tyr	His	Gln	Asn	Glu	Lys	Asp	Arg
A	2.8	1.9	4.5	3.8	4.2	2.5	-0.9	1.8	-0.7	-0.4	-0.8	-1.6	-1.3	-3.2	-3.5	-3.5	-3.5	-3.9	-3.5	-4.5
B	3.7	3.4	3.1	2.8	2.6	2.0	1.9	1.6	1.2	1.0	0.6	-0.2	-0.7	-3.0	-4.1	-4.8	-8.2	-8.8	-9.2	-12.3

Row A is from J. Kyte and R.F. Doolittle; row B, from D.A. Engelman, T.A. Steitz, and A. Goldman.

hydrophilic surface of the molecule, in the transmembrane helices of the reaction center there are hydrophilic side chains (which are often important for function) among the hydrophobic. However, hydrophobic residues are in a clear majority in transmembrane helices, and such residues occur less frequently in other continuous regions of the polypeptide chain. Therefore, we need some method to measure the amount of hydrophobicity in a segment of the amino acid sequence in order to be able to predict whether or not it is likely to be a transmembrane helix.

Hydrophobicity scales measure the degree of hydrophobicity of different amino acid side chains

Each amino acid side chain within a transmembrane helix has a different hydrophobicity. It is easy to state that side chains such as Val, Met, and Leu are the most hydrophobic and that charged residues such as Arg and Asp are at the other end of the scale. However, to order all side chains according to hydrophobicity and to assign actual numbers that represent their degree of hydrophobicity is not trivial. Many such **hydrophobicity scales** have been developed over the past decade on the basis of solubility measurements of the amino acids in different solvents, vapor pressures of side-chain analogs, analysis of side-chain distributions within soluble proteins, and theoretical energy calculations. In Table 12.1 two of these hydrophobicity scales are listed. The most frequently used scale, which was introduced by J. Kyte and R.F. Doolittle at University of California, San Diego, is based on experimental data. A more refined scale was developed by D.A. Engelman, T.A. Steitz, and A. Goldman at Yale University. They used a semitheoretical approach to calculating the hydrophobicity, taking into account the fact that the side chains are attached to an α-helical framework.

Hydropathy plots identify transmembrane helices

These hydrophobicity scales are frequently used to identify those segments of the amino acid sequence of a protein that have hydrophobic properties consistent with a transmembrane helix. For each position in the sequence, a hydropathy index is calculated. The **hydropathy index** is the mean value of the hydrophobicity of the amino acids within a "window," usually 19 residues long, around each position. In transmembrane helices the hydropathy index is high for a number of consecutive positions in the sequence. Charged amino acids are usually absent in the middle region of transmembrane helices because it would cost too much energy to have a charged residue in the hydrophobic lipid environment. It might be possible, however, to have two residues of opposite charge close together inside the lipid membrane because they neutralize each other. Such charge neutralization has been observed in the hydrophobic interior of soluble globular proteins.

When the hydropathy indices are plotted against residue numbers, the resulting curves, called hydropathy plots, identify possible transmembrane helices as broad peaks with high positive values. Such hydropathy plots are shown in Figure 12.23 for the L and M chains of the reaction center.

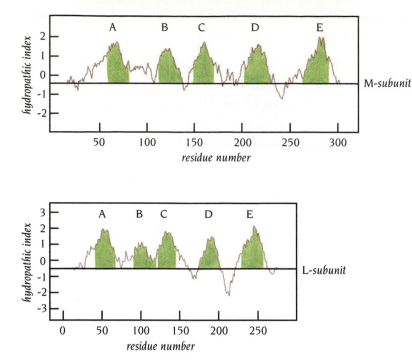

Figure 12.23 Hydropathy plots for the polypeptide chains L and M of the reaction center of *Rhodobacter sphaeroides*. A window of 19 amino acids was used with the hydrophobicity scales of Kyte and Doolittle. The hydropathy index is plotted against the tenth amino acid of the window. The positions of the transmembrane helices as found by subsequent x-ray analysis by the group of G. Feher, La Jolla, California, are indicated by the green regions.

Reaction center hydropathy plots agree with crystal structural data

The hydropathy plots in Figure 12.23 were calculated and published several years before the x-ray structure of the reaction center was known. It is therefore of considerable interest to compare the predicted positions of the transmembrane-spanning helices with those actually observed in the x-ray structure. These observed positions are indicated in green in Figure 12.23.

It is immediately apparent that these plots correctly predict the number of transmembrane helices, five each in the L and M chains, and also their approximate positions in the polypeptide chain. This gives us confidence in the hydropathy plot method. Transmembrane helices are the only secondary structure elements that can be predicted from novel amino acid sequences with a high degree of confidence using current knowledge and methods. The exact ends of transmembrane helices, however, cannot be predicted, essentially because they are usually inserted within the polar head-groups of the membrane lipids and therefore contain charged and polar residues. The transmembrane helices in the reaction center, for example, contain a number of charged residues at their ends (Table 12.2), most of which are at the cytoplasmic side. All of the helices, however, have a segment of at least 19 consecutive amino acids that contain no charged side chains. The majority of residues in these segments are hydrophobic, but there are a number of polar residues, such as Ser, Thr, Tyr, and Trp, among them. The presence of histidine residues in the D and E helices of subunits L and M is accounted for by their special function in the reaction center; they are ligands to the magnesium atoms of chlorophyll molecules and to the Fe atom.

Membrane lipids have no specific interaction with protein transmembrane α helices

Comparison of the amino acid sequences of the L and M subunits of the reaction centers from three different bacterial species shows that about 50% of all residues in those two subunits are conserved in all three species. In the transmembrane helices, sequence conservation varies. Residues that are buried and have contacts either with pigments or with other transmembrane helices are about 60% conserved. In contrast, residues that are fully exposed to the membrane lipids are only 16% conserved. Clearly, fewer restrictions

Table 12.2 Amino acid sequences of the transmembrane helices of the photosynthetic reaction center in *Rhodobacter sphaeroides*

	Sequence
LA	G F F G V A T F F F A A L G I I L I A W S A V L
LB	L K R C I E V E R L A W S V F A G T A C I T I I Q W L G G
LC	H I P F A F A F A I L A Y L T L V L F R P V M
LD	A A S L V L A G H L A L A L A N T F F S I A I M H A P
LE	G T L G I H R L G L L L S L S A V F F S A L C M I I
MA	S L G V L S L F S G L M W F F T I G I W F W Y Q A
MB	A Q A R L Y T R G W W S W V A V F M F F S A I L W L G G E K L
MC	A W A F L S A I W L W M V L G F I R P I L M
MD	V A L I T A G H M A F L L A S G Y L F A I S L G H F P
ME	M E G I H R W A I W M A V L V T L T G G I G I L L
HA	M N E T Q L Y Y I L G A L F I W F S Y I A L S A L

The helices are aligned according to approximate positions within the membrane and with respect to the photosynthetic pigments. LA is the first helix of subunit L, ME is the last helix of subunit M, HA is the only transmembrane helix of subunit H. Charged residues are colored red, polar residues are blue, hydrophobic residues are green, and glycine is yellow. (From T.O. Yeates et al., *Proc. Natl. Acad. Sci. USA* 84: 6438–6442, 1987.)

are placed on residues that are exposed to the membrane lipids than to residues having contact with polypeptide chains or pigments. This implies that there are relatively few specific interactions between these transmembrane helices and the fatty acid side chains of the membrane that require the presence of specific residues. This is consistent with the observation that membrane proteins can move within the plane of the membrane, by lateral diffusion, and are not at fixed positions.

Conclusion

The x-ray structure of a bacterial photosynthetic reaction center and the associated light-harvesting complexes has given insight into the mechanism for the primary reaction of photosynthesis: the capture and conversion of light energy to chemical energy by a stable separation of negatively and positively charged molecules. In addition, the structure has provided geometrical constraints for theoretical calculations on the electron flow through pigments across the membrane during this charge separation.

Important novel information has thus been obtained for the specific biological function of those molecules, but disappointingly few general lessons have been learned that are relevant for other membrane-bound proteins with different biological functions. In that respect the situation is similar to the failure of the structure of myoglobin to provide general principles for the construction of soluble protein molecules as described in Chapter 2.

The most important general lesson is that there are hydrophobic transmembrane helices, the positions of which within the amino acid sequence can be predicted with reasonable accuracy. This applies both to the single transmembrane-spanning helix within the H polypeptide chain of the reaction center and the five transmembrane helices of the L and M chains that are connected by loop regions. This does not imply, however, that it is equally simple to predict the positions of transmembrane helices in different classes of proteins by hydropathy plots. The transmembrane helices of ion channels, for example, contain charged residues facing the channel. The latter would give quite different hydropathy plots from a single transmembrane helix in a receptor protein.

The structure of the reaction center also established that membrane-spanning helices can be tilted with respect to the plane of the membrane and that their relative positions within the membrane might be determined by the way they are anchored to the loop regions. Finally, several structures provide examples of how binding pockets for ligands are formed between such transmembrane-spanning helices.

The three-dimensional structure of the bacterial membrane protein, bacteriorhodopsin, was the first to be obtained from electron microscopy of two-dimensional crystals. This method is now being successfully applied to several other membrane-bound proteins.

Unlike the other membrane proteins discussed here, the porins in the outer membrane of Gram-negative bacteria have transmembrane regions that are β strands, not α helices. The β strands, 16 in some porins, 18 in others, are arranged into up and down antiparallel β barrels. These barrels, because they have so many strands in their walls, do not have a tightly packed hydrophobic core. Their center is a channel, which is partially blocked by loop regions between two of the strands, leaving an eyelet about 9 Å long and 8 Å in diameter. The complete functional porin molecule comprises three of these barrels; loop regions from the upper surface of the three barrels are in contact with the extracellular space and form a broad funnel leading to the barrels and via their eyelets to the periplasmic space. The porins in this way form size-restricted channels for the passive diffusion of molecules in and out of the periplasmic space.

Like the photosynthetic reaction center and bacteriorhodopsin, the bacterial K^+ ion channel also has tilted transmembrane helices, two in each of the subunits of the homotetrameric molecule that has fourfold symmetry. These transmembrane helices line the central and inner parts of the channel but do not contribute to the remarkable 10,000-fold selectivity for K^+ ions over Na^+ ions. This crucial property of the channel is achieved through the narrow selectivity filter that is formed by loop regions from the four subunits and lined by main-chain carbonyl oxygen atoms, to which dehydrated K^+ ions bind.

Selected readings

General

Barber, J., Andersson, B. Revealing the blueprint of photosynthesis. *Nature* 370: 31–34, 1994.

Cowan, S.W. Bacterial porins: lessons from three high-resolution structures. *Curr. Opin. Struct. Biol.* 3: 501–507, 1993.

Cowan, S.W., Rosenbusch, J.P. Folding pattern diversity of integral membrane proteins. *Science* 264: 914–916, 1994.

Deisenhofer, J., Michael, H. Nobel lecture. The photosynthetic reaction center from the purple bacterium *Rhodopseudomonas viridis*. *EMBO J.* 8: 2149–2169, 1989.

Engelman, D.M., Steitz, T.A., Goldman, A. Identifying nonpolar transbilayer helices in amino acid sequences of membrane proteins. *Annu. Rev. Biophys. Biophys. Chem.* 15: 321–353, 1986.

Fasman, G.D., Gilbert, W.A. The prediction of transmembrane protein sequences and their conformation: an evaluation. *Trends Biochem. Sci.* 15: 89–95, 1990.

Garavito, R.M., Rosenbusch, J.P. Isolation and crystallization of bacterial porin. *Methods Enzymol.* 125: 309–328, 1986

Gust, D., Moore, T.A. Mimicking photosynthesis. *Science* 244: 35–41, 1989.

Huber, R. Nobel lecture. A structural basis of light energy and electron transfer in biology. *EMBO J.* 8: 2125–2147, 1989.

Kovari, L.C., Momany, C., Rossmann, M.C. The use of antibody fragments for crystallization and structure determinations. *Structure* 3: 1291–1293, 1995.

Kühlbrandt, W. Structure and function of bacterial light-harvesting complexes. *Structure* 3: 521–525, 1995.

Landau, E.M., Rosenbuch, J.P. Lipid cubic phases: a concept for the crystallization of membrane proteins. *Proc. Natl. Acad. Sci. USA* 93: 14532–14535, 1996.

Lanyi, J.K. Bacteriorhodopsin as a model for proton pumps. *Nature* 375: 461–464, 1995.

Michel, H. Crystallization of membrane proteins. *Trends Biochem. Sci.* 8: 56–59, 1983.

Michel, H., Deisenhofer, J. Relevance of the photosynthetic reaction center from purple bacteria to the structure of photosystem II. *Biochemistry* 27: 1–7, 1988.

Nikaido, H. Porins and specific diffusion channels in bacterial outer membranes. *J. Biol. Chem.* 269: 3905–3908, 1994.

Norris, J.R., Schiffer, M. Photosynthetic reaction centers in bacteria. *Chem. Eng. News* 68(31): 22–37, 1990.

Rees, D.C., et al. The bacterial photosynthetic reaction center as a model for membrane proteins. *Annu. Rev. Biochem.* 58: 607–633, 1989.

Unwin, N., Henderson, R. The structure of proteins in biological membranes. *Sci. Am.* 250(2): 78–95, 1984.

Youvan, D.C., Marrs, B.L. Molecular mechanisms of photosynthesis. *Sci. Am.* 256(6): 42–49, 1987

Specific structures

Allen, J.P., et al. Structure of the reaction center from *Rhodobacter sphaeroides* R-26: the cofactors. *Proc. Natl. Acad. Sci. USA* 84: 5730–5734, 1987.

Allen, J.P., et al. Structure of the reaction center from *Rhodobacter sphaeroides* R-26: the protein subunits. *Proc. Natl. Acad. Sci. USA* 84: 6162–6166, 1987.

Brisson, A., Unwin, P.N.T. Quaternary structure of the acetylcholine receptor. *Nature* 315: 474–477, 1985.

Cowan, S.W., et al. Crystal structures explain functional properties of two *E. coli* porins. *Nature* 358: 727–733, 1992.

Deisenhofer, J., et al. Structure of the protein subunits in the photosynthetic reaction center of *Rhodopseudomonas viridis* at 3 Å resolution. *Nature* 318: 618–624, 1985.

Deisenhofer, J., et al. X-ray structure analysis of a membrane protein complex. Electron density map at 3 Å resolution and a model of the chromophores of the photosynthetic reaction center from *Rhodopseudomonas viridis*. *J. Mol. Biol.* 180: 385–398, 1984.

Doyle, D.A., et al. The structure of the potassium channel: molecular basis of K$^+$ conduction and selectivity. *Science* 280: 69–77, 1998.

Dutzler, R., et al. Crystal structures of various maltooligosaccharides bound to maltoporin reveal a specific sugar translocation pathway. *Structure* 4: 127–134, 1996.

Eisenberg, D., Weiss, R.M., Terwilliger, T.C. The hydrophobic moment detects periodicity in protein hydrophobicity. *Proc. Natl. Acad. Sci. USA* 82: 140–144, 1984.

Garavito, R.M., Rosenbusch, J.P. Three-dimensional crystals of an integral membrane protein: an initial x-ray analysis. *J. Cell. Biol.* 86: 327–329, 1980.

Henderson, R., et al. Model for the structure of bacteriorhodopsin based on high-resolution electron cryo-microscopy. *J. Mol. Biol.* 213: 899–929, 1990.

Henderson, R., Unwin, P.N.T. Three-dimensional model of purple membrane obtained by electron microscopy. *Nature* 257: 28–32, 1975.

Iwata, S., Ostermeier, C., Ludwig, B., Michel, H. Structure at 2.8 Å resolution of cytochrome c oxidase from *Paracoccus denitrificans*. *Nature* 376: 660–669, 1995.

Karrasch, S., Bullough, P.A., Ghosh, R. 8.5-Å projection map of the light-harvesting complex I from *Rhodospirillum rubrum* reveals a ring composed of 16 subunits. *EMBO J.* 14: 631–638, 1995.

Koepke, J., et al. The crystal structure of the light-harvesting complex II (B800–850) from *Rhodospirillum molischianum*. *Structure* 4: 581–597, 1996.

Kühlbrandt, W., Wang, D.A., Fujiyoshi, Y. Atomic model of the plant light-harvesting complex. *Nature* 367: 614–621, 1994.

Kyte, J., Doolittle, R.F. A simple method for displaying the hydropathic character of a protein. *J. Mol. Biol.* 157: 105–132, 1982.

Leifer, D., Henderson, R. Three-dimensional structure of orthorhombic purple membrane at 6.5 Å resolution. *J. Mol. Biol.* 163: 451–466, 1983.

MacKinnon, R., et al. Structural conservation in prokaryotic and eukaryotic potassium channels. *Science* 280: 106–109, 1998.

McDermott, G., et al. Crystal structure of an integral membrane light-harvesting complex from photosynthetic bacteria. *Nature* 374: 517–521, 1995.

Michel, H. Three-dimensional crystals of a membrane protein complex. The photosynthetic reaction center from *Rhodopseudomonas viridis*. *J. Mol. Biol.* 158: 567–572, 1982.

Michel, H., et al. The "heavy" subunit of the photosynthetic reaction center from *Rhodopseudomonas viridis*: isolation of the gene, nucleotide and amino acid sequence. *EMBO J.* 4: 1667–1672, 1985.

Michel, H., et al. The "light" and "medium" subunits of the photosynthetic reaction center from *Rhodopseudomonas viridis*: isolation of the genes, nucleotide and amino acid sequence. *EMBO J.* 5: 1149–1158, 1986.

Michel, H., Epp, O., Deisenhofer, J. Pigment–protein interactions in the photosynthetic reaction center from *Rhodopseudomonas viridis*. *EMBO J.* 5: 2445–2451, 1986.

Picot, D., Loll, P.J., Garavito, R.M. The x-ray crystal structure of the membrane protein prostaglandin H2 synthase-1. *Nature* 367: 243–249, 1994.

Rees, D.C., DeAntonio, L., Eisenberg, D. Hydrophobic organization of membrane proteins. *Science* 245: 510–513, 1989.

Schirmer, T., Keller, T.A., Wang, Y.-F., Rosenbusch, J.P. Structural basis for sugar translocation through malto-porin channels at 3.1 Å resolution. *Science* 267: 512–514, 1995.

Tsukihara, T., et al. The whole structure of the 13-subunit oxidized cytochrome c oxidase at 2.8 Å resolution. *Science* 272: 1136–1144, 1996.

Unwin, N. Acetylcholine receptor channel imaged in the open state. *Nature* 373: 37–43, 1995.

Valpuesta, J.M., Henderson, R., Frey, T.G. Electron cryo-microscopic analysis of crystalline cytochrome oxidase. *J. Mol. Biol.* 214: 237–251, 1990.

Weiss, M.S., et al. Molecular architecture and electrostatic properties of a bacterial porin. *Science* 254: 1627–1630, 1991.

Yeates, T.O., et al. Structure of the reaction center from *Rhodobacter sphaeroides* R-26: membrane–protein interactions. *Proc. Natl. Acad. Sci. USA* 84: 6438–6442, 1987.

Signal Transduction

13

Many signal-transducing receptors are plasma membrane proteins that bind specific extracellular molecules, such as growth factors, hormones, or neurotransmitters, and then transmit to the cell's interior a signal that elicits a specific response. These responses are usually cascades of enzymatic reactions giving rise to many different effects within the cell, including changes in gene expression. Interference with receptor-signaling systems can therefore have drastic consequences. For example, oncogene products that simulate the function of receptors or their associated signal transmitters cause the loss of cellular growth control.

The DNA sequences of membrane-bound signal-transducing receptor molecules have shown that many are structurally and evolutionarily related, and they can be grouped into a few families making up three major classes: ion-channel linked receptors homologous to the K^+ channel molecule described in Chapter 12, G protein-linked receptors, and enzyme-linked receptors. These last two classes of receptor molecules have an extracellular domain that recognizes a specific molecular signal, a transmembrane region through which the signal is transmitted, and an intracellular domain that produces a response (Figure 13.1). These proteins provide yet another example of the now familiar story that there are only a limited number of different types of domains and that protein molecules with different functions have evolved either by accumulation of point mutations or by combining domains in different ways, by gene shuffling.

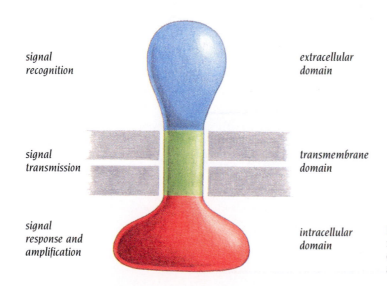

signal recognition

signal transmission

signal response and amplification

extracellular domain

transmembrane domain

intracellular domain

Figure 13.1 The basic organization of a membrane receptor molecule consists of an extracellular domain, a transmembrane region, and an intracellular domain.

No three-dimensional structure is yet available for any complete receptor because they are large and membrane bound, and hence difficult to crystallize, and are too large to study by NMR. However, by recombinant DNA techniques it has been possible to crystallize isolated extracellular and intracellular domains of receptor molecules. In this chapter we will give examples of signal recognition by the binding of growth hormone to the extracellular domain of its receptor, and of the intracellular signal response and amplification by G protein and protein tyrosine kinase-linked receptors. We have no detailed structural knowledge on how signals are transmitted through the membrane, nor on how receptors are linked to ion channel signaling.

G proteins are molecular amplifiers

Several important physiological responses, including vision, smell, and stress response, involve large metabolic effects produced from a small number of input signals. The receptors that trigger these responses have two things in common. First, their transmembrane regions contain seven helices each spanning the lipid bilayer of the plasma membrane. Second, the signals transmitted to the intracellular domain of these receptors are amplified, and the amplifiers are members of a common family of proteins with homologous amino acid sequences called **G proteins**. G proteins bind guanine nucleotides (hence the term G protein) and act as molecular switches that are activated by binding GTP and are inactivated when the GTP is hydrolyzed to GDP. The hydrolysis of GTP is catalyzed by the G protein itself, but G proteins on their own are very slow GTPases, and switching off the G protein is normally accelerated by regulatory molecules, known as **RGS** (regulators of GTP hydrolysis), which bind to the active G protein and increase the rate of GTP hydrolysis. When in the active GTP-bound state, the G protein can activate many downstream effectors, greatly amplifying the signal, before RGS binds and the signal is switched off. The structural basis of this important switch mechanism is now understood.

Most G proteins are heterotrimers consisting of one copy each of α (45 kDa), β (35 kDa), and γ (8 kDa) subunits. It is the α subunit that contains the GTPase activity. When the G protein is activated by binding GTP, the heterotrimer dissociates into the α subunit and a βγ heterodimer. Both the α subunit and the βγ heterodimer can independently transmit the signal received from the receptor to different effector proteins. The human genome is estimated to contain over 1000 different genes for these receptors with seven transmembrane helices and, in addition, enough genes for different α, β and γ subunits to allow the formation of hundreds of different G proteins. Consequently, there is a very large number of possible combinations of these receptors and G proteins, which provide the potential for a cell to respond to a wide variety of external signals. Table 13.1 gives some examples of physiological responses by G proteins.

Both the α and the γ subunits of G proteins are anchored to the membrane by lipids covalently bound to the N-terminal region of the G_α chain

Table 13.1 Examples of physiological processes mediated by G proteins

Stimulus	Receptor	Effector	Physiological response
Epinephrine	β-adrenergic receptor	adenylate cyclase	glycogen breakdown
Light	rhodopsin	c-GMP phosphodiesterase	visual excitation
IgE–antigen complexes	mast cell IgE receptor	phospholipase C	histamine secretion in all allergic reactions
Acetylcholine	muscarinic receptor	potassium channel	slowing of pacemaker activity that controls the rate of the heartbeat

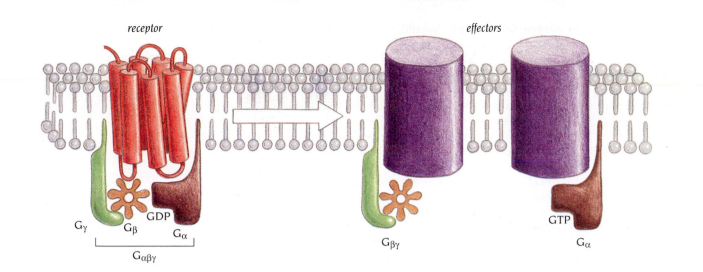

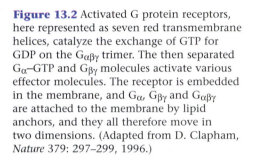

Figure 13.2 Activated G protein receptors, here represented as seven red transmembrane helices, catalyze the exchange of GTP for GDP on the $G_{\alpha\beta\gamma}$ trimer. The then separated G_{α}–GTP and $G_{\beta\gamma}$ molecules activate various effector molecules. The receptor is embedded in the membrane, and G_{α}, $G_{\beta\gamma}$ and $G_{\alpha\beta\gamma}$ are attached to the membrane by lipid anchors, and they all therefore move in two dimensions. (Adapted from D. Clapham, *Nature* 379: 297–299, 1996.)

and the C-terminal region of the G_{γ} chain (Figure 13.2). The β subunit lacks such lipid modification but remains at the membrane because it is always complexed with the γ subunit, forming the heterodimeric species $G_{\beta\gamma}$. All three species that exist in the cell—G_{α}, $G_{\beta\gamma}$, and $G_{\alpha\beta\gamma}$—are consequently attached to the cell membrane through the lipid modification of subunits α and γ. On their own both the β and the γ subunits are unstable and easily degraded, whereas the α subunit is stable as a monomer. When GDP is bound to G_{α}, it forms a heterotrimeric complex with $G_{\beta\gamma}$ but when GTP is bound, G_{α} remains monomeric. Because they are attached to the cell membrane, G proteins are correctly oriented with respect to each other and to receptors and effector molecules as they move about in two dimensions (see Figure 13.2). Thus, the lipid anchors markedly enhance the search for physiological partners.

In the inactive state, the G protein consists of a G_{α}–GDP complex combined with a dimeric $G_{\beta\gamma}$ molecule to form a stable heterotrimeric $G_{\alpha\beta\gamma}$, which binds to the cytosolic domain of a receptor (Figure 13.3). When the external domain of a receptor binds its specific ligand (or, as in the case of the visual receptor, rhodopsin, when the ligand is photoisomerized, as discussed in Chapter 12), a signal passes through the membrane causing the cytosolic domain to become activated by a conformational change. The G protein is activated when the changed cytosolic surface of the receptor catalyzes the exchange of bound GDP for GTP. The activated G protein is then released from the receptor and dissociates into the two active components $G_{\beta\gamma}$ and G_{α}–GTP. These transmit the signal to effector proteins that in turn produce second messenger molecules, such as cyclic AMP, which give rise to physiological responses.

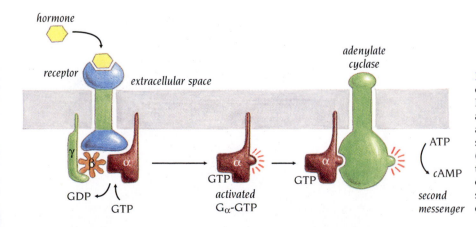

Figure 13.3 G protein–mediated activation of adenylate cyclase by hormone binding. Hormone binding on the extracellular side of a receptor such as the β adrenergic receptor activates a G protein on the cytoplasmic side. The activated form of the G protein (α subunit with bound GTP) dissociates from the β and γ subunits and activates adenylate cyclase to produce cyclic AMP, which is a second messenger that affects a diverse range of cellular processes.

253

As long as the ligand remains on the extracellular domain of the receptor, its cytoplasmic domain can continue to trigger the production of active G_α–GTP molecules. Each active G_α–GTP, by complexing with an effector molecule, in turn triggers the production of many second messenger molecules. For example, one molecule of the hormone epinephrine bound to the receptor in this way produces a cascade of molecules of the second messenger, cyclic AMP, through the action of G proteins and adenylate cyclase. The GTPase activity of the G_α subunit determines the length of time that the signal remains on. Once hydrolysis to GDP has occurred, the α subunit is inactivated and able to bind to free $G_{\beta\gamma}$, ready for the cycle to recommence.

A failure to turn off GTP-activated G_α has dire consequences. For example, in the disease cholera, cholera toxin produced by the bacterium *Vibrio cholerae* binds to G_α and prevents GTP hydrolysis, resulting in the continued excretion of sodium and water into the gut.

To illustrate the structural basis of signal transduction by G proteins, we shall discuss the following: (i) the structural similarities and differences of G_α and the regulator molecule Ras; (ii) the structural changes that convert resting G_α–GDP to active G_α–GTP; (iii) the mechanism of the intrinsic GTPase activity of G_α and Ras, and how the rate of GTP hydrolysis is changed by binding to regulators; (iv) the structure of the complex of G_α with $G_{\beta\gamma}$, in which the GTPase activity is inhibited and the G protein is prepared for binding to the receptor; (v) the formation of complexes of $G_{\beta\gamma}$ with regulators, which prevents the formation of $G_{\alpha\beta\gamma}$.

Ras proteins and the catalytic domain of G_α have similar three-dimensional structures

The **Ras** proteins, regulators of signal transduction processes leading to cell multiplication and differentiation, are molecular switches activated in response to protein tyrosine kinase receptors. They are GTP-binding proteins with essentially no GTPase activity, and like G_α they are activated as molecular switches by binding GTP and are inactive when complexed with GDP. To be switched off they require the action of a GTPase-activating protein called **GAP**. About 25% of all tumor cells have mutant Ras proteins that cannot be

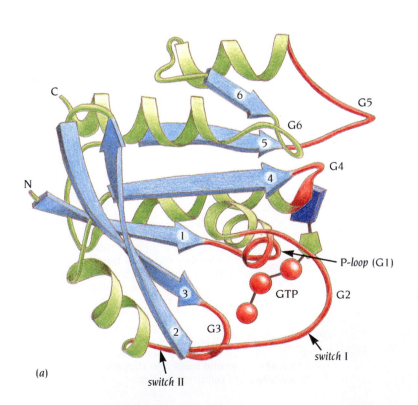

(a)

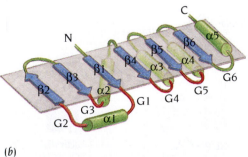

(b)

Figure 13.4 Schematic diagram (a) and topology diagram (b) of the polypeptide chain of cH-ras p21. The central β sheet of this α/β structure comprises six β strands, five of which are parallel; α helices are green, β strands are blue, and the adenine, ribose, and phosphate parts of the GTP analog are blue, green, and red, respectively. The loop regions that are involved in the activity of this protein are red and labeled G1–G5. The G1, G3, and G4 loops have the consensus sequences G-X-X-X-X-G-K-S/T, D-X-X-E, and N-K-X-D, respectively. (Adapted from E.F. Pai et al., *Nature* 341: 209–214, 1989.)

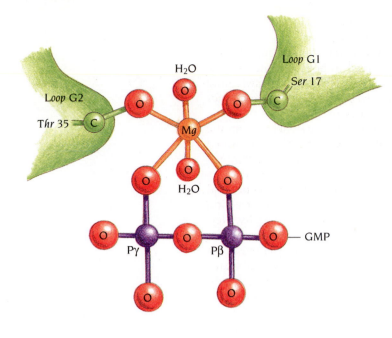

Figure 13.5 A Mg^{2+} atom links GTP to the Ras protein. Mg is coordinated to one oxygen atom each from the β and γ phosphates of GTP as well as to the side chains of Ser 17 and Thr 35 of Ras. Two water molecules complete the octahedral coordination of Mg.

activated to hydrolyze GTP; as a result, these mutant Ras molecules remain bound to GTP and activated, and this leads to uncontrolled cell growth.

The three-dimensional structure of Ras was determined by the groups of Sung-Hou Kim at the University of California, Berkeley, and Ken Holmes at the Max-Planck Institute, Heidelberg. Ras has an α/β-type structure in which the central β sheet comprises six β strands, five of which are parallel (Figure 13.4). There are five α helices positioned on both sides of the β sheet. The GTP is bound in a pocket at the carboxy ends of the β strands in a way similar to the binding of nucleotides to other nucleotide-binding proteins. Loop regions that connect the β strands with the α helices form the binding pocket. The topology of this fold, shown in Figure 13.4b, as well as the mode of nucleotide binding, is exactly the same as that observed earlier by Jens Nyborg in Brian Clark's laboratory at Aarhus University, Denmark, and Frances Jurnak at the University of California, Riverside, in their crystal structures of the elongation factor Tu, which is involved in protein synthesis on the ribosome.

Almost all nucleotide triphosphate hydrolyzing enzymes, including Ras and G_α, require magnesium ions for their catalytic activities. Mg^{2+} is presumably required both for proper positioning of the γ phosphate and for weakening the P–O bond that is split during catalysis, as well as for maintaining the stability of the guanine nucleotide complex. In the structure of Ras and all GTPases bound to a GTP analog, a magnesium ion is coordinated to two phosphate oxygen atoms, one each from the β and γ phosphates. In addition the Mg^{2+} coordinates two water molecules and two protein oxygen ligands, in Ras one from the side chain of Ser 17 in loop G1 and one from Thr 35 in loop G2 (Figure 13.5).

Five of the six loop regions (G1–G5 in Figure 13.4) that are present at the carboxy end of the β sheet in the Ras structure participate in the GTP binding site. Three of these loops, G1 (residues 10–17), G3 (57–60), and G4 (116–119), contain regions of amino acid sequence conserved among small GTP-binding proteins and the G_α subunits of trimeric G proteins.

The first loop region, G1, of Ras, which is also called the diphosphate-binding loop or P-loop, connects the β1 strand with the α1 helix (see Figure 13.4) and is essential for proper positioning of the phosphate groups by binding to the α and β phosphates of the guanine nucleotide. Loop G3, which connects β3 with α2, forms a link between the subsites that bind Mg^{2+} and the γ phosphate of GTP. The consensus amino acid sequences of these two loop regions have been found in other nucleotide-binding proteins such as myosin, described in Chapter 14, and are frequently used as a fingerprint motif to identify from amino acid sequences ATP- and GTP- binding proteins. Loop G2 contains the threonine residue conserved in both Ras and G_α proteins that binds Mg^{2+} and which is important in both structural switching and GTP hydrolysis. Loops G4 and G5 are involved in recognizing and binding the guanine base of the nucleotide. Two regions in the Ras molecule undergo large conformational changes when the molecule switches from the active GTP-bound form to the inactive GDP-bound form. They are called switch regions and are found in loop G2 (switch I) and the region comprising loop G3 and helix $\alpha 2$ (switch II) (Figure 13.4a).

X-ray structures of G_α protein complexes with GDP and GTP analogs that simulate both the ground state and transition states of the GTPase reaction have been determined by the groups of Paul Sigler at Yale University, New Haven and Stephen Sprang at the University of Texas, Dallas. These studies have given a detailed picture of the structural aspects of the mechanism for GTP hydrolysis by G_α. Paul Sigler's group has studied the G_α subunit of **transducin**, a G protein that mediates vision in the retina by transmitting light signals detected by rhodopsin, the photon receptor in the eye, to cyclic GMP phosphodiesterase, the effector. Both groups have obtained essentially similar results with similar conclusions and we will here focus on the G_α subunit of transducin, from which Sigler's group removed the covalently bound lipid at the N-terminus in order to promote crystallization.

The polypeptide chain of transducin G_α is about twice as long as that of Ras and is divided in two domains, a GTPase domain with a structure similar to Ras and an α-helical domain with a unique topology (Figure 13.6). The helical domain is connected to the GTPase domain by two linker regions, one of which follows helix $\alpha 1$ and the other precedes strand $\beta 2$ in the GTPase domain. The whole helical domain including parts of the linker regions can be regarded as a large insertion in the GTPase domain at a position corresponding to the G2 loop in Ras (compare Figures 13.4 and 13.6). The bound nucleotide, GTP-γS, which is a chemically modified GTP with a sulfur atom instead of oxygen in the γ phosphate to prevent hydrolysis, is completely buried between the two domains. The inaccessibility of the GTP analog in G_α

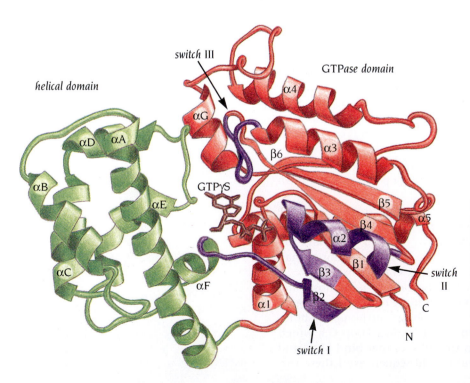

Figure 13.6 Schematic diagram of G_α from transducin with a bound GTP analog. The polypeptide chain is organized into two domains; a catalytic domain (red) with a structure similar to Ras, and a helical domain (green) which is an insert in the loop between $\alpha 1$ and $\beta 2$. There are three switch regions (violet) that have different conformations in the different catalytic states of G_α. The GTP analog (brown) is bound to the catalytic domain in a cleft between the two domains. (Adapted from J. Noel et al., *Nature* 366: 654–663, 1993.)

suggests that in order to exchange guanine nucleotides there may be a conformational change that opens the cleft.

The helical domain is built up from one long α helix of 28 residues that acts as a scaffold, supporting five smaller helices through hydrophobic interactions. These interactions provide a rigid internal structure consistent with the notion that the whole domain functions as a rigid lid on the nucleotide binding site, and that conformational changes might operate through hinge movements opening and closing the whole lid.

Compared to Ras, G_α subunits are extended at their N-termini by about 30 residues that in free G_α are disordered but, as we will see, obtain a helical conformation in association with $G_{\beta\gamma}$.

G_α is activated by conformational changes of three switch regions

The transduction of signals by G proteins depends on the ability of G_α to cycle between the resting (GDP-bound) conformation primed for interaction with stimulated receptors, and the active (GTP-bound) conformation capable of activating or inhibiting a variety of effector molecules. How can the presence or absence of a phosphate in the bound nucleotide cause such drastic effects? Both Sigler's and Sprang's groups have revealed details of the conformational differences between these two states of G_α by comparing high-resolution structures of G_α complexed with GDP (resting form) and GTP-γS (active form).

The structural differences are essentially localized to three regions in the GTPase domain (see Figure 13.6), two of which are the same as switch regions I and II described for the Ras structure, while the third, switch region III, is another loop region linking β4 with α3, corresponding to G4 in Ras (see Figures 13.4 and 13.6). The structural changes of these regions are considerable, as can be seen in Figure 13.7. The trigger for the conformational change from the GDP form to the GTP form seems to be the formation of hydrogen bonds between the protein and the γ-phosphate of GTP, coupled to the formation of new bonds to the Mg^{2+} ion from protein ligands.

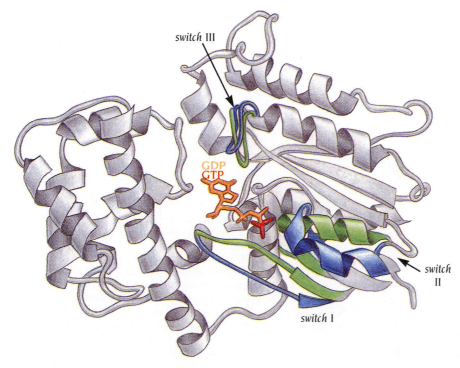

Figure 13.7 Structural differences between the inactive and active forms of G_α. The major part of the structure is unchanged (gray regions) but the three switch regions change conformation. In the active GTP-bound form (green) the switch regions are closer to the catalytic site than they are in the inactive GDP-bound form (blue). (Adapted from D. Lambright et al., *Nature* 369: 621–628, 1994.)

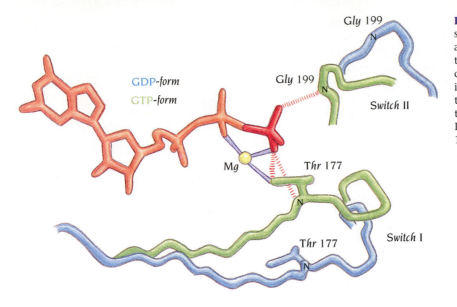

Figure 13.8 Interactions involved in the switch from the inactive GDP- (blue) to the active GTP- (green) bound forms of G_α from transducin. The diagram illustrates the local changes required in the switch I and II regions in order to bring the side chain of Thr 177 and the main chain N of Gly 199 into contact with the γ phosphate of GTP. (Adapted from D. Lambright et al., *Nature* 369: 621–628, 1994.)

In both structures the Mg^{2+} ion is coordinated to six ligands with octahedral geometry. Four water molecules as well as the side chain oxygen atom of a serine residue from the P-loop and one oxygen atom from the β-phosphate bind to Mg^{2+} in the GDP structure. Two of the water molecules are replaced in the GTP structure by a threonine residue from switch I and an oxygen atom from the γ phosphate (similar to the arrangement shown in

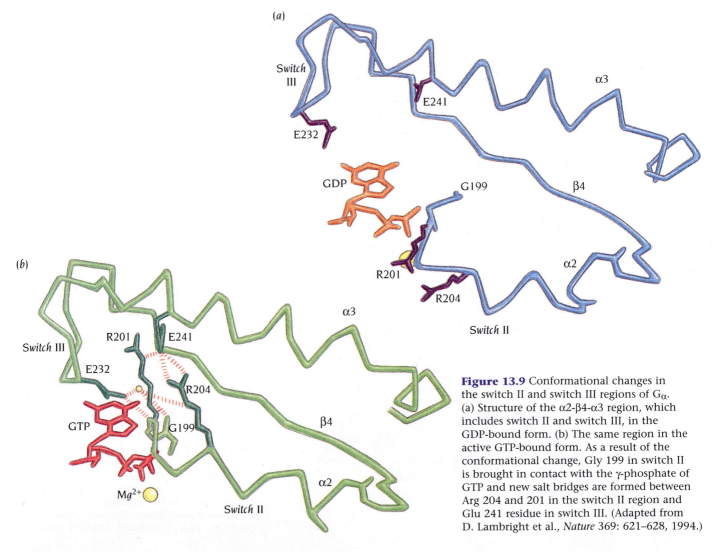

Figure 13.9 Conformational changes in the switch II and switch III regions of G_α. (a) Structure of the α2-β4-α3 region, which includes switch II and switch III, in the GDP-bound form. (b) The same region in the active GTP-bound form. As a result of the conformational change, Gly 199 in switch II is brought in contact with the γ-phosphate of GTP and new salt bridges are formed between Arg 204 and 201 in the switch II region and Glu 241 residue in switch III. (Adapted from D. Lambright et al., *Nature* 369: 621–628, 1994.)

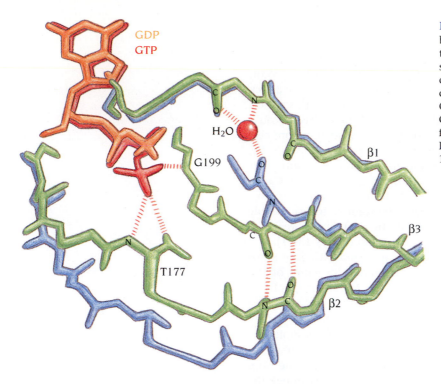

Figure 13.10 Rearrangements of the hydrogen bond network between strands 1, 2, and 3 in the β sheet of G_α as a consequence of the switch from the GDP (blue) to the GTP (green) conformation. Strand β3 pulls away from β1 and disrupts two hydrogen bonds in order to bring Gly 199 into contact with the γ-phosphate of GTP. As a consequence new hydrogen bonds are formed between β2 and β3. (Adapted from D. Lambright et al., *Nature* 369: 621–628, 1994.)

Figure 13.5). Structural changes in the switch I region are induced by hydrogen bonds to the γ phosphate from both the main chain N and side chain OH of the threonine residue, Thr 177, the oxygen of which is also coordinated to the Mg^{2+} ion (Figure 13.8). These structural changes are conceptually simple: after exchange of GDP for GTP, switch region I is drawn essentially as a whole towards the γ phosphate, bringing the side chain of Thr 177 into position for hydrogen bonding with the oxygen of the phosphate, and also into the vicinity of the Mg^{2+} where it replaces one of the water molecules observed in the GDP form (see Figure 13.8).

Changes in the switch II region are initiated by the formation of a hydrogen bond between the main chain NH of Gly 199 and the γ phosphate of GTP (Figures 13.8 and 13.9). Switch II comprises both the loop containing Gly 199 and helix α2. Movement in this loop is coupled to α2 so that to bring Gly 199 into this position, the region is stretched and rotated relative to the conformation in the GDP form (see Figure 13.7). As a consequence the end of β3 pulls away from β1, disrupting two hydrogen bonds (Figure 13.10). The new conformation is stabilized by the formation of additional hydrogen bonds between β3 and β2 and also by new salt bridges involving arginine residues in the switch II region and a glutamate residue in the switch III region (Figure 13.9).

The switch III region has no direct interactions with the γ phosphate, but forms a network of new highly conserved interactions with residues in switch II. The structural changes in switches I and II, which are triggered by the γ phosphate, have thus indirectly propagated to switch III. These three switches are clustered in one area of the G_α molecule that structural studies show is involved both in interactions with the β subunit of $G_{\beta\gamma}$ and with GTPase activator molecules. Structural changes in this region can therefore explain why the GTP form of G_α, but not the GDP form, binds to effectors and vice versa for binding to $G_{\beta\gamma}$.

GTPases hydrolyze GTP through nucleophilic attack by a water molecule

Some G proteins are slow GTP hydrolases with turnover numbers around two per minute, others such as Ras are only marginally catalytic. Kinetic experiments in solution have shown that in both cases the most likely mechanism of hydrolysis of GTP to GDP and phosphate is direct transfer of the γ phosphate to water without the formation of a covalently bound intermediate with the protein, as described for serine proteases in Chapter 11. This mechanism requires the GTPase to provide a catalytic base to remove a proton from a water molecule, producing a hydroxyl group that can directly attack the phophorous atom of the γ phosphate. GTP is not hydrolyzed in solution by negatively charged hydroxyl groups in the absence of the GTPase because the γ phosphate is negatively charged too. The GTPase must therefore also reduce the negative charge on the phosphate to catalyze hydrolysis. Ideally, to identify groups on the protein that carry out these two crucial functions we would determine the structure of the protein in complex with the transition state, but because transition states are very short-lived, even for slow enzymatic reactions, their structures cannot easily be studied. Instead, stable analogs that simulate the transition state are used; the complex between G_α, GDP and AlF_4^- is such an analog for the GTPase reaction (Figure 13.11).

The structures of two different G_α molecules complexed with AlF_4^- show only minor conformational changes compared with the structure of the ground state, represented by the complex with GTP-γS. These changes bring a glutamine side chain from the switch II region deeper into the active site closer to the γ phosphate. This glutamine residue acts as the catalytically required base when its oxygen atom accepts a proton from a water molecule, with the glutamine nitrogen atom donating a proton to the γ phosphate. This glutamine residue together with an arginine and a threonine residue from the switch I region reduce the negative charge of the γ phosphate through the formation of hydrogen bonds, thereby stabilizing the presumed transition state complex, as shown in Figure 13.12. The hydrogen atom taken from the water molecule to generate the attacking hydroxyl group is a part of this stabilizing network of hydrogen bonds. The phosphorous atom in the γ phosphate is assumed to be pentacoordinated, with the hydrogen bonds firmly stabilizing the geometry of the pentacoordinate γ phosphate.

All the residues involved in important functions in the catalytic mechanism are strictly conserved in all homologous GTPases with one notable exception. Ras does not have the arginine in the switch I region that stabilizes the transition state. The assumption that the lack of this catalytically important residue was one reason for the slow rate of GTP hydrolysis by Ras was confirmed when the group of Alfred Wittinghofer, Max-Planck Institute,

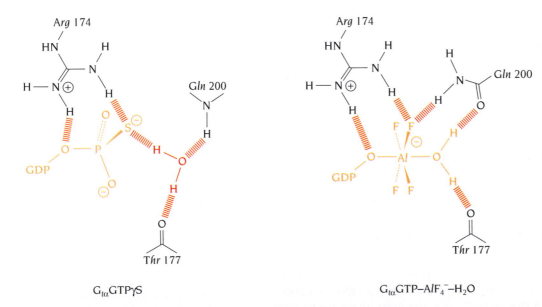

$G_{t\alpha}GTP\gamma S$

$G_{t\alpha}GTP–AlF_4^−–H_2O$

Figure 13.11 Interactions that are relevant for the GTPase mechanism, involving GTP-γS and GDP–AlF$_4^-$–H$_2$O at the active site of transducin G_α. GTP-γS mimics the ground state of bound GTP and GDP–AlF$_4^-$–H$_2$O mimics the transition state during catalysis. The positively charged guanidinium group of Arg 174 would increase the transfer of negative charge from the attacking hydroxyl to the β–γ bridging oxygen thereby facilitating hydrolysis of the terminal phosphate. (Adapted from J. Sondek et al., *Nature* 372: 276–279, 1994.)

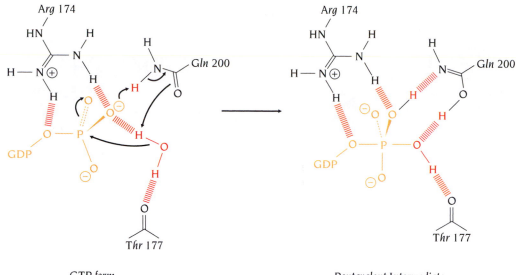

Figure 13.12 A concerted mechanism for GTP hydrolysis by G_α in which transfer of a proton to the γ phosphate is coupled to deprotonation of the attacking water by Gln 200. (Adapted from J. Sondek et al., *Nature* 372: 276–279, 1994.)

GTP *form*   Pentavalent *Intermediate*

Dortmund, determined the structure of a complex between Ras, a transition state analog GDP. AlF$_3$ and the GTPase-activating protein GAP. GAP proteins accelerate the GTPase reaction of Ras nearly 100,000-fold. The structure of the Ras–GAP complex shows that GAP provides an arginine residue in the active site of Ras with the guanidinium group occupying a very similar position to that of the arginine in the switch I region in G_α. This arginine of GAP can increase the rate of the reaction by a factor of about 1000. The remainder of the rate increase results from the stabilization of the active conformations of switch regions I and II of Ras in the complex with GAP. The GAP binds to these regions such that the switches are locked in the structure that stabilizes the transition state of the reactants. The RGS molecules that regulate G protein signaling bind to G_α and enhance its rate of GTP hydrolysis by a similar mechanism. RGS molecules do not provide additional residues in the active site of G_α but rather bind to all three switch regions and stabilize their conformations.

Disruption of the Ras–GAP relationship can have profound consequences for a cell, leading to unregulated cell division. For instance, about 25% of human tumors contain either of two mutated forms of Ras which resist GAP stimulation. In one of these oncogenic forms of Ras, the glutamine residue in the switch II region that stabilizes the transition state is mutated. In the other mutant the glycine residue in the P-loop, which is intimately packed against both the glutamine residue in Ras and the arginine residue from GAP, is changed. Any mutation of this residue changes the geometry of the GAP stabilized transition state, leading to loss of GAP-mediated activation of the catalytic activity of Ras.

In summary, structural studies of Ras and G_α with GTP-γS and a transition state analog have illuminated the catalytic mechanism of their GTPase activity, as well as the mechanism by which GTP hydrolysis is stimulated by GAP and RGS. In addition, these structural studies have shown how tumor-causing mutations affect the function of Ras and G_α.

The G_β subunit has a seven-blade propeller fold, built up from seven WD repeat units

The G_β polypeptide chain has about 350 amino acid residues whereas the γ chain has only about 70 residues. There are small variations in these lengths between species and between different gene products within the same species, but all known β chains and γ chains show significant sequence

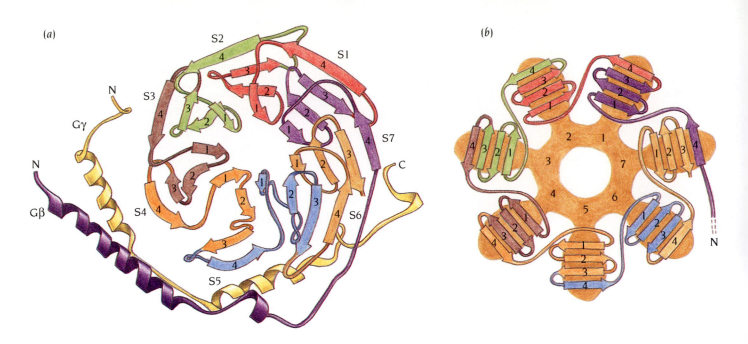

homology among each other. Most of a β chain comprises seven repeats of a unit of approximately 40 amino acids. These repeated motifs are called WD repeats because they often end with a tryptophan, W, and an aspartate, D, residue. Such WD repeats occur not only in G proteins but also in effector molecules for signal transduction as well as in proteins involved in RNA processing, gene regulation, vesicular traffic, cell cycle regulation, and in a number of proteins of unknown function.

Structure determinations of $G_{\alpha\beta\gamma}$ by both Sigler's and Sprang's groups have shown that WD repeats form the blades of a β propeller structure, as described in Chapter 5 for neuraminidase, but with seven blades instead of the six β sheet blades of neuraminidase (Figure 13.13). Each blade consists of four antiparallel β strands forming a four-stranded β sheet. The seven blades are arranged like the blades of a propeller with the first β strand (β1) of each sheet close to the center of the structure and the last strand (β4) at the outer edge of the blade. Intuitively one would assume that each blade of four β strands would correspond to each WD sequence repeat. However, this is not the case. Instead, each WD sequence repeat forms the outermost β strand of one blade followed by the inner three strands of the following blade (Figure 13.13b). The outer strand of the seventh blade is provided by the N-terminal strand of the first WD repeat, thus linking the beginning and the end of the propeller. This arrangement of the β strands elegantly links the blades so that seven sequence repeat units form a circular superbarrel.

Analysis of the sequences of the seven repeats in G_β shows that there are five residues that are almost totally invariant (Table 13.2). These residues

Figure 13.13 (a) Schematic diagram of the $G_{\beta\gamma}$ heterodimer from transducin. The view is along the central tunnel. The seven four-stranded β sheets that form the seven blades of the propeller-like structure are labeled S1 to S7. The strands are colored to highlight the seven WD sequence repeats. The N-terminal α helices of the β and γ chains form a coiled coil. (b) Topological diagram of the seven-blade propeller domain of G_β. The strands are colored to highlight the seven WD sequence repeats. [(a) Adapted from J. Sondek et al., *Nature* 379: 369–374, 1996. (b) Adapted from D. Lambright et al., *Nature* 379: 311–319, 1996.]

Table 13.2 Sequence alignments of the seven WD repeats in the G_β subunit of transducin

	1	5	10		20		30		40	42
WD-1	- - - - - - - - - R T R R R T L R G	H	L A K - -	T Y A M H W G T D S R L L L	S	A S Q	D	G K L I I	W D
WD-2	S Y T - - - - T N K V H A I P L	R	S S W - -	V M T C A Y A P S G N Y V A C		G G L	D	N I C S I	Y N
WD-3	L K T R E G N V R V S R E L A G	H	T G Y - -	L S C C R F L D D - N Q I V	T	S S G	D	T T C A L	W D
WD-4	I E T - - - - G Q Q T T T F T G	H	T G D - -	V M S L S L A P D T R L F V	S	G A C	D	A S A K L	W D
WD-5	V R E - - - - G M C R Q T F T G	H	E S D - -	I N A I C F F P N G N A F A	T	G S D	D	A T C R L	F D
WD-6	L R A - - - - D Q E L M T Y S	H D	N I I C G I T S V S F S K S G R L L L	A	G Y D	D	F N C N V	W D
WD-7	A L K - - - - A D R A G V L A G	H	D N R - -	V S C L G V T D D G M A V A	T	G S W	D	S F L K T	W N

Arrows denote β strands. Colored columns correspond to the five almost invariant residues described in the text and shown in Figure 13.14.

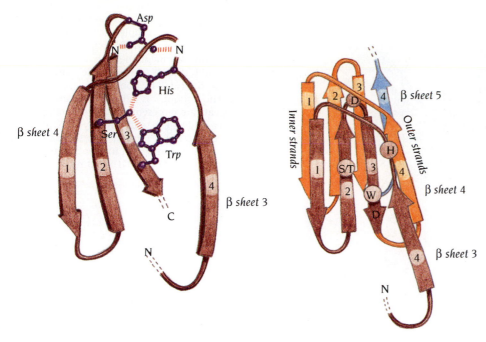

Figure 13.14 (a) Schematic diagram of the main chain and four almost invariant residues of the fourth WD repeat of G_β from transducin. The view is roughly perpendicular to the central tunnel and the plane of the sheet. The red stripes denote hydrogen bonds. (b) Schematic view of two WD repeats illustrating the structural relationships between two consecutive repeats. The first repeat is brown and the second repeat is orange. The positions of the four almost invariant residues in the first repeat are circled. (Adapted from J. Sondek et al., *Nature* 379: 369–374, 1996.)

occupy strategic positions to couple the outer strand of each blade to the inner three strands of the next blade (Figure 13.14). An Asp residue, which is present in all seven strands, is located in a tight turn between the middle two β strands. Side chain oxygens of this Asp form hydrogen bonds to main chain nitrogen atoms both in the tight turn and in the loop connecting the first and second strand of the same sequence repeat, which is the loop between two adjacent blades. This hydrogen bonding arrangement stabilizes and effectively couples the tight turn of one β sheet to the outer strand of the previous β sheet (Figure 13.14). A conserved His in the same loop between the blades participates in a hydrogen bond network to a Ser or Thr in the second strand which in turn forms a hydrogen bond to a Trp in the third strand. This spatial arrangement of conserved residues preserves a network of interactions that link the blades together in proper orientation. However, these conserved residues are not invariant, so there are small structural differences when one or the other conserved residue is absent.

The structure of $G_{\beta\gamma}$ (see Figure 13.13a) also explains why the dimer is stable whereas G_β alone is unstable and G_γ alone is unfolded *in vitro*. G_γ, which in the dimer is folded into two α helices in an extended arrangement lacking intrachain tertiary interactions (see Figure 13.13a), forms extensive, mainly hydrophobic interactions with G_β. The N-terminal α helix of G_γ forms a parallel coiled coil (see Chapter 3) with a long N-terminal α helix of G_β that precedes the seven WD repeats in the β-polypeptide chain. While these coiled-coil helices interact with one edge of the propeller region, the C-terminal half of G_γ is bound at the bottom side of the propeller structure as shown in Figure 13.13a. This unusual structure of G_γ is stabilized by the interactions with G_β; in the absence of these interactions the γ chain is unfolded and has no stable structure. The major side-chain interactions between G_β and G_γ involve residues that are conserved in different β and γ chains, so that the arrangement shown in Figure 13.13a is likely to be valid for all types of $G_{\beta\gamma}$ dimers.

The GTPase domain of G_α binds to G_β in the heterotrimeric $G_{\alpha\beta\gamma}$ complex

G_α-GDP binds to the $G_{\beta\gamma}$ dimer through its GTPase domain in a region of the β propeller opposite to where G_γ is bound (Figure 13.15). There are, therefore, no contacts between G_α and G_γ in the heterotrimeric $G_{\alpha\beta\gamma}$ complex. The

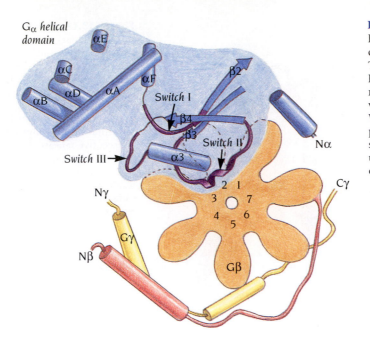

Figure 13.15 Schematic diagram of the heterotrimeric $G_{\alpha\beta\gamma}$ complex based on the crystal structure of the transducin molecule. The α subunit is blue with some of the α helices and β strands outlined. The switch regions of the catalytic domain of G_α are violet. The β subunit is light red and the seven WD repeats are represented as seven orange propeller blades. The γ subunit is yellow. The switch regions of G_α interact with the β sub-unit, thereby locking them into an inactive conformation that binds GDP but not GTP.

interactions between G_α and G_β occur at two distinct interfaces; these surfaces do not involve the helical domain of G_α, which is far from the β subunit and has no interactions with it. The more extensive interface involves residues in or adjacent to switch regions I and II of the GTPase domain of G_α, which interact with loops and turns on the top side of the β propeller. The second interface is formed between an N-terminal α helix of G_α and one edge of the β propeller. The N-terminal region in free G_α is disordered but it becomes stabilized into an α helix by interactions with the β subunit. Most of the contacts in the two interfaces involve residues that are highly conserved in both G_α and G_β. Substitutions at these positions are conservative even in sequences from distantly related organisms. It is, therefore, likely that the arrangement of subunits seen in Sigler's and Sprang's structures can be generalized to other members of the heterotrimeric G protein family.

The structure of $G_{\beta\gamma}$ is not changed at all on formation of the complex with G_α. The overall structure of G_α is also preserved except for the ordering of the N-terminal region into an α helix and significant conformational changes within the switch I and II regions (violet in Figure 13.15). The largest local structural changes in the switch interface occur in the switch II region, and involve those residues that interact with G_β directly. Some of these are conserved residues that are important to the function of G_α, for example Gly 199, which triggers conformational changes in the switch II region through hydrogen bonding with the γ phosphate of GTP, and Gln 200, which stabilizes the transition state for GTP hydrolysis (see Figures 13.8 and 13.9).

Binding to $G_{\beta\gamma}$ locks the flexible switch regions I and II of G_α into a conformation that firmly binds GDP but is nonproductive for GTP binding and hydrolysis. The replacement of GDP with GTP causes local but dramatic conformational changes to switch regions I and II, as shown in the G_α–GTP-γS structure, which disrupt nearly all of the contacts between $G_{\beta\gamma}$ and G_α in the switch interface, thereby triggering release of G_α from $G_{\beta\gamma}$ (see Figures 13.10 and 13.11).

The $G_{\beta\gamma}$ dimer appears to function as a rigid unit with critical residues positioned to interact with G_α–GDP. Whereas G_α activation proceeds through nucleotide-dependent structural reorganization, activation of $G_{\beta\gamma}$ occurs solely as a function of its release from G_α. As we will see, the G_α subunit acts as a negative regulator of $G_{\beta\gamma}$ by masking sites on the surface of $G_{\beta\gamma}$ that interact with downstream effector molecules.

Phosducin regulates light adaptation in retinal rods

One of the best characterized heterotrimeric G protein coupled pathways is the visual transduction system in rod cells, where rhodopsin absorbs photons through its retinal chromophore in a similar way as bacteriorhodopsin (see Chapter 12). Instead of initiating proton transfers through the membrane, as occurs in bacteriorhodopsin, rhodopsin activates the G protein transducin by catalyzing nucleotide exchange, leading to an activation of cyclic GMP phosphodiesterase, which degrades cyclic GMP. In a dark-adapted rod cell the absorption of one photon is sufficient to activate one rhodopsin molecule, leading to the degradation of more than 100,000 cyclic GMP molecules, enough to cause a neural impulse. This level of amplification is too high for daytime vision and a G protein regulator molecule called **phosducin** plays an important role in long-term light adaptation of rod cells.

Phosducin is expressed at high levels in retinal rods, and reaches a concentration similar to that of the G protein transducin. In light-adapted rods, unphosphorylated phosducin binds the $G_{\beta\gamma}$ of transducin with high affinity, sequestering $G_{\beta\gamma}$ and translocating it from the membrane to the cytosol. This prevents reassociation with G_α, thereby reducing signal amplification by transducin. In dark-adapted rods, phosducin is phosphorylated at serine 73, which reduces the stability of the phosducin–$G_{\beta\gamma}$ complex, allowing G_α to reassociate with $G_{\beta\gamma}$ and increase signal amplification. The crystal structure of a phosducin–$G_{\beta\gamma}$ complex, determined by Sigler's group explains some of these biological effects.

Phosducin binding to $G_{\beta\gamma}$ blocks binding of G_α

The phosducin polypeptide chain, of some 240 amino acids, is folded into two domains (Figure 13.16). The N-terminal domain is mostly α-helical and appears to be quite flexible since only a weak electron density is obtained in the structure determination. The actual path of the polypeptide chain from the end of helix A_α to the beginning of helix B_α is tentative due to slight disorder. This region is close to serine 73 at the beginning of B_α, which also becomes disordered on phosphorylation.

The C-terminal domain of phosducin is a five-stranded mixed β sheet with α helices on both sides, similar to the thioredoxin fold of disulfide isomerase DsbA described in Chapter 6. Despite significant sequence homology to thioredoxin, the phosducin domain, unlike other members of this family,

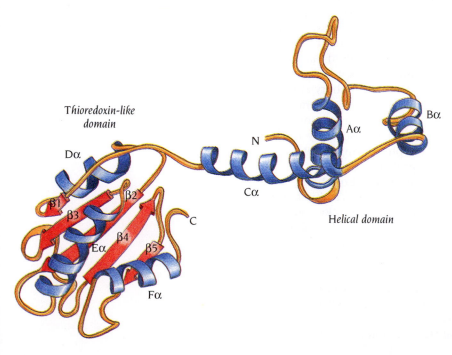

Figure 13.16 Schematic diagram of the phosducin molecule. Helices are blue, β strands are red and loop regions are orange. The structure folds into two separate domains, a N-terminal helical domain and a C-terminal domain that has the thioredoxin fold. Some of the loop regions in the helical domain are not well defined. (Adapted from R. Gaudet et al., *Cell* 87: 577–588, 1996.)

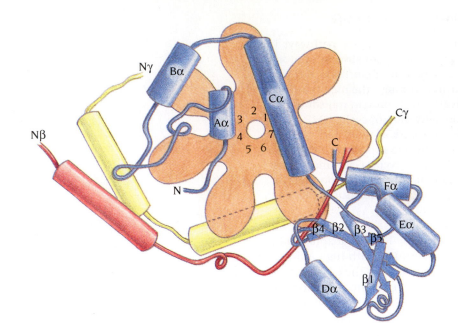

Figure 13.17 Schematic diagram of the structure of a complex between phosducin and the transducin $G_{\beta\gamma}$ dimer. The β subunit of transducin is light red and the seven WD repeats are represented as seven orange blades of a propeller. The γ subunit is yellow and the phosducin molecule is blue. The helical domain of phosducin interacts with G_β in the same region that G_α binds, thereby blocking the formation of a trimeric $G_{\alpha\beta\gamma}$ complex.

does not have a redox function because it contains only one of the two conserved cysteine residues that form the redox-active disulfide bridge.

In the complex with $G_{\beta\gamma}$ the two phosducin domains do not interact with each other, instead they wrap around the edge and the top side of the β propeller, to form an extensive interaction surface (Figure 13.17). The N-terminal domain of phosducin interacts with all of the top loops of the β propeller including part of the surface of $G_{\beta\gamma}$ that interacts with G_α (Figures 13.15 and 13.17). This interface between phosducin's N-terminal domain and $G_{\beta\gamma}$ clearly precludes association of the latter with G_α.

The C-terminal domain of phosducin binds to the edge of the β propeller opposite the β-γ coiled-coil region and close to the C-terminal region of the γ chain that carries the lipid anchoring $G_{\beta\gamma}$ to the membrane (see Figure 13.2). This surface of $G_{\beta\gamma}$ has also been implicated in receptor binding, raising the possibility that, by disrupting the orientation of $G_{\beta\gamma}$ relative to the membrane and the receptor, phosducin can interfere with the assembly and activation of the $G_{\alpha\beta\gamma}$ heterotrimer, in addition to blocking G_α binding to $G_{\beta\gamma}$.

In the structure of unphosphorylated phosducin that binds to $G_{\beta\gamma}$, Ser 73 points towards the flexible loop of phosducin and not towards $G_{\beta\gamma}$; it is, therefore, accessible on the surface for phosphorylation. Phosphorylation of Ser 73 cannot lead to the direct disruption of the phosducin/$G_{\beta\gamma}$ interaction. Rather, the structure suggests that phosphorylation may lead to conformational changes in the N-terminal domain of phosducin, especially in the flexible loop region, that could weaken or alter the phosducin/$G_{\beta\gamma}$ interface.

Two types of regulators of heterotrimeric G proteins have been identified. The RGS (regulators of G protein signaling) proteins act on the G_α subunits like GAP acts on Ras proteins, increasing their rate of GTP hydrolysis and thereby decreasing signal amplification as discussed above. By contrast, phosducin downregulates G protein function by acting on $G_{\beta\gamma}$ subunits, binding to $G_{\beta\gamma}$ and translocating it from the membrane to the cytosol, thereby preventing $G_{\alpha\beta\gamma}$ reassociation and subsequent rounds of receptor-mediated G_α activation. $G_{\beta\gamma}$ is an active component of the signal transduction system, interacting with numerous effector molecules so phosducin binding also directly inhibits the signaling pathway. As phosducin is present in various tissues and binds $G_{\beta\gamma}$ through conserved G_β residues, it is likely to be an important modulator of $G_{\beta\gamma}$ in several signal transduction pathways.

The human growth hormone induces dimerization of its cognate receptor

Peptide hormone receptors are sensory machines that direct cells to proliferate, differentiate, or even die. The growth hormone receptor belongs to one class of the cytokine or hematopoietic superfamily of hormone receptors that mediate signals from more than 20 known cytokines such as interleukin, erythropoietin and prolactin, as well as growth factors. Members of this receptor superfamily have a three-domain organization comprising an extracellular ligand-binding domain, a single transmembrane segment, and an intracellular domain that is quite diverse in sequence among the family members. The intracellular domain is structurally not well characterized, but it is known to associate reversibly with different cytoplasmic tyrosine kinases, which transmit the hormone-induced signal to activate a specific family of transcription factors. We shall describe here the crystal structure of the complex between the **human growth hormone**, **GH**, and the extracellular domain of its specific receptor, **GHR**, and compare this structure with that of growth hormone complexed with the **prolactin receptor**, **PLR**. These structures were determined by the group of Anthony Kossiakoff, Genentech, San Francisco.

The complex between GH and GHR contains one molecule of the hormone and two molecules of the receptor, even though the hormone does not have a pseudosymmetric structure with two similar binding sites. Instead, there are two completely different binding sites on the hormone, each of which binds to similar sites on the receptor molecules. These interactions are so far unique.

Like other hormones in this class of cytokines, GH has a four-helix bundle structure as described in Chapter 3 (see Figures 3.7 and 13.18). Two of the α helices, A and D, are long (around 30 residues) and the other two are about 10 residues shorter. Similar to other four-helix bundle structures, the internal core of the bundle is made up almost exclusively of hydrophobic residues. The topology of the bundle is up-up-down-down with two cross-over connections from one end of the bundle to the other, linking helix A with B and helix C with D (see Figure 13.18). Two short additional helices are in the first cross-over connection and a further one in the loop connecting helices C and D.

The two receptor extracellular domains in the complex are of course chemically identical and have very similar structures. Each polypeptide chain is arranged in two immunoglobulin-like domains joined by a linker region (Figure 13.19). Compared with an immunoglobulin constant domain (see Chapter 15) there is a "strand-switching" such that the edge of strand D, which forms hydrogen bonds to strand E in the immunoglobulin, is instead hydrogen bonded to strand C in the receptor and forms part of the sheet GFCD. This modified immunoglobulin fold is called a fibronectin III domain, since it also occurs in fibronectin.

The two domains are arranged with an acute angle between them so that loop regions from both domains come close together, forming the hormone-binding site (yellow in Figure 13.19). The loop regions are from different ends of the two domains; loops A–B, C–D and E–F in the N-terminal domain and loops B–C and F–G in the C-terminal domain. In contrast to the TIM-barrel domains (see Chapter 4), where the active site is always found at the same place in relation to the fold, immunoglobulin-like domains are much more promiscuous and utilize different areas of the fold for specific interactions with different molecules.

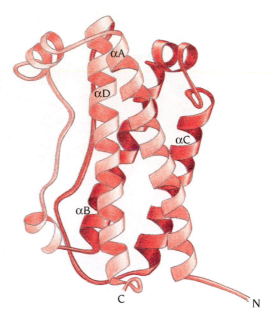

Figure 13.18 Ribbon diagram of the structure of human growth hormone. The fold is a four-helix bundle with up-up-down-down topology, and consequently there are two long cross-connections between helices A and B as well as between helices C and D. (Adapted from J. Wells et al., *Annu. Rev. Biochem.* 65: 609–634, 1996.)

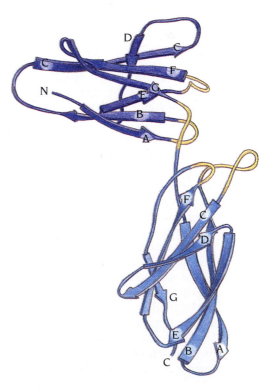

Figure 13.19 Ribbon diagram of the structure of the extracellular domain of the human growth hormone receptor. The hormone-binding region is formed by loops (yellow) at the hinge region between two fibronectin type III domains. (Adapted from J. Wells et al., *Annu. Rev. Biochem.* 65: 609–634, 1996.)

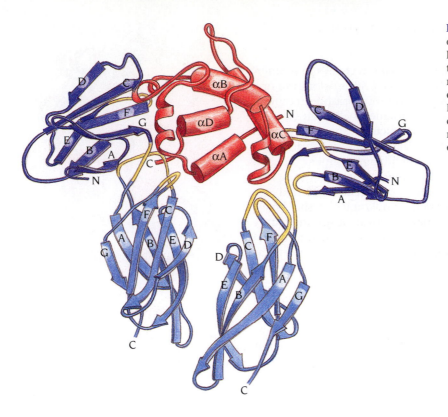

Figure 13.20 Ribbon diagram of the structure of a 1:2 complex between the human growth hormone and the extracellular domains of two receptor molecules. The two receptor molecules (blue) bind the hormone (red) with essentially the same loop regions (yellow). In contrast, the hormone molecule uses totally different surface regions to bind the two receptor molecules. (Adapted from Somers et al., *Nature* 372: 478–481, 1994.)

In the complex with the hormone, the two receptor domains form a dimer through direct interactions between their C-terminal regions and indirect interactions mediated by the hormone (see Figure 13.20). The two C-terminal domains are almost parallel, each having its C-terminus pointing away from the hormone towards the membrane. Biochemical and genetic studies have shown that dimerization of the receptor is the essential step responsible for activation of the intracellular domain. A reasonable speculation is that the close joining of the extracellular domains immediately above the membrane forces the transmembrane and intracellular domains into close proximity within and below the membrane, so as to initiate signaling.

Dimerization of the growth hormone receptor is a sequential process

The crystal structure of the GH–GHR complex shows that the receptor has a single ligand-binding site that can interact with either of two different sites on the hormone. Both receptor molecules in the complex use their A–B, C–D and E–F loops of the N-terminal domain as well as the linker and loops B–C and F–G of the C-terminal domain to form the binding site. Not only are essentially the same amino acid residues on both receptors used for binding, but the three-dimensional structures of these loop regions are also very similar. In contrast, the two binding sites on the hormone are very different; one site comprises residues from α helices A and D, and the other site is formed by residues from helices A and C (see Figure 13.20).

There is no structural similarity to these two binding sites; the A–C binding site is flat whereas the A–D site is concave, neither is there any similarity in the nature of the amino acids that form these two different binding sites. Nevertheless they bind to the same site on the receptor. However, the binding strengths of these two sites on the hormone are different. At high concentrations of the hormone a 1:1 complex is formed with receptor instead of the normal 1:2 complex. Experiments with mutants of the hormone show that this complex is formed by interactions to the A–D site on the hormone. Mutants in which binding to the A–C site is blocked still form the 1:1 complex whereas mutants that block binding to the A–D site do not form such complexes. These results nicely correlate to the structure of the complex. The A–D binding site involves 31 residues and the receptor covers an area of 1300 Å² of the hormone surface. The A–C site on the other hand only involves 24 residues and the binding area is 850 Å². Even though binding strength depends on the specific interactions involved there is in general a direct correlation between the size of the binding area and the strength of binding when proteins interact.

These results suggest that activation of the receptor by dimerization through binding the hormone is a sequential process. A 1:1 complex is first formed by hormone binding firmly to one receptor molecule through its A–D site. Once this complex is formed the binding of a second receptor to the A–C site of the hormone is facilitated by the interactions between the C-terminal domains of the two receptor molecules (see Figure 13.20). This latter interaction area is 500 Å², which together with the 850 Å² of the A–C site, provides a total binding area of 1350 Å², sufficient to stabilize the binding of the second receptor molecule to form the active complex.

The growth hormone also binds to the prolactin receptor

The prolactin receptor, PLR, which regulates milk production in mammals, belongs to the same receptor class as the growth hormone receptor. In addition to binding the hormone prolactin, PLR also binds and is activated by growth hormone. The extracellular domain of PLR forms a very stable 1:1 complex with growth hormone in solution: this complex has been crystallized and its structure determined (Figure 13.21). We shall compare this structure with the 1:2 complex of the same hormone with GHR.

The structure of the growth hormone in these two complexes is essentially the same, differing only in some of the loop regions and cross-over connections. The hormone interacts with the prolactin receptor through the same area that forms the strong A–D binding site to the growth hormone receptor. The structures of the individual domains of the prolactin receptor are very similar to the corresponding domains of the growth hormone receptor (even though their amino acid sequences share only 28% identity). The orientation of the domains with respect to each other is, however, significantly different. In spite of these differences the same loop regions of both

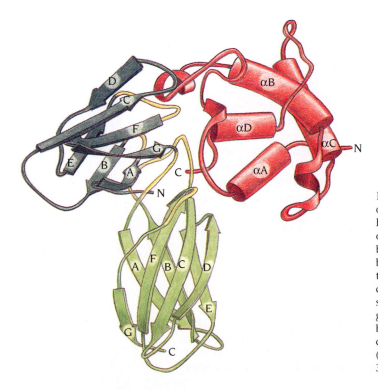

Figure 13.21 Ribbon diagram of the structure of a 1:1 complex between the human growth hormone (red) and the extracellular domain of the prolactin receptor (green). The hormone binds to the prolactin receptor using the same binding area as in the strong binding site to the growth hormone receptor. The two domains of the prolactin receptor have similar structures to the corresponding domains of the growth hormone receptor, but the orientation between the domains is different, causing differences in the details of the binding area. (Adapted from W. Somers et al., *Nature* 372: 478–481, 1994.)

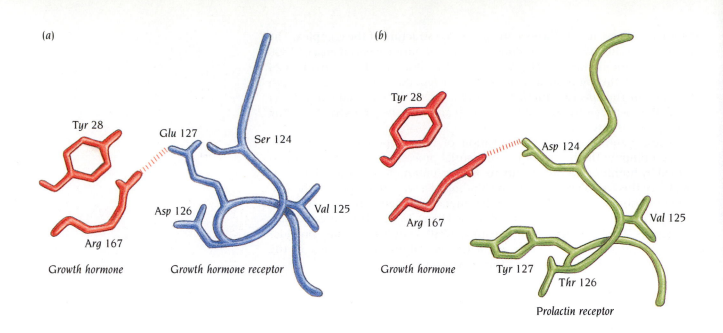

(a)

Tyr 28

Glu 127

Ser 124

Asp 126

Val 125

Arg 167

Growth hormone

Growth hormone receptor

(b)

Tyr 28

Asp 124

Arg 167

Val 125

Growth hormone

Tyr 127

Thr 126

Prolactin receptor

Figure 13.22 Hormone–receptor interactions involving the domain–domain linker region in the receptor. (a) Interactions between the growth hormone (red) and the growth hormone receptor (blue) linker region. Glu 127 of the receptor forms a salt bridge to Arg 167 in the hormone. (b) The same interaction area in the growth hormone (red)–prolactin receptor (green) complex. The displacement of the linker region due to differences in the domain orientations have brought Asp 124 in the prolactin receptor into contact with Arg 167 of the hormone. (Adapted from W. Somers et al., *Nature* 372: 478–481, 1994.)

receptors interact with the hormone. An example is given in Figure 13.22, which shows part of the interactions between growth hormone and the linker regions of the two receptors. The first four residues (124–127) of the linker regions are Ser-Val-Asp-Glu in GHR and Asp-Val-Thr-Tyr in PLR. The remaining seven residues of the linker regions are identical in the two molecules. The orientation of the linker in GHR is such that Glu 127 forms a salt linkage with Arg 167 of the hormone (Figure 13.22a). In PLR the corresponding Tyr 127 residue cannot form such a salt bridge. However, the different orientation of the linker region in the PLR complex removes the tyrosine from the interaction area and instead brings Asp 124 into a position where it forms a salt linkage with Arg 167 (Figure 13.22b).

A unique feature of the interaction of the hormone and PLR is at the beginning of the F-G loop in the C-terminal domain. In HGR the sequence is Arg-Asn-Ser whereas in PLR it is Asp-His-deletion. This loop interacts with His 18 and Glu 174 of the hormone. In PLR the orientation of this loop is such that the Asp and His residues, in combination with His and Glu from the hormone, form a strong binding site for a zinc atom that links the hormone and the receptor (Figure 13.23b). The presence of zinc increases the affinity of the hormone for the receptor *in vitro* by a factor of 10,000. As shown by mutagenesis studies His 18 and Glu 174 of the hormone are important for the tight binding to PRL but not to GHR.

These two examples demonstrate that both sequence differences and differences in the relative orientation of receptor domains are important factors for forming ligand-binding sites. This has considerable consequences for modeling binding sites of a receptor based on sequence homology with another receptor of known structure. Such modeling studies are frequently done, especially by drug companies, since many of these receptors are important targets for drugs. It is reasonably simple to model amino acid substitutions based on a structure with fixed main chain conformations. It is considerably more difficult to predict possible changes in the orientations of the domains due to sequence differences or ligand-induced structural changes.

Tyrosine kinase receptors are important enzyme-linked receptors

There are five known classes of **enzyme-linked receptors**: (1) receptor tyrosine kinases, which phosphorylate specific tyrosine residues on intracellular signaling proteins; (2) tyrosine kinase-associated receptors, such as the prolactin and growth hormone receptors we have already discussed, which

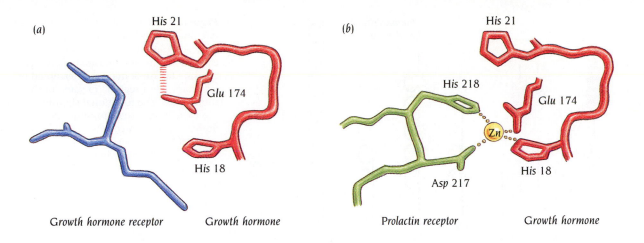

Figure 13.23 The F–G loop in the C-terminal domain of the prolactin receptor is involved in a unique interaction. (a) The F-G loop of the growth hormone receptor (blue) is not involved in any specific interactions with the growth hormone (red). (b) The F-G loop in the prolactin receptor forms a strong zinc-binding site that links the receptor (green) to the hormone (red). (Adapted from W. Somers et al., *Nature* 372: 478–481, 1994.)

associate with proteins that have tyrosine kinase activity; (3) receptor tyrosine phosphatases, which remove phosphate groups from tyrosine residues of specific intracellular signaling proteins; (4) transmembrane receptor serine/threonine kinases, which add a phosphate group to serine and threonine side chains on target proteins; and (5) transmembrane guanyl cyclases, which catalyze the production of cyclic GMP in the cytosol.

The protein kinase domains of groups 1, 2, and 4 all have homologous amino acid sequences and belong to one of the largest superfamilies known; the cyclin-dependent protein kinase CDK2 described in Chapter 6 (see Figure 6.16a) is one such member. In the complete yeast genome sequence more than 100 genes for members of this superfamily have been identified by sequence similarities, and it has been estimated that there are about 1000 members in humans. The first two types of receptors are by far the most numerous and they are thought to work in a similar way: ligand binding induces the receptors to oligomerize, which activates the tyrosine kinase activity of either the receptor itself or its associated nonreceptor tyrosine kinase. When activated, receptor tyrosine kinases usually cross-phosphorylate themselves on multiple tyrosine residues, with the phosphotyrosine residues then serving as docking sites for intracellular proteins. This results in the formation of an ensemble of proteins immobilized at the cell membrane, bringing together a diverse range of proteins such as kinases, phosphatases, phospholipases and G proteins such as Ras. In this way, a multisignaling complex is activated from which the signal spreads from the membrane to the cell interior.

Signaling through tyrosine kinase domains is involved in a variety of diverse biological processes including cell growth, cell shape, cell cycle control, transcription, and apoptosis (programmed cell death). Receptors regulating cell growth and differentiation were among the first to be studied, and they show a similar overall structural organization (Figure 13.24). The cytosolic region has a tyrosine kinase domain of 250–300 amino acid residues and, in addition, regions that contain tyrosine residues, in different sequence environments, which bind to different adaptor proteins when phosphorylated. The extracellular domains are different for different subclasses of these receptors, but are in many cases built up from immunoglobulin or fibronectin domains. A single transmembrane α helix links the two domains.

The second group, the tyrosine kinase associated receptors, have cytosolic domains that lack a defined catalytic function. This large and heterogeneous group includes receptors for cytokines as well as for some hormones including growth hormone and prolactin, discussed earlier. These receptors work through associated cytosolic tyrosine kinases, which phosphorylate various target proteins when the receptor binds its ligand. The best characterized of these associated tyrosine kinases are members of the **Src** family, so called because the first identified member of this family, the transforming agent of Rous sarcoma virus, induces sarcoma tumors of connective tissues. The structure and regulation of Src will be described later in this chapter.

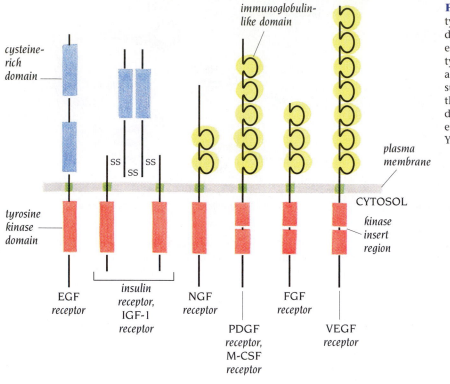

immunoglobulin-like domain

cysteine-rich domain

tyrosine kinase domain

plasma membrane

CYTOSOL

kinase insert region

ss ss ss

EGF receptor

insulin receptor, IGF-1 receptor

NGF receptor

FGF receptor

VEGF receptor

PDGF receptor, M-CSF receptor

Figure 13.24 Six subfamilies of receptor tyrosine kinases involved in cell growth and differentiation. Only one or two members of each subfamily are indicated. Note that the tyrosine kinase domain is interrupted by a "kinase insert region" in some of the subfamilies. The functional significance of the cysteine-rich and immunoglobulin-like domains is unknown. (Adapted from B. Alberts et al, *Molecular Biology of the Cell*, 3rd ed. New York: Garland Publishing, 1994.)

Small protein modules form adaptors for a signaling network

How are the internal pathways controlled and organized in the complex network of signaling molecules involving more than a thousand different protein kinase domains and probably even more target molecules? A set of protein modules, which are covalently attached either to their associated protein kinases or to their target molecules, play an important role in regulating signal pathways (Figure 13.25a). These modules function as adaptors that bring together a kinase domain with its proper targets. Three important such modules are the **SH2** (Src-homology-2) and **SH3** (Src-homology-3) domains, whose names derive from the Src protein kinase molecule where they were first discovered, and the **PH** (pleckstrin-homology) domain, which was first defined as two repeated sequences in pleckstrin, the major substrate for Ser/Thr phosphorylation by protein kinase C in platelets. The SH2 and SH3 domains recognize short peptide motifs containing phosphotyrosine (pTyr)

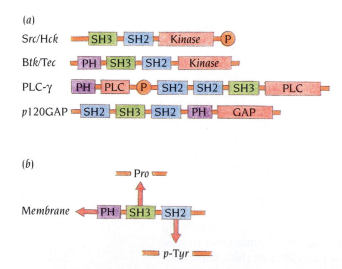

(a)

Src/Hck   SH3   SH2   Kinase   P

Btk/Tec   PH   SH3   SH2   Kinase

PLC-γ   PH   PLC   P   SH2   SH2   SH3   PLC

p120GAP   SH2   SH3   SH2   PH   GAP

(b)

Pro

Membrane   PH   SH3   SH2

p-Tyr

Figure 13.25 (a) Receptor-associated tyrosine kinases frequently contain within their polypeptide chains small modules such as Src homology domains, SH2 or SH3 and pleckstrin homology domains, PH. These modules function as adaptors to bring the kinase to its correct target molecule. (b) SH2 domains bind phosphotyrosine-containing peptide regions of the target molecules, SH3 domains bind proline-rich peptide regions of the target molecules and PH domains are involved in membrane association of kinases with their target molecules.

in the case of SH2, and proline-rich motifs in the case of SH3 (Figure 13.25b). The PH domains appear to mediate membrane association of kinases with their target molecules, in some cases by interaction with the headgroup of a phosphoinositide lipid.

SH2 domains bind to phosphotyrosine-containing regions of target molecules

The structures of many SH2 domains have been determined both by x-ray crystallography and by NMR. The overall fold of these domains is very similar, but residues involved in peptide binding differ, giving recognition specificity to SH2 domains from different signaling molecules. Figure 13.26 shows the crystal structure of the Src SH2 domain complexed with a peptide, determined by the group of John Kuriyan at the Rockefeller University, New York. The basic structure of the SH2 domain is an essentially antiparallel β sheet flanked by two helices. The β sheet comprises three long antiparallel β strands βB–βD and in addition two short strands βA and βG, which are hydrogen bonded in a parallel fashion to βB. These secondary structure elements are connected by relatively short loops, except the loop between βD and αB, which includes a small antiparallel β sheet. Hydrophobic cores are formed by packing the two α helices against each side of the β sheet.

The peptide binds to a relatively flat surface created by the two helices and the edge of the β sheet. Screening different phosphotyrosine-containing peptides showed the motif p-Tyr-Glu-Glu-Ile to be optimal for binding to Src. The peptide that Kuriyan studied contains this motif, and the structure shows that the phosphotyrosine moiety is firmly anchored to SH2 by three positively charged residues (green in Figure 13.26): two arginines from αA and βB and a lysine from βD. The two Glu residues of the peptide make weak

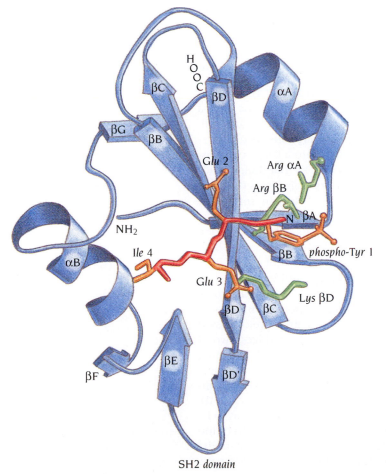

Figure 13.26 Schematic diagram of the SH2 domain from the Src tyrosine kinase with bound peptide. The SH2 domain (blue) comprises a central β sheet surrounded by two α helices. Three positively charged residues (green) are involved in binding the phosphotyrosine moiety of the bound peptide (red). (Adapted from G. Waksman et al., *Cell* 72: 779–790, 1993.)

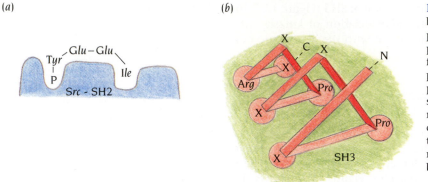

(a)   (b)

Figure 13.27 (a) All SH2 domains have a binding pocket with strong affinity for phosphotyrosine. Src-SH2 binds strongly to peptides containing two polar side chains followed by an aliphatic residue, after the phosphotyrosine. (b) SH3 domains bind proline-rich peptide regions with the consensus sequence X-Pro-X-X-Pro-X-Arg where X is any residue. These peptide regions bind in a helical conformation called polyproline II, which has three residues per turn. Such a helix has three ridges of side chains, two of which contact the binding surface of SH3 (green).

yet specific interactions with polar side chains of SH2, but the Ile residue is bound tightly to SH2 inside a hydrophobic pocket. This so-called pY+3 pocket is critical for determining peptide specificity and is formed by side chains from αB and βD, with loop regions from αB–βG and the small sheet βE–βF forming "jaws" that residue pY+3 fits between. The general hydrophobic character of the pocket is conserved in all SH2 domains, but the precise nature of the specific side chains that form the pocket in different SH2 domains determines the specificity of binding to target molecules.

The Src SH2 domain typifies a large number of those characterized to date. The pTyr fits into a pocket on the opposite side of the central sheet to the pY+3 pocket (Figure 13.27a). All known SH2 domains bind pTyr in essentially the same way, but some have a different pattern of contacts for the residues that follow. For example, in the Grb2 SH2 domain, a tryptophan side chain from the small sheet fills the pY+3 pocket, and the bound peptide takes a different course, with important interactions to an asparagine at pY+2. Screens of peptide libraries have detected the importance of this asparagine. The SH2 domain from PLC-γ1 contacts five mainly hydrophobic residues that follow pTyr.

SH3 domains bind to proline-rich regions of target molecules

The structures of many SH3 domains have also been determined; as an example, Figure 13.28a shows the SH3 domain from the tyrosine kinase Fyn, a member of the Src family, as determined by the group of Ian Campbell, Oxford University, using NMR methods. The SH3 domain consists of a five-stranded up-and-down antiparallel β structure, twisted into a barrel comprising two antiparallel β sheets that pack against each other so that their strands are nearly orthogonal. Strand β2 takes part in both sheets. This fold is common to all SH3 modules. The loop regions differ, however, and in some SH3 domains they contain small regions of secondary structure.

The peptide-binding site is a hydrophobic groove flanked by the RT loop between β1 and β2 and the n-Src loop between β3 and β4 (see Figure 13.28a). The latter is so named because neuronal Src has an insertion of six residues in this loop. The groove is lined with conserved aromatic residues.

SH3 binding elements in target molecules consist of proline-rich peptides. Upon binding, these segments adopt a left-handed polyproline type II helix with three residues per turn (see Figure 13.27b)—a structure discussed further in Chapter 14. The peptide ligand therefore has three ridges, two of which contact the SH3 domain (Figure 13.28b). Combinatorial screening studies (see Chapter 17) have shown that SH3 domains recognize sequences of about 10 residues. In the case of the Src SH3 domain, two classes of consensus motif were found: RXLPPLPXX (class I) and XXXPPLPXR (class II), where X is any residue. These were shown by subsequent NMR analysis to lie in opposite directions in the SH3 binding groove. Because proline is an imino acid, polyproline II helices with opposite N to C polarities present very similar surfaces. The ridges of these surfaces fit into notches formed by the aromatic

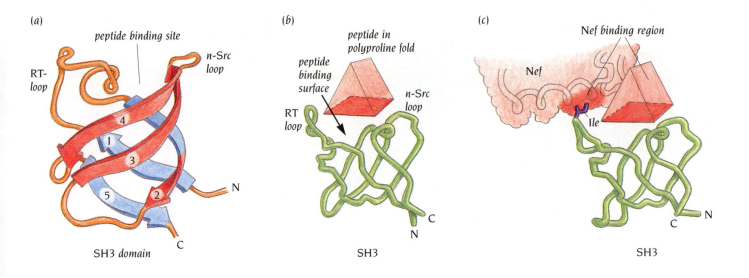

(a) *peptide binding site*
RT-loop
n-Src loop
4
1
3
5
2
N
C
SH3 *domain*

(b) *peptide in polyproline fold*
peptide binding surface
RT loop
n-Src loop
C
N
SH3

(c) Nef *binding region*
Nef
Ile
C
N
SH3

side chains in the SH3 groove. The polarity of binding is determined by additional interactions at one end of the groove. In the case of Src SH3, a negatively charged pocket helps to select the orientation of the bound peptide by receiving an arginine residue at one end or the other of the proline-rich segment.

The interactions of SH3 domains with their targets are strikingly promiscuous and relatively weak. Specificity is provided either by parallel contacts involving additional interacting modules (Grb2, for example, contains two SH3 domains, which bind to neighboring proline-rich sequences on its target), or by SH3–target interactions outside the polyproline helix, elegantly illustrated by the HIV protein, Nef. Nef is a regulatory protein encoded in the HIV genome; it is critical for viral pathogenesis and progression to AIDS. Nef has a high affinity for the SH3 domain of the Src family member Hck but not for the SH3 domain of the closely related kinase, Fyn. A key determinant of this specificity is an isoleucine residue, present in Hck but not in Fyn. Mutation of the corresponding Fyn residue (Arg 96) to Ile produces high specificity for Nef. John Kuriyan has determined the crystal structure of a complex between Nef and this mutant Fyn-SH3. As expected, a P-X-X-P element in Nef lies in the binding groove of the SH3 domain (Figure 13.29). The key Ile residue, which lies in the RT-loop of the SH3, forms a hydrophobic contact with side chains from two α helices in Nef (Figures 13.28b,c). Moreover, the polyproline helix itself is stabilized within the Nef molecule by packing against an α helix. Thus, the polyproline element is pre-formed, saving the free-energy "cost" of folding a disordered segment and increasing its affinity for the SH3 domain. Interactions of SH3 domains with their cellular targets are generally weak in order to be able to regulate the signal-transduction pathways in which they participate (which we discuss below). That is, the interaction must be readily reversed. Nef appears to have evolved to intervene strongly and irreversibly in a signaling pathway, but its physiological target is still not known. Identification of its partner might lead to a new target for anti-HIV drugs.

Figure 13.28 Schematic diagrams of the structure and binding modes of SH3 domains. (a) The polypeptide chain of the SH3 domain is folded into a five-stranded antiparallel β barrel with orthogonal packing of the β sheets. The peptide-binding site is at one end of the barrel, flanked by two loop regions, the RT-loop and the n-Src loop. (b) Isolated peptides (red triangular prism) bind rather weakly to SH3 even in the form of a polyproline helix. (c) Complete target molecules with the same peptide region bind much more strongly to the same SH3 module. The target molecule, Nef (red), which is involved in AIDS, interacts strongly with an Ile residue (violet) in the RT-loop of SH3, in addition to the proline-rich peptide binding (red triangular prism). [(a) Adapted from C.J. Morton et al., *Structure* 4: 705–714, 1996. (c) Adapted from W. Lim, *Structure* 4: 657–659, 1996.]

Src tyrosine kinases comprise SH2 and SH3 domains in addition to a tyrosine kinase

The polypeptide chain of **Src tyrosine kinase**, and related family members, comprises an N-terminal "unique" region, which directs membrane association and other as yet unknown functions, followed by a SH3 domain, a SH2 domain, and the two lobes of the protein kinase. Members of this family can be phosphorylated at two important tyrosine residues—one in the "activation loop" of the kinase domain (Tyr 419 in c-Src), the other in a short

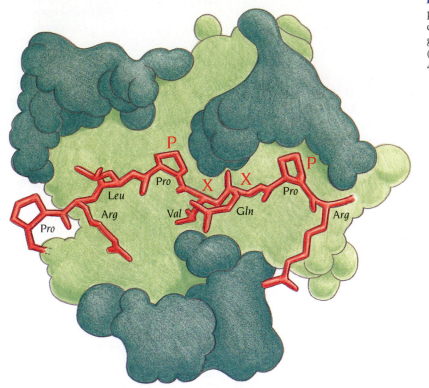

Figure 13.29 Schematic diagram of the proline-rich region of Nef bound to an SH3 domain. The peptide region (red) binds in a groove that extends across the surface of SH3 (green). (Adapted from W. Lim, *Structure* 4: 657–659, 1996.)

C-terminal tail (Tyr 527 in c-Src). Phosphorylation of Tyr 419 activates the kinase; phosphorylation of Tyr 527 inhibits it. Crystal structures of a fragment containing the last four domains of two members of this family were reported simultaneously in 1997—cellular Src by the group of Stephen Harrison and Hck by the group of John Kuriyan. The two structures are very similar, as expected since the 440 residue polypeptide chains have 60% sequence identity. The crucial C-proximal tyrosine that inhibits the activity of the kinases was phosphorylated in both cases; the activation loop was not.

The structures of each of the domains in the fragment (Figure 13.30) are very similar to previously described structures; the small N- and the large

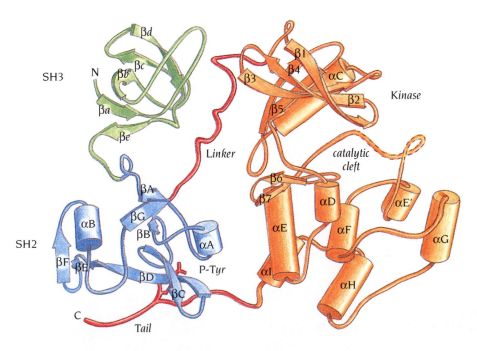

Figure 13.30 Ribbon diagram of the structure of Src tyrosine kinase. The structure is divided in three units starting from the N-terminus; an SH3 domain (green), an SH2 domain (blue), and a tyrosine kinase (orange) that is divided into two domains and has the same fold as the cyclin dependent kinase described in Chapter 6 (see Figure 6.16a). The linker region (red) between SH2 and the kinase is bound to SH3 in a polyproline helical conformation. A tyrosine residue in the carboxy tail of the kinase is phosphorylated and bound to SH2 in its phosphotyrosine-binding site. A disordered part of the activation segment in the kinase is dashed. (Adapted from W. Xu et al., *Nature* 385: 595–602, 1997.)

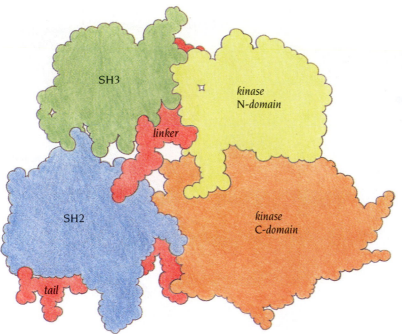

Figure 13.31 Space-filling diagram of Src tyrosine kinase in the same view as Figure 13.30. The SH2 domain makes only a few contacts with the rest of the molecule except for the tail region of the kinase. The SH3 domain contacts the N-domain of the kinase in addition to the linker region. There are extensive contacts between the N- and C-domains of the kinase. (Adapted from W. Xu et al., *Nature* 385: 596–602, 1997.)

C-terminal lobes of the tyrosine kinase are similar to those of cyclin-dependent kinase described in Chapter 6 (see Figure 6.16a), while the SH2 and SH3 domains of Src and Hck have structures very similar to those of the isolated domains (see Figures 13.26 and 13.28a).

The inactive state of the kinase is a compact ensemble of the four domains. The SH2 and SH3 domains lie respectively beside the large and small domains of the tyrosine kinase, on the opposite side to the catalytic cleft. The phosphorylated C-terminal tail extends from the base of the large catalytic domain of the kinase into SH2, where the phosphotyrosine is bound in the usual SH2 binding site for phosphotyrosine. The mode of interaction between other residues of this tail and the binding groove of SH2 suggest a low-affinity interaction, because there is no side chain of the tail in the pY+3 pocket that binds Ile in high-affinity complexes of Src-SH2 (Figure 13.27a). The phosphorylated tail is a short but flexible tether that anchors the SH2 domain to the kinase. Apart from this tail, the SH2 domain makes only a few contacts with the catalytic domain of the kinase (Figure 13.31).

The linker region between the SH2 domain and the N-terminal domain of the kinase interacts with the groove of the SH3 domain. Residues 249–253 of the linker adopt a polyproline II helix structure in the SH3-binding groove even though the sequence around this region, -Pro-Gln-Thr-Gln-Gly-Leu-, deviates significantly from the consensus SH3 recognition sequence -P-X-X-P-X-R-. The proline residue occupies the position of the first proline in the consensus sequence, but the Gln residue cannot be accommodated in the binding pocket for the second proline, and so the course of the linker deviates from that of proline-rich peptides at this point. Contacts between the SH2 and SH3 domains are very limited (see Figure 13.31).

The two domains of the kinase in the inactive state are held in a closed conformation by assembly of the regulatory domains

A number of kinase structures have been determined in various catalytic states. For example, structures of the cyclin-dependent kinase, CDK2, in its inactive state and in a partially active state after cyclin binding have been discussed in Chapter 6. The most thoroughly studied kinase is the cyclic AMP-dependent protein kinase; the structure of both the inactive and the active

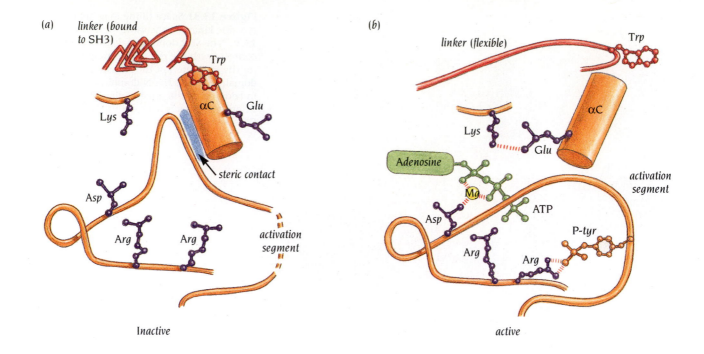

(a)

linker (bound
to SH3)

Trp

αC   Glu

Lys

steric contact

Asp

Arg   Arg

activation
segment

Inactive

(b)

linker (flexible)   Trp

αC

Lys

Adenosine   Glu

Mg

Asp   ATP

Arg   Arg   P-tyr

activation
segment

active

Figure 13.32 Regulation of the catalytic activity of members of the Src family of tyrosine kinases. (a) The inactive form based on structure determinations. Helix αC is in a position and orientation where the catalytically important Glu residue is facing away from the active site. The activation segment has a conformation that through steric contacts blocks the catalytically competent positioning of helix αC. (b) A hypothetical active conformation based on comparisons with the active forms of other similar protein kinases. The linker region is released from SH3, and the activation segment changes its structure to allow helix αC to move and bring the Glu residue into the active site in contact with an important Lys residue. In the activation segment an Arg binds to phosphotyrosine, another Arg binds to the γ phosphate of ATP and an Asp binds to the Mg atom in the active site. (Adapted from F. Sicheri et al., *Nature* 385: 602–609, 1997.)

states have been determined by the group of Susan Taylor at the University of California, San Diego. In the inactive state the two domains of the kinase are wide apart, in an open conformation. In the active form the domains have moved toward each other and are in a closed conformation. A comparison of the phosphorylated Src kinase structure with these other kinases shows that the Src kinase domains are in a closed conformation but that their close approach is accompanied by a displacement and a rotation of the α helix, αC, in the N-terminal domain, away from its position in active kinases (Figure 13.32). Helix αC of Src is equivalent to the PSTAIRE helix in CDK2, where a similar rearrangement of the helix and the activation segment is seen in the inactive form (compare Figures 6.17 and 13.32). Because αC (like PSTAIRE) contributes to the active site of the enzyme, the observed, closed form of Src is catalytically inactive. What causes this inactivation, and how is the kinase activated?

The interaction of the phosphotyrosine tail with SH2 is clearly important for inactivation of the tyrosine kinase, because regulation *in vivo* is lost upon dephosphorylation or mutation of Tyr 527. The observed interactions of the SH2 and SH3 domains with the linker, the phosphorylated C-terminal tail and the kinase itself, all serve to hold the kinase lobes together, thereby stabilizing ejection of the αC from the catalytic site (see Figure 13.32) as well as holding the activation loop in a conformation that protects Tyr 419 from phosphorylation. A key feature of the assembled regulatory apparatus is the position of the SH3 domain, which binds the linker regions between SH2 and the N-terminal lobe of the kinase and may affect the conformation of αC. Dephosphorylation of tyrosine 527 or recruitment of the SH2 or SH3 domains by strongly binding target molecules abolishes the intermolecular interaction of the linker region and presumably allows helix αC and the activation segment to shift into their active conformations. The SH2 and SH3 domains apparently not only function as adaptors between the kinase and target molecules but are also important for autoregulation of the intrinsic catalytic activity of Src-tyrosine kinases.

Conclusion

Signal-transducing receptors are plasma membrane proteins that bind specific molecules, such as growth factors, hormones, or neurotransmitters, and then transmit a signal to the cell's interior, causing the cell to respond in a specific manner. All membrane-bound receptor molecules have an extracellular domain that receives and recognizes a specific signal, a transmembrane region through which the signal is transmitted, and an intracellular domain that elicits a response. In many cases, the signal is amplified by a network of intracellular signaling pathways.

G proteins are molecular amplifiers for a large number of seven-transmembrane helix receptors that regulate responses like vision, smell and stress response. They are heterotrimeric molecules, $G_{\alpha\beta\gamma}$, that dissociate into membrane-bound G_{α} and $G_{\beta\gamma}$ signal transmitters upon activation of the receptor.

G_{α} has two domains: a helical domain and a catalytically active GTPase domain, which is similar to the small G protein, Ras. Both Ras and G_{α} switch between two conformations, an active GTP-bound form and an inactive GDP-bound form. Only the inactive GDP-bound form binds to $G_{\beta\gamma}$. Three switch regions have different conformations in the two forms. The cellular response of G_{α} and Ras depends on the length of time these molecules stay in the active GTP form—that is, on the catalytic rate of their GTPase activity. The rate of Ras GTPase activity is enhanced by binding to GAP, which provides a catalytically important arginine to the GTPase active site and stabilizes the switch regions in their active conformations. Mutations of Ras that abolish this rate enhancement are oncogenic because they render Ras constitutively active, promoting undifferentiated cell growth. The activity of G_{α}, which has a far higher catalytic rate than Ras due to the presence of an intrinsic corresponding Arg, is regulated by proteins that affect the conformation of the switch regions.

G_{β} has an N-terminal α helix that forms a coiled-coil structure with an N-terminal helix of G_{γ} in the $G_{\beta\gamma}$ dimer. The remaining part of G_{β} contains seven WD sequence repeats, which build up a circular seven-blade propeller structure, where each blade is an antiparallel β sheet of four strands. Each blade is formed by regions from two WD repeats, facilitating the formation of a circular structure. The GTPase domain of G_{α} binds to G_{β} in trimeric $G_{\alpha\beta\gamma}$ in such a way that the switch region of G_{α} is locked into the inactive form that binds GDP. The negative regulator of $G_{\beta\gamma}$ signaling, phosducin, binds to the same region of $G_{\beta\gamma}$ as G_{α}, thereby both preventing $G_{\beta\gamma}$ activating downstream signaling molecules and preventing reactivation of G_{α} by blocking the formation of the $G_{\alpha\beta\gamma}$ complex. Essentially all of the molecular mechanisms discussed (switch region conformational changes, hydrolysis, association and dissociation) are conserved throughout the trimeric G protein superfamily.

Growth hormones activate their cognate receptors by inducing the formation of receptor dimers. The structure of a complex between human growth hormone and the extracellular domains of the growth hormone receptor shows that one hormone molecule can bind two receptors. The hormone uses quite different surface areas for binding the same regions of the two receptor molecules. The structure of a complex between the growth hormone and the prolactin receptor has shown that subtle conformational differences can provide novel binding interactions, thereby increasing the range of possible interactions between a hormone and receptors.

Signaling through tyrosine kinase domains is involved in such diverse biological activities as cell growth, cell shape, cell cycle control, transcription and apoptosis. Receptor-associated cytosolic tyrosine kinases contain a set of protein modules such as SH2, SH3, and PH domains that function as adaptors to bring together a kinase domain and its appropriate target. SH2 and SH3 domains recognize short peptide sequences containing phosphotyrosine and proline-rich regions, respectively; some PH domains recognize lipid head groups. SH2 domains fold into a central β sheet surrounded by two helices and have a specific phosphotyrosine-binding pocket. SH3 domains fold into a β barrel structure with the peptide-binding site at one end of the barrel, wedged between two loop regions. The peptide region binds in a polyproline type II helix conformation with three residues per turn. Proline-rich sequences favor such a conformation. Additional specificity for target proteins can be provided by interactions between the loop regions and residues of the target protein outside the proline-rich sequence.

Src tyrosine kinase contains both an SH2 and an SH3 domain linked to a tyrosine kinase unit with a structure similar to other protein kinases. The phosphorylated form of the kinase is inactivated by binding of a phosphotyrosine in the C-terminal tail to its own SH2 domain. In addition the linker region between the SH2 domain and the kinase is bound in a polyproline II conformation to the SH3 domain. These interactions lock regions of the active site into a nonproductive conformation. Dephosphorylation or mutation of the C-terminal tyrosine abolishes this autoinactivation.

Selected readings

General

Bohm, A., Gaudet, R., Sigler, P.B. Structural aspects of heterotrimeric G-protein signaling. *Curr. Opin. Struct. Biol.* 8: 480–487, 1997.

Bos, J.L. *Ras* oncogenes in human cancer: a review. *Cancer Res.* 49: 4682–4689, 1989.

Clapham, D.E. The G-protein nanomachine. *Nature* 379: 297–299, 1996.

Harrison, S.C. Peptide-surface association: the case of PDZ and PTB domains. *Cell* 86: 341–343, 1996.

Hunter, T., Plowman, G.D. The protein kinases of budding yeast: six score and more. *Trends Biochem. Sci.* 22: 18–22, 1997.

Johnson, L.N., Noble, M.E.M., Owen, D.J. Active and inactive protein kinases: structural basis for regulation. *Cell* 85: 149–158, 1996.

Kjeldgaard, M., Nyborg, J., Clark, B.F.C. The GTP binding motif: variations on a theme. *FASEB J.* 10: 1347–1368, 1996.

Kuriyan, J., Cowburn, D. Structures of SH2 and SH3 domains. *Curr. Opin. Struct. Biol.* 3: 828–837, 1993.

Murzin, A.G. Structural principles for the propeller assembly of β sheets: the preference for seven-fold symmetry. *Proteins* 14: 191–201, 1992.

Neer, E.J., et al. The ancient regulatory-protein family of WD-repeat proteins. *Nature* 371: 297–300, 1994.

Pawson, T. Protein modules and signaling networks. *Nature* 373: 573–580, 1995.

Polakis, P., McCormick, F. Structural requirements for the interaction of p21[ras] with GAP, exchange factors, and its biological effector target. *J. Biol. Chem.* 268: 9157–9160, 1993.

Sprang, S.R. G protein mechanisms: insights from structural analysis. *Annu. Rev. Biochem.* 66: 639–678, 1997.

Sprang, S.R., Bazan, J.F. Cytokine structural taxonomy and mechanisms of receptor engagement. *Curr. Opin. Struct. Biol.* 3: 815–827, 1993.

Staub, O., Rotin, D. WW domains. *Structure* 4: 495–499, 1996.

Wedegaertner, P.B., Wilson, P.T., Bourne, H.R. Lipid modifications of trimeric G proteins. *J. Biol. Chem.* 270: 503–506, 1995.

Wells, J.A., De Vos, A.M. Hematopoietic receptor complexes. *Annu. Rev. Biochem.* 65: 609–634, 1996.

Specific structures

Berghuis, A.M., et al. Structure of the GDP-Pi complex of Gly203-Ala $G_{i\alpha 1}$: a mimic of the ternary product complex of G_α-catalyzed GTP hydrolysis. *Structure* 4: 1277–1290, 1996.

Coleman, D.E., et al. Structures of active conformations of $G_{i\alpha 1}$ and the mechanism of GTP hydrolysis. *Science* 265: 1405–1412, 1994.

De Vos, A.M., Ultsch, M., Kossiakoff, A.A. Human growth hormone and extracellular domain of its receptor: crystal structure of the complex. *Science* 255: 306–312, 1992.

Eck, M.J., Shoelson, S.E., Harrison, S.C. Recognition of a high-affinity phosphotyrosyl peptide by the Src homology-2 domain of p56[lck]. *Nature* 362: 87–91, 1993.

Ferguson, K.M., et al. Structure of the high affinity complex of inositol triphosphate with a phospholipase C pleckstrin homology domain. *Cell* 83: 1037–1046, 1995.

Gaudet, R., Bohm, A., Sigler, P.B. Crystal structure at 2.4 Å resolution of the complex of transducin βγ and its regulator, phosducin. *Cell* 87: 577–588, 1996.

Holbrook, S.R., Kim, S.-H. Molecular model of the G protein, a subunit based on the crystal structure of the H-ras protein. *Proc. Natl. Acad. Sci. USA* 86: 1751–1755, 1989.

Jurnak, F. Structure of the GDP domain of EF-Tu and location of the amino acids homologous to *ras* oncogene proteins. *Science* 230: 32–36, 1985.

LaCour, T.F.M., et al. Structural details of the binding of guanosine diphosphate to elongation factor Tu from *E. coli* as studied by x-ray crystallography. *EMBO J.* 4: 2385–2388, 1985.

Lambright, D.G., et al. Structural determinants for activation of the α-subunit of heterotrimeric G protein. *Nature* 369: 621–628, 1994.

Lambright, D.G., et al. The 2.0 Å crystal structure of a heterotrimeric G protein. *Nature* 379: 311–319, 1996.

Lee, C.-H., et al. Crystal structure of the conserved core of HIV-1 Nef complexed with a Src-family SH3 domain. *Cell* 85: 931–942, 1996.

Lim, W.A. Reading between the lines: SH3 recognition of an intact protein. *Structure* 4: 657–659, 1996.

Mixon, M.B., et al. Tertiary and quaternary structural changes in $G_{i\alpha1}$ induced by GTP hydrolysis. *Science* 270: 954–960, 1995.

Morton, C.J., et al. Solution structure and peptide binding of the SH3 domain from human Fyn. *Structure* 4: 705–714, 1996.

Noel, J.P., Hamm, H.E., Sigler, P.B. The 2.2 Å crystal structure of transducin-α complexed with GTPγS. *Nature* 366: 654–663, 1993.

Pai, E.F., et al. Refined crystal structure of the triphosphate conformation of H-*ras* at 1.35 Å resolution: implications for the mechanism of GTP hydrolysis. *EMBO J.* 9: 2351–2360, 1990.

Pai, E.F., et al. Structure of the guanine-nucleotide-binding domain of the Ha-*ras* oncogene product p21 in the triphosphate conformation. *Nature* 341: 209–214, 1989.

Rittinger, K. et al. Crystal structure of a small G protein in complex with GTPase-activating protein rhoGAP. *Nature* 388: 693–697, 1997.

Rittinger, K., et al. Structure at 1.65 Å of RhoA and its GTPase-activating protein in complex with a transition-state analogue. *Nature* 389: 758–762, 1997.

Scheffzek, K., et al. The Ras–RasGAP complex: structural basis for GTPase activation and its loss in oncogenic Ras mutants. *Science* 277: 333–338, 1997.

Schlichting, I., et al. Time-resolved x-ray crystallographic study of the conformational change in Ha-ras p21 protein on GTP hydrolysis. *Nature* 345: 309–315, 1990.

Schreuder, H., et al. A new cytokine-receptor binding mode revealed by the crystal structure of the IL-1 receptor with an antagonist. *Nature* 386: 194–200, 1997.

Sicheri, F., Moarefi, I., Kuriyan, J. Crystal structure of the Src family tyrosine kinase Hck. *Nature* 385: 602–609, 1997.

Somers, W., Ultsch, M., De Vos, A.M., Kossiakoff, A.A. The X-ray structure of a growth hormone-prolactin receptor complex. *Nature* 372: 478–481, 1994.

Sondek, J., et al. Crystal structure of a G_A protein βγ dimer at 2.1 Å resolution. *Nature* 379: 369–374, 1996.

Sondek, J., et al. GTPase mechanism of G proteins from the 1.7 Å crystal structure of transducin α.GDP.AlF$_4^-$. *Nature* 372: 276–279, 1994.

Sunahara, R.K., et al. Crystal structure of the adenylyl cyclase activator $G_{s\alpha}$. *Science* 278: 1943–1947, 1997.

Tesmer, J.J.G., et al. Structure of RGS4 bound to AlF$_4$—activated $G_{i\alpha1}$: Stabilization of the transition state for GTP hydrolysis. *Cell* 89: 251–261, 1997.

Tesmer, J.J.G., et al. Crystal structure of the catalytic domains of adenylyl cyclase in a complex with $G_{s\alpha}$.GTPγS. *Science* 278: 1907–1916, 1997.

Vigers, G.P.A., et al. Crystal structure of the type-I interleukin-1 receptor complexed with interleukin-1β. *Nature* 386: 190–194, 1997.

Waksman, G., et al. Binding of a high affinity phosphotyrosyl peptide to the Src SH$_2$ domain: crystal structures of the complexed and peptide-free forms. *Cell* 72: 779–790, 1993.

Wall, M.A., et al. The structure of the G protein heterotrimer $G_{i\alpha1\beta1\gamma2}$. *Cell* 83: 1047–1058, 1995.

Xu, W., Harrison, S.C., Eck, M.J. Three-dimensional structure of the tyrosine kinase c-Src. *Nature* 385: 595–602, 1997.

Yamaguchi, H., Hendrickson, W. Structural basis for activation of human kinase Lck upon tyrosine phosphorylation. *Nature* 384: 484–489, 1996.

Yu, H., et al. Solution structure of the SH3 domain of Src and identification of its ligand-binding site. *Science* 258: 1665–1668, 1992.

Zhang, G., Ruoho, A.E., Hurley, J.H. Structure of the adenylyl cyclase catalytic core. *Nature* 386: 247–253, 1997.

Fibrous Proteins

14

Proteins are usually separated into two distinct functional classes: passive structural materials, which are built up from long fibers, and active components of cellular machinery in which the protein chains are arranged in small compact domains, as we have discussed in earlier chapters. In spite of their differences in structure and function, both these classes of proteins contain α helices and/or β sheets separated by regions of irregular structure. In most cases the fibrous proteins contain specific repetitive amino acid sequences that are necessary for their specific three-dimensional structure.

Fibrous proteins can serve as structural materials for the same reason that other polymers do; they are long-chain molecules. By cross-linking, interleaving and intertwining the proper combination of individual long-chain molecules, bulk properties are obtained that can serve many different functions. Fibrous proteins are usually divided in three different groups dependent on the secondary structure of the individual molecules: coiled-coil α helices present in keratin and myosin, the triple helix in collagen, and β sheets in amyloid fibers and silks.

The α-helical fibers of wool are flexible, can be extended up to twice their length, and are also elastic so that the fibers return to their original length when the tension is released. Collagen fibers by contrast are strong, resistant to stretching and relatively rigid. The β sheet fibers are both strong and very flexible. A prime example are the fibers of spiders' webs, some of which are stronger than steel wires of the same dimension, and at the same time are flexible enough not to break when an insect is caught in the web. Fibrous β sheets can also be formed *in vivo* by misfolding of soluble globular proteins, thereby causing conditions such as Alzheimer's disease and prion diseases.

Fibrous proteins often form protofilaments or protofibrils that assemble into structurally specific, higher-ordered filaments and fibrils. Examples include collagen, amyloids, intermediate filaments, tubulin, myosin, and fibrinogen. Such filaments or fibrils cannot be crystallized because they can only be ordered in two dimensions. However, ordered fibers give two-dimensional diffraction patterns from which information on their overall structure can be deduced (see Chapter 18). On the other hand, some fibrous proteins, or fragments thereof, such as fibrinogen, tropomyosin, collagens, spectrin, and myosin, have been crystallized and their detailed atomic structures determined. Such information has been useful for interpretation of diffraction patterns or electron micrographs of ordered filaments or fibers.

The assembly of fibrous proteins into higher-ordered structures is in some cases a dynamic process. Microtubules, which are long, hollow polymers of globular proteins, tubulins, continually depolymerize and re-polymerize, thereby governing the location of organelles and other cell components. In fibroblasts the half-life of an individual microtubule is about 10 minutes, while the average lifetime of a tubulin molecule is more than 20 hours. Thus each tubulin molecule will be recycled through many microtubules during its lifetime.

Collagen is a superhelix formed by three parallel, very extended left-handed helices

The term **collagen** derives from the Greek word for glue and was defined in the 1893 edition of the Oxford Dictionary as "that constituent of connective tissue which yields gelatin on boiling." A modern definition, based on sequence determinations of different family members, would be that collagens are proteins that assemble into fibrous supramolecular aggregates in the extracellular space and which comprise three polypeptide chains with a large number of repeat sequences Gly-X-Y where X is often proline and Y is often hydroxyproline. Hydroxyproline is formed by the posttranslational modification of proline by a specific enzyme, prolyl hydroxylase, and is rarely found in other proteins. Each collagen polypeptide chain contains about 1000 amino acid residues and the entire triple chain molecule is about 3000 Å long.

Collagen chains are synthesized as longer precursors, called procollagens, with globular extensions—propeptides of about 200 residues—at both ends. These procollagen polypeptide chains are transported into the lumen of the rough endoplasmic reticulum where they undergo hydroxylation and other chemical modifications before they are assembled into triple chain molecules. The terminal propeptides are essential for proper formation of triple

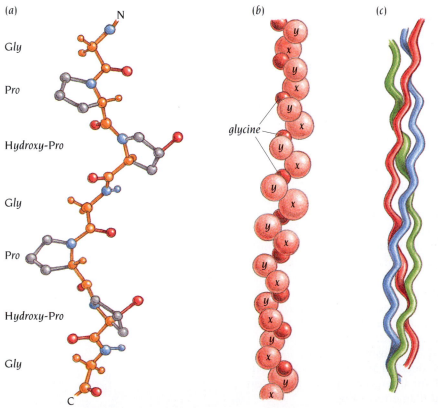

Figure 14.1 Each polypeptide chain in the collagen molecule folds into an extended polyproline type II helix with a rise per turn along the helix of 9.6 Å comprising 3.3 residues. In the collagen molecule three such chains are supercoiled about a common axis to form a 3000-Å-long rod-like molecule. The amino acid sequence contains repeats of -Gly-X-Y- where X is often proline and Y is often hydroxyproline. (a) Ball and stick model of two turns of one polypeptide chain. (b) A model of one collagen chain in which each amino acid is represented by a sphere. (c) A small part of the collagen superhelix in which the three chains have different color. [(a) Adapted from R. Fraser et al., *J. Mol. Biol.* 129: 463–481, 1979. (b) Adapted from B. Alberts et al, *Molecular Biology of the Cell*, 3rd ed. New York: Garland Publishing, 1994.]

284

chains: they form interchain disulfide bonds that align the three chains in register before formation of the triple chain structure. Only following exocytosis are the propeptides cleaved off, by extracellular enzymes. Excision of both propeptides allows the triple chain molecules to polymerize into fibrils several micrometers long and 50–200 nanometers in diameter. These fibrils then pack side by side into parallel bundles, the collagen fibers, which are stronger than steel of the same size. When mature collagen is denatured, since there are no propeptides present, the polypeptides associate at many different places other than their ends, forming triple chains that are out of register. These polymerize to form a gel, gelatin.

Early interpretations of fiber diffraction studies of collagen by, among others, Linus Pauling, Francis Crick, and Alexander Rich, established that each of the three polypeptide chains is folded into an extended left-handed helix with 3.3 residues per turn and a rise per residue along the helical axis of 2.9 Å. In contrast the more compact right-handed α helix has 3.6 residues per turn and a rise per residue of 1.5 Å (see Chapter 2). The rise per turn in the collagen helix is therefore 9.6 Å, compared with 5.4 Å for the α helix, and this gives such an extended chain that it must aggregate to form a stable structure (Figure 14.1). Synthetic polymers of proline or glycine fold into similar extended, left-handed helices and so this helix type is called a polyproline type II helix. SH3 domains, described in Chapter 13, recognize short proline-rich segments in other proteins and these segments have a similar helical conformation when bound to SH3.

The three polyproline type II helices in collagen form a trimeric molecule by coiling about a central axis to form a right-handed superhelix with a repeat distance of about 100 Å (see Figure 14.1c). In this superhelix the side chain of every third residue is close to the central axis where there is no space available for a side chain, consequently every third residue must be a glycine. Any other residue deforms the superhelix and certain inherited connective tissue diseases are due to mutations in codons for these glycine residues. This sequence requirement is a hallmark of triple helix collagen-like domains that is used in sequence analyses of proteins of unknown structure. Triple-helical domains are found in a variety of supramolecular aggregates, ranging from the collagen fibrils in tendons and cartilage, to network forms seen in basement membranes, and to the parallel clusters of short triple helices seen in the blood complement component C1q or in the sugar-binding collectins.

A detailed picture of the collagen triple helix has been obtained by the group of Helen Berman at Rutgers University, who have determined the crystal structure to 1.9 Å resolution of a synthetic collagen-like peptide (Pro-Hydroxypro-Gly)$_{10}$ with one Gly substituted by Ala. This collagen-like peptide did not form fibers but single crystals. The specific sequence was chosen in order to study both the details of the regular collagen triple helix structure and the effect of mutation at a glycine position. The structure shows the importance of direct as well as water-mediated hydrogen bonds in stabilizing the triple helix structure. In addition it shows that the alanine side chain can be accommodated inside the triple helix by a local small change of the helix geometry (Figure 14.2), which allows the incorporation of interstitial water molecules to link the chains. Such conformational shifts may help to accommodate sequence variations that deviate from the consensus.

(a)

(b)

Figure 14.2 Models of a collagen-like peptide with a mutation Gly to Ala in the middle of the peptide (orange). Each polypeptide chain is folded into a polyproline type II helix and three chains form a superhelix similar to part of the collagen molecule. The alanine side chain is accommodated inside the superhelix causing a slight change in the twist of the individual chains. (a) Space-filling model. (b) Ribbon diagram. Compare with Figure 14.1c for the change caused by the alanine substitution. (Adapted from J. Bella et al., *Science* 266: 75–81, 1994.)

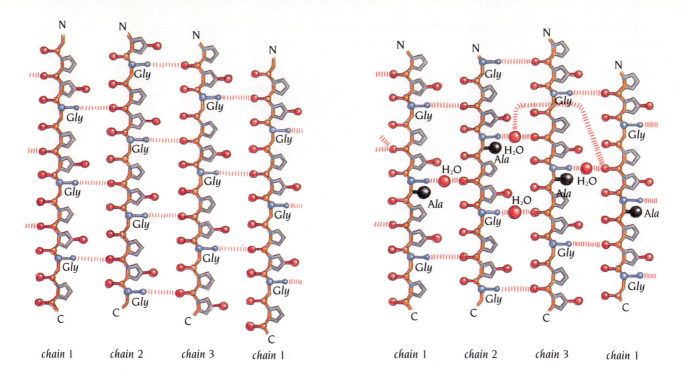

chain 1   chain 2   chain 3   chain 1   chain 1   chain 2   chain 3   chain 1

In the regular, triple helix collagen structure the three chains are held close together by direct hydrogen bonds between proline C=O groups of one chain and the glycine NH groups of another (Figure 14.3a). In the region around the alanine residues the three polypeptide chains are forced apart by the alanine side chains and four water molecules are incorporated between the chains; this allows the direct hydrogen bonds to be replaced by water-mediated hydrogen bonds (Figure 14.3b).

All side chains as well as the C=O group of glycines in all three chains are on the outside of the triple helix molecule and in contact with water molecules. These water molecules mediate hydrogen bonds between the hydroxyl groups of hydroxyproline and the peptide C=O and NH groups both within each chain and between different chains (Figure 14.4). These water mediated hydrogen bonds are essential for the stability of the triple helix and are presumably the reason for the presence of hydroxyproline in collagen.

Coiled coils are frequently used to form oligomers of fibrous and globular proteins

We described in Chapter 3 the basic features of α-helical coiled coils whose amino acid sequences are recognized by heptad repeats *a* to *g* in which positions *a* and *d* frequently are hydrophobic residues (see Figures 3.2 and 3.3).

Figure 14.3 Schematic diagrams of interchain hydrogen bonding (stripes) in the collagen triple helix showing a cylindrical projection with chain 1 repeated at the right side to provide a clear picture of the chain 3 to chain 1 hydrogen bonds. The three chains are staggered by one residue and go clockwise from 1 to 3. (a) Interchain hydrogen bonds in regular collagen triple helix between proline C=O groups and glycine NH groups. (b) Water-mediated hydrogen bonds in the middle of the Gly-Ala substituted peptide structure. Four water molecules are inserted in the interior of the triple helix to mediate hydrogen bonds between the polypeptide chains, which are displaced due to the alanine side chains in this region. Interchain hydrogen bonds are also shown. (Adapted from J. Bella et al., *Science* 266: 75–81, 1994.)

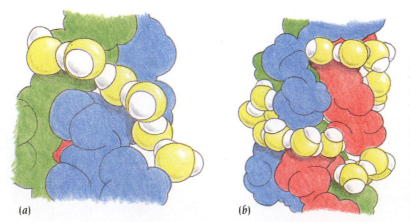

(a)   (b)

Figure 14.4 Space-filling models of intrachain water bridges observed in the crystal structure of a collagen-like peptide. The three peptide chains are green, blue and red. The oxygen atoms of the water molecules are yellow and the hydrogen atoms are white. (a) A bridge of three water molecules hydrogen bonded to two C=O groups and linked to other water molecules. (b) Long bridges of water molecules establish a network that surrounds the triple helix. (Adapted from J. Bella et al., *Structure* 3: 893–906, 1995.)

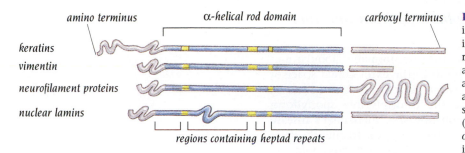

amino terminus | α-helical rod domain | carboxyl terminus

keratins

vimentin

neurofilament proteins

nuclear lamins

regions containing heptad repeats

Figure 14.5 The domain organization of intermediate filament protein monomers. Most intermediate filament proteins share a similar rod domain that is usually about 310 amino acids long and forms an extended α helix. The amino-terminal and carboxy-terminal domains are non-α-helical and vary greatly in size and sequence in different intermediate filaments. (Adapted from B. Alberts et al, *Molecular Biology of the Cell*, 3rd ed. New York: Garland Publishing, 1994.)

The leucine zipper DNA-binding proteins, described in Chapter 10, are examples of globular proteins that use coiled coils to form both homo- and heterodimers. A variety of fibrous proteins also have heptad repeats in their sequences and use coiled coils to form oligomers, mainly dimers and trimers. Among these are myosin, fibrinogen, actin cross-linking proteins such as spectrin and dystrophin as well as the intermediate filament proteins keratin, vimentin, desmin, and neurofilament proteins.

Keratins, which are the major structural protein in skin, hair, and feathers, belong to the superfamily of **intermediate filaments** that comprise about 60 genes of greatly differing sizes in the human genome. All members of this family contain a homologous central region of about 300 amino acid residues that exhibit heptad repeats typical for coiled coil formation (Figure 14.5). It is assumed that these central regions form a double-stranded coiled coil giving a homodimer with a rod-like central domain and two globular domains, one at each end. In assembling into an intermediate filament, the rod-like domains interact with one another to form the uniform core of the filament, while the globular domains, which vary in size in different intermediate filament proteins, project from the filament surface. A hypothetical view of the assembly of the peptide chains into intermediate filaments is shown in Figure 14.6.

Figure 14.6 A model of intermediate filament construction. The monomer shown in (a) pairs with an identical monomer to form a coiled-coil dimer (b). The dimers then line up to form an antiparallel tetramer (c). Within each tetramer the dimers are staggered with respect to one another, allowing it to associate with another tetramer (d). In the final 10-nm rope-like intermediate filament, tetramers are packed together in a helical array (e). (Adapted from B. Alberts et al, *Molecular Biology of the Cell*, 3rd ed. New York: Garland Publishing, 1994.)

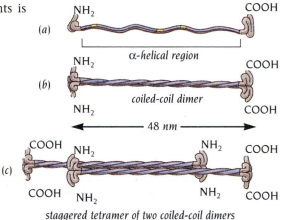

(a) NH_2 COOH
α-helical region

(b) NH_2 COOH
coiled-coil dimer
NH_2 COOH

← 48 nm →

(c) COOH NH_2 NH_2 COOH

COOH NH_2 NH_2 COOH

staggered tetramer of two coiled-coil dimers

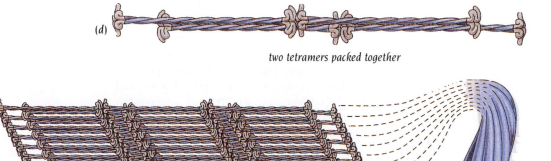

(d)

two tetramers packed together

(e)

10 nm

287

Amyloid fibrils are suggested to be built up from continuous β sheet helices

Amyloidosis are diseases in which proteins that are normally globular and soluble are deposited as stable, insoluble fibrils in the extracellular space of tissues such as the brain or the eye. Deposition of such fibrils is associated with several amyloid diseases including the transmissable spongiform encephalopathies (BSE and Creutzfeld–Jacob diseases), Alzheimer's disease and type II diabetes. **Amyloid fibrils** from different sources frequently share a common ultrastructure; they are long, straight, unbranched rods about 100 Å in diameter. A unifying model for fibril structures has been proposed on the basis of amyloid fibers of the protein transthyretin. Colin Blake and Louise Serpell at Oxford University have obtained fiber diffraction patterns to high resolution, using synchrotron radiation, of such amyloid fibers from patients suffering from familial amyloidotic polyneuropathy. These patients have an inherited single-point mutation, Val to Met, in the protein **transthyretin** which carries the hormone thyroxine in the bloodstream. This seemingly innocuous mutation causes the soluble, globular transthyretin molecule to polymerize and gradually deposit as fibrous material in the heart or eye. The group of Colin Blake had earlier determined the crystal structure of the soluble form of transthyretin and showed it to be a homotetramer with the subunits folded into an antiparallel β structure (Figure 14.7).

Earlier fiber diffraction studies to lower resolution of amyloid fibers had established that they have a β structure, with fibers from different sources giving similar diffraction patterns. Structural models had been suggested based on planar pleated β sheets, similar to a model suggested for silk fibers, which we shall discuss later. Blake and Serpell's high resolution data, however, showed a distinct, repeating pattern corresponding to structural repeats of 115.5 Å along the fiber axis. Because almost all β sheets in globular proteins are twisted and not planar, they proposed, therefore, that the 115.5 Å repeats they observed are compatible with a twisted β sheet completing one revolution of 360° every 115.5 Å (Figure 14.8). Since the mean distance between β strands in globular proteins is about 4.8 Å, the repeating unit along the fibril would correspond to 24 β strands with an average twist of 15° between each successive β strand, a value that is in good agreement with the values observed in β sheets in globular proteins. Finally, they suggested that each repeat unit is built up from four transthyretin molecules, each contributing six β strands to the β sheet helix. The model includes the idea that a partially unfolded form that has lost two strands forms the fibril. This idea is supported by the absence of fibrils in the brain, where hormone binding stabilizes the native state.

Electron microscopy of transthyretin fibrils shows that they have a uniform width of about 130 Å and are composed of four parallel protofilaments each 50–60 Å in diameter. Blake and Serpell propose that each protofilament is built up of four continuous twisted β sheets of indefinite length running parallel to the axis of the fibril, with their constituent β strands, each containing about 10 amino acid residues, running perpendicular to the axis. The four β sheets are arranged in pairs, each pair being stabilized by a hydrophobic core between the two sheets, as is usually the case in globular proteins with β structure. The two pairs of β helices coil about each other. The twisted β helix only comprises the regular, diffracting, core structure of the protofilament and must be surrounded by other material such as loops of polypeptide chains that do not contribute to the x-ray pattern.

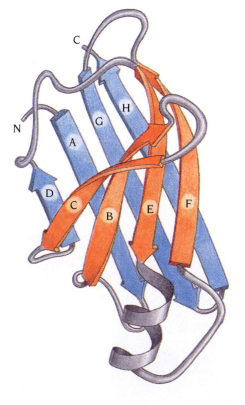

Figure 14.7 Ribbon diagram of one subunit of the globular form of transthyretin. The β strands are labeled A to H from the amino end. Strands C and D are thought to be unfolded to produce the conformation that forms amyloid fibrils. (Adapted from C.C.F. Blake et al., *J. Mol. Biol.* 121: 339–356, 1978.)

Figure 14.8 Proposed model of β sheet helix of the fibrous form of transthyretin. The repeating unit of the β helix comprises 24 β strands with an average twist of 15° between each strand giving a complete turn of 360°. Four transthyretin polypeptide chains contribute to the repeat unit and are shown here in different colors. (Adapted from C. Blake and L. Serpell, *Structure* 4: 989–998, 1996.)

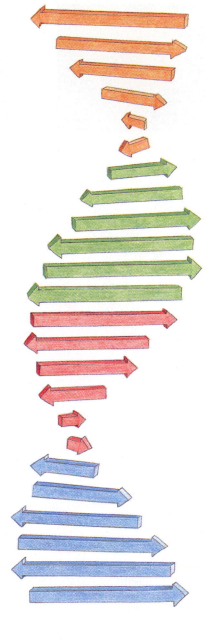

This model implies a considerable refolding of the globular state of transthyretin to form the fibrous state. However, large refolding events are known to occur in other proteins which form amyloid β sheet fibers from subunits that are mainly α-helical in their globular state, for example prions and α lactalbumin. Even though not every detail of this model may be correct, it not only is compatible with the fiber diffraction model but also incorporates knowledge gained from structural studies of β sheets in globular proteins.

Spider silk is nature's high-performance fiber

Individual spiders generate up to seven distinct silk fibers by drawing liquid-crystalline proteins from a set of separate gland-spinneret complexes. The proteins, called **silk fibroins**, are produced inside the glands from a family of homologous genes and stored as a highly concentrated (up to 50%) solution of mainly α-helical globular molecules. This solution is passed through the spinning machinery of the spider and mixed with other components, producing a variety of different fibers in which the proteins have adopted a β structure. Spiders can produce different types of silk with different mechanical properties for different functions. The most remarkable properties are exhibited by the dragline fiber that the spider spins rapidly and uses to climb from the web to its hiding place. This fiber is stronger than steel, yet it can contract, stretch out and bend without breaking. The structure of silk fibers has therefore attracted the attention of materials scientists who are increasingly willing to turn to biology for lessons that might be applied to synthetic systems.

Several silk fibroin genes have been cloned and sequenced and they all show a similar sequence pattern: variable domains at the N- and C-termini flank a large region of repetitive short sequences of alternating poly-Ala (8 to 10 residues) and Gly-Gly-X repeats (where X is usually Ser, Tyr, or Gln). This middle region varies in length and may comprise up to 800 residues.

Early fiber diffraction studies in the 1950s established that at least some of the proteins in the spider silk fibers adopt a β sheet structure, similar to the β sheets that had been observed in common silk. Simple models of planar β sheets were suggested to explain these diffraction patterns. They were, however, at very low resolution due to the difficulty in aligning thousands of fibers into a large enough bundle to obtain visible diffraction spots. Bundles of fibers that are not precisely aligned give broad and diffuse diffraction patterns that do not reveal structural details (see Chapter 18). Not until very recently has it been possible to obtain x-ray diffraction patterns from individual spider dragline fibers that are only a few micrometers thick, using the very intense x-ray beams from the new generation of synchrotrons.

Progress in deducing more structural details of these fibers has instead been achieved using NMR, electron microscopy and electron diffraction. These studies reveal that the fibers contain small microcrystals of ordered regions of the polypeptide chains interspersed in a matrix of less ordered or disordered regions of the chains (Figure 14.9). The microcrystals comprise about 30% of the protein in the fibers, are arranged in β sheets, are 70 to 100 nanometers in size, and contain trace amounts of calcium ions. It is not yet established if the β sheets are planar or twisted as proposed for the amyloid fibril discussed in the previous section.

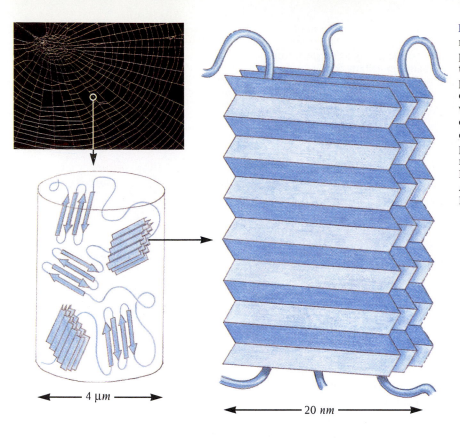

Figure 14.9 Spider fibers are composite materials formed by large silk fibroin polypeptide chains with repetitive sequences that form β sheets. Some regions of the chains participate in forming 100-nm crystals, while other regions are part of a less-ordered meshwork in which the crystals are embedded. The diagram shows a model of the current concepts of how these fibers are built up, which probably will be modified and extended as new knowledge is gained. (Adapted from F. Vollrath, *Sci. Am.* p. 54–58, March 1992 and A.H. Simmons, *Science* 271: 84–87, 1996. Photograph courtesy of Science Photo Library.)

An elegant NMR experiment by the group of Lynn Jelinski at Cornell University has established that at least part of the microcrystals is built up from the polyalanine repeats in the protein chains. These experiments, which were made on ^{13}C-enriched proteins produced by feeding the spiders ^{13}C-labeled alanine, showed that there were two populations of alanine side chains, one ordered and oriented perpendicular to the fiber axis and a second less ordered. Jelinski's interpretation is that parts of the polyalanine sequences are incorporated as β strands in the microcrystals with an orientation parallel to the fiber axis. Whether or not the Gly-Gly-X repeats also form β strands in the microcrystals remains an open question.

Even if much remains to be done to clarify the relations between structure and mechanical properties of the spider fibers, the knowledge obtained so far gives two important messages. First, the process of spinning converts the silk proteins from a soluble α-helical form to a fibrous form containing β sheets, analogous to the conversion in humans and other animals of soluble amyloid and prion proteins to insoluble fibrous forms. Second, the mechanical properties of silk fibers, like those of synthetic fibers, are dependent on having well-ordered crystalline regions which give strength, within a matrix of less-ordered regions, which give flexibility. An understanding of the structural details of these regions, as well as the interplay between them, might pave the way for fascinating new materials.

Muscle fibers contain myosin and actin which slide against each other during muscle contraction

Muscle contraction takes place by the mutual sliding of two sets of interdigitating filaments made of fibrous proteins: **thick filaments** (containing **myosin**) and **thin filaments** (containing **actin**). The thick and thin filaments are organized in basic contractile units called sarcomeres, each 2–3 μm long, which also contain other proteins and give muscle its cross-striated appearance in the microscope (Figure 14.10). Another fibrous component of the sarcomere, titin, is the largest known polypeptide chain, with a molecular

(a)

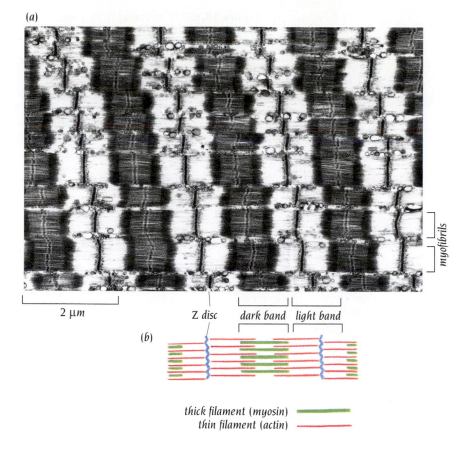

2 μm

Figure 14.10 A muscle viewed under the microscope is seen to contain many myofibrils that show a cross-striated appearance of alternating light and dark bands, arranged in repeating units called sarcomeres. The dark bands comprise myosin filaments and are interupted by M (middle) lines, which link adjacent myosin filaments to each other. The light bands comprise actin filaments that are attached to disks at each end of the sarcomeres. (Courtesy of Roger Craig, adapted from B. Alberts et al, *Molecular Biology of the Cell*, 3rd ed. New York: Garland Publishing, 1994.)

Z disc   dark band   light band

(b)

thick filament (myosin) ——————
thin filament (actin) ——————

weight of approximately 3000 kDa. Titin is thought to help measure the length of the sarcomere and return the stretched muscle to the correct length. Figure 14.11 shows how the sarcomeres contract as the actin and myosin filaments slide past each other.

Myosin heads form cross-bridges between the actin and myosin filaments

Within each sarcomere the relative sliding of thick and thin filaments is brought about by "**cross-bridges**," parts of the myosin molecules that stick out from the myosin filaments and interact cyclically with the thin actin filaments, transporting them by a kind of rowing action. During this process, the hydrolysis of ATP to ADP and phosphate couples the conformational

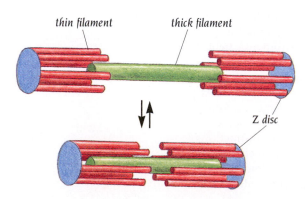

thin filament   thick filament

Z disc

Figure 14.11 The sliding filament model of muscle contraction. The actin (red) and myosin (green) filaments slide past each other without shortening. (Adapted from B. Alberts et al, *Molecular Biology of the Cell*, 3rd ed. New York: Garland Publishing, 1994.)

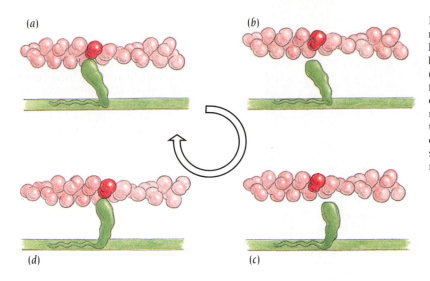

(a) (b) (d) (c)

Figure 14.12 The swinging cross-bridge model of muscle contraction driven by ATP hydrolysis. (a) A myosin cross-bridge (green) binds tightly in a 45° conformation to actin (red). (b) The myosin cross-bridge is released from the actin and undergoes a conformational change to a 90° conformation (c), which then rebinds to actin (d). The myosin cross-bridge then reverts back to its 45° conformation (a), causing the actin and myosin filaments to slide past each other. This whole cycle is then repeated.

changes in myosin to actin binding and release. Myosin in the thick filaments is a fibrous protein with individual chains arranged in helical coiled coils. Actin by contrast is a fibrous protein formed by linking together globular monomeric subunits. The first molecular theories of muscle contraction, which appeared in the 1930s, were based on polymer science. They proposed that there was a rubber-like shortening of myosin filaments brought about by altering the state of ionization of myosin. This aberration was corrected in 1954 by the work of H.E. and A.F. Huxley, which showed that sarcomeres contain two sets of filaments that glide over each other without altering their length. The question naturally arose: what makes them glide?

Following the discovery by electron microscopy of the myosin cross-bridges to the actin filaments, two conformations of the cross-bridge were observed in insect flight muscle. This seminal finding led to the **swinging cross-bridge model** to explain the sliding of the actin filaments into myosin filaments (Figure 14.12). The myosin cross-bridge was thought to bind to actin in an initial (90°) conformation and then go over to an angled (45°) conformation, followed by release from the actin. For each complete cycle one molecule of ATP would be hydrolyzed. The actual movement per cycle of ATP hydrolysis was measured in physiological experiments to be about 80–100 Å and since the cross-bridge was an elongated structure such a movement could be accommodated by swinging the cross-bridge. However, this model basically rested on the observation of two conformations of cross-bridges in fixed and sectioned insect flight muscle under the electron microscope. It required the development of synchrotron radiation x-ray beams to provide more direct evidence.

Time-resolved x-ray diffraction of frog muscle confirmed movement of the cross-bridges

Excised living frog muscles will continue to contract for many hours if bathed in Ringer solution and stimulated electrically. H.E. Huxley showed that living frog muscles give detailed low-angle x-ray fiber diffraction patterns (see Chapter 18) with a series of layer-lines arising from the helical arrangement of cross-bridges. A strong reflection on one of these layer-lines indicates that the distance between cross-bridges along the myosin filament helix is 143.5 Å. If the cross-bridges tilt during contraction, then this reflection should get weaker. Muscles contract fast, so millisecond time resolution is necessary. Initial attempts to observe this effect with conventional x-ray sources failed: the signal in the short times available for the diffractions was too weak. The first beam lines for x-ray diffraction experiments using synchrotron radiation were set up at DESY, Hamburg, by Ken Holmes and his colleagues from Heidelberg; these provided the requisite power to allow data collection over only a few milliseconds. The key experiments, carried out in 1980 by H.E. Huxley, showed the anticipated changes in intensity of the diffraction data. If a contracting muscle is allowed to relax, the intensity of the strong reflection arising from the cross-bridges drops within a few milliseconds to a fraction of its initial value. If one waits some time with the muscle at its new length, the intensity recovers. These and subsequent time-resolved experiments at synchrotrons are fully consistent with the swinging cross-bridge hypothesis.

Structures of actin and myosin have been determined

Structural studies to atomic resolution of both actin and myosin have clarified a number of long-standing questions concerning the mechanism of cross-bridge movements in muscle. Fibrous actin, F-actin, is a helical polymer of a globular polypeptide chain, G-actin, comprising about 375 amino acid residues. The first crystal structure of a monomeric G-actin molecule was determined by the group of Ken Holmes, Max-Planck Institute, Heidelberg, who prevented polymerization by forming a complex with the enzyme DNAse. Subsequently the structures of several other G-actin molecules have been determined in complex with other inhibitors of actin polymerization. The structure comprises four domains, two of which are similar α/β domains that contain an ATPase catalytic site (Figure 14.13). Related structures have also been found in such diverse ATP-requiring enzymes as hexokinase and the chaperone hsc70. These molecules probably form an evolutionary family, but their specific functions and amino acid sequences have completely diverged. Only their ATPase activity and three-dimensional structures are retained.

The F-actin helix has 13 molecules of G-actin in six turns of the helix, repeating every 360 Å. Oriented gels of actin fibers yield x-ray fiber diffraction patterns to about 6 Å resolution. Knowing the atomic structure of G-actin it was possible for the group of Ken Holmes to determine its orientation in the F-actin fiber, and thus arrive at an atomic model of the actin filament that best accounted for the fiber diffraction pattern.

C

N

Figure 14.13 Structure of G-actin. Two α/β-domains, (dark and light red and dark and light green) bind an ATP molecule (brown) between them. This ATP is hydrolyzed when the actin monomer polymerizes to F-actin.

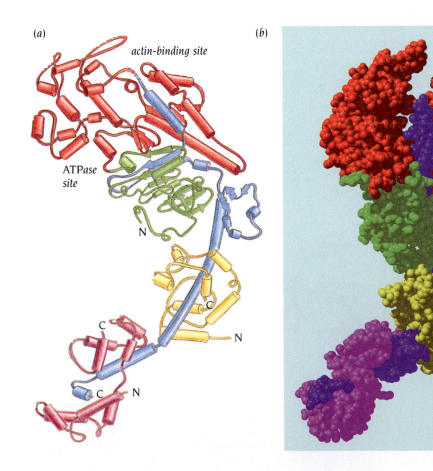

Figure 14.14 Schematic diagram of the myosin molecule, comprising two heavy chains (green) that form a coiled-coil tail with two globular heads and four light chains (gray) of two slightly differing sizes, each one bound to each heavy-chain globular head.

coiled coil of two α helices

N-terminus

light chains

2 nm

C-terminus

150 nm

The myosin molecule, which is a dimer consisting of two heavy chains and four light chains, forms a 1400-Å-long tail and two heads, each of 120,000 kDa molecular weight (Figure 14.14). The C-terminal regions of the heavy chains are folded into long α helices that form the tail by dimerizing through parallel coiled coils. Fragments of myosin called subfragment 1, or S1, can be cleaved off; S1 consists of the two light chains and the N-terminal region of one heavy chain that includes the globular head and a short part of the helical tail. The structure of S1 from chicken myosin has been determined by the group of Ivan Rayment at the University of Wisconsin, Madison. The fragment is tadpole-like in form, with an elongated head and a tail (Figure 14.15). The head contains a seven-stranded β sheet (green) and numerous associated α helices (red) which form a deep cleft, with the actin-binding site and the nucleotide-binding site on opposite sides of the β sheet. The ATP-binding site contains a typical switch II region, which is also found in the G-proteins (see Chapter 13) that changes conformation depending on whether ATP or ADP is bound. The cleft separates two parts of the head, which are referred to as the 50K upper and 50K lower domains (see Figure 14.15b). The tail, which also provides the connection to the thick filament, has a long extended α helix formed by the heavy chain that binds two light chains (yellow and magenta in Figure 14.15).

(a)

actin-binding site

ATPase site

N

C

C

N

C

N

(b)

Figure 14.15 Structure of the S1 fragment of chicken myosin as a Richardson diagram (a) and a space-filling model (b). The two light chains are shown in magenta and yellow. The heavy chain is colored according to three proteolytic fragments produced by trypsin: a 25-kDa N-terminal domain (green); a central 50-kDa fragment (red) divided by a cleft into a 50K upper and a 50K lower domain; and a 20-kDa C-terminal domain (blue) that links the myosin head to the coiled-coil tail. The 50-kDa and 20-kDa domains both bind actin, while the 25-kDa domain binds ATP. [(b) Courtesy of I. Rayment.]

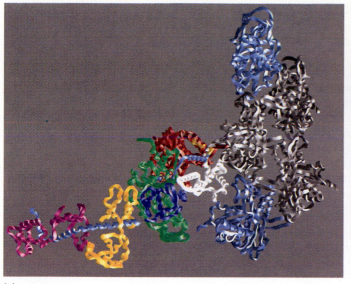
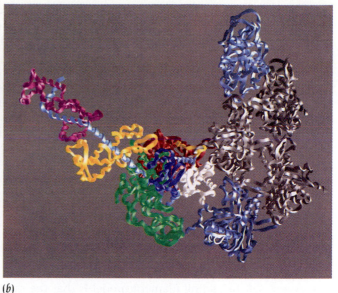

(a)   (b)

The groups of Rayment and Holmes fitted the atomic structures of F-actin and the S1 fragment of myosin into a low-resolution three-dimensional cryoelectron microscopy image of the actin–myosin complex determined by Ronald Milligan, Scripps Research Institute, California and obtained a detailed model of the complex (Figure 14.16a). This model establishes the spatial orientation of the S1 myosin fragment and shows that a cleft in myosin extends from the ATP-binding site to the actin-binding site. These sites are likely to be coupled by the opening and closing of this cleft. Furthermore, the very extended C-terminal α-helical tail of S1 is ideally placed to be a lever arm. The lever arm joins onto the bulk of the molecule via a small compact "converter domain" that lies just distal to a kinked α helix containing two reactive thiol groups. Numerous experiments point to the putative "hinge" for the lever arm being in the region of these thiol groups.

The structure of myosin supports the swinging cross-bridge hypothesis

According to the swinging cross-bridge scheme (see Figure 14.12) the myosin cross-bridge is envisaged as having two discrete conformations: (1) when it first attaches to actin with the ADP still bound and with the lever at the beginning of the working stroke; and (2) at the end of the working stroke when the ADP is released. The switch from state 1 to state 2 is often referred to as the "power stroke." The end state is referred to as "rigor," since it is the state muscles enter on ATP depletion when they become locked in *rigor mortis*. It is also called "strong" because it binds to actin quite tightly. The initial state is called the "weak binding state" because of its low affinity for actin. We might anticipate that these two states of the myosin cross-bridge exist independently from actin and indeed protein crystallography shows this to be the case.

The crystal structure of chicken S1, which was solved without bound nucleotide, fits well into the electron micrograph reconstructions of the strong actin–myosin, nucleotide-free interaction. In addition, Rayment and his colleagues at the University of Wisconsin have studied a crystalline truncated fragment of the *Dictyostelium* myosin II cross-bridge. The truncation eliminates the lever arm and the associated light chains, helping crystallization. However, the converter domain is still present. The crystal structures of this fragment have been determined with a number of ATP analogs bound, particularly ADP–vanadate, which is considered to be an analog of the transition state or possibly of the ADP–Pᵢ state. The ADP–vanadate structure

Figure 14.16 Structure of the actin–myosin complex, in the absence of bound nucleotide (a) and with ADP–vanadate bound (b). Five actin subunits (two pale blue and three gray) are shown on the right of each panel. The myosin on the left of each panel is colored similarly to Figure 14.15 with the light chains yellow and magenta and the 25-kDa domain green. The 50-kDa domain is colored red and white to show the 50K upper and 50K lower domains, respectively; a portion of the 20-kDa fragment, containing the converter region, is dark blue; a short linker region between the 20-kDa and the 50-kDa domains is shown in yellow; and the first part of the 20-kDa domain that is proposed to form part of the "hinge" region of the myosin head is shown in pale blue between the red, green and white domains. Note that the view is rotated with respect to Figure 14.15. (Courtesy of Ken Holmes.)

shows dramatic changes in the shape of S1 compared with the chicken structure. There is a closing of the cleft associated with a movement of the helix corresponding to the switch II region in G-proteins (see Chapter 13), which is coupled to a rotation of the converter domain by 70°.

A model of this new state is shown in Figure 14.16b and compared with the chicken S1 structure. The coordinates of the missing lever arm have been generated from the chicken coordinates by superimposing the converter domains. Since the converter domain has been rotated 70° the end of the lever arm has moved through 120 Å along the actin helix axis, which is greater than most estimates of the size of the power stroke. The mechanism for coupling the movement of the lever arm with the status of the nucleotide-binding pocket that is revealed by this structure suggests that the two events are tightly coupled: cleft closed, lever up (beginning); cleft open, lever down (end).

The role of ATP in muscular contraction has parallels to the role of GTP in G-protein activation

The essence of the sliding filament model is that the myosin head binds to the actin filament in one position relative to its anchor point on the myosin filament, changes this relative position about 100 Å along the fiber axis, and during this process the two filaments slide relative to each other by about the same distance. The myosin head then detaches from the actin filament ready to repeat the process. We will now discuss the molecular mechanisms of the power stroke suggested by Ken Holmes, from the known structures of myosin and the actin–myosin complex, that couple ATP binding and hydrolysis, myosin–actin interactions and the repositioning of the myosin heads by conformational changes.

In the absence of nucleotides, the myosin nucleotide-binding cleft is open, the lever arm is "down," the actin-binding site is intact and this form binds strongly to actin (Figure 14.17a). This is the rigor state into which in the absence of nucleotides the muscle is locked as in *rigor mortis*. If ATP is added, the myosin head, bound to actin, will bind ATP, and then dissociate from the actin (Figure 14.17b). Binding of ATP to the nucleotide-binding domain in the cleft causes the P loop, which corresponds to the switch II region in G-proteins, described in Chapter 13, to change its conformation. The γ-phosphate of ATP plays the same role in this conformational change as the γ-phosphate of GTP in G-proteins. This change in the loop conformation is coupled to a major conformational change of parts of the head protein; as a result the cleft closes and the region that binds actin releases the actin filament. The bound ATP is then hydrolyzed to ADP and phosphate (Figure 14.17c).

It seems likely, therefore, that as the bound phosphate molecule is released, the cleft starts to open and the myosin head binds to actin (Figure 14.17d). Release of ADP coincides with a conformational change that fully opens the myosin cleft, causing actin to be tightly bound, and moves the lever arm to the "up" position. Since the myosin head is now strongly bound to actin at one end and covalently linked to the myosin fibril at the other

Figure 14.17 A sequence of events combining the swinging cross-bridge model of actin and myosin filament sliding with structural data of myosin with and without bound nucleotides. (a) Myosin in the absence of nucleotide binds tightly to actin in a 45° conformation. (b) The binding of ATP induces myosin to dissociate from actin. (c) As a result of ATP hydrolysis the myosin undergoes a conformational switch to the 90° state. (d) The release of phosphate coincides with myosin binding weakly to actin, and the subsequent release of ADP returns the myosin to the initial 45° state. This conformational switch completes the power stroke that drives the sliding of the actin and myosin filaments relative to each other by about 100 Å. (Adapted from B. Alberts et al, *Molecular Biology of the Cell*, 3rd ed. New York: Garland Publishing, 1994.)

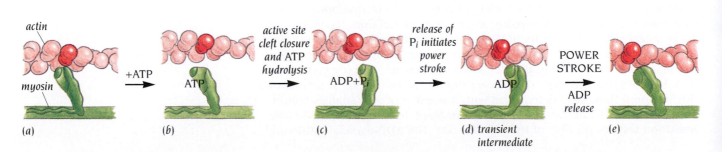

actin

myosin

+ATP

active site cleft closure and ATP hydrolysis

ATP

ADP+P$_i$

release of P$_i$ initiates power stroke

ADP

POWER STROKE

ADP release

(a)   (b)   (c)   (d) transient intermediate   (e)

end, the conformational change leads to a relative displacement of the actin and myosin filaments; they slide past each other (Figure 14.17e).

The conformational change of myosin that leads to the sliding of the filaments thus consists of switching a lever arm that changes the relative position of the two ends of the elongated heads by about 100 Å. Part of the energy of ATP hydrolysis must be stored as a mechanical "spring" to drive this conformational change. The two major states described here are not the only states of the actin–myosin complex; additional intermediate states are now being defined but it is unlikely that we shall have to change the current overall picture of the mechanism of muscle contraction.

Conclusion

Fibrous proteins are long-chain polymers that are used as structural materials. Most contain specific repetitive amino acid sequences and fall into one of three groups: coiled-coil α helices as in keratin and myosin; triple helices as in collagen; and β sheets as in silk and amyloid fibrils.

The coiled-coil fibrous proteins have heptad repeats in their amino acid sequence and form oligomers—usually dimers or trimers—through their coiled coils. These oligomeric units then assemble into fibers.

Collagen is a large trimeric superhelical molecule about 3000 Å long. Each of the three polypeptide chains has about 1000 amino acids, and is folded into an extended left-handed helix with a rise per turn of 9.6 Å. Proper assembly of the chains requires propeptide regions that are cleaved upon maturation of the triple helix. The three helical subunits coil about a central axis to form a right-handed superhelix with a 100-Å repeat distance. Steric constraints within this superhelix account for the large number of repeat sequences Gly-X-Y, where X is proline and Y is hydroxyproline, in the polypeptide chains. The crystal structure of a synthetic collagen-like peptide has shown the importance of both direct and water-mediated hydrogen bonds in stabilizing the triple helix.

Amyloid fibers, which are deposited in cells of the eye, brain, and heart, amongst others, are associated with a variety of inherited or infectious diseases. The amyloid fibers are formed as a result of the refolding of normally globular soluble proteins, which leads to their polymerization into long, straight, unbranched fibrils, about 100 Å in diameter, that aggregate into fibers. In a rare and fatal disease, familial amyloidotic polyneuropathy, point mutations in transthyretin, a carrier of the hormone thyroxine, causes the normally soluble globular transthyretin to deposit in the heart and brain. The structure of normal transthyretin and the fibrillar form have been studied by high-resolution crystallography and fiber diffraction, respectively. A model for the fibrillar form has been proposed, with four parallel protofilaments packed together to form a fibril. Each protofilament is built up from four continuous twisted β sheets running parallel to the fiber axis, with the constituent β strands in each at right angles to the axis.

Silk fibers, which have incredible strength, comprise well-ordered microcrystals of β-sheets that make up about 30% of the protein mass, interspersed in a matrix of polypeptide chains without order. The β strands of the sheets are oriented parallel to the fiber axis.

Muscle is composed of two sets of interdigitating filaments: thick filaments containing the fibrous protein myosin and thin filaments containing F-actin, a helical polymer of the soluble, globular, monomeric G-actin. In addition there are a number of other muscle proteins that regulate muscle activity. When muscles contract, actin filaments and myosin filaments slide with respect to each other. This movement is brought about by rounds of conformational changes of the heads of myosin molecules that project as cross-bridges from the myosin thick filaments to actin subunits in F-actin. Each round requires the hydrolysis of one molecule of ATP. Detailed structural studies of fragments of myosin and of actin at atomic resolution have clarified many of the molecular changes that occur during the movement of myosin cross-bridges.

Selected readings

General

Cohen, C., Parry, D.A.D. α-Helical coiled coils and bundles: how to design an α-helical protein. *Proteins* 7: 1–15, 1990.

Cohen, C., Parry, D.A.D. α-Helical coiled coils: more facts and better predictions. *Science* 263: 488–489, 1993.

Crick, F.H.C. The packing of α helices: simple coiled coils. *Acta Cryst.* 6: 689–697, 1953.

Fuchs, E. Keratins and the skin. *Annu. Rev. Cell Dev. Biol.* 11: 123–153, 1995.

Harbury, P.H., et al. A switch between two-, three-, and four-stranded coiled coils in GCN4 leucine zipper mutants. *Science* 262: 1401–1407, 1993.

Holmes, K.C. Muscle proteins—their actions and interactions. *Curr. Opin. Struct. Biol.* 6: 781–789, 1996.

Huxley, A.F., Simmons, R. Proposed mechanism of force generation in striated muscle. *Nature* 233: 533–538, 1971.

Huxley, E. The mechanism of muscular contraction. *Science* 164: 1356–1366, 1969.

Kielty, C.M., Hopkinson, I., Grant, M.E. Collagen: structure, assembly and organization in the extracellular matrix. In *Connective Tissue and its Heritable Disorders* Wiley-Liss, Chichester, pp. 103–147, 1993.

Lymn, R.W. Taylor, E.W. Mechanism of adenosine triphosphate hydrolysis of actomyosin. *Biochemistry* 10: 4617–4624, 1971.

Matsudaira, P. Modular organization of actin crosslinking proteins. *TIBS* 16: 87–92, 1991.

Prockop, D.J., Kivirikko, K.I. Collagens: molecular biology, diseases and potentials for therapy. *Annu. Rev. Biochem.* 64: 403–434, 1995.

Van der Rest, M., et al. Collagen family of proteins. *FASEB J.* 5: 2814–2823, 1991.

Vollrath, F. Spider webs and silks. *Sci. Am.*, pp. 52–58, March 1992.

Specific structures

Bella, J., Brodsky, B., Berman, H.M. Hydration structure of a collagen peptide. *Structure* 3: 893–906, 1995.

Bella, J., et al. Crystal and molecular structure of a collagen-like peptide at 1.9 Å resolution. *Science* 266: 75–81, 1994.

Blake, C., Serpell, L. Synchrotron x-ray studies suggest that the core of the transthyretin amyloid fibril is a continuous β sheet helix. *Structure* 4: 989–998, 1996.

Blake, C.C.F., et al. Structure of human plasma prealbumin at 2.5 Å resolution. A preliminary report on the polypeptide chain conformation, quaternary structure and thyroxine binding. *J. Mol. Biol.* 88: 1–12, 1974.

Bram, A., et al. X-ray diffraction from single spider silk fibers. *J. Appl. Cryst.* 30: 390–394, 1997.

Guerette, P.A., et al. Silk properties determined by a gland-specific expression of a spider fibroin gene family. *Science* 272: 112–115, 1996.

Hijirida, D.H., et al. ^{13}C NMR of *Nephila clavipes* major ampullate gland. *Biophys. J.* 71: 3442–3447, 1996.

Holmes, K.C., et. al. Atomic model of the actin filament. *Nature* 347: 44–49, 1990.

Kabsch, W., et. al. Atomic structure of the actin: DNase 1 complex. *Nature* 347: 37–44, 1990.

North, A.C.T., Steinert, P.M., Parry, D.A.D. Coiled-coil stutter and link segments in keratin and other intermediate filament molecules: a computer modeling study. *Proteins* 20: 174–184, 1994.

Rayment, I., et al. Structure of the actin–myosin complex and its implications for muscle contraction. *Science* 261: 58–65, 1996.

Rayment, I., et al. Three-dimensional structure of myosin subfragment–1: a molecular motor. *Science* 261: 5–58, 1996.

Schroeder, R.R., et. al. Three-dimensional atomic model of F-actin decorated with *Dictyostelium myosin* S1. *Nature* 364: 171–174, 1993.

Simmons, A.H., Michal, C.A., Jelinski, L.W. Molecular orientation and two-component nature of the crystalline fraction of spider dragline silk. *Science* 271: 84–87, 1996.

Smith, C. A., Rayment, I. X-ray structure of the magnesium (I I). ADP–vanadate complex of the *Dictyostelium discoideum* myosin motor domain to 1.9 Å resolution. *Biochemistry* 35: 5404–5407, 1996.

Spraggon, G., et al. Crystal structures of fragment D from human fibrinogen and its crosslinked counterpart from fibrin. *Nature* 389: 455–462, 1997.

Thiel, B.L., Kunkel, D.D., Viney, C. Physical and chemical microstructure of spider dragline: a study by analytical transmission electron microscopy. *Biopolymers* 34: 1089–1097, 1994.

Warwicker, J.O. Comparative studies of fibroins. II. The crystal structure of various fibroins. *J. Mol. Biol.* 2: 350–362, 1960.

Wess, T.J., et al. Type I collagen packing, conformation of the triclinic unit cell. *J. Mol. Biol.* 248: 487–493, 1995.

Xu, M., Lewis, R.V. Structure of a protein superfiber: spider dragline silk. *Proc. Natl. Acad. Sci. USA* 87: 7120–7124, 1990.

Recognition of Foreign Molecules by the Immune System

15

The immune system in vertebrates provides a defense mechanism against foreign parasites such as viruses and bacteria. Three main properties are essential to its successful operation: specific recognition of foreign molecules, the ability to destroy the foreign parasite, and a memory mechanism that allows a more rapid response to a second infection by the same microorganism.

Foreign invaders are recognized through specific and tight binding of the proteins of the immune system to molecules specific to the foreign organisms. The sites on foreign molecules that are recognized by the immune system are called **antigenic determinants**, and they interact with two different classes of antigen receptors produced by two major cell types of the immune system. **Antibodies**, also known as **immunoglobulins**, are produced by B cells, which are stimulated by antigen binding to secrete antibodies into the bloodstream (Figure 15.1a). Related molecules known as **T-cell receptors**, found on the surface of T cells, recognize and destroy virus-infected cells and play an important role in coordinating the immune response (Figure 15.1b).

Antibodies secreted by B cells bind to foreign material (antigen) and serve as tags or identifiers for such material. Antibody-tagged bacteria,

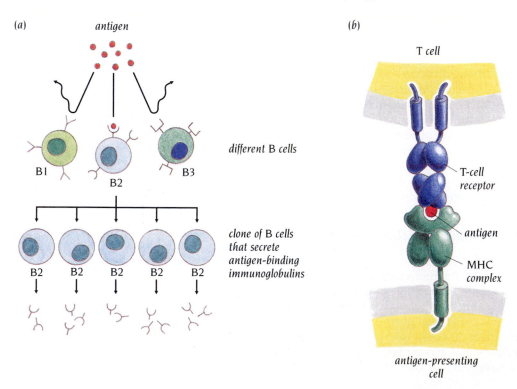

Figure 15.1 (a) The clonal selection theory. An antigen (red) activates those B cells that have immunoglobulin molecules on their surfaces that can recognize and bind the antigen. This binding triggers production of a clone of identical B cells that secrete soluble antigen-binding immunoglobulins into the bloodstream. (b) T cells recognize foreign viral antigens (red), through a T-cell receptor (blue) that can bind degraded fragments of the antigen when they are associated with an MHC molecule (green). The polypeptide chains of both T-cell receptors and MHC molecules are folded into immunoglobulin-like domains, represented as ellipsoids in the diagram.

pathogens, or other foreign objects are recognized and disposed of by macrophages and other effector cells in the immune system; in addition, antibody-tagged bacteria can be directly lysed by a set of nonspecific serum proteins collectively called complement (because they are required in addition to antibody in triggering cell lysis). B cells rearrange the genes that code for their antibody proteins, so that each cell makes a unique antibody. B cells can display on their surface a membrane-bound form of their antibody. Binding of antigen to this form triggers a process that eventually activates the B cell to synthesize and secrete many copies of the antibody. In this way a small amount of antigen can elicit an amplified and specific immune response that helps to clear the body of the source of the antigen.

By contrast, T-cell receptors recognize antigenic determinants only when presented as part of a complex with a third important group of proteins, the **MHC molecules**, which fall into two classes, I and II. These proteins are encoded by the *m*ajor *h*istocompatibility gene *c*omplex, a region of chromosome six (in humans) originally identified as the genetic element that controls transplant rejection. Each individual carries a discrete set of MHC molecules, and differences in this set from one individual to another cause a transplanted organ to appear foreign to the immune system. MHC molecules bind degraded fragments of peptide antigens, generated inside infected cells or taken up by specialized phagocytotic cells, and display the fragments at the cell surface for recognition by T cells. T cells also rearrange their receptor genes so that the receptor carried by each T cell is unique. Binding of a T-cell receptor to an MHC–antigen complex triggers activation of the T cell to kill the infected cell, or elicits help from other cells in generating an immune response against the phagocytosed material.

In this chapter we will see that all three of the molecules directly involved in the specific recognition of antigen—immunoglobulins, the T-cell receptor, and the molecules of the MHC—belong to a family of proteins that seems to have evolved by duplication and diversification from a single ancestral domain. Because antibodies are naturally produced in very large quantities as soluble molecules, the crystal structure of an immunoglobulin was solved long before the structure of either of the other two molecules could be experimentally approached: hence the presumed ancestral domain structure is known as an immunoglobulin or immunoglobulin-like domain. MHC proteins and T-cell receptors are cell surface proteins that occur naturally in low abundance. Since the development of recombinant sources of protein for crystallization, structures have been determined for soluble forms of MHC proteins and T-cell receptors. We shall examine in detail the three-dimensional structures of antibodies and their bound antigens, and discuss the relationship between the genetic basis for antibody diversity and the structural basis for specific recognition. We shall also examine the structure of an MHC molecule and show that the structural principles for recognition and binding of antigens to MHC are quite different from those of antibodies and that they reflect the special requirement for presentation of bound antigen to the T-cell receptor. We will then look at the structure of a T-cell receptor and its special adaptations for binding to its ligand, which is the complex of an MHC molecule with bound antigen. Finally, we shall discuss the use of the immunoglobulin-like domain in other proteins of immunological interest.

The polypeptide chains of antibodies are divided into domains

The basic structure of all immunoglobulin (Ig) molecules comprises two identical **light chains** and two identical **heavy chains** linked together by disulfide bonds (Figure 15.2a). There are two different classes, or isotypes, of light chains, λ and κ, but there is no known functional distinction between them. Heavy chains, by contrast, have five different isotypes that divide the immunoglobulins into different functional classes: IgG, IgM, IgA, IgD, and IgE, each with different effector properties in the elimination of antigen

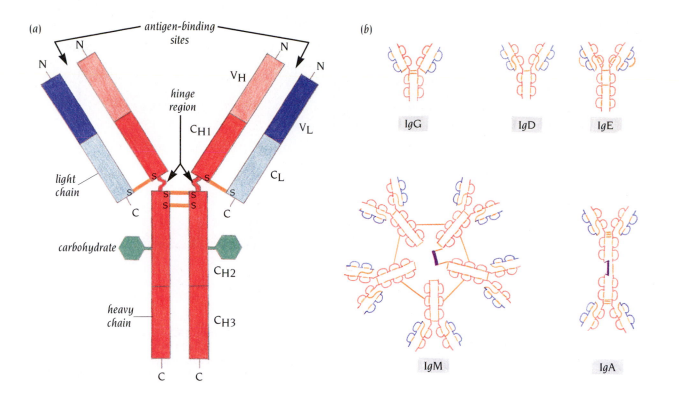

(a)

antigen-binding sites

N

N

N

V_H

hinge region

C_{H1}

V_L

C_L

light chain

S S S S S S S S

C C

carbohydrate

C_{H2}

heavy chain

C_{H3}

C C

(b)

IgG   IgD   IgE

IgM   IgA

(Figure 15.2b). Each class of heavy chains can combine with either of the two different classes of light chains.

In this chapter we will discuss immunoglobulins of the **IgG** class, which is the major type of immunoglobulin in normal human serum, and which has the simplest structure. Each chain of an IgG molecule is divided into domains of about 110 amino acid residues. The light chains have two such domains, and the heavy chains have four.

The most remarkable feature of the antibody molecule is revealed by comparing the amino acid sequences from many different immunoglobulin IgG molecules. This comparison shows that between different IgGs the amino-terminal domain of each polypeptide chain is highly variable, whereas the remaining domains have constant sequences. A light chain is thus built up from one amino-terminal **variable domain** (V_L) and one carboxy-terminal **constant domain** (C_L), and a heavy chain from one amino-terminal variable domain (V_H), followed by three constant domains (C_{H1}, C_{H2}, and C_{H3}).

The variable domains are not uniformly variable throughout their lengths; in particular, three small regions show much more variability than the rest of the domain: they are called **hypervariable regions** or **complementarity determining regions**, CDR1–CDR3 (Figure 15.3). They vary both in size and in sequence among different immunoglobulins. These are the regions that determine the specificity of the antigen–antibody interactions. The remaining parts of the variable domains have quite similar amino acid sequences. In fact, all variable domains show significant overall sequence homology and, hence, evolutionary relationships among each other. Similarly, the different constant domains—C_L, C_{H1}, C_{H2}, and C_{H3}—show significant (30–40%) sequence identity with each other and with constant domains from immunoglobulin chains of different classes. This strongly suggests that the genes that code for these constant domains arose by successive gene duplication of an ancestral antibody gene. In contrast, there is no significant sequence homology between constant and variable domains, even though, as we will see, there are striking structural similarities. Before describing the structures of the immunoglobulin domains, however, we pause to outline the genetic mechanisms underlying the variability of immunoglobulin molecules.

Figure 15.2 (a) The immunoglobulin molecule, IgG, is built up from two copies each of two different polypeptide chains, heavy (H) and light (L). The L chain folds into two domains: V_L with variable sequence between different IgG molecules and C_L with constant sequence. The H chain folds into four domains: one variable V_H and three constant domains, C_{H1}, C_{H2}, and C_{H3}. Disulfide bridges connect the four chains. The antigen-binding sites are at the ends of the variable domains. (b) Schematic polypeptide chain structure of different immunoglobulin molecules. IgM is a pentamer where the heavy chains have one variable domain and four constant domains. An additional polypeptide chain, the J chain, (black bar) is associated with the pentameric molecule. The J chain also links two units to form the dimeric IgA. IgG, IgD, and IgE are monomeric immunoglobulin molecules.

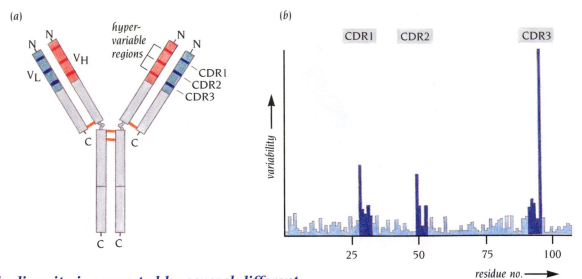

(a)

N N
hyper-
variable
regions
VL VH N N
CDR1
CDR2
CDR3
C C
C C

(b)

CDR1 CDR2 CDR3

variability

25   50   75   100

residue no. ⟶

Antibody diversity is generated by several different mechanisms

The human body generates as many as 50 million new antibody-producing B cells every day. The vast majority of these cells produce an antibody molecule unique to that cell and its progeny. Genetic information for this diversity is contained in more than 1000 small segments of DNA. One of the outstanding achievements of modern biology has been the elucidation of the mechanisms that generate antibody diversity.

The gene segments encoding the variable and constant regions of antibodies are clustered in three gene pools, one for the heavy chain and one for each of the two light-chain isotypes, on separate chromosomes. In the heavy-chain gene pool the variable domain is encoded by three types of segments, V, D, and J (Figure 15.4). V codes for approximately the first 90 residues, D for the hypervariable region CDR3, and J for the remaining 15 residues of the variable domain.

There are about 1000 different V segments, about 10 different D segments of variable lengths, and about 4 different J segments in the heavy-chain gene pool. The DNA for the variable domain of a new B cell is assembled by random joining of one of each of these segments into a single continuous exon (Figure 15.5), a process called combinatorial joining that can occur in about 40,000 different ways. The joining of these segments is not precise at the junction of D with V and J, and this creates additional junction diversity. Extra nucleotides can be added during this joining procedure, giving still further diversity. The D segment and nearby junctional regions encodes CDR3, which is thus a focus of particular diversity in the heavy chain.

The newly created V–D–J exon becomes joined to one of the eight C segments that encode the constant regions of the various classses of heavy chain. Usually, a variable region is first expressed in an IgM molecule. Later, further genetic rearrangements will delete the mu (μ) exons encoding the IgM constant region, and the variable region will become joined to a second set of constant-region exons encoding IgA, IgD, IgE, or most commonly IgG. This exchange of variable regions among heavy chains is called **class-switching**.

The genetic mechanisms underlying light chain diversity are similar except that there are no D segments, and the diversity of CDR3 therefore depends largely upon junctional diversity generated during V–J joining.

Figure 15.3 Certain regions within the 110 amino acid variable domains show a high degree of sequence variation. These regions determine the antigen specificity and are called hypervariable regions or complementarity determining regions, CDR (a). There are three such regions, CDR1–CDR3, in each variable domain. The sequences of a large number of variable domains in L chains have been compared, and the sequence variability is plotted as a function of residue number along the polypeptide chain in (b). The three large peaks in this diagram correspond to the three hypervariable regions in (a): CDR1–CDR3. [(b) Adapted from T.T. Wu and E.A. Kabat, *J. Exp. Med.* 132: 211–250, 1970.]

Figure 15.4 Variable domains in immunoglobulins (and T-cell receptors) are made by combinatorial joining of gene segments. Three such segments, V, D, and J, are joined to make the variable domain of a heavy chain. In the mouse the gene-segment pool for an H chain contains about 1000 V segments, 12 D segments, 4 J segments, and a cluster of C segments, each encoding a different class of H chain.

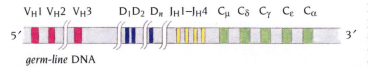

V_H1 V_H2 V_H3 D_1D_2 D_n J_H1–J_H4 C_μ C_δ C_γ C_ϵ C_α

5′   3′

germ-line DNA

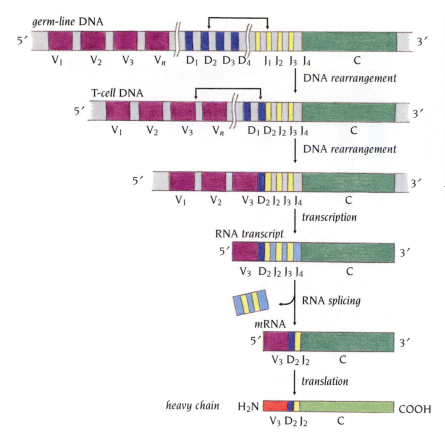

Figure 15.5 The V–D–J joining process involved in making a heavy chain in the mouse. In the germ-line DNA the cluster of about 4 J gene segments and about 10 D segments are separated from the C gene segment by a short intron and from the about 1000 V gene segments by long regions of DNA. During B cell development the chosen V gene segment (V_3) is moved to lie precisely next to the chosen D and J segments (D_2 and J_2). The "extra" J genes (J_3 and J_4) and the intron sequence are transcribed along with the V_3, D_2, J_2, and C segments and then removed by RNA splicing to generate mRNA molecules.

By these genetic mechanisms we can estimate that at least 90,000 different heavy chains and 3000 different light chains can be produced. We will see that it is reasonable to assume from structural considerations that almost any heavy chain can combine with almost any light chain to produce a functional antibody. Since the antigen-binding sites are built up from the CDR regions of both chains, this ability of light and heavy chains to combine and form functionally viable pairs generates considerable additional diversity. By this combinatorial association, the number of different antibody molecules a new B cell can choose from is increased to 3000 × 90,000 = 270 million. The actual number of antibody molecules with different antigen binding sites is, however, several orders of magnitude higher because of yet one more mechanism for increasing diversity in mature B cells. In these cells, point mutations can be introduced into the exons for the variable domains by somatic hypermutation.

All immunoglobulin domains have similar three-dimensional structures

Complete IgG molecules have been difficult to crystallize because of the conformational flexibility in the **hinge region** between C_{H1} and C_{H2} of the heavy chains (see Figure 15.2). When an IgG molecule is partially digested with the proteolytic enzyme papain, however, the heavy chains are cleaved in the hinge region to give two identical **Fab** fragments (for *Fragment, antigen binding*) and one **Fc** fragment (for *Fragment, crystallizes easily*) (Figure 15.6). The Fab molecule comprises one complete light chain linked by a disulfide bridge to a fragment of the heavy chain consisting of V_H and C_{H1}. The Fc molecule comprises C_{H2} and C_{H3} from both heavy chains, which are also linked by disulfide bridges. Until recently, all detailed high-resolution x-ray structural information on immunoglobulins has been obtained from structure determinations of Fab, Fc, or similar fragments including those of complexes with antigen.

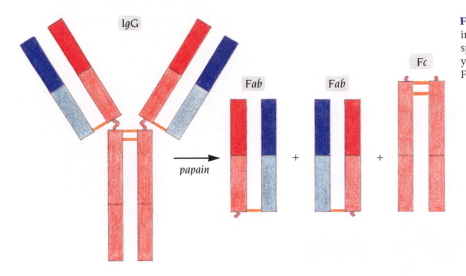

IgG

Fab   Fab   Fc

papain   +   +

Figure 15.6 Enzymatic cleavage of immunoglobulin IgG. The enzyme papain splits the molecule in the hinge region, yielding two Fab fragments and one Fc fragment.

Before the advent of monoclonal antibody technology, Allen Edmundson and collaborators at Argonne National Laboratories, US, obtained the first detailed description of an immunoglobulin domain. The high-resolution structure they determined in 1973 was that of a Bence-Jones protein, a tumor-derived molecule that is a dimer of two light chains. In the same year Roberto Poljak and collaborators at Johns Hopkins Medical School, Baltimore, determined the first high-resolution structure of an intact immunoglobulin Fab fragment, and David Davies and collaborators at the National Institutes of Health demonstrated the binding of an antigenic determinant to the antigen-binding site in a Fab fragment. Since then the structures of over one hundred various other proteolytic fragments of immunoglobulin molecules, alone or in complex with antigen, have been solved to high resolution. In all cases both constant and variable domains from both heavy and light chains have proved to have similar structures, which has become known as the **immunoglobulin fold**.

The immunoglobulin fold is best described as two antiparallel β sheets packed tightly against each other

All constant domains are built up from seven β strands arranged so that four strands form one β sheet and three strands form a second sheet (Figure 15.7). The sheets are closely packed against each other and joined together by a disulfide bridge from β strand F in the three-stranded sheet, to β strand B in the four-stranded sheet. The topology of the sheets is quite simple, as can be seen from the topological diagram in Figure 15.7b. Even though the actual structure consists of two separate β sheets, it is convenient to regard the topology as a form of a Greek key barrel.

The loops are short, and as a result the majority of the residues of the domain are in the two β sheets. These framework residues have the same structure in every constant domain studied, and the structure is the same in both heavy and light chains. Most of the invariant residues of the constant domain, including the disulfide bridge, are found in the sheets of the framework region. These invariant residues have two important functions. One is to form and stabilize the framework by packing the β sheets through hydrophobic interactions to give a hydrophobic core between the sheets; the other is to provide interactions between constant domains of different chains to form a complete immunoglobulin molecule.

The residues not in the framework region form the loops between the β strands. These loops may vary in length and sequence among immunoglobulin chains of different classes but are constant within each class; the sequence of the loops is invariant. The functions of these loops are not known, but they are probably involved in the effector functions of antibodies. When an antibody–antigen complex has been formed, signals are

(a)

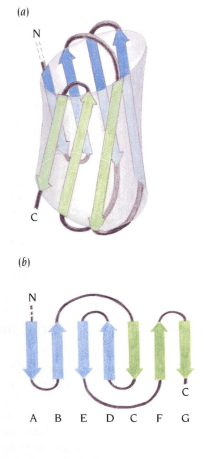

(b)

Figure 15.7 The constant domains of immunoglobulins are folded into a compressed antiparallel β barrel built up from one three-stranded β sheet packed against a four-stranded sheet (a). A topological diagram (b) shows the connected Greek key motifs of this fold.

(a)

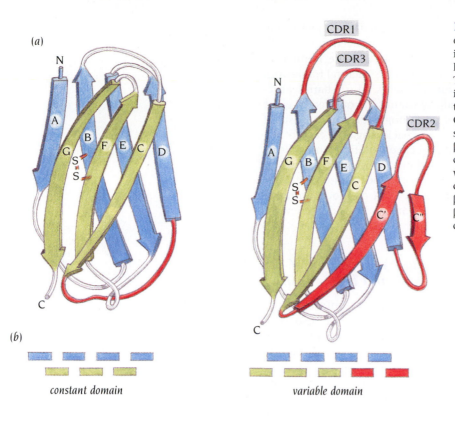

(b)

constant domain

variable domain

Figure 15.8 (a) Comparison of the structures of the constant and variable domains of immunoglobulins. Beta strands labeled A–G have the same topology and similar structures. There are two extra β strands, C′ and C″ (red) in the variable domain. The loop between these strands contains the hypervariable region CDR2. The remaining CDR regions are at the same end of the barrel in the loops connecting β strands B and C and strands F and G. A disulfide bond bridges strand B in one sheet with strand F in the other sheet in both the constant and the variable domains. (b) The β strands viewed end on, illustrating that one β sheet has the same four β strands in the two domains.

mediated through the Fc region to different systems, such as phagocytic cells or the complement system, which will destroy the complex. These activities are triggered by specific interactions of ligands with the constant domains of immunoglobulins.

The overall structure of the variable domain is very similar to that of the constant domain, but there are nine β strands instead of seven. The two additional β strands are inserted into the loop region that connects β strands C and D (red in Figure 15.8). Functionally, this part of the polypeptide chain is important since it contains the hypervariable region CDR2. The two extra β strands, called C′ and C″, provide the framework that positions CDR2 close to the other two hypervariable regions in the domain structure (Figure 15.8).

The hypervariable regions are clustered in loop regions at one end of the variable domain

The specificity of immunoglobulins is determined by the sequence and size of the hypervariable regions in the variable domains. From amino acid sequence comparisons of several hundred different variable domains, these hypervariable regions have been defined as amino acids 24–34, 50–56, and 89–97 in the light chains and 31–35, 50–65, and 95–102 in the heavy chains. In the three-dimensional structures these residues occur in the loop regions that connect β strands B–C, C′–C″, and F–G. We can see in Figure 15.8 that these loop regions are clustered together at one end of the β sheets. CDR2 and CDR3 are hairpin loops between β strands C′–C″ and F–G in the five-stranded β sheet. The third hypervariable region CDR1 is a cross-over connection from β strand B in the four-stranded sheet to strand C in the five-stranded sheet. All strands in the five-stranded sheet thus contribute loops to the hypervariable regions, whereas only one strand from the four-stranded sheet is involved. The additional loop at this end of the sheet, between β strands D and E in the four-stranded sheet, does not have a variable sequence and does not participate in antigen binding. It is thus evident that loops from the five-stranded β sheet contribute more residues to the complete antigen-binding site than the four-stranded sheet.

The variable domains of immunoglobulins are excellent examples of the important structural principle discussed in Chapter 4 in connection with α/β-barrel structures. Functional residues in protein molecules are frequently provided by loop regions attached to one end of a stable structural framework that is built up mainly by secondary structure elements. In α/β-barrel structures this framework is built from eight parallel β strands surrounded by eight helices, whereas in the variable domains of immunoglobulins it consists of nine antiparallel β strands arranged in two sheets packed against each other. The fact that there exists a great number of sequence variations in the loop regions of antibodies demonstrates that the folding of the structural framework is quite insensitive to changes in the functional loop regions. Thus, it has been possible to construct, using recombinant DNA methods, chimeric antibodies carrying CDR loop regions from one antibody, and framework regions from another antibody.

The antigen-binding site is formed by close association of the hypervariable regions from both heavy and light chains

Fab fragments contain one arm of the IgG molecule with an intact antigen binding site. In these molecules, as in intact IgG, the light chain is associated with the heavy chain in such a way that C_L associates with C_{H1} and V_L with V_H (Figure 15.9). These associations are very tight and extensive. The segments of the light and heavy polypeptide chains that join V_L to C_L and V_H to C_{H1} have few contacts. As a result, the Fab fragment consists of two globular regions, one formed by the two constant domains and the other by the two variable domains. Together, they make an elongated molecule. The hypervariable regions of both the light and heavy chains are close together

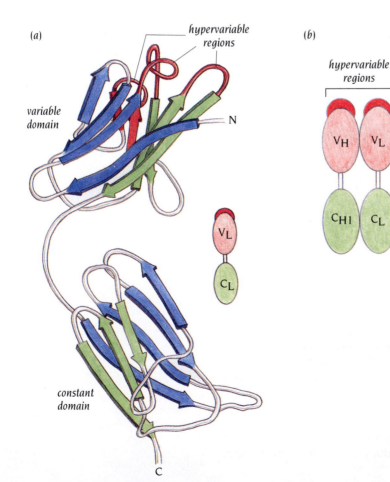

(a)   *hypervariable regions*   (b)

hypervariable regions

variable domain

N

V_H V_L

V_L

C_L

C_{H1} C_L

constant domain

C

Figure 15.9 (a) The variable and constant domains in the light chain of immunoglobulins are folded into two separate globular units. In both domains the four-stranded β sheet is blue and the other sheet is green. The hypervariable CDR regions are at one end of this elongated molecule. (b) In the Fab fragment, as well as in the intact immunoglobulin molecule, the domains associate pairwise so that V_H interacts with V_L and C_{H1} with C_L. By this interaction the CDR regions of both variable domains are brought close to each other and together form the antigen-binding site.

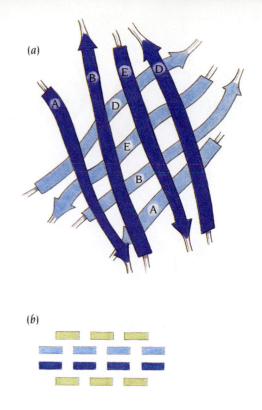

(a)

(b)

Figure 15.10 Schematic diagrams of the packing of the four-stranded β sheets of the constant domains C_{H1} and C_L in an Fab fragment of IgG. The sheets are viewed perpendicular to the β strands in (a) and end-on in (b), where the four-stranded β sheets are blue.

at one end of this elongated molecule. We will now examine more closely how these domains associate.

The constant domains associate by interactions between the four-stranded β sheets in both C_L and C_{H1}, which are closely packed with mainly hydrophobic residues in the interface. The directions of the β strands in the two domains are almost at right angles to each other (Figure 15.10a). This mode of packing β sheets against each other occurs frequently within the core of single domains that are built up from two β sheets—for example, in retinol-binding protein discussed in Chapter 5—but they are rather unusual for domain–domain associations. The effect of this packing is to bury the central regions of the four-stranded β sheets away from the solvent. Each four-stranded β sheet has the three-stranded β sheet of the same domain on one side and the four-stranded β sheet from the second domain on its other side (Figure 15.10b). This places rather stringent constraints on the sequence of these regions and most of the residues that occur in the interface are, in fact, conserved. In addition, residues that form the hydrophobic core within each domain are generally conserved.

The variable domains associate in a strikingly different manner. It is obvious from Figure 15.11 that if they were associated in the same way as the constant domains, via the four-stranded β sheets, the CDR loops, which are linked mainly to the five-stranded β sheet, would be too far apart on the outside of each domain to contribute jointly to the antigen-binding site. Thus in the variable domains the five-stranded β sheets form the domain–domain interaction area (Figure 15.11). Furthermore, the relative orientation of the β strands in the two domains is closer to parallel than in the constant domains and the curvature of the five-stranded β sheets is such that they do not pack

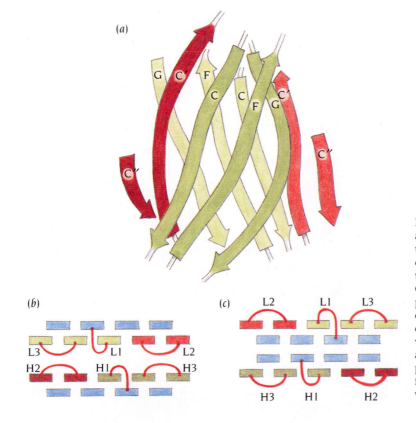

Figure 15.11 The two variable domains V_H and V_L are packed against each other so that the six hypervariable regions are close to each other. (a) Schematic diagram of the packing of the five-stranded β sheets of V_H and V_L in IgG. Only four of the β strands are involved in packing the variable domains against each other. Strand C″ is not involved. (b) Schematic diagram of the strands viewed end-on in the variable domain. Hypervariable loop regions are red. (c) Diagram illustrating a hypothetical packing of the variable domains through their four-stranded sheets. The hypervariable regions would be far apart.

307

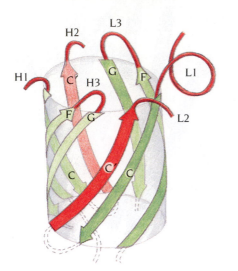

Figure 15.12 Schematic diagram of the barrel arrangement of four β strands from each of the variable domains in Fab. The six hypervariable regions, CDR1–CDR3 from the light chain (L1–L3) and from the heavy chain (H1–H3), are at one end of this barrel. (From J. Novotny et al., *J. Biol. Chem.* 258: 14433–14437, 1983.)

tightly against each other. Instead, each sheet forms half a barrel. When the two sheets associate, four of the five strands from both domains complete a barrel structure of eight antiparallel β strands with the six CDR loops at the same end of the barrel (Figure 15.12). Hydrophobic side chains fill the inside of this barrel.

The antigen-binding site binds haptens in crevices and protein antigens on large flat surfaces

The six CDR loops from the two variable chains at the rim of the eight-stranded barrel provide an ideal arrangement for generating antigen-binding sites of different shapes depending on the size and sequence of the loops (Figure 15.13). They can create flat extended binding surfaces for protein antigens or specific deep binding cavities for small hapten molecules.

Haptens, a special class of antigen, are small molecules that induce specific antibody production when they are attached to a protein that acts as a carrier. Phosphorylcholine is one such hapten that has been widely used in the investigation of immune responses. The specific binding of this hapten

Figure 15.13 (a) Drawing of a space-filling model of the hypervariable regions of an Fab fragment. The superpositions of five sections are shown, cut through a model as shown in (b). It is clearly seen that all six hypervariable regions (L1–L3, H1–H3) contribute to the surface shown here. (From C. Chothia and A. Lesk, *J. Mol. Biol.* 196: 901–917, 1987.)

(a)

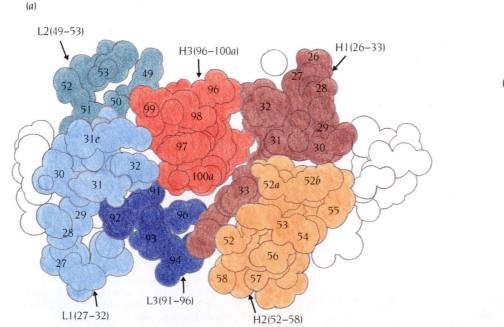

(b)

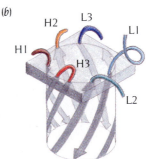

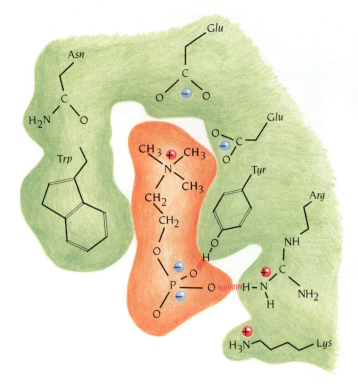

Figure 15.14 Schematic representation of the specific interactions between phosphorylcholine (orange) and the protein side groups (green) in Fab. The binding cavity is in a cleft between the light and the heavy chains. Choline binds in the interior while the phosphate group is toward the surface. (Adapted from E.A. Padlan et al., *Immunochemistry* 13: 945–949, 1976.)

to an Fab fragment has been studied crystallographically by David Davies and coworkers at National Institutes of Health. The binding cavity is about 15 Å wide at the mouth and 12 Å deep. It is lined by residues from all CDR loops except the short CDR2 from the light chain. A number of side chains from these loops bind phosphorylcholine in a manner that is very similar to that of a substrate or inhibitor binding to an enzyme (Figure 15.14).

In enzyme-catalyzed reactions, substrates in their transition state bind very tightly to the enzyme and are positioned close to polar or charged side chains that participate in catalyzing the enzymatic reaction. It has recently been possible to induce production of antibodies with enzymatic activity by choosing haptens that are analogs of the transition state of the substrate for a chemical reaction. When an animal is immunized with a transition-state analog as a hapten, an array of different antibodies that bind the hapten specifically are produced. Some of these antibodies happen to have charged or polar side chains suitably placed to promote catalytic reactions involving the corresponding substrate. These catalytic antibodies can be isolated and improved upon by recombinant DNA techniques.

The analysis of hapten binding to antibodies was possible before monoclonal antibodies became available because some myeloma proteins bind haptens. Thus hapten binding was studied as early as 1974. Once monoclonal antibodies with any chosen specificity could be produced, it became possible in principle to study the interactions of antibodies with protein antigens. However, the path from a cell line to a high-resolution three-dimensional antigen-antibody structure is long, and not until 1986 was the first such investigation reported by Roberto Poljak and coworkers at the Pasteur Institute, France. They chose a lysozyme–Fab complex because the structure of the enzyme lysozyme was known and its antigenic properties had been extensively studied.

The shape of the interaction area between lysozyme and the CDR loops of the antibody is easily distinguished from the hapten-binding crevice. The interaction extends over a large area with maximum dimensions of about 20 × 30 Å (Figure 15.15). The interaction surface is irregular but relatively flat, with small protuberances and depressions that are complementary in the antigen and the antibody. Residues from all six CDR loops contribute to the

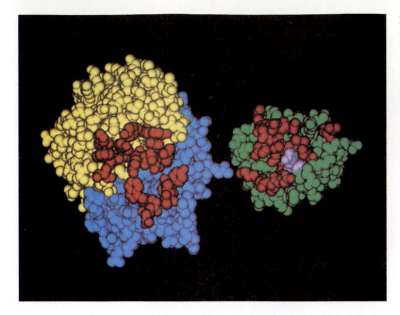

Figure 15.15 Space-filling representation of a complex between lysozyme (green) and the Fab fragment of a monoclonal antilysozyme (blue and yellow). The Fab fragment and the antigen (lysozyme) have been separated in this diagram, and their combining surfaces are viewed end-on. Atoms that are in contact in the complex are colored red both in Fab and lysozyme, except Gln 121 in lysozyme, which is violet. The diagram illustrates the large size of the interaction surfaces. (After A.G. Amit et al., *Science* 233: 747–753, 1986; courtesy of R. Poljak.)

antibody surface, and residues from two stretches of the polypeptide chain of lysozyme, 18–27 and 116–129, form the antigen surface. Seventeen antibody residues make close contact with 16 lysozyme residues. These residues pack against each other with a density similar to that found in the interior of protein molecules. For example, Gln 121 in lysozyme (violet in Figure 15.15) protrudes from the lysozyme surface and fits snugly into a cleft between the CDR3 loops of V_H and V_L (Figure 15.16). On the other hand, there are some imperfections, in the form of holes, in the interaction area. Appropriate changes in the CDR loops could fill these holes and produce an even better fit. Presumably, such adjustment could occur through somatic mutations in the antibody genes to give rise to higher-affinity antibodies.

The interaction area is centered on CDR3 of the heavy chain, which makes a proportionally greater contribution to the formation of this area than any of the other CDR loops. Four residues in the heavy-chain CDR3 participate in interactions with lysozyme, and they are all encoded in the D segment, the special significance of which now becomes clear. CDR3 of the heavy chain is the principal focus of the somatic mechanisms for generating antibody diversity and is central to the specific recognition of protein antigens. It is also the loop that eludes the structural predictive formula of Chothia and Lesk that we describe in the following section.

The structure of lysozyme in the complex is the same as that in crystals of free lysozyme, and no conformational changes are seen even in the regions

Figure 15.16 Detailed views of the environment of Gln 121 in the lysozyme–antilysozyme complex. Gln 121 in lysozyme is colored green both in the space-filling representation to the left and in the ball and stick model to the right. This side chain of the antigen fits into a hole between CDR3 regions of both the heavy (Tyr 101) and the light (Trp 92) chains as well as CDR1 from the light chain (Tyr 32). (After A.G. Amit et al., *Science* 233: 747–753, 1986.)

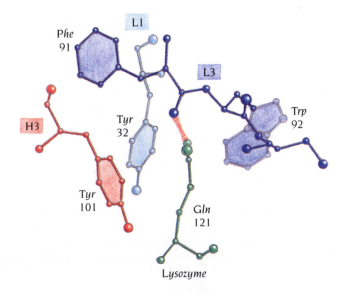

that bind to the antibody. The structure of the Fab fragment complexed to lysozyme is also similar to its structure when not bound to antigen. However, in other cases, conformational changes that increase the complementarity between the interacting surfaces of antigen and antibody can accompany antigen binding. Adjustments ranging from simple side chain rotations, to CDR loop reorientations of up to 6 Å, to rigid body domain motions of up to 15°, have been observed for various complexes. Although antibody–antigen interactions have been always at the forefront of "lock-and-key" versus "induced-fit" debate about the physical origin of biological specificity, no general rules have yet emerged.

The CDR loops assume only a limited range of conformations, except for the heavy chain CDR3

We have discussed how antibodies are able to bind a wide range of antigens and even catalyze chemical reactions through variations in their hypervariable CDR loops, which are attached to the essentially invariant β-sheet framework of the immunoglobulin domain. Therefore, in order to engineer an antibody with a predetermined binding specificity, or to predict the binding specificity of an antibody with known sequence but unknown function, we must be able to predict and model accurately the conformations of CDR loops. Thus the question arises: can the structure of CDR loops vary randomly, or are there certain preferred conformations that can be deduced from the lengths and sequences of these regions?

Cyrus Chothia and Arthur Lesk at MRC, Cambridge, UK, have attempted to answer this question by an analysis of the CDR-loop conformations in known structures. For each observed conformation they identified those residues that were mainly responsible for maintaining that conformation, either through interactions with framework residues or through their ability to assume unusual conformations. Examples of the latter are Gly or Pro. Examination of the sequences of variable domains of unknown structure showed that many have CDR loops that are similar in size to those of one of the known structures, with identical residues at the sites responsible for the observed conformation. Thus, for these loops, conformations could reliably be predicted. In addition to their value in the modeling and design of new antibodies, these results have interesting implications for the molecular mechanisms involved in the generation of antibody diversity. For five of the six hypervariable regions of most immunoglobulins (CDR3 of the heavy chain is an exception) there appears to be only a small repertoire of main-chain conformations, most of which are known from the set of immunoglobulin structures so far determined. Sequence variations within the hypervariable regions modulate the surface that these canonical loop structures present to the antigens. Sequence variations within both the framework and the hypervariable regions shift the canonical structures relative to each other by small but significant amounts.

Three examples from the analysis by Chothia and Lesk, which is further described in Chapter 17, demonstrate these phenomena. The loop between β strands C' and C" in CDR2 of the light chain is three residues long in all known sequences. This loop has very similar conformations in the known structures, and 95% of 250 known sequences of mouse and human light chains fit the sequence constraints of this structure. There is thus a high probability that this loop has the same conformation in almost all immunoglobulins. Four different conformations of CDR2 from the heavy chain have been found in the 10 x-ray structures that were analyzed. For each of these conformations there are specific constraints at a few residues. For 236 of 300 known sequences of heavy chains, the sequence of the CDR2 loop fits one of these constraints and would therefore be expected to have one of these four canonical CDR2-loop structures. The CDR3 loop of the heavy chain, however, varies in length from 6 to 14 residues in the known sequences and forms a long hairpin loop. For such large loops many conformations are possible. The one actually found will depend both on the size and sequence of the loop and on its packing against the rest of the protein. This loop, therefore, is difficult to predict and model correctly.

An IgG molecule has several degrees of conformational flexibility

Early physicochemical and electron microscopy studies suggested that in the intact IgG molecule the Fab arms were flexible and could move relative to each other and to the Fc stem. The region of the heavy chains that connects C_{H1} with $C_{H2,}$ (and therefore Fab with Fc) is called the hinge region. The amino acid sequence of this region includes the cysteine residues that form disulfide bridges between heavy chains, and has a high proportion of proline residues frequently found in flexible regions of proteins. The flexibility of the hinge region is important for the function of immunoglobulins, but it complicates crystallization, and in many cases crystallized intact antibodies have disordered Fc and hinge regions that are not visualized in the crystal structure.

The structure of an intact antibody, an antitumor IgG that forms unusually well-ordered crystals, was determined by Alexander McPherson and colleagues in 1992 (Figure 15.17). All segments of the antibody, including the hinge domain and attached carbohydrate, are visible in the electron density maps. As expected from earlier studies of Fc domains, the C_{H2} and C_{H3} domains of each of the heavy chains associate in C_{H2}–C_{H2} and C_{H3}–C_{H3} pairs. The C_{H3} pair is associated in the same way as the C_{H1}–C_L pair discussed previously. In contrast, there is no protein–protein interaction between the C_{H2} domains: instead, a carbohydrate chain is attached to each C_{H2} domain in the interface region and forms a weak bridge between them. Polypeptide segments between the Fab and Fc portions are extended (compare Figures 15.2a and 15.17) and exhibit large temperature factors indicative of a high degree of mobility. Thus, the hinge "domain" is better considered as a flexible tether approximately 50 Å long than as a rigid hinge. There is no overall symmetry to the molecule: the local twofold rotation axis relating the Fc C_{H3}–C_{H3} domains is different from that relating the Fab C_{H1}–C_L domains. The orientation of the two Fab arms is such that the intact IgG molecule resembles a "T". However, studies of other intact antibodies by crystallography and NMR spectroscopy indicate that there is no preferred orientation of the arms. Because the two Fab arms can move relative both to each other and to Fc, the two antigen-binding sites at the tips of the Fab arms can move to bind antigenic determinants separated by different distances and in different orientations, and the Fc region can move relative to the Fab to facilitate interaction with effector molecules that bind Fc.

Structures of MHC molecules have provided insights into the molecular mechanisms of T-cell activation

Few protein crystal structures have had such a direct impact in biology as that of the human class I major histocompatibility complex (MHC) molecule, **human lymphocyte antigen A2 (HLA-A2)**, determined in 1987 by Pamela Björkman and others in Don Wiley's group at Harvard University. MHC molecules are crucial to the activation of T cells, which play a major role in immune defense. Many years of detailed molecular genetic analysis had established that T-cell activation is initiated when the antigen receptors on the surface of these cells come into contact with foreign peptide antigen bound to MHC molecules on the surface of antigen-presenting cells. Several major questions arose from these observations. How can a few different MHC molecules recognize and bind an almost unlimited number of antigens? How are these antigens presented to T-cell receptors so that they discriminate between self and nonself? Do the T cells see antigen and MHC molecules, which are firmly attached to the surface of the antigen-presenting cells, as a single structure or as two separate entities?

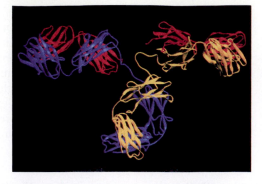

Figure 15.17 The three-dimensional structure of an intact IgG. Hinge regions connecting the Fab arms with the Fc stem are relatively flexible, despite the presence of disulfide bonds in this region linking the heavy and light chains. Carbohydrate residues that bridge the two C_{H2} domains are not shown. (Courtesy of A. McPherson and L. Harris, *Nature* 360: 369–372, 1992, by copyright permission of Macmillan Magazines Limited.)

The answering of these questions began with the structure determination of the human class I MHC protein HLA-A2 (see Figure 15.18a). Subsequently, as methods became available for producing MHC proteins in recombinant expression systems and loading them *in vitro* with defined peptides, structures were determined for a class II MHC protein, and for single peptide complexes of both class I and class II proteins. These structures elegantly demonstrated that T cells recognize antigens as peptides bound in a groove in the MHC protein, and that peptide and MHC are recognized as a single surface. They have also helped to answer some important questions about MHC polymorphism and T-cell recognition and activation.

MHC molecules are composed of antigen-binding and immunoglobulin-like domains

Class I MHC molecules are plasma membrane proteins expressed in all cells and are composed of two polypeptide chains, a heavy chain of about 360 residues that spans the membrane bilayer and a noncovalently attached light chain of 99 residues. Class I MHC molecules present antigen to cytotoxic or "killer" T cells that destroy infected cells. The light chain, which is called β_2 microglobulin (β_2m), is homologous to the immunoglobulin constant domains and has essentially the same three-dimensional structure. The extracellular portion of the heavy chain is divided into three domains, $\alpha1$, $\alpha2$, and $\alpha3$, each about 90 amino acids long. The carboxy-terminal domain $\alpha3$ also shows sequence and structural similarities to immunoglobulin constant domains, whereas $\alpha1$ and $\alpha2$ have very different structures. For structural analysis by x-ray crystallography, the extracellular portion of HLA-A2, composed of $\alpha1$, $\alpha2$, $\alpha3$, and β_2m, was removed from the cells by proteolytic cleavage of a short flexible linker region that connects $\alpha3$ to the transmembrane domain.

The class I MHC structure is divided into two globular regions (Figure 15.18). One region is built up from the two immunoglobulin-like domains $\alpha3$

Figure 15.18 (a) Schematic representation of the path of the polypeptide chain in the structure of the class I MHC protein HLA-A2. Disulfide bonds are indicated as two connected spheres. The molecule is shown with the membrane proximal immunoglobulin-like domains ($\alpha3$ and β_2m) at the bottom and the polymorphic $\alpha1$ and $\alpha2$ domains at the top. (b) The domain arrangement in class I and class II MHC proteins. The domain structures of the MHC class II molecule are similar to those of the class I molecule shown in (a). [(a) Adapted from P.J. Björkman et al., *Nature* 329: 506–512, 1987.]

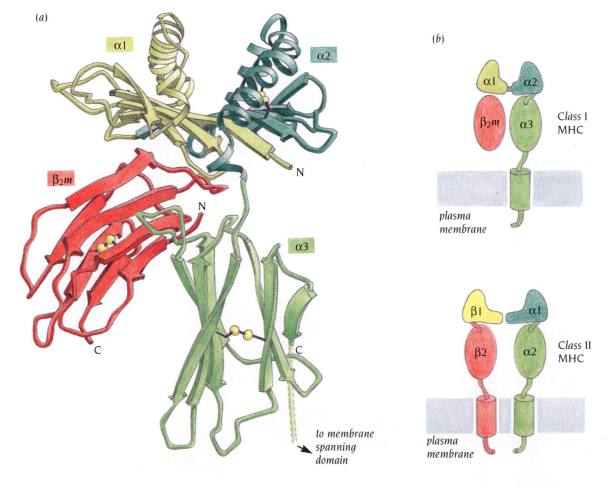

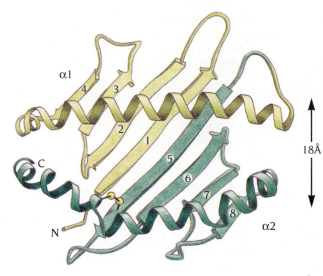

Figure 15.19 Schematic representation of the peptide-binding domain of a class I MHC protein. The α1 and α2 domains are viewed from the top of the molecule, showing the empty antigen-binding site as well as the surface that is contacted by a T-cell receptor. (Adapted from P.J. Björkman et al., *Nature* 329: 506–512, 1987.)

and β_2m. Since α3 is attached to the membrane in the intact HLA molecule, this region is closest to the cell surface. These domains are packed together through their four-stranded β sheets as for the pairing of the constant domains C_{H1}–C_L and C_{H3}–C_{H3} in IgG, but the relative orientation of the domains and the details of the packing are different. The second region, which contains the antigen binding site and presumably faces away from the membrane surface, is composed of domains α1 and α2. Each domain has a similar simple structure: starting from the N-terminus each region of the chain forms four up-and-down antiparallel β strands followed by a helical region across the β strands on one side of the β sheet. The two domains associate in such a way that their β sheets are hydrogen-bonded to each other in an antiparallel fashion. By this association, the structure of the complete region has a "floor" of a continuous eight-stranded antiparallel β sheet (Figure 15.19). This floor sits on top of and is stabilized by the immunoglobulin-like domains α3 and β_2m. Two helical regions are above the floor almost parallel to each other, and separated by a large distance, about 18 Å from center to center. A large crevice that faces the solution is thus formed with the floor forming its bottom and the helices its sides. This crevice is the antigen-binding site.

Recognition of antigen is different in MHC molecules compared with immunoglobulins

In the first structures determined for class I MHC proteins, the antigen binding site was occupied by a heterogeneous mixture of tightly bound peptides that copurified with the MHC. Subsequent structures for complexes with single, defined peptides have provided much information about the molecular details of peptide binding, which appears to be similar for all class I MHC proteins. Peptides bind with their N- and C-termini buried in the peptide-binding site, and their central regions exposed and bulging out from the cleft (Figure 15.20a). Sets of residues conserved in all class I MHC proteins form hydrogen bonds with these buried termini (Figure 15.20b). Peptides of 7 to 11 residues bind in this way to class I MHC, with varying conformations in the bulged region. Some of the side chains of the peptide are bound into small pockets or subsites within the overall binding cleft. These side chains are important to the overall peptide-binding affinity, while residues with side chains that project away from the MHC do not contribute substantially to the binding. In many cases the pockets accommodate the penultimate and the last residues of the peptide, but some class I MHC proteins have pockets into which central residues of the peptide bind.

The genes for MHC molecules, unlike immunoglobulin genes, do not undergo rearrangements to create structural diversity. The β_2m light chain is invariant, but the class I MHC heavy chain is the most genetically polymorphic

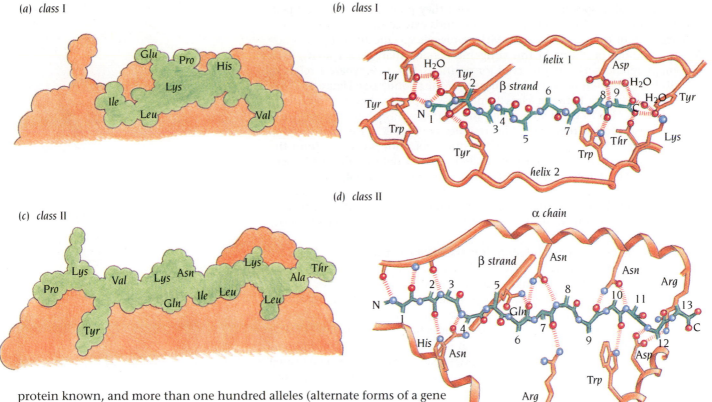

(a) class I

(b) class I

(c) class II

(d) class II

α chain

β chain

Figure 15.20 Peptide binding by MHC class I and class II molecules. (a) A cross-section of a peptide antigen (green) bound to pockets in the class I molecule (orange). (b) Hydrogen bonds (red) form between the ends of the bound peptide and conserved residues of the class I molecule (orange). (c) Class II molecules bind longer peptides than class I molecules, with the ends of the peptide extending beyond the peptide-binding site. (d) Hydrogen bonds between bound peptide and the class II molecule occur along the length of the binding site, also in contrast to the case with class I complexes.

protein known, and more than one hundred alleles (alternate forms of a gene within the species) have been characterized for the three human class I MHC genes. Because of this polymorphism and the expression of MHC genes from both copies of chromosome 6, any given individual is likely to have a set of six different class I molecules that is almost unique to that individual. Differences among the MHC genes of different (mismatched) individuals account for transplantation rejection, and may contribute to differences in susceptibility to infection and autoimmune disease. Different MHC alleles have different binding preferences, as determined by analysis of the mixtures of endogenously bound peptides. How MHC gene polymorphism influences the selection of peptides by MHC proteins is clear from examination of MHC crystal structures: the polymorphic residues cluster in the side chain binding pockets.

Thus the mechanism of antigen recognition by MHC proteins differs from that of antibodies. Antibodies recognize a wide variety of different molecular species and conformations, and are sensitive to the three-dimensional structure of their antigen. MHC proteins recognize peptide antigens based on their sequence, and force the antigen into an extended conformation that displays many of the peptide side chains to the T-cell receptor (see below). Because MHC proteins almost always recognize peptides, some parts of their binding sites are conserved, while the pockets in the MHC molecule that bind the side chains of peptide antigens can vary between alleles to provide a range of binding specificity.

Peptides are bound differently by class I and class II MHC molecules

Class II MHC proteins are expressed on specialized antigen-presenting cells, such as macrophages or B cells, and are important in presenting antigen to helper T cells that regulate and coordinate the immune response to foreign antigen. Class II MHC proteins have the same basic tertiary structure as class I MHC, but the domains are distributed differently between the α and β subunits (see Figure 15.18b). In class II MHC, the subunits are of equal size, around 320 residues, and each chain folds into one immunoglobulin-like domain and one antigen-binding domain. Class II MHC proteins have a different binding specificity to class I MHC; they prefer to bind longer peptides (greater than 10 residues) with their ends extending from the binding site (see Figure 15.20c). The class II MHC molecule specifically recognizes side chains at more positions in the antigenic peptide than class I MHC, with strong preferences usually observed for positions 1, 4, 6, 9, where 1 refers to an arbitrary position near the N-terminus of the peptide (see Figure 15.20d).

This difference between class I and class II peptide-binding specificity appears to be due to small alterations and changes in the binding site. Additional conserved hydrogen-bonding residues are found in the center of the class II peptide-binding site, and the peptide does not bulge out from the MHC (compare Figures 15.20a and 15.20c). Interestingly, in all class II–peptide complexes determined to date, the peptide adopts a polyproline type II-like structure similar to polypeptides bound to SH3 domains as described in Chapter 13. This conformational homogeneity is in dramatic contrast to the conformational flexibility observed for peptides bound to class I MHC proteins.

The differences in peptide-binding specificity between short peptides with defined length for class I MHC and longer peptides for class II MHC may be related to the different mechanisms and intracellular locations for loading class I and class II MHC proteins with peptides. As membrane glycoproteins, both class I and class II MHC proteins are synthesized by membrane-bound ribosomes and assemble on the endoplasmic reticulum membrane. Class I MHC proteins generally bind peptides that have been derived from cytoplasmic proteins and transported into the endoplasmic reticulum by a dedicated transmembrane peptide pump (a mechanism used to alert the immune system to intracellular pathogens, for example, viruses that replicate in the cytoplasm). In contrast, class II MHC proteins move from the endoplasmic reticulum to a specialized endosomal compartment where they bind peptides derived from cell surface and extracellular proteins, a mechanism that detects extracellular or microsomal pathogens such as bacteria. These two mechanisms for peptide generation and loading, together with the ability of any particular MHC protein to bind many different peptide sequences, allow MHC molecules to display antigens derived from all types of cellular pathogens to the immune system.

T-cell receptors have variable and constant immunoglobulin domains and hypervariable regions

T-cell receptors (TCR) are heterodimeric transmembrane glycoproteins found exclusively in T cells, with extracellular domains that closely resemble antibody Fab structures. Each of the TCR α and β chains forms half of an extracellular antigen-binding domain, and in addition has one transmembrane

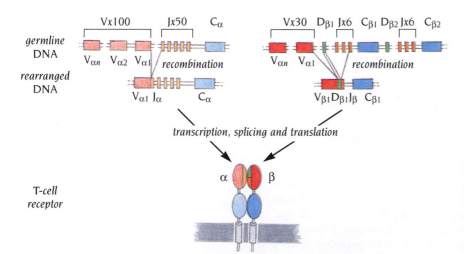

Figure 15.21 T-cell receptor α and β genes undergo rearrangments much like immunoglobin genes (see Figure 15.5). The $V_\alpha J_\alpha$ and $V_\beta D_\beta J_\beta$ regions correspond to the CDR loops. Nucleotides can also be added or deleted from the junctions, adding further amino acid sequence diversity.

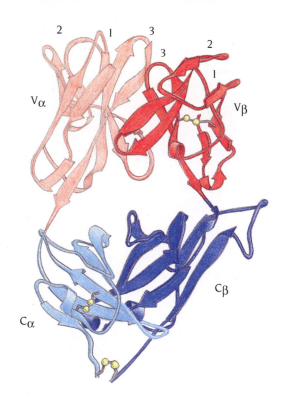

Figure 15.22 T-cell stucture shown as a ribbon diagram. The antigen-binding site is formed by CDR loops (labeled 1 to 3) from the V_α and V_β domain, as for antibodies. A disulfide bond (yellow) links the two peptide chains. (Courtesy of A.I. Wilson.)

domain and a short cytoplasmic tail. Like the antibody light chains, TCR α and β chains both have one variable and one constant domain as well as hypervariable regions that are involved in antigen recognition. In the T-cell plasma membrane, TCR assembles with other proteins to form a signaling complex that transmits information across the T-cell membrane in response to antigen binding. TCR $\alpha\beta$ genes are rearranged during T cell development much as are antibody genes, except that the number of J genes is larger, and nucleotides are added to both V_α–D_α–J_α and V_β–J_β junctions (Figure 15.21). Unlike antibody genes, however, TCR genes are not subject to somatic mutation after rearrangement.

In 1996, crystal structures were determined for a murine TCR (by the group of Ian Wilson at Scripps Institute) and for a human TCR–MHC–peptide complex (by the group of Don Wiley at Harvard University), which together with earlier structures for isolated TCR α and β chains (by Roy Mariuzza and coworkers, University of Maryland), provided a coherent view of TCR structure. As predicted much earlier by Cyrus Chothia on the basis of homology with immunoglobulins, each TCR chain forms two immunoglobulin-like domains (Figure 15.22). The membrane proximal constant domains C_α and C_β are similar to C_{H1}, C_{H2}, and C_{H3}, and the membrane distal variable domains V_α and V_β are similar to V_H and V_L. Packing between the V_α and V_β domains and between the C_α and C_β domains is similar to that found in antibody Fab fragments. A single disulfide bond at the end of the protein links the C_α and C_β domains, and is equivalent to the bond that links C_{H1} with C_L. Finally, the antigen combining site is formed by hypervariable CDR loops in the variable domains, as it is in antibodies.

In spite of these similarities, some alterations to the basic immunoglobulin fold are found in the TCR subunits and these may be important for function. The C″ strand in the V_β domain is switched from the inner G-F-C-C′ sheet to the outer A-B-E-D sheet, resulting in a different shape for the CDR2 loop. There is also an unusual insertion in the C_β F-G connection that places a large loop near the $V_\beta C_\beta$ interface, which extends away from the protein surface (see the right edge of Figure 15.22). This loop may well be important in interactions with other components of the overall TCR complex. Finally, the C_α domain is unusual and structurally divergent from other Ig domains.

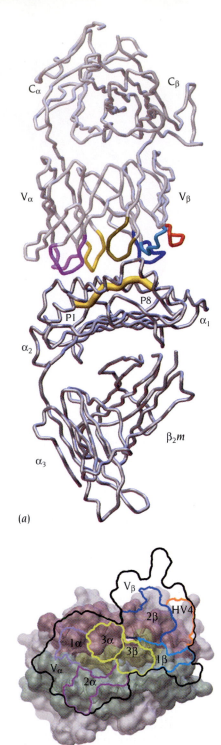

Figure 15.23 Interactions between the class I MHC–peptide complex with the T-cell receptor. (a) The T-cell receptor (top) binds to the MHC–peptide complex with its hypervariable CDR loops effectively burying the eight-residue foreign peptide (yellow, with the N- and C-terminal residues labeled P1 and P8, respectively). The TCR α subunit loops 1–3 are colored light purple (CDR1α), dark purple (CDR2α), and yellow (CDR3α); the β chain loops 1–3 are colored light blue (CDR2β), dark blue (CDR3β), and green (CDR3β). The β subunit hypervariable loop that has no counterpart in immunoglobulins, HV4, is shown in red. (b) A finger-print of how the TCR hypervariable loops contact the MHC–peptide complex, as seen looking down onto the MHC surface as oriented in (a). The TCR sits diagonally across the peptide-binding site, contacting both the MHC molecule (gray and green space-filling model) and its bound peptide (pale green space-filling model). (Courtesy of A.I. Wilson.)

MHC–peptide complexes are the ligands for T-cell receptors

The physiological role of TCR is to bind MHC–peptide complexes presented by antigen-presenting cells. During T cell development, cells bearing TCR that bind to MHC complexed with self peptides are removed from the immune system; thus mature T cells carry only receptor that can bind to MHC proteins with foreign antigens bound in their peptide binding groove. The structures of several class I MHC–peptide complexes bound to TCR $\alpha\beta$ have been determined. These all show the MHC and TCR proteins interacting diagonally, with the TCR contacting the peptide and both helices of the MHC protein such that the peptide is virtually buried in the interface (Figure 15.23a).

In the TCR, packing of the V_α and V_β domains brings together CDR1, CDR2, and CDR3 to form a relatively flat combining site well-suited to interact with the surface of the MHC–peptide complex. Among these three loops, CDR3 is the largest and most variable as it is in immunoglobulins, and appears strongly to determine peptide specificity. In the TCR–MHC–peptide complex, the CDR loops lie over the MHC peptide-combining site, with the CDR3 loops from V_α and V_β (CDR3α and CDR3β) packed together over the center of the bound peptide (see Figure 15.23b). CDR1α and CDR2α extend diagonally towards the N-terminal end of peptide, and also contact MHC residues from the α_2 subunit helix. CDR1β and CDR2β are oriented correspondingly toward the peptide's C-terminus and the MHC α_1 subunit helix. Thus, all the CDR loops of the TCR can potentially contact the MHC molecule and/or bound peptide. Small changes in the CDR loops of the receptor, and in exposed regions of the peptide and MHC protein, have been observed upon complex formation, but no large conformational changes that could account for signal transduction have been seen.

To date, there has not been a reported structure determination of a class II MHC–peptide–TCR complex. However, T-cell receptors that recognize class I MHC–peptide complexes and those that recognize class II MHC–peptide complexes utilize the same set of V_α and V_β genes, and the principal feature that defines the site of class I MHC–TCR interaction, the cleft formed by the α_1 and α_2 subunits of the MHC molecule, is shared by both class I and class II proteins (see Figure 15.20b and c). Thus, the general mode of class II MHC interaction with TCR is expected to be similar to that determined for class I MHC.

Many cell-surface receptors contain immunoglobulin-like domains.

Many other cell-surface proteins involved in immunological recognition utilize immunoglobulin-like domains as structural elements. Immunoglobulin domains have been classified into five types, namely V (like antibody variable

(a)

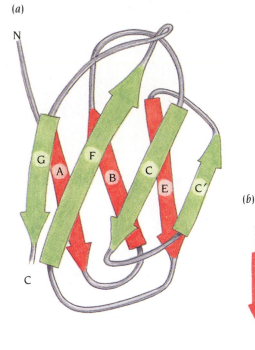

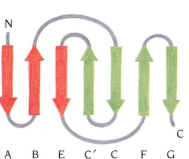

(b)

A   B   E   C′   C   F   G

Figure 15.24 Ribbon diagram (a) and topology diagram (b) of the fibronectin type III domain, which is composed of a three-stranded and a four-stranded β sheet packed together as a compressed barrel.

domains), C1 (like antibody constant domains) C2 (strand-switched variant of C1), I, and E, based on structural and sequence homology. A close cousin of the immunoglobulin domain is the fibronectin type III domain, which has a β-sandwich fold closely related to the immunoglobulin fold (Figure 15.24). Hundreds of proteins appear to be made up of combinations of these domains in an essentially modular fashion. Interaction sites on immunoglobulin domains are sometimes found in the CDR loops or equivalent regions, but can also involve other parts of the domain. In many cases, the immunoglobulin domains appear to play a structural role not directly involved in ligand binding.

An interesting adaptation is found in CD4, a T-cell coreceptor that binds class II MHC along with the TCR during T cell activation. In addition to its role in T cell activation, CD4 also serves as the major binding site for the human immunodeficiency virus (HIV), through interactions with the HIV surface glycoprotein receptor, gp120. CD4 is a monomer with four extracellular Ig domains. The structure of the first and second extracellular domains, which include the binding sites for MHC and for gp120, was determined simultaneously in 1990 by the groups of Stephen Harrison at Harvard Univerisity, and Wayne Hendrickson at Columbia University. The first domain is a V-type immunoglobulin with strand A completely switched to the G-F-C-C′-C″ sheet: the second domain is a conventional C2 type. At the junction of the first and second domains, the last strand from one domain continues uninterrupted to form the first strand of the next domain (Figure 15.25). Presumably this arrangement provides rigidity to the molecule. A similar arrangement has also been observed between the third and fourth domains of CD4. Sharing β strands between domains is one way to confer rigidity to a modular protein.

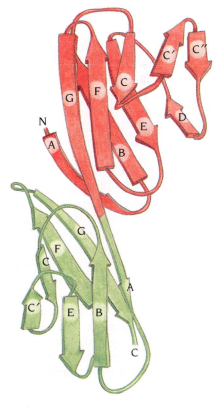

Figure 15.25 Ribbon digram for the first two domains of CD4. The G strand of domain 1 is contiguous with the A strand of domain 2. The C″ ridge has been implicated in binding to the HIV surface protein, gp120.

Conclusion

IgG antibody molecules are composed of two light chains and two heavy chains joined together by disulfide bonds. Each light chain has one variable domain and one constant domain, while each heavy chain has one variable and three constant domains. All of the domains have a similar three-dimensional structure known as the immunoglobulin fold. The Fc stem of the molecule is formed by constant domains from each of the heavy chains, while two Fab arms are formed by constant and variable domains from both heavy and light chains. The hinge region between the stem and the arms is flexible and allows the arms to move relative to each other and to the stem.

The constant domain has a stable framework structure composed of two antiparallel sheets comprising seven β strands, four in one sheet and three in the other. The variable domains have a similar framework structure but comprising nine β strands, five in one sheet and four in the other. The three hypervariable regions are in loops at one end of the variable domain. The variable domains from the heavy and light chains associate through their five-stranded β sheets to form a barrel with the hypervariable loop regions from both domains close together at the top of the barrel.

By this arrangement the hypervariable loops can form antigen-binding sites of different shapes depending on the size and sequence of the loops. Haptens induce antibodies with binding cavities, and protein antigens induce antibodies with extended binding surfaces, which contain grooves or pockets for protruding antigen residues. In general, no gross structural alterations occur in the antibody or antigen when they bind together, although a variety of small conformational changes may be seen. The third hypervariable region, CDR3, of the heavy chain is in the center of the interaction area and makes a large number of contacts with the antigen. This region corresponds to the D gene segment, which, through junctional diversity and addition of nucleotides during the recombination events, has the largest sequence variation of the hypervariable regions.

Class I and class II MHC molecules bind peptide antigens and present them at the cell surface for interaction with receptors on T cells. The extracellular portion of these molecules consists of a peptide-binding domain formed by two helical regions on top of an eight-stranded antiparallel β sheet, separated from the membrane by two lower domains with immunoglobulin folds. These domains are differently disposed between the two protein subunits in class I and class II molecules.

Class I MHC molecules bind short peptides of 8–11 residues, with the peptide's termini buried within the protein and the central regions exposed for interactions with the T-cell receptor. Conserved residues interact with main chain groups of the peptide's termini, and two or three peptide side chains at particular positions bind into pockets within the overall binding groove. These pockets vary between MHC class I alleles, and define the peptide-binding specificity for different class I MHC molecules. Class II MHC molecules bind longer peptides, the ends of which are exposed and extend beyond the peptide-binding pocket. Conserved residues in the MHC molecules interact with main-chain atoms along the peptide. Class II peptide-binding motifs generally bind their ligands at more positions than those of class I. Both class I and class II MHC proteins are extremely polymorphic, with hundreds of alleleic variants for some MHC genes.

T-cell receptors are cell surface proteins with extracellular domains composed of two similarly sized subunits, each with a variable and a constant immunoglobulin-like domain, arranged similarly to the Fab arms of antibodies. Like antibodies, T-cell receptors have an antigen-binding site formed by hypervariable loop regions from both variable domains, with the CDR3 loops playing a key role in antigen recognition. Unlike antibodies, T-cell receptors recognize antigen only when it is bound in the peptide-binding site of an MHC protein. In ternary complexes of T-cell receptor, MHC molecule, and peptide antigen that have been determined to date, the T-cell receptor lies diagonally across the top of the MHC peptide-binding site, and contacts exposed regions of the peptide as well as the MHC helices.

The immune system can use a few different types of protein modules to bind to an almost infinite number of antigens, achieving binding diversity by using variable loops or pockets supported by a constant framework. The immunoglobulin-like fold often forms the basis of this framework. If a particular type of antigen is to be recognized, the binding site will contain conserved regions that contact invariant portions of the antigen; if the antigens are diverse, the binding regions will be variable. For antibodies and T-cell receptors, the required variability exceeds the number of genes in the genome, and gene recombination and mutation are utilized to generate diversity.

Selected readings

General

Alzari, P.N., Lascombe, M.-B., Poljak, R.J. Three-dimensional structure of antibodies. *Annu. Rev. Immunol.* 6: 555–580, 1988.

Chothia, C. Jones, E.Y. The molecular structure of cell adhesion molecules. *Annu. Rev. Biochem.* 66: 823–862, 1997.

Colman, P.M. Structure of antibody-antigen complexes: implications for immune recognition. *Adv. Immunol.* 43: 99–132, 1988.

Davies, D.R., Metzger, H. Structural basis of antibody function. *Annu. Rev. Immunol.* 1: 87–117, 1983.

Fremont, D.H., Rees, W.A., Kozono, H. Biophysical studies of T-cell receptors and their ligands. *Curr. Opin. Struct. Immunol.* 8: 93–100, 1996.

Kennedy, R.C., Melnick, J.L., Dreesman, G.R. Anti-idiotypes and immunity. *Sci. Am.* 255(1): 40–63, 1986.

Leder, P. The genetics of antibody diversity. *Sci. Am.* 247(5): 72–83, 1982.

Lerner, R.A., Tramontano, A. Catalytic antibodies. *Sci. Am.* 258(3): 42–53, 1988.

Madden, D.R. The three-dimensional structure of peptide–MHC complexes. *Annu. Rev. Immunol.* 13: 587–622, 1995.

Mariuzza, R.A., Phillips, S.E.V., Poljak, R.J. The structural basis of antigen–antibody recognition. *Annu. Rev. Biophys. Biophys. Chem.* 16: 139–159, 1987.

Marrack, P., Kappler, J. The T cell and its receptor. *Sci. Am.* 254(2): 28–37, 1986.

Raghavan M., Bjorkman P.J. Fc receptors and their interactions with immunoglobulins. *Annu. Rev. Cell Dev. Biol.* 12: 181–220, 1996.

Rammensee, H.-G. Chemistry of peptides associated with class I and class II molecules. *Curr. Opin. Struct. Immunol.* 7: 85–96, 1995.

Riechmann, L., et al. Reshaping human antibodies for therapy. *Nature* 332: 323–327, 1988.

Shokat, K.M., Schultz, P.G. Catalytic antibodies. *Annu. Rev. Immunol.* 8: 335–364, 1990.

Stern, L.J., Wiley, D.C. Antigenic peptide binding by class I and class II histocompatibility proteins. *Structure* 2: 245–251, 1994.

Tonegawa, S. Somatic generation of antibody diversity. *Nature* 302: 575–581, 1983.

Wilson, I.A., Bjorkman, P.J. Unusual MHC-like molecules: CD1, Fc receptor, the hemochromatosis gene product, and viral homologs. *Curr. Opin. Struct. Biol.* 10: 67–73, 1998.

Wilson, I.A., Garcia, K.C. T-cell receptor structure and TCR complexes. *Curr. Opin. Struct. Biol.* 7: 839–848, 1997.

Wilson, I.A., Stanfield, R.L. Antibody–antigen interactions: new structures and new conformational changes. *Curr. Opin. Struct. Biol.* 4: 857–867, 1994.

Specific structures

Amit, A.G., et al. Three-dimensional structure of an antigen–antibody complex at 2.8 Å resolution. *Science* 233: 747–753, 1986.

Bjorkman, P.J., et al. Structure of the human class I histocompatibility antigen, HLA-A2. *Nature* 329: 506–512, 1987.

Braden, B.C., et al. Crystal structure of an Fv–Fv idiotope–anti-idiotope complex at 1.9 Å resolution. *J. Mol. Biol.* 264: 137–151, 1996.

Brown, J.H., Wiley, D.C. The three-dimensional structure of the human class II histocompatibility antigen HLA-DR1. *Nature* 364: 33–39, 1993.

Bruccoleri, R.E., Haber, E., Novotny, J. Structure of antibody hypervariable loops reproduced by a conformational search algorithm. *Nature* 335: 564–568, 1988.

Burnmeister, W.P., Gastinel, L.N., Sinister, N.E., Blum M.L., Bjorkman, P.J. Crystal structure of the complex of rat neonatal Fc receptor with Fc. *Nature* 372: 379–383, 1994.

Chothia, C., et al. Conformations of immunoglobulin hypervariable regions. *Nature* 343: 877–883, 1989.

Chothia, C., et al. Domain association in immunoglobulin molecules. The packing of variable domains. *J. Mol. Biol.* 186: 651–663, 1985.

Chothia, C., Boswell, D.R., Lesk, A.M. The outline structure of the T-cell αβ receptor. *EMBO J.* 7: 3745–3755, 1988.

Colman, P.M., et al. Three-dimensional structure of a complex of antibody with influenza virus neuraminidase. *Nature* 326: 358–363, 1987.

Dall'Acqua, W., Goldman, E.R., Eisenstien, E., Mariuzza, R.A. A mutational analysis of the binding of two different proteins to the same antibody. *Biochemistry* 25: 9667–9676, 1996.

Deisenhofer, J. Crystallographic refinement and atomic models of a human Fc fragment and its complex with fragment B of protein A from *Staphylococcus aureus* at 2.9 and 2.8 Å resolution. *Biochemistry* 20: 2361–2369, 1981.

Fields, B.A., et al. Crystal stucture of the V_α domain of a T-cell antigen receptor. *Science* 270: 1821–1824, 1995.

Fields, B.A., et al. Molecular basis of antigen mimicry by an anti-idiotope. *Nature* 374: 739–742, 1995.

Fremont, D.H., Hendrickson, W.A., Marrack, P., Kappler, J. Structures of an MHC class II molecule with covalently bound single peptides. *Science* 272: 1001–1004, 1996.

Fremont, D.H., Matsumura, M., Stura, E.A., Peterson, P.A., Wilson, I.A. Crystal structures of two viral peptides in complex with murine MHC class I H2-K^b. *Science* 257: 919–927, 1992.

Fremont, D.H., Monnaie, D., Nelson, C.A., Hendrickson, W.A., Unanue, E.R. Crystal structure of I-A^k in complex with a dominant epitope of lysozyme. *Immunity* 8: 305–317, 1998.

Furey, W., Jr., et al. Structure of a novel Bence-Jones protein (Rhe) fragment at 1.6 Å resolution. *J. Mol. Biol.* 167: 661–692, 1983.

Garboczi, D.N., Ghosh, P., Utz, U., Fan, Q.R., Biddison, W.E., Wiley, D.C., Structure of the complex between human T-cell receptor, viral peptide, and HLA-A2. *Nature* 384: 134–141, 1996.

Garcia, K.C., et al. An αβ T cell receptor structure at 2.5 Å and its orientation in the TCR–MHC complex. *Science,* 274: 209–219, 1996.

Guo, H.-C., et al. Comparison of the P2 specificity pocket in three human histocompatibility antigens: HLA-A*6801, HLA-A*0201, and HLA-B*2705. *Proc. Natl. Acad. Sci. USA* 90: 8053–8057, 1993.

Harris, L.J., Larson, S.B., Hasel, K.W., McPherson, A. Refined structure of an intact IgG2a monoclonal antibody. *Biochemistry* 36: 1581–1597, 1997.

Huber, R., et al. Crystallographic structure studies of an IgG molecule and an Fc fragment. *Nature* 264: 415–420, 1976.

Janda, K.D., Benkovic, S.J., Lerner, R.A. Catalytic antibodies with lipase activity and R or S substrate selectivity. *Science* 244: 437–440, 1989.

Jardetzky, T.S., Wiley, D.C. Crystallographic analysis of endogenous peptides associated with HLA-DR1 suggests a common, polyproline II-like conformation for bound peptides. *Proc. Natl. Acad. Sci. USA* 93: 734–728, 1996.

Lascombe, M.-B., et al. Three-dimensional structure of Fab R 19.9, a monoclonal murine antibody specific for the p-azobenzenearsonate group. *Proc. Natl. Acad. Sci. USA* 86: 607–611, 1989.

Leahy, D.J., Hendrickson, W.A., Aukhil, A., Erickson, H.P. Structure of a fibronectin type II domain from tenascin phased by MAD analysis of the selenomethionyl protein. *Science* 258: 987–991, 1992.

Madden, D.R., Garboczi, D.N., Wiley, D.C. The antigenic identity of peptide–MHC complexes: a comparison of the conformation of five viral peptides presented by HLA-A2. *Cell* 75: 693–708, 1993.

Madden, D.R., Gorga, J.C., Strominger, J.L., Wiley, D.C. The three-dimensional structure of HLA-B27 at 2.1 Å resolution suggests a general mechanism for tight peptide binding to MHC. *Cell* 70: 1035–1048, 1992.

Murthy, V.L., Stern, L.J. The class II MHC protein HLA-Dr1 in complex with an endogenous peptide: implications for the structural basis of the specificity of peptide binding. *Structure* 5: 1385–1396, 1997.

Novotny, J., et al. Molecular anatomy of the antibody binding site. *J. Biol. Chem.* 258: 14433–14437, 1983.

Padlan, E.A., et al. Structure of an antibody–antigen complex. Crystal structure of the HyHEL–10 Fab–lysozyme complex. *Proc. Natl. Acad. Sci. USA* 86: 5938–5942, 1989.

Poljak, R.J., et al. Three-dimensional structure of the Fab fragment of a human immunoglobulin at 2.8 Å resolution. *Proc. Natl. Acad. Sci. USA* 70: 3305–3310, 1973.

Pollack, S.J., Jacobs, J.W., Schultz, P.G. Selective chemical catalysis by an antibody. *Science* 234: 1570–1573, 1986.

Ryu, S.-E., et al. Crystal structure of an HIV-binding recombinant fragment of human CD4. *Nature* 348: 419–426, 1990.

Saper, M.A., Bjorkman, P.J., Wiley, D.C. Refined structure of the human histocompatibility antigen HLA-A2 at 2.6 Å resolution. *J. Mol. Biol.* 219: 277–319, 1991.

Satow, Y., et al. Phosphorylcholine-binding immunoglobulin Fab McPC603. An x-ray diffraction study at 2.7 Å. *J. Mol. Biol.* 190: 593–604, 1986.

Scott, C.A., Peterson. P.A., Teyton, L., Wilson, I.A. Crystal structures of two I-A^d-peptide complexes reveal that high affinity can be achieved without large anchor residues. *Immunity* 8: 319–329, 1998.

Segal, D.M., et al. The three-dimensional structure of a phosphorylcholine-binding mouse immunoglobulin Fab and the nature of the antigen-binding site. *Proc. Natl. Acad. Sci. USA* 71: 4298–4302, 1974.

Sheriff, S., et al. Three-dimensional structure of an antibody–antigen complex. *Proc. Natl. Acad. Sci. USA* 84: 8075–8079, 1987.

Stanfield, R.L., et al. Crystal structures of an antibody to a peptide and its complex with peptide antigen at 2.8 Å. *Science* 248: 712–719, 1990.

Stanfield, R.L., et al. Major antigen-induced domain rearrangements in an antibody. *Structure* 1: 83–93, 1993.

Stern, L.J., Brown, J.H., Jardetzky, T.S., Gorga, J.C., Urban, R.G., Strominger, J.L., Wiley, D.C. Crystal structure of the human class II MHC protein HLA-DR1 complexed with an influenza virus peptide. *Nature* 368: 215–221, 1994.

Wang, J., et al. Atomic structure of a fragment of human CD4 containing two immunoglobulin-like domains. *Nature* 348: 411–418, 1990.

Wang, J., et al. Atomic structure of an αβ T-cell receptor (TCR) heterodimer in complex with an anti-TCR Fab fragment derived from a mitogenic antibody. *EMBO J.* 17: 10–26, 1988.

Zheng, Z.-H., Cataño, A. R., Segelke, B.W., Stura, E.A., Peterson, P.A., Wilson, I.A. Crystal structure of mouse CD1: an MHC-like fold with a large hydrophobic binding groove. *Science* 277: 339–345, 1997.

The Structure of Spherical Viruses

16

All viruses depend for their existence on their ability to infect cells, causing them to make more virus particles. The infected cells generally die in the process. Even though the infection cycles of different types of viruses vary in detail, they all follow the same basic pattern. First, the viruses deliver their nucleic acid genome into the host cell, often in association with viral proteins; the mechanism of delivery varies depending on the virus, but the end result is the same. Second, the host cell's biosynthetic machinery is subverted for the replication, transcription, and translation of the viral genes at the expense of cellular gene expression. Finally, progeny virus particles assemble in the infected cell and by one route or another leave it to infect a fresh host cell. Individual viruses can only infect a restricted range of hosts, and the host organisms usually have defense mechanisms, ranging from restriction enzymes in bacteria to the immune system of vertebrates. The combination of restricted host ranges and host defense mechanisms keeps in check the extent of viral infections.

Viruses come in many different sizes, shapes, and compositions. They all have a nucleic acid genome, but this is DNA in some viruses and RNA in others, double stranded in some but single stranded in others, a single molecule in some but segmented into several different molecules in others. The nucleic acid genome is always surrounded by a protein shell, a **capsid**. Some viruses also have a lipid bilayer membrane, an **envelope**, enclosing the capsid; one example is the influenza viruses discussed in Chapter 5. For animal viruses the distinction between nonenveloped and enveloped viruses is important because it corresponds to different strategies for entry into and exit from a cell. Enveloped viruses bud out through a cellular membrane, acquiring the lipid bilayer in the process. Their entry into cells involves a membrane fusion event. Nonenveloped viruses have evolved more complex strategies and some have elaborately differentiated structures. For example, bacteriophage T4, which must penetrate the bacterial cell wall at specific sites and inject its DNA into the cell, is seen in electron micrographs to resemble a hypodermic syringe.

A nucleic acid can never code for a single protein molecule that is big enough to enclose and protect it. Therefore, the protein shell of viruses is built up from many copies of one or a few polypeptide chains. The simplest viruses have just one type of capsid polypeptide chain, which forms either a rod-shaped or a roughly spherical shell around the nucleic acid. The simplest such viruses whose three-dimensional structures are known are plant and insect viruses: the rod-shaped tobacco mosaic virus, the spherical satellite tobacco necrosis virus, tomato bushy stunt virus, southern bean mosaic virus, and flock house virus. The mammalian parvoviruses also have a simple, single-subunit design. The mammalian picornaviruses have a spherical shell built up of four different polypeptide chains. The structures of several picornaviruses are known: common cold virus, Mengo virus, poliovirus, and foot-and-mouth disease virus. The largest virus particles whose structures have been determined are SV40 and polyomavirus, which are closely related double-stranded DNA viruses, and the inner capsid particle of bluetongue virus, a double-stranded RNA virus. These viruses all infect species of mammals. Schematic diagrams illustrating the size and shape of some of the viruses mentioned above are given in Figure 16.1.

The protein shells, or capsids, of the simple spherical animal and plant viruses are made up of symmetrically arranged protein subunits. This symmetry of subunits has been exploited by crystallographers to facilitate the determination of the structure of the capsids at atomic resolution by x-ray methods. Consequently, these viruses are the largest aggregates of biological macromolecules whose structures have been determined at high resolution. The nucleic acid genome, which is inside the protein shell of the virus, lacks symmetry and is unlikely to have a unique structure. However, we are beginning to understand packaging signals and the unique protein–RNA or protein–DNA interactions that initiate virus particle assembly. As we will see later, the unique interactions in the interior of an RNA phage have been seen crystallographically by making a complex of the MS2 capsid with 90 identical RNA fragments.

Detailed knowledge of the structure of viral capsids alone is valuable for both practical and theoretical reasons. In principle, it should be possible to exploit this information for the rational design of antiviral drugs. For instance, the entry of mammalian viruses into cells depends upon specific recognition by the viral capsid of proteins or carbohydrates that occur on the surface of mammalian cells. This interaction thus plays a crucial part in determining the viral host range, and compounds that interfere with it should block infection. Understanding the capsid structure should help in designing such compounds. Because viral binding sites for surface molecules of host cells are also a target for protective antibodies, the immune system of the host has in all probability played a major role in the evolution of animal viruses. Detailed knowledge of the surface regions of these viruses should therefore teach us more about the structural adaptations that viruses have evolved to evade host defences.

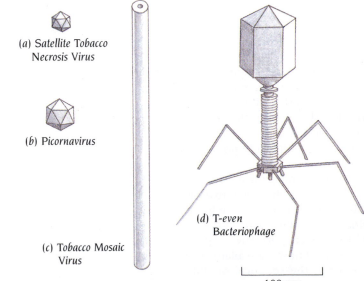

(a) Satellite Tobacco Necrosis Virus

(b) Picornavirus

(c) Tobacco Mosaic Virus

(d) T-even Bacteriophage

100 nm

Figure 16.1 Viruses vary in size and shape from the simplest satellite viruses (a) that need another virus for their replication to the T-even bacteriophages (d) that have developed sophisticated mechanisms for injecting DNA into bacteria. Four different virus particles are shown to scale.

In this chapter we will examine the construction principles of spherical viruses, the structures of individual subunits and the host cell binding properties of the surface of one of the picornaviruses, the common cold virus.

The protein shells of spherical viruses have icosahedral symmetry

Two basic principles govern the arrangement of protein subunits within the shells of spherical viruses. The first is specificity; subunits must recognize each other with precision to form an exact interface of noncovalent interactions because virus particles assemble spontaneously from their individual components. The second principle is genetic economy; the shell is built up from many copies of a few kinds of subunits. These principles together imply symmetry; specific, repeated bonding patterns of identical building blocks lead to a symmetric final structure.

The first question then is, what types of symmetry can form roughly spherical objects that are built up from identical units. In other words, how can one make a sphere by arranging identical objects symmetrically on its surface?—a problem also faced by producers of soccer balls. This question was answered more than 2000 years ago by the classical Greek mathematicians, who showed that there is only a very limited number of ways to build such objects. Among these, the icosahedron and dodecahedron have the highest possible symmetry, generally called **icosahedral symmetry**, and therefore allow a maximum number of identical objects to form a closed symmetrical shell (Figure 16.2). Do these geometric principles apply to viral structures?

Fortunately, virus particles are large enough to be studied by electron microscopy using negative staining. In the 1950s and 1960s such studies, mainly by Aaron Klug and his collaborators in Cambridge, England, showed that indeed all the spherical viruses examined exhibited icosahedral symmetry. They could also demonstrate from the x-ray diffraction patterns of virus crystals that not only does the external surface obey this symmetry, but so also does the arrangement of subunits that make up this surface. Since icosahedral symmetry is central to an understanding of the architecture of spherical viruses, it will be examined in detail.

The icosahedron has high symmetry

The icosahedron is a roughly spherical object that is built up from 20 identical equilateral triangles. These triangular tiles are arranged side by side in such a way that they enclose the volume inside the icosahedron. Figure 16.3 shows a view of an icosahedron where there are 5 tiles at the top, 5 at the bottom, and 10 in a band around the middle region. These faces are identical: in other words, identical icosahedra are obtained irrespective of which of the vertices is at the top in Figure 16.3a.

The symmetry of the icosahedron is described by the different types of rotations that bring it into self-coincidence (see Figure 16.3). There are 12 vertices, each of which has a fivefold rotation axis that passes through it and through the center. Rotation by a fifth of a revolution about one of these axes brings the icosahedron to an orientation indistinguishable from the starting orientation. In addition, there are 20 threefold symmetry axes through the center of the icosahedron and the middle of each of the 20 faces, and 30 twofold axes through the middle of each of the 30 edges. All this symmetry has important consequences for the way in which the shell of spherical viruses are constructed.

Any symmetric object is built up from smaller pieces that are identical and that are related to each other by symmetry. An icosahedron can therefore be divided into a number of smaller identical pieces called symmetry-related units. Protein subunits are asymmetric objects; hence, a symmetry axis cannot pass through them. The minimum number of protein subunits that can form a virus shell with icosahedral symmetry is therefore equal to

Figure 16.2 The icosahedron (*top*) and dodecahedron (*bottom*) have identical symmetries but different shapes. Protein subunits of spherical viruses form a coat around the nucleic acid with the same symmetry arrangement as these geometrical objects. Electron micrographs of these viruses have shown that their shapes are often well represented by icosahedra. One each of the twofold, threefold, and fivefold symmetry axes is indicated by an ellipse, triangle, and pentagon, respectively.

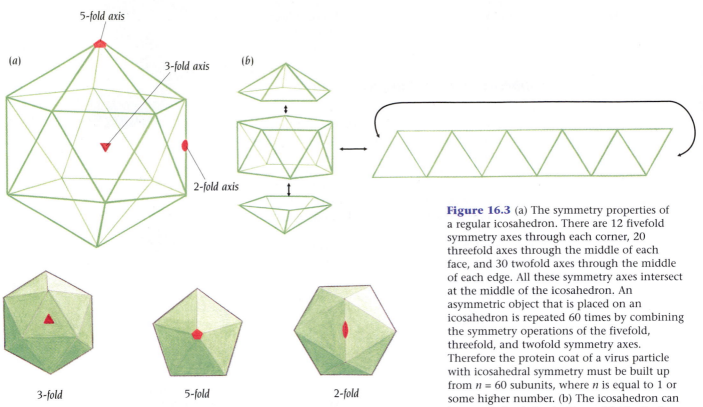

5-fold axis

3-fold axis

2-fold axis

(a)

(b)

3-fold   5-fold   2-fold

Figure 16.3 (a) The symmetry properties of a regular icosahedron. There are 12 fivefold symmetry axes through each corner, 20 threefold axes through the middle of each face, and 30 twofold axes through the middle of each edge. All these symmetry axes intersect at the middle of the icosahedron. An asymmetric object that is placed on an icosahedron is repeated 60 times by combining the symmetry operations of the fivefold, threefold, and twofold symmetry axes. Therefore the protein coat of a virus particle with icosahedral symmetry must be built up from $n = 60$ subunits, where n is equal to 1 or some higher number. (b) The icosahedron can be regarded as being built up of 5 identical equilateral triangular tiles at the top, 5 at the bottom, and 10 in a band around the middle region. (c) The icosahedron viewed along each of its different symmetry axes.

the number of symmetry-related units in an icosahedron. This number can easily be deduced by regarding the 20 triangular tiles on the surface of the icosahedron. These tiles have threefold symmetry, and therefore, by definition, three identical objects are required to form one tile. One obvious way to regard such tiles is shown in Figure 16.4a, which illustrates the division of one such triangle into three equal parts, each of which contains an object. Since there are 20 tiles the total number of units in the icosahedron is $20 \times 3 = 60$. We could equally well have regarded the 12 different corners with fivefold symmetry or the 30 edges with twofold symmetry. Such units are called **asymmetric units**; thus the icosahedron and, by inference, spherical viruses have 60 asymmetric units (Figure 16.4b). Hence at least 60 protein subunits are required to form a virus shell with icosahedral symmetry.

The symmetry properties of an icosahedron are not restricted to the surface but extend through the whole volume. An asymmetric unit is therefore a part of this volume; it is a wedge from the surface to the center of the icosahedron. Sixty such wedges completely fill the volume of the icosahedron.

The simplest virus has a shell of 60 protein subunits

The asymmetric unit of an icosahedron can contain one or several polypeptide chains. The protein shell of a spherical virus with icosahedral symmetry

(a)

(b)

Figure 16.4 The division of the surface of an icosahedron into asymmetric units. (a) One triangular face is divided into three asymmetric units into which an object is placed. These are related by the threefold symmetry axis. (b) Triangular tiles with the three symmetry-related objects have been positioned on the surface of the icosahedron in such orientations that the objects are related by the fivefold and twofold symmetry axes.

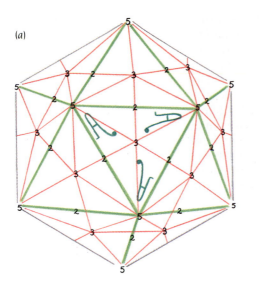

(a)

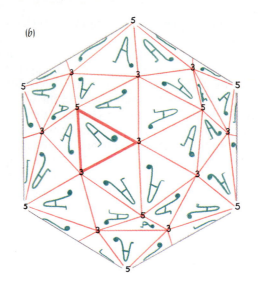

(b)

must therefore have a minimum number of 60 identical units. Such a unit can be a single protein chain, a dimer of two identical protein chains as in the inner shell of bluetongue virus or several different protein chains as in the picornaviruses to be described later. Is there any virus known with only 60 protein chains, or is the volume inside such a shell too small to enclose the viral genome?

Very few self-sufficient viruses have only 60 protein chains in their shells. The satellite viruses do not themselves encode all of the functions required for their replication and are therefore not self-sufficient. The first satellite virus to be discovered, **satellite tobacco necrosis virus**, which is also one of the smallest known with a diameter of 180 Å, has a protein shell of 60 subunits. This virus cannot replicate on its own inside a tobacco cell but needs a helper virus, tobacco necrosis virus, to supply the functions it does not encode. The RNA genome of the satellite virus has only 1120 nucleotides, which code for the viral coat protein of 195 amino acids but no other protein. With this minimal genome the satellite viruses are obligate parasites of the viruses that parasitize cells.

The structure of satellite tobacco necrosis virus has been determined to 2.5 Å resolution by the group of Bror Strandberg in Uppsala, Sweden. As expected, the viral capsid has icosahedral symmetry with one polypeptide chain in the asymmetric unit. This is schematically illustrated in Figure 16.5a. The subunits are arranged so that they are quite close together around the fivefold axes forming tight contacts at the surface of the particle (Figure 16.5b). On the inside of the shell, on the other hand, the polypeptide chains are clustered around the threefold axes, one α helix from each subunit forming a bundle of three α helices. These tight interactions around the fivefold and threefold axes form a scaffold that links all subunits together to complete the shell.

The outer diameter of the shell is approximately 180 Å and the inner diameter about 125 Å except around the threefold axes, where the N-terminal α helices project about 20 Å into the core. The RNA molecule that is present in the core is not visible in the electron density map.

Complex spherical viruses have more than one polypeptide chain in the asymmetric unit

We have seen in the structure of this simple satellite virus that 60 subunits are sufficient to form a shell around an RNA molecule that codes for the subunit protein, but there is little room for additional genetic information.

Figure 16.5 Schematic illustration of the way the 60 protein subunits are arranged around the shell of satellite tobacco necrosis virus. Each subunit is shown as an asymmetric A. The view is along one of the threefold axes, as in Figure 16.3a. (a) Three subunits are positioned on one triangular tile of an icosahedron, in a similar way to that shown in 16.4a. The red lines represent a different way to divide the surface of the icosahedron into 60 asymmetric units. This representation will be used in the following diagrams because it is easier to see the symmetry relations when there are more than 60 subunits in the shells. (b) All subunits are shown on the surface of the virus, seen in the same orientation as 16.4a. The shell has been subdivided into 60 asymmetric units by the red lines. When the corners are joined to the center of the virus, the particle is divided into 60 triangular wedges, each comprising an asymmetric unit of the virus. In satellite tobacco necrosis virus each such unit contains one polypeptide chain.

Self-sufficient viruses have longer genomes, which code for enzymes essential for the replication of the viral nucleic acid in addition to the structural proteins of the capsid. These genomes need a larger volume inside the capsid and therefore need a larger capsid. How are such larger shells constructed within the constraints of icosahedral symmetry? Very little might be gained by increasing the size of each subunit because the shell could become thicker as it is enlarged. The only way out is to increase the number of subunits. To preserve icosahedral symmetry, the asymmetric unit itself must then contain more than one subunit. Since there are 60 asymmetric units, the total number of subunits in the shell must therefore be a multiple of 60. These subunits can be either identical or different. Genetic economy is achieved by using many identical subunits; only one gene is needed to code for the entire protein shell. There are many examples of viruses of this type, and we will examine one example before describing viruses that contain different types of subunits in their capsids.

Can any number of identical subunits be accommodated in the asymmetric unit while preserving specificity of interactions within an icosahedral arrangement? This question was answered by Don Caspar then at Children's Hospital, Boston, and Aaron Klug in Cambridge, England, who showed in a classical paper in 1962 that only certain multiples (1, 3, 4, 7...) of 60 subunits are likely to occur. They called these multiples **triangulation numbers**, T. Icosahedral virus structures are frequently referred to in terms of their triangulation numbers; a T = 3 virus structure therefore implies that the number of subunits in the icosahedral shell is $3 \times 60 = 180$.

Caspar and Klug based their arguments on the principle of specificity, or that the protein–protein contacts should be approximately similar. When T is larger than one, it is no longer possible to pack the protein subunits into an icosahedral shell in a strictly equivalent way as it is for a T = 1 structure like satellite tobacco necrosis virus where all subunits have the same environment and hence the same packing interactions. What Caspar and Klug showed is that for certain specific values of T above one (3, 4, 7...) it is possible to pack the subunits with only slightly different bonding patterns, in a **quasi-equivalent** way.

As examples of such quasi-equivalent arrangement of subunits, we will examine the T = 3 and T = 4 packing modes, both of which are found in known virus particles. In the T = 3 structure, which has 180 subunits (3×60),

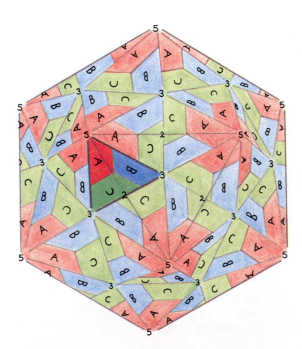

Figure 16.6 A T = 3 icosahedral virus structure contains 180 subunits in its protein shell. Each asymmetric unit (one such unit is shown in thick lines) contains three protein subunits A, B, and C. The icosahedral structure is viewed along a threefold axis, the same view as in Figure 16.5. One asymmetric unit is shown in dark colors.

330

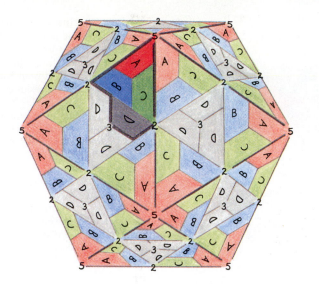

Figure 16.7 A T = 4 icosahedral virus structure contains 240 subunits in its protein shell. Each asymmetric unit (one such unit is shown in dark colors) contains four protein subunits, A, B, C, and D. We have now twisted the icosahedron somewhat, compared with Figure 16.6, so that we view the structure along one of the twofold axes, the same view as in Figure 16.3c.

each asymmetric unit contains three protein subunits with different environments, which we will call A, B, and C. These are arranged so that the A subunits interact around the fivefold axes, and the B and C subunits alternate around the threefold axes (Figure 16.6). There are thus six subunits, three each of B and C, arranged in a pseudosymmetric way around the threefold axes, which therefore are also pseudosixfold axes. Caspar and Klug argued that the subunits could pack around a fivefold and a pseudosixfold axis in a rather similar fashion with only minor alterations of the packing mode. We will see in the structure of tomato bushy stunt virus that they were right. This viral capsid is made from chemically identical proteins that are able to accommodate the necessary differences in packing mode by changing their conformation according to their different environments.

In the T = 4 structure there are 240 subunits (4 × 60) in four different environments, A, B, C, and D, in the asymmetric unit. The A subunits interact around the fivefold axes, and the D subunits around the threefold axes (Figure 16.7). The B and C subunits are arranged so that two copies of each interact around the twofold axes in addition to two D subunits. For a T = 4 structure the twofold axes thus form pseudosixfold axes. The A, B, and C subunits interact around pseudothreefold axes clustered around the fivefold axes. There are 60 such pseudothreefold axes. The T = 4 structure therefore has a total of 80 threefold axes: 20 with strict icosahedral symmetry and 60 with pseudosymmetry.

Structural versatility gives quasi-equivalent packing in T = 3 plant viruses

The molecular basis for quasi-equivalent packing was revealed by the very first structure determination to high resolution of a spherical virus, **tomato bushy stunt virus**. The structure of this T = 3 virus was determined to 2.9 Å resolution in 1978 by Stephen Harrison and co-workers at Harvard University. The virus shell contains 180 chemically identical polypeptide chains, each of 386 amino acid residues. Each polypeptide chain folds into distinct modules: an internal domain R that is disordered in the structure, a region (a) that connects R with the S domain that forms the viral shell, and, finally, a domain P that projects out from the surface. The S and P domains are joined by a hinge region (Figure 16.8).

When they form the three subunits A, B, and C of the asymmetric unit, the identical polypeptides adopt different three-dimensional structures. The C subunit in particular is distinct from the A and B structures, its hinge region assuming a different conformation so that the S and P domains are

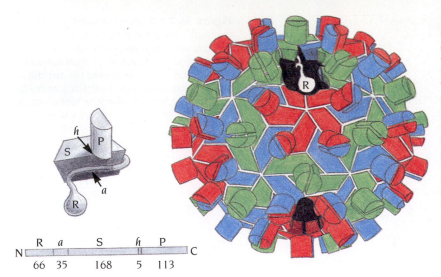

Figure 16.8 Architecture of the tomato bushy stunt virus particle. The polypeptide chain of each subunit folds into three domains (R, S, P) with a 35-residue connecting arm (a) between R and S and a hinge (h) between S and P. The number of amino acid residues in each structural module is indicated. The subunits pack into the virus particle in one of three conformations, colored red (A), blue (B), and green (C). The S domains of the red subunits pack around fivefold axes; the S domains of the blue and green subunits alternate around the threefold axes. The P domains project out from the surface of the particle. The cutaway region in the large diagram shows the connecting arm (a) between domains S and R and the region that the disordered domain R might occupy. (Adapted from S.C. Harrison et al., *Nature* 276: 368–373, 1978.)

quite differently oriented. In addition, the arm region (a) is ordered in C but disordered in A and B. There are no gross conformational differences between A and B, but there are significant local structural differences, especially in the interaction areas. The subunits thus accommodate to the three different environments and preserve quasi-equivalence in most contact regions by conformational differences of the protein chain; structural diversity has been achieved without sacrificing genetic economy.

The S domains form the viral shell by tight interactions in a manner predicted by the Caspar and Klug theory and shown in Figure 16.8. The P domains interact pairwise across the twofold axes and form protrusions on the surface. There are 30 twofold axes with icosahedral symmetry that relate the P domains of C subunits (green) and in addition 60 pseudotwofold axes relating the A (red) and B (blue) subunits (Figure 16.9). By this arrangement the 180 P domains form 90 dimeric protrusions.

One of the remarkable features of this structure as well as other T = 3 plant virus structures is the way in which the ordered connecting arms of C subunits interdigitate to form an internal framework (Figure 16.10). These arms (a in Figure 16.8) extend along the inner face of the S domain and loop around icosahedral threefold axes. Three such C subunit arms contact each other in this way, and all 60 C subunits form a coherent network (see Figure 16.10). The function of this framework is to determine the size of the particle—that is, to ensure that the viral shell closes round on itself correctly during assembly. It does so by serving as a molecular switch. The curvature of the shell along the line joining two threefold axes is much less than the curvature along a line joining a threefold and a fivefold symmetry axis. This difference results from insertion of the ordered C-subunit arms into the C–C contact thereby separating the inner parts of the S domains and flattening the surface lattice.

The size of this viral particle is of course larger than that of a virus with only 60 subunits. The diameter of tomato bushy stunt virus is 330 Å compared with 180 Å for satellite tobacco necrosis virus. The increase in volume of the capsid means that a roughly four times larger RNA molecule can be accommodated.

The protein subunits recognize specific parts of the RNA inside the shell

The N-terminal part of the tomato bushy stunt virus polypeptide chain (the R-segment in Figure 16.8) is disordered in all the subunits. As in the core of many other single-strand RNA viruses this region of the polypeptide chain

Figure 16.9 Contacts between P domains in tomato bushy stunt virus. S_A, S_B, and S_C are the shell domains of subunits A, B, and C, respectively. P_A, P_B, and P_C are the protruding domains of subunits A, B, and C, respectively.

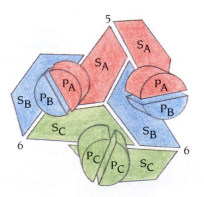

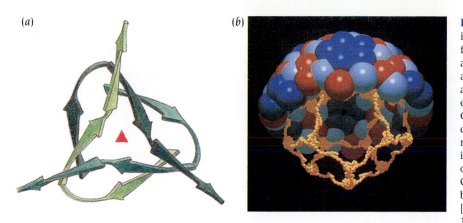

(a)   (b)

Figure 16.10 The arms of all 60 C subunits in tomato bushy stunt virus form an internal framework. (a) Configuration of interdigitated arms from the three C subunits, viewed down a threefold axis. The β strands are shown as arrows. (b) Cutaway view of the virus particle, emphasizing the framework function of the C-subunit arms. These arms are shown as chains of small balls, one per residue. The region where three arms meet and interdigitate is shown schematically in (a). The main part of each subunit is represented by large balls. Only about one hemisphere of these is drawn, but all the C-subunit arms are included. [(a) Adapted from A.J. Olson et al., *J. Mol. Biol.* 171: 61–93, 1983. (b) Courtesy of A.J. Olson.]

contains many positively charged residues. They appear to serve two functions. One is to bind RNA "nonspecifically" and neutralize its negative charge; the other is to recognize specifically a unique "packaging sequence" in the viral genome. The precise structure of this packaging interaction has yet to be worked out, but it is likely to involve a specific RNA structure and some part of the R-segment bound in a defined way by the RNA.

The protein capsid of picornaviruses contains four polypeptide chains

A protein capsid built up from identical subunits is an economical way for the virus to protect its RNA. However, more elaborate schemes have also evolved, such as that found in **picornaviruses**. These viruses constitute a large family of animal viruses with single-stranded RNA genomes. They have been classified into four groups: (1) cardioviruses, such as encephalomyocarditis virus and Mengo virus; (2) enteroviruses such as poliovirus and hepatitis A virus; (3) rhinoviruses, which cause common colds and of which there are about 100 serotypes; and (4) aphtoviruses, which include foot-and-mouth disease virus. The three-dimensional structures of representative examples from each of these groups are known. The group of Michael Rossmann at Purdue University, Indiana, has determined the structure of Mengo virus to 3.0 Å resolution and that of human rhinovirus strain 14 to 2.6 Å resolution. The structure of poliovirus was determined by the group of James Hogle at Scripps Clinic, La Jolla, California, to 2.9 Å resolution, and the structure of foot-and-mouth disease virus, by the group of David Stuart at Oxford University to 2.5 Å resolution.

All these viruses have a molecular mass of around 8.5 million daltons; one long RNA molecule of around 8000 nucleotides accounts for 30% of this molecular mass. The RNA chain occupies the interior of the spherical virion and extends from its center to a radius of about 110 Å (Figure 16.11). A continuous protective shell of protein subunits surrounds the RNA chain. This continuous shell has an average thickness of about 30 Å, giving the virus particle an approximate total diameter of 300 Å. The loops of the polypeptide chain that project from the main part of the protein shell are recognized by the immune defense mechanism of the host, and they contain most of the binding sites for host antibodies. The surface cavities bounded by these protrusions are believed in some cases to be specific sites for receptors on the surface of the host cell. The structural properties of the protein molecules that form the shell must thus include not only the usual requirements for assembly into an icosahedron but also the ability to preserve the receptor sites that render the virus infectious, while accommodating mutations in neighboring antigenic sites in order to escape the immune defenses of the host.

Figure 16.11 Schematic diagram of a picornavirus particle, illustrating the volume occupied by RNA and protein. The surface of the particle contains protrusions and depressions.

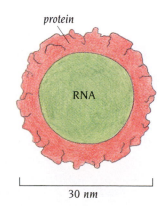

protein

RNA

30 *nm*

There are four different structural proteins in picornaviruses

The shell of all picornaviruses is built up from 60 copies each of four polypeptide chains, called VP1 to VP4. These are translated from the viral RNA into a single polypeptide, which is posttranslationally processed by stepwise proteolysis involving virally encoded enzymes. First, the polypeptide chain is cleaved into three proteins VP0 (which is the precursor for VP2 and VP4), VP1 and VP3. These proteins begin the assembly process. The last step of the processing cascade occurs during completion of the virion assembly; the precursor protein VP0 is cleaved into VP2 and VP4 by a mechanism that is probably autocatalytic but may also involve the viral RNA. VP1, VP2, and VP3 have molecular masses of around 30,000 daltons, whereas VP4 is small, being 7000 daltons, and is completely buried inside the virion.

The arrangement of subunits in the shell of picornaviruses is similar to that of T = 3 plant viruses

The capsids of many plant viruses, like tomato bushy stunt virus, have 180 polypeptide subunits (T = 3) that are chemically identical. Picornaviruses also have 180 polypeptide subunits in their capsids; these, however, are 60 copies each of three different types, VP1, VP2, and VP3, with no significant amino acid sequence homology between them. The 60 copies of the fourth small subunit, VP4, we consider as a continuation of VP2 that has become detached. Given this diversity of subunits, the picornaviruses could, in principle, form an icosahedral shell with a completely different subunit arrangement and different subunit structures from those of the plant viruses. This is not the case, however. The 180 large subunits of the picornaviruses are arranged in an extraordinarily similar way to the 180 subunits of the T = 3 plant viruses. This packing is achieved not by conformational changes, as is the case with the single polypeptide of the plant viruses, but by having within the shell three chemically different polypeptide chains with local structural differences.

The asymmetric unit contains one copy each of the subunits VP1, VP2, VP3, and VP4. VP4 is buried inside the shell and does not reach the surface. The arrangement of VP1, VP2, and VP3 on the surface of the capsid is shown in Figure 16.12a. These three different polypeptide chains build up the virus shell in a way that is analogous to that of the three different conformations A, C, and B of the same polypeptide chain in tomato bushy stunt virus. The viral coat assembles from 12 compact aggregates, or pentamers, which contain five of each of the coat proteins. The contours of the outward-facing surfaces of the subunits give to each pentamer the shape of a molecular mountain: the VP1 subunits, which correspond to the A subunits in T = 3 plant viruses, cluster at the peak of the mountain; VP2 and VP3 alternate around the foot; and VP4 provides the foundation. The amino termini of the five VP3 subunits of the pentamer intertwine around the fivefold axis in the interior of the virion to form a β structure that stabilizes the pentamer and in addition interacts with VP4.

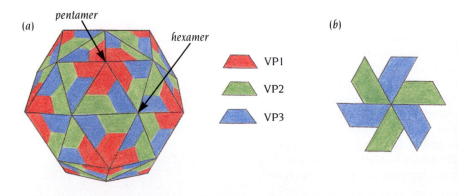

Figure 16.12 (a) Schematic diagram of the surface of a picornavirus. Each triangle represents an asymmetric unit that contains three protein subunits on its surface: VP1 is red, VP2 is green, and VP3 is blue. The diagram illustrates that, at the surface, VP1 subunits are clustered around the fivefold axes, forming pentamers, whereas VP2 and VP3 alternate around the threefold axes, forming hexamers. (b) The hexameric arrangement of VP2 and VP3.

Subunits VP2 and VP3 from different pentamers alternate around the threefold symmetry axes like subunits B and C in the plant viruses (Figure 16.12b). Since VP2 and VP3 are quite different polypeptide chains, they cannot be related to each other by strict symmetry, or even by quasi-symmetry in the original sense of the word. To a first approximation, however, they are related by a quasi-sixfold symmetry axis, since the folded structures of the cores of the subunits are very similar.

The fact that spherical plant viruses and some small single-stranded RNA animal viruses build their icosahedral shells using essentially similar asymmetric units raises the possibility that they have a common evolutionary ancestor. The folding of the main chain in the protein subunits of these viruses supports this notion.

The coat proteins of many different spherical plant and animal viruses have similar jelly roll barrel structures, indicating an evolutionary relationship

One of the most striking results that has emerged from the high-resolution crystallographic studies of these icosahedral viruses is that their coat proteins have the same basic core structure, that of a jelly roll barrel, which was discussed in Chapter 5. This is true of plant, insect, and mammalian viruses. In the case of the picornaviruses, VP1, VP2, and VP3 all have the same jelly roll structure as the subunits of satellite tobacco necrosis virus, tomato bushy stunt virus, and the other T = 3 plant viruses. Not every spherical virus has subunit structures of the jelly roll type. As we will see, the subunits of the RNA bacteriophage, MS2, and those of alphavirus cores have quite different structures, although they do form regular icosahedral shells.

The canonical jelly roll barrel is schematically illustrated in Figure 16.13. Superposition of the structures of coat proteins from different viruses show that the eight β strands of the jelly roll barrel form a conserved core. This is illustrated in Figure 16.14, which shows structural diagrams of three different coat proteins. These diagrams also show that the β strands are clearly arranged in two sheets of four strands each: β strands 1, 8, 3, and 6 form one sheet and strands 2, 7, 4, and 5 form the second sheet. Hydrophobic residues from these sheets pack inside the barrel.

In all jelly roll barrels the polypeptide chain enters and leaves the barrel at the same end, the base of the barrel. In the viral coat proteins a fairly large number of amino acids at the termini of the polypeptide chain usually lie outside the actual barrel structure. These regions vary considerably both in size and conformation between different coat proteins. In addition, there are three loop regions at this end of the barrel that usually are quite long and that also show considerable variation in size in the plant viruses and the

Figure 16.13 The known subunit structures of plant, insect, and animal viruses are of the jelly roll antiparallel β barrel type, described in Chapter 5. This fold, which is schematically illustrated in two different ways, (a) and (b), forms the core of the S domain of the subunit of tomato bushy stunt virus (c). [(b), (c) Adapted from A.J. Olson et al., *J. Mol. Biol.* 171: 61–93, 1983.]

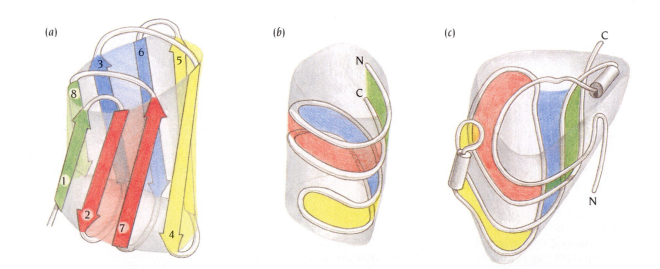

(a)　　　　　　　(b)　　　　　　　(c)

(a)

(b)

(c)

(d)

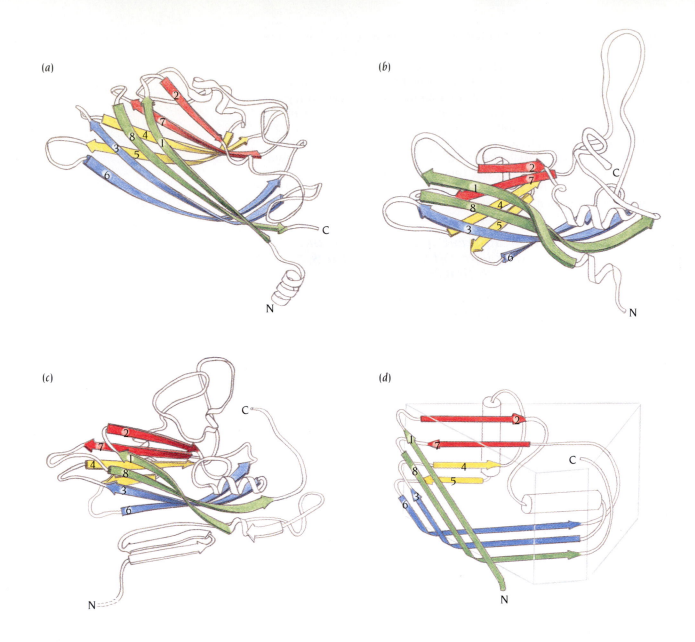

picornaviruses. In contrast, the four loop regions at the other end of the barrel, the top, are short and exhibit only minor variations in size. As a consequence all coat proteins are shaped like a wedge or flattened cone with a broad base formed by the long loops and a narrow tip where the short loops keep the β strands together (see Figure 16.14d).

The finding that the coat proteins of these spherical viruses have the same general topology and that the subunit arrangements of picornaviruses and T = 3 plant viruses are very similar means it is highly likely that all these viruses are evolutionarily related. Amino acid sequence comparisons reveal relationships only within the same class of viruses. The three major protein chains of a picornavirus, VP1, VP2, and VP3, which range in size from 230 to 280 amino acids, have different sequences and show no obvious homology with each other, either at the gene or protein level. Furthermore, there are no sequence similarities to the coat proteins of spherical viruses of other families such as the plant viruses. In contrast, the same protein from different picornaviruses show significant sequence similarity. For example, VP2 from Mengo virus and rhinovirus have about 30% sequence identity. Similar sequence similarity is found in other proteins that are coded by the

Figure 16.14 Schematic diagrams of three different viral coat proteins, viewed in approximately the same direction. Beta strands 1 through 8 form the common jelly roll barrel core. (a) Satellite tobacco necrosis virus coat protein. (b) Subunit VP1 from poliovirus. (c) Subunit VP2 from human rhinovirus. (d) Canonical jelly roll barrel, illustrating the wedge-shaped arrangement of the β strands in the subunit structures of viruses. [(a) Adapted from L. Liljas et al., *J. Mol. Biol.* 159: 93–108, 1982. (b) Adapted from J. Hogle et al., *Science* 229: 1358–1365, 1985. (c) Adapted from M.G. Rossmann et al., *Nature* 317: 145–153, 1985.]

picornaviral genome, such as RNA polymerase. The picornaviruses obviously are an evolutionarily related family.

From sequence data alone no one would suggest that picornaviruses and T = 3 plant viruses are related. However, in order to obtain both the same topology of the subunits and a similar packing arrangement, there must be constraints on a large number of amino acids. Even though they are not the same in the different proteins, they must fulfill the same functional role. Despite their different sequences, the common topology and packing properties suggest that they all derive from the same ancestor. The alternative, that both the same topology and packing arrangements of the coat proteins have arisen independently many times during evolution, seems much less likely.

Drugs against the common cold may be designed from the structure of rhinovirus

Most drugs are effective because they bind to a specific receptor site and block the physiological function of a protein. Classical drug design has been based on this concept for several decades. Compounds that are structural variants of substrates for selected target proteins are synthesized and tested in binding studies. In this way it is usually possible to obtain inhibitors that bind better than the physiological substrates, although several thousands of compounds may have to be synthesized and tested before a suitable drug is found. It has, therefore, been the dream of many pharmaceutical chemists to be able to design new drugs in a more rational way based on a knowledge of the three-dimensional structure of the binding sites of relevant proteins. Understandably, Michael Rossmann attracted considerable attention in 1986 by reporting a detailed crystallographic analysis of the binding of an antiviral drug called WIN 51711 to **human rhinovirus**. This work raised hopes that it might eventually be possible to design efficient drugs against the common cold, as well as many other viral diseases.

The cleft where this drug binds is inside the jelly roll barrel of subunit VP1. Most spherical viruses of known structure have the tip of one type of subunit close to the fivefold symmetry axes (Figure 16.15a). In all the picornaviruses this position is, as we have described, occupied by the VP1 subunit. Two of the four loop regions at the tip are considerably longer in VP1 than in the other viral coat proteins. These long loops at the tips of VP1 subunits protrude from the surface of the virus shell around its 12 fivefold axes (Figure 16.15b).

In rhinoviruses there are depressions, or "canyons," which are 25 Å deep and 12 to 30 Å wide and which encircle the protrusions (Figure 16.15b). One wall of the canyons is lined by residues from the base of VP1. The structure of VP1 is such that the barrel is open at the base and permits access to the hydrophobic interior of the barrel, as in the up-and-down barrel structure of the retinol-binding protein described in Chapter 5.

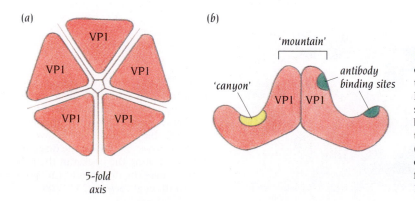

(a)

(b)

'mountain'

'canyon'

antibody binding sites

VP1 VP1

5-fold axis

Figure 16.15 Subunits VP1 in picornaviruses cluster around the fivefold axis (a) so that their tips form a protrusion on the viral surface. In rhinoviruses a depression or "canyon" surrounds the protrusion and contains a binding site for initial attachment of the virus to its receptor, the adhesion molecule ICAM-1. (b) Antiviral agents bind below the floor of this canyon in a cavity within VP1. [(b) Adapted from M. Luo et al., *Science* 235: 182–191, 1987.]

The antiviral compound that Rossmann studied binds inside the barrel of VP1 (Figure 16.16a). It appears to gain access to this site through an opening of the floor of the canyon. One end of the compound is a 3-methylisoxazole group that is inserted into the hydrophobic interior of the VP1 barrel (Figure 16.16b). This end is connected by an aliphatic chain to a 4-oxazolinyl phenoxy group, OP. Both the aliphatic chain and the OP group are contacted by side chains of the amino acids that line the bottom of the canyon. It is quite apparent by looking at the geometries of the cleft and the anti viral compound that the fit is not optimal and that it should be possible to design compounds with stronger binding to this site. In fact, such compounds have been designed and synthesized and are now subject to clinical trials.

Rossmann suggested that the canyons form the binding site for the rhinovirus receptor on the surface of the host cells. The receptor for the major group of rhinoviruses is an adhesion protein known as ICAM-1. Cryoelectron microscopic studies have since shown that ICAM-1 indeed binds at the canyon site. Such electron micrographs of single virus particles have a low resolution and details are not visible. However, it is possible to model components, whose structure is known to high resolution, into the electron microscope pictures and in this way obtain rather detailed information, an approach pioneered in studies of muscle proteins as described in Chapter 14.

(a)

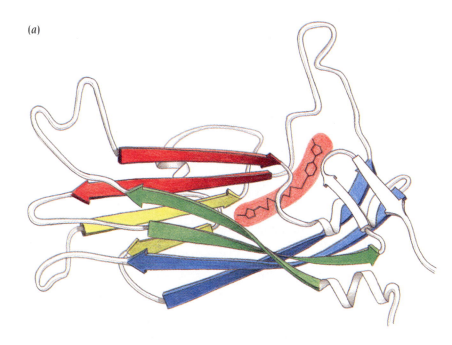

(b)

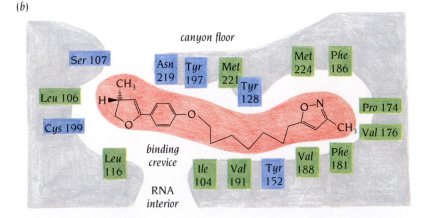

Figure 16.16 Schematic diagrams illustrating the binding of an antiviral agent to human rhinovirus strain 14. (a) The drug binds in a hydrophobic pocket of VP1 below the floor of the canyon. (b) Schematic diagram of VP1 illustrating the pocket in the jelly roll barrel where the drug binds. (Adapted from T.J. Smith et al., *Science* 233: 1286–1293, 1986.)

Bacteriophage MS2 has a different subunit structure

The structures of many different plant, insect, and animal spherical viruses have now been determined to high resolution, and in most of them the subunit structures have the same jelly roll topology. However, a very different fold of the subunit was found in **bacteriophage MS2**, whose structure was determined to 3 Å resolution by Karin Valegård in the laboratory of Lars Liljas, Uppsala.

MS2 belongs to a family of small single-stranded RNA phages—MS2, R17, f2, and Qβ—that infect *Escherichia coli*. These phages are about 250 Å in diameter and contain only four genes in their 4000-nucleotide-long chromosome. One of these genes codes for the 129 amino acid coat protein that is present in 180 copies in the T = 3 virus particle. A second gene encodes a protein 393 amino acids long that is present in a single copy in each virus particle and is needed both for attachment of the phage to its host and penetration of the viral RNA into the bacterium.

The x-ray structure of the MS2 particle was determined by similar methods to those applied to other viral structures; the icosahedral symmetry of the particle was used to compute an average electron density for the subunits. The electron densities for both the attachment protein and the RNA, one copy of which are present in each virus, are lost in this process and only the structure of the coat protein is obtained. The subunit folds into a five-stranded up-and-down antiparallel β sheet with an additional short hairpin at the amino terminus and two α helices at the other end of the chain (Figure 16.17). The α helices are responsible for interactions with a second subunit to form a tight dimer. The two β sheets, one from each subunit in the dimer, are aligned at their edges so that a large continuous β sheet of 10 adjacent antiparallel β strands is formed, with α helices from one subunit packing against the β sheets of the second subunit (Figure 16.18). This structure is quite different from the jelly roll found in all other spherical viruses so far.

Since all members of this family of RNA phages have homologous coat proteins, their subunits are expected to have the same three-dimensional structure. It remains to be seen if the MS2 fold is also present in any other unrelated viruses. The fold is so far unique for the MS2 subunit, but similar structures have been observed in other proteins such as the major histocompatibility antigen, HLA, which was discussed in Chapter 15.

A dimer of MS2 subunits recognizes an RNA packaging signal

The coat protein of MS2 recognizes and binds to a sequence of 19 nucleotides in the viral genome that contains the start codon for the phage replicase gene. This recognition controls two processes: translation of the replicase protein and packaging of the correct RNA molecule, the phage genome, into the protein shell. In solution, the MS2 subunits associate as stable dimers, and encapsidation of viral genomic RNA is initiated when a dimer binds to the 19-nucleotide recognition sequence. Once this assembly initiation complex has been formed, capsid growth occurs, with coat protein dimers adding to the complex until the complete virus particle has been formed.

Capsids can also be formed in the absence of RNA as well as in the presence of small RNA molecules comprising the 19 nucleotides of the

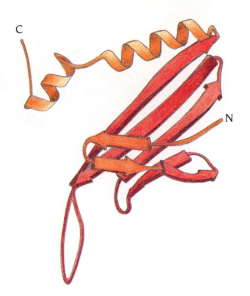

Figure 16.17 The subunit structure of the bacteriophage MS2 coat protein is different from those of other spherical viruses. The 129 amino acid polypeptide chain is folded into an up-and-down antiparallel β sheet of five strands, β3–β7, with a hairpin at the amino end and two C-terminal α helices. (Adapted from a diagram provided by L. Liljas.)

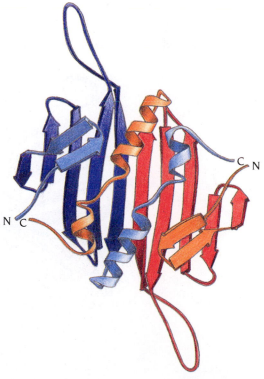

Figure 16.18 A dimer is the basic unit that builds up the capsid of bacteriophage MS2. The two subunits (red and blue) are arranged so that the dimer has a β sheet of 10 antiparallel strands on one side and the hairpins and α helices on the other side. The helices from one subunit pack against β strands from the other subunit and vice versa. (Adapted from a diagram provided by L. Liljas.)

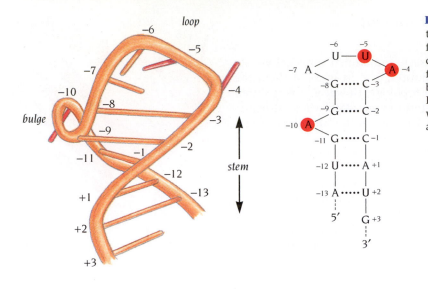

Figure 16.19 Schematic drawing illustrating the structure and sequence of the RNA fragment that is recognized and bound by the coat protein of bacteriophage MS2. The RNA fragment forms a base-paired stem with a bulge at base –10 and a loop of four bases. Bases that form sequence-specific interactions with the coat protein are red. (Adapted from a diagram provided by L. Liljas.)

recognition sequence. The structure of such a capsid complex has been determined by the group of Lars Liljas, and it has given insight into the recognition of viral RNA by the MS2 coat protein. The RNA molecule forms a stem of five base pairs with a loop of four nucleotides at the top (Figure 16.19). There is a bulge in the middle of the stem causing an adenine base to project out from the base-paired double helix of the stem. This adenine in combination with two bases in the loop region form sequence-specific interactions with amino acid residues from the extended β sheet of the coat protein dimer (Figure 16.20). These interactions require a complementarity between the structures of the RNA and of the protein. Consequently, recognition involves not only the proper sequence of the three interacting bases in the RNA but also the formation of a stem-loop structure, as well as a rigidly folded and assembled protein dimer.

The core protein of alphavirus has a chymotrypsin-like fold

Alphaviruses, such as Sindbis virus and Semliki Forest virus, are a group of mosquito-borne, enveloped RNA viruses that can cause encephalitis, fever, arthritis and rashes in mammals. These viruses have two protein shells—an outer glycoprotein layer and an inner core—which are separated by a lipid bilayer, a membrane. Studies by cryoelectron microscopy have shown that

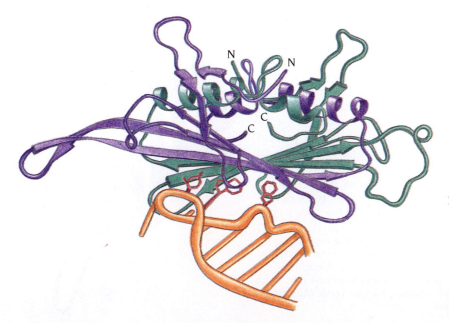

Figure 16.20 The structure of the complex between a dimer of the coat protein of bacteriophage MS2 and the RNA fragment shown in Figure 16.19. One subunit of the coat protein dimer is green, the other is violet and the RNA fragment is orange. Bases that form sequence specific interactions with the protein are red. (Adapted from a diagram provided by L. Liljas.)

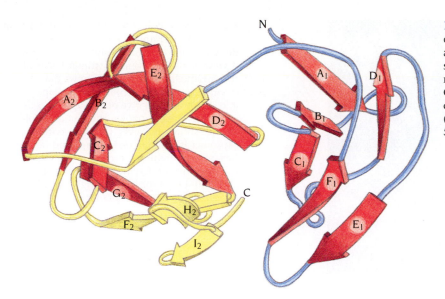

Figure 16.21 Structure of one subunit of the core protein of Sindbis virus. The protein has a similar fold to chymotrypsin and other serine proteases, comprising two Greek key motifs separated by an active site cleft. The C-terminus of the protein is bound in the catalytic site, making the coat protein inactive. (Adapted from S. Lee et al., *Structure* 4: 531–541, 1996.)

both the core and the outer glycoprotein shell conform to T = 4 icosahedral lattices. No high-resolution structure of an alphavirus has yet been obtained, but the subunit structure of the core protein of **Sindbis virus** has been determined by x-ray crystallography by the group of Michael Rossmann. It differs from both the MS2 coat protein and the jelly roll fold.

The core protein of alphaviruses is generated from a longer precursor protein by an autocatalytic cleavage; in other words the precursor protein has protease activity. This activity resides in the part of the precursor that after cleavage forms the inner shell of the virus, the core. Unassembled subunits of this core protein have two domains: a flexible N-terminal domain of roughly 110 residues rich in positive charges, which binds to RNA, and a 150-residue C-terminal domain with a structure similar to chymotrypsin and other serine proteases described in Chapter 11. In Rossmann's crystals the flexible N-terminal domain is not visible. The C-terminal domain comprises two Greek key β barrel motifs with the active site in a cleft between them (compare Figures 16.21 and 11.7). In contrast to chymotrypsin, there is only one polypeptide chain, and there are no disulfide bridges. Like chymotrypsin, however, the coat protein protease domain has a catalytic triad of Asp-His-Ser residues in the active site, and therefore the catalytic mechanism of cleavage is similar to chymotrypsin (see Chapter 11).

Each precursor protein molecule is cleaved only once to generate one molecule of the coat protein, and catalytic activity is restricted to the precursor protein. Why is the coat protein itself catalytically inactive? The structure of the coat protein shows that its C-terminus is bound in the active site cleft and thereby prevents other proteins entering the cleft and being cleaved. This arrangement allows the precursor protein to fulfill its function to generate the coat protein and prevents the coat protein from destroying other proteins in the infected cell, including other coat proteins.

SV40 and polyomavirus shells are constructed from pentamers of the major coat protein with nonequivalent packing but largely equivalent interactions

Polyomavirus, simian virus 40 (SV40), and the papilloma viruses are members of the **papovavirus** family and were for a long time thought to have a simple T = 7 icosahedral design. Electron micrographs by Klug and co-workers of negatively stained particles showed strongly contrasted projections, or capsomers, that would correspond to clusters of five and six subunits around the 12 fivefold and the 60 pseudosixfold positions of an icosahedral T = 7 lattice. This arrangement would give 72 capsomers and 420 subunits (7 × 60 = 12 × 5 + 60 × 6), per virus particle.

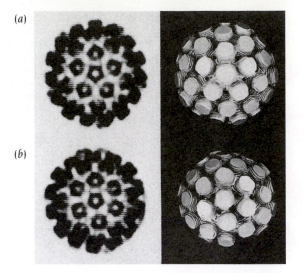

(a)

(b)

Figure 16.22 Polyomavirus was believed from electron micrographs to have a simple T = 7 design until x-ray studies showed otherwise. A view down the fivefold symmetry axis of the icosahedral structure (a) shows that the central capsomer is pentameric in shape and surrounded by five other capsomers as expected. The view down the pseudosixfold axis (b) shows, however, that the central capsomer is pentameric in shape and not hexameric as required for a T = 7 structure. (Adapted from I. Rayment et al., *Nature* 295: 110–115, 1982, by copyright permission of Macmillan Magazines Limited.)

X-ray studies at 22.5 Å resolution of murine **polyomavirus** by I. Rayment and D.L.D. Caspar at Brandeis University confirmed the presence of these 72 capsomers at the expected positions, but even at low resolution the pentagonal shape of all 72 capsomers was evident (Figure 16.22). They concluded that each capsomer must be a pentameric assembly of the major viral subunit, known as viral protein 1 (VP1). Each of the 60 icosahedral asymmetric units contains 6 VP1 subunits, not 7, and the complete shell contains 360 VP1 subunits. The 12 VP1 pentamers centered on icosahedral fivefold axes are identically related to their five neighbors, but the 60 pentamers centered on pseudosixfold positions "see" each of their 6 neighbors quite differently (Figure 16.23). How can such diversity of interaction be incorporated into the bonding properties of just one type of protein subunit, without compromising specificity and accuracy of assembly?

The structures of **SV40** and polyomavirus, determined by Stephen Harrison and coworkers at 3.1 Å and 3.6 Å resolution, respectively, provide a striking answer to this question. The major part of each VP1 subunit folds into a jelly roll domain, which together with four others make up the core of the pentameric capsomers (Figures 16.24 and 16.25). About 45 N-terminal residues and 60 C-terminal residues are not part of the jelly-roll domain. The C-terminal segment projects from each VP1 subunit and "invades" a neighboring pentamer. Part of this C-terminal arm forms a β strand, which augments one β sheet of the jelly roll in a VP1 subunit in the invaded pentamer (yellow in Figure 16.25). This β strand is clamped in place by an additional short strand made by residues in the N-terminal segment of the target subunit (see Figure 16.25). Thus, the pentamers are building blocks tied together by the C-terminal arms of their subunits rather than cemented together across preformed complementary interfaces (Figure 16.26).

The interactions of all 360 arms with their targets are essentially identical. Most of the variability is in how the arms span the interpentamer gap—that is in how they get from their subunit of origin to the pentamer they invade. By expanding our notion of functional contacts between subunits to include such linkages, rather than restricting ourselves to the extensive surface contacts implied by diagrams such as those in Figures 16.6 or 16.7, we can readily explain the high degree of specificity seen in these assemblies.

The links between VP1 pentamers are quite tenuous. With one exception, (across the twofold axis between the γ subunits in Figure 16.26), all

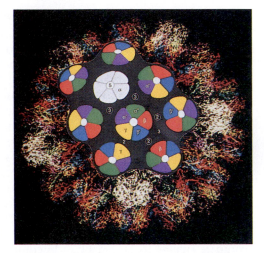

Figure 16.23 Overview of the structure of the SV40 virus particle, showing the packing of pentamers. The subunits of pentamers on fivefold positions are shown in white; those of pentamers in six-coordinated positions are shown in colors. The six colors indicate six quite different environments for the subunit. (Courtesy of S. Harrison.)

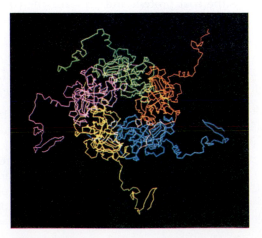

Figure 16.24 A six-coordinated pentamer of SV40 "extracted" from the model shown in Figure 16.23. The extended carboxy-terminal arms are shown in the conformations they adopt in the assembled particle; in the free pentamer they are disordered and flexible. (Courtesy of S. Harrison.)

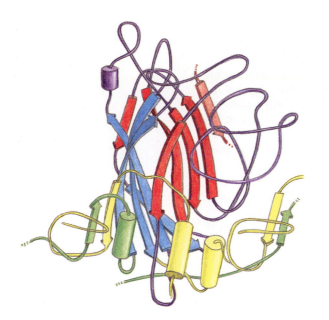

Figure 16.25 Schematic diagram of the SV40 subunit, β strands are represented as ribbons; α helices as cylinders and loops as narrow tubes. The two sheets of the β barrel domain are colored blue and red, the intervening loops are violet; the carboxy-terminal arms are yellow and the amino-terminal arms are green. The β strand in pale red is from a neighboring subunit in the pentamer. The complete yellow carboxy-terminal arm emanates from a subunit in another pentamer; only the initial, α-helical segment of the arm from this subunit is shown, since it extends out of the page. Note how the green amino-terminal segment clamps the invading yellow arm in place. (Courtesy of S. Harrison.)

interpentamer contacts involve the C-terminal arms—a single segment of polypeptide chain rather than a complex surface. Rigidity is imparted to the shell by short α helices that are formed by the proximal segments of the C-terminal arms (see Figure 16.26). These helices act like wedges, stabilizing the curvature of the capsid. We can think of the C-terminal arms as ropes that tie together the pentameric building blocks. We can then extend the analogy by imagining that the ropes are drawn tight and that these helices act like small, rigid spacers, which bear against each other and against the pentamer cores.

The augmentation of a β sheet in one protein by a strand emanating from another is a mode of protein association not restricted to viral shells. Small domains involved in intracellular signal transduction bind to "arms" of other proteins by presenting the edge of a sheet on which those arms can form an additional strand.

Quasi-equivalence was conceived to explain why icosahedral symmetry should be selected for the design of closed containers built of a large number of identical structural units with conserved binding specificity. In simple viruses preformed complementary surfaces provide quasi-equivalent interactions between the units. In the papovaviruses on the other hand, the pentameric units are packed in a nonequivalent way and linked together by extended polypeptide chains that become ordered only on association with targets. The actual contacts between different pentamers are largely equivalent, differing primarily by the conformations of the linker regions.

Conclusion

Small spherical viruses have a protein shell around their nucleic acid that is constructed according to icosahedral symmetry. Objects with icosahedral symmetry have 60 identical units related by fivefold, threefold, and twofold symmetry axes. Each such unit can accommodate one or several polypeptide chains. Hence, virus shells are built up from multiples of 60 polypeptide chains. To preserve quasi-equivalent symmetry when packing subunits into the shell, only certain multiples (T = 1, 3, 4, 7...) are allowed.

Satellite tobacco necrosis virus is an example of a T = 1 virus structure. The 60 identical subunits interact tightly around the fivefold axes on the surface of the shell and around the threefold axes on the inside. These interactions form a scaffold that links all subunits together to complete the shell.

Tomato bushy stunt virus is a T = 3 plant virus with 180 chemically identical subunits. Each polypeptide chain is divided into several domains. The subunits preserve quasi-equivalent packing in most contact regions by conformational differences of the protein chains, especially a large change in

Figure 16.26 Schematic diagram illustrating the different conformations of the flexible arms that link the pentamers in SV40. The color scheme corresponds to Figure 16.23. The body of a pentamer is shown as a five-petaled flower; the arms are cylinders (α helices) and lines.

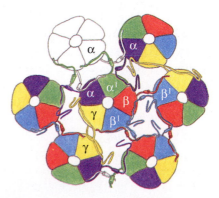

the orientation of the domains. A 35-residue region in 60 of the subunits forms a connected network of interactions around the icosahedral threefold axis that determines the size and shape of the particle.

Picornaviruses construct their shells from 60 copies each of three different polypeptide chains. These 180 subunits are arranged within the shell in a manner very similar to the 180 identical subunits of bushy stunt virus. In some picornaviruses there are protrusions around the fivefold axes, which are surrounded by deep "canyons." In rhinoviruses, the canyons form the virus's attachment site for protein receptors on the surface of the host cells, and they are adjacent to cavities that bind antiviral drugs.

Most subunits of plant and animal spherical viruses have the same topology in their core, that of a jelly roll antiparallel β barrel structure. This implies evolutionary relationships even though there is no significant sequence similarity. The barrel is flattened with short loop regions at the top and long at the base, giving the barrel a wedge shape. The subunit structures of alphaviruses and of the bacteriophage MS2 have different arrangements of antiparallel β strands. The coat protein of the core of Sindbis virus has a chymotrypsin fold and its precursor protein has an autocatalytic protease activity. Dimers of the MS2 coat protein bind to a specific recognition structure of RNA, initiating assembly of the bacteriophage particle.

The capsids of polyoma virus and the related SV40 have icosahedral symmetry, with 72 pentameric assemblies of the major capsid protein. The pentamers are linked to their neighbors by flexible arms, with a β strand that augments a β sheet in the invaded pentamer. These flexible arms allow the pentamers to be linked together with both fivefold and sixfold symmetry.

Selected readings

General

Baltimore, D. Picornaviruses are no longer black boxes. *Science* 229: 1366–1367, 1985.

Caspar, D.L.D., Klug, A. Physical principles in the construction of regular viruses. *Cold Spring Harbor Symp. Quant. Biol.* 27: 1–24, 1962.

Harrison, S.C. Finding the receptors. *Nature* 338: 205–206, 1989.

Harrison, S.C. Multiple modes of subunit association in the structures of simple spherical viruses. *Trends Biochem. Sci.* 9: 345–351, 1984.

Harrison, S.C. Structure of simple viruses: specificity and flexibility in protein assemblies. *Trends Biochem. Sci.* 3: 3–7, 1978.

Harrison, S.C., Skehel, J.J., Wiley, D.C. Virus structure. In *Fields Virology*, 3rd edn. (eds. Fields, B.N. et al.), pp 59–99. New York: Raven Press, 1996.

Hogle, J.M., Chow, M., Filman, D.J. The structure of poliovirus. *Sci. Am.* 256(3): 28–35, 1987.

Hurst, C.J., Benton, W.H., Enneking, J.M. Three-dimensional model of human rhinovirus type 14. *Trends Biochem. Sci.* 12: 460, 1987.

Liljas, L. Viruses. *Curr. Opin. Struct. Biol.* 6: 151–156, 1996.

McKinlay, M.A., Rossmann, M.G. Rational design of antiviral agents. *Annu. Rev. Pharmacol. Toxicol.* 29: 111–122, 1989.

Rossmann, M.G. The canyon hypothesis. *J. Biol. Chem.* 264: 14587–14590, 1989.

Rossmann, M.G. Virus structure, function, and evolution. *Harvey Lectures*, Series 83: 107–120, 1989.

Rossmann , M.G., Johnson, J.E. Icosahedral RNA virus structure. *Annu. Rev. Biochem.* 58: 533–573, 1989.

Specific structures

Abad-Zapatero, C., et al. Structure of southern bean mosaic virus at 2.8 Å resolution. *Nature* 286: 33–39, 1980.

Acharya, R., et al. The three-dimensional structure of foot-and-mouth disease virus at 2.9 Å resolution. *Nature* 337: 709–716, 1989.

Arnold, E., et al. Implications of the picornavirus capsid structure for polyprotein processing. *Proc. Natl. Acad. Sci. USA.* 84: 21–25, 1987.

Arnold, E., Rossmann, M.G. Analysis of the structure of a common cold virus, human rhinovirus 14, refined at a resolution of 3.0 Å. *J. Mol. Biol.* 211: 763–801, 1990.

Badger, J., et al. Structural analysis of a series of antiviral agents complexed with rhinovirus 14. *Proc. Natl. Acad. Sci. USA* 35: 3304–3308, 1988.

Badger, J., et al. Three-dimensional structures of drug-resistant mutants of human rhinovirus 14. *J. Mol. Biol.* 207: 163–174, 1989.

Chen, Z., et al. Protein–nucleic acid interactions in a spherical virus: the structure of beanpod mottle virus at 3.0 Å resolution. *Science* 245: 154–159, 1989.

Choi, H.-K., et al. Structure of Sindbis virus core protein reveals a chymotrypsin-like serine proteinase and the organization of the virion. *Nature* 354: 37-43, 1991.

Filman, D.J., et al. Structural factors that control conformational transitions and serotype specificity in type 3 poliovirus. *EMBO J.* 8: 1567–1579, 1989.

Harrison, S.C., et al. Tomato bushy stunt virus at 2.9 Å resolution. *Nature* 276: 368–373, 1978.

Hogle, J.M., Chow, M., Filman, D.J. Three-dimensional structure of poliovirus at 2.9 Å resolution. *Science* 229: 1358–1365, 1985.

Hogle, J.M., Maeda, A., Harrison, S.C. Structure and assembly of turnip crinkle virus. I. X-ray crystallographic structure analysis at 3.2 Å resolution. *J. Mol. Biol.* 191: 625–638, 1986.

Hosur, M.V., et al. Structure of an insect virus at 3.0 Å resolution. *Protein: Struct. Funct. Gen.* 2: 167–176, 1987.

Jones, T.A., Liljas, L. Structure of satellite tobacco necrosis virus after crystallographic refinement at 2.5 Å resolution. *J. Mol. Biol.* 177: 735–768, 1984.

Kim, S., et al. Crystal structure of human rhinovirus serotype 1A (HRV1A). *J. Mol. Biol.* 210: 91–111, 1989.

Krishnaswamy, S., Rossmann, M.G. Structural refinement and analysis of mengo virus. *J. Mol. Biol.* 211: 803–844, 1990.

Liddington, R.C., et al. Structure of simian virus 40 at 3.8 Å resolution. *Nature* 354: 278–284, 1991.

Luo, M., et al. The atomic structure of mengo virus at 3.0 Å resolution. *Science* 235: 182–191, 1987.

Olson, A.J., Bricogne, G., Harrison, S.C. Structure of tomato bushy stunt virus IV. *J. Mol. Biol.* 171: 61–93, 1983.

Rayment, I., et al. Polyoma virus capsid structure at 22.5 Å resolution. *Nature* 295: 110–115, 1982.

Robinson, I.K., Harrison, S.C. Structure of the expanded state of tomato bushy stunt virus. *Nature* 297: 563–568, 1982.

Rossmann, M.G. Antiviral agents targeted to interact with viral capsid proteins and a possible application to human immunodeficiency virus. *Proc. Natl. Acad. Sci. USA*, 85: 4625–4627, 1988.

Rossmann, M.G., et al. Structure of a human common cold virus and functional relationship to other picornaviruses. *Nature* 317: 145–153, 1985.

Rossmann, M.G., et al. Subunit interactions in southern bean mosaic virus. *J. Mol. Biol.* 166: 37–72, 1983.

Smith, T.J., et al. The site of attachment in human rhinovirus 14 for antiviral agents that inhibit uncoating. *Science* 233: 1286–1293, 1986.

Stehle, T., Gamblin, S.J., Yan, Y., Harrison, S.C. The structure of simian virus 40 refined at 3.1 Å resolution. *Structure* 4: 165–182, 1996.

Valegård, K., et al. The three-dimensional structure of the bacterial virus MS2. *Nature* 345: 36–41, 1990.

Valegård, K., et al. The three-dimensional structures of two complexes between recombinant MS2 capsids and RNA operator fragments reveal sequence-specific protein–RNA interactions. *J. Mol. Biol.* 270: 724–738, 1997.

Prediction, Engineering, and Design of Protein Structures

17

Over a period of more than 3 billion years, a large variety of protein molecules has evolved to become the complex machinery of present-day cells and organisms. These molecules have evolved by random changes of genes by point mutations, exon shuffling, recombination and gene transfer between species, in combination with natural selection for those gene products that have conferred some functional advantage contributing to the survival of individual organisms.

Long before Darwin and Wallace proposed the theory of evolution and Mendel discovered the laws of genetics, plant and animal breeders had begun to interfere with the process of evolution in the species that gave rise to domesticated animals and cultivated plants. Considering their total lack of knowledge of both evolutionary theory and genetics, their achievements, brought about by forcing the pace of and subverting natural selection, were impressive albeit very gradual. With the advent of molecular genetics and in particular techniques for gene manipulation, we have now entered an era of genetic exploitation of organisms undreamed of only 50 years ago. We can now design genes to produce, in host organisms, novel gene products for the benefit of human beings; we are no longer restricted to selecting useful genes that arise by mutation. We are, however, only at the beginning of this new era, and so far we have only scratched the surface of the knowledge that is required for true engineering and design of protein molecules. We distinguish **protein engineering**, by which we mean mutating the gene of an existing protein in an attempt to alter its function in a predictable way, from **protein design**, which has the more ambitious goal of designing *de novo* a protein to fulfill a desired function.

Genome projects have now provided us with a description of the complete sequences of all the genes in more than a dozen organisms, and they will provide many more complete genome sequences within the next decade, including that of the human genome. These databases provide great opportunities for the analysis and exploitation of genes and their corresponding proteins. Central to reaping the intellectual and commercial benefits of this genetic information is the ability to find out the function of individual gene products. Almost all functional assignments to date have been based on sequence similarity to proteins of known function.

Knowledge of a protein's tertiary structure is a prerequisite for the proper understanding and engineering of its function. Unfortunately, in spite of recent significant technological advances, the experimental determination of tertiary structure is still slow compared with the rate of accumulation of amino acid sequence data. This makes the **folding problem**, the successful prediction of a protein's tertiary structure from its amino acid sequence, central to rapid progress in post-genomic biology. We will, therefore, in this chapter first briefly describe implications of protein homology and methods for the prediction of secondary and tertiary structure before giving some examples of protein engineering and protein design.

Homologous proteins have similar structure and function

The term **homology** as used in a biological context is defined as similarity of structure, physiology, development and evolution of organisms based upon common genetic factors. The statement that two proteins are homologous therefore implies that their genes have evolved from a common ancestral gene.

Homologous proteins are mostly recognized by statistically significant similarities in their amino acid sequences. Usually, they also have similar functions although there are some known exceptions, where genes for ancient enzymes have been recruited at a later stage in evolution to produce proteins with quite different functions. An example is provided by one of the structural components in the eye lens that is homologous to the ancient glycolytic enzyme lactate dehydrogenase. Once a novel gene has been cloned and sequenced, a search for amino acid sequence similarity between the corresponding protein and other known protein sequences should be made. Usually, this is done by comparison with databases of known protein sequences using one of the standard sequence alignment computer programs.

Two proteins are considered to be homologous when they have identical amino acid residues in a significant number of sequential positions along the polypeptide chains. Using statistical methods based on comparisons of computer-generated random sequences, it is relatively straightforward to assess how many positions need to be identical for a statistically significant identity between two sequences. However, it is frequently found that two proteins with sequence identity below the level of statistical significance have similar functions and similar three-dimensional structures. In these cases, functionally important residues are identical and usually such residues form sequence patterns or motifs that can be used to identify other proteins that belong to the same functional family. Frequently, members of such families are also considered to be homologous, even though the identities are not statistically significant, only functionally significant. Databases for such families, based on identical or similar sequence motifs, are available on the World Wide Web (see pp. 393–394) and they are very useful for assigning function to a novel protein.

If significant amino acid sequence identity is found with a protein of known crystal structure, a three-dimensional model of the novel protein can be constructed, using computer modeling, on the basis of the sequence alignment and the known three-dimensional structure. This model can then serve as an excellent basis for identifying amino acid residues involved in the active site or in antigenic epitopes, and the model can be used for protein engineering, drug design, or immunological studies.

Since the sequence databases are large and growing exponentially, currently comprising more than 500,000 known protein sequences, the standard sequence alignment programs have been designed to provide a compromise between the speed and the accuracy of the search. As a result, they work well only when there is a reasonably high degree of sequence identity, usually of the order of 30% or more. Much more sensitive programs have been written that search for both identity and conserved structural properties and also for relatedness in different physical properties, but these inevitably require far more computing time. Carefully used, such programs can identify structural and functional similarity where the standard programs fail to do so.

Homologous proteins have conserved structural cores and variable loop regions

Homologous proteins always contain a core region where the general folds of the polypeptide chains are very similar. This core region contains mainly the secondary structure elements that build up the interior of the protein: in other words, the scaffolds of homologous proteins have similar three-dimensional structures. Even distantly related proteins with low sequence identity have similar scaffold structures, although minor adjustments occur in the positions of the secondary structure elements to accommodate differences in the arrangements of the hydrophobic side chains in the interior of the protein. The greater the sequence identity, the more closely related are the scaffold structures (Figure 17.1). This has important implications for model building of homologous proteins; the more distantly related two proteins are, the more the scaffold must be adjusted to model the new structure.

Loop regions that connect the building blocks of scaffolds can vary considerably both in length and in structure. The problem of predicting the three-dimensional structure of a protein that is homologous to a protein of known three-dimensional structure is therefore mainly a question of predicting the structure of loop regions and side-chain conformations, after the scaffold has been adjusted. As mentioned in Chapter 2, loop regions do not have random structures, and their main-chain conformations cluster in sets of similar structures. The conformation of each set depends more on the number of amino acids in the loop and the type of secondary structure elements that it connects, whether they are α-α, β-β, α-β, or β-α connections, than on the actual amino acid sequences. Therefore it is possible to use a database of loop regions from proteins of known structure to obtain a preliminary model of the loops of an unknown structure. To model a protein structure, suitable main-chain loop conformations from this database are attached to the scaffold modeled to have a structure similar to that of the known homologous protein. Finally, the conformations of the side chains are predicted by energy refinement of the model, which minimizes the free energy of the protein by maximizing the interaction energies of the amino acids. Analysis of structures determined to high resolution has shown that only a few side-chain conformations frequently occur. These are called rotamers and model building of side chains employs databases of such rotamers.

An instructive example of the use of such procedures has been in modeling antigen-binding sites in immunoglobulins. These binding sites are built up from three hypervariable loop regions, CDR1–CDR3, from the variable domains of both the light and the heavy chains of immunoglobulins as described in Chapter 15. There is usually high sequence identity within the scaffolds of the variable domains in different immunoglobulin molecules. Consequently, the scaffold of variable domains of known three-dimensional structures can be used in modeling a new monoclonal antibody with a known amino acid sequence. However, the CDR regions of a new antibody are usually very different in sequence from those of any other known antibody, and their three-dimensional structures must be predicted. By comparing

Figure 17.1 The relation between the divergence of amino acid sequence and three-dimensional structure of the core region of homologous proteins. Known structures of 32 pairs of homologous proteins such as globins, serine proteinases, and immunoglobulin domains have been compared. The root mean square deviation of the main-chain atoms of the core regions is plotted as a function of amino acid homology (red dots). The curve represents the best fit of the dots to an exponential function. Pairs with high sequence homology are almost identical in three-dimensional structure, whereas deviations in atomic positions for pairs of low homology are of the order of 2 Å. (From C. Chothia and A. Lesk, *EMBO J.* 5: 823–826, 1986.)

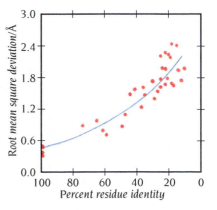

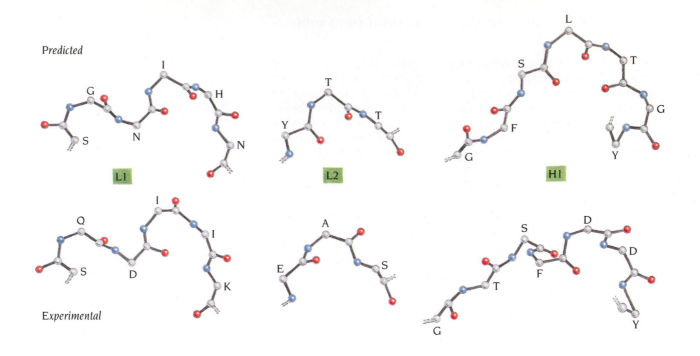

Predicted

L1   L2   H1

Experimental

known antibody structures and sequences, it has been shown that there is only a small repertoire of main-chain conformations for at least five of the six CDR regions and that the particular conformation adopted is determined by a few key conserved residues for each loop conformation. For example, three different conformations were found for the CDR3 regions of the light chains in nine known x-ray structures. More than 90% of the known sequences of light-chain CDR3 regions obey the sequence constraints of one or other of these three conformations. By using this repertoire of loop conformations, considerable success has been achieved in correctly predicting the structure of antigen-binding surfaces. An example of such a prediction compared with the actual structure, subsequently determined, is given in Figure 17.2.

Knowledge of secondary structure is necessary for prediction of tertiary structure

What can be done by predictive methods if the sequence search fails to reveal any homology with a protein of known tertiary structure? Is it possible to model a tertiary structure from the amino acid sequence alone? There are no methods available today to do this and obtain a model detailed enough to be of any use, for example, in drug design and protein engineering. This is, however, a very active area of research and quite promising results are being obtained; in some cases it is possible to predict correctly the type of protein, α, β, or α/β, and even to derive approximations to the correct fold.

Today's predictive methods rely on prediction of secondary structure: in other words, which amino acid residues are α-helical and which are in β strands. We have emphasized in Chapter 12 that secondary structure cannot in general be predicted with a high degree of confidence with the possible exceptions of transmembrane helices and α-helical coiled coils. This imposes a basic limitation on the prediction of tertiary structure. Once the correct secondary structure is known, we know enough about the rules for packing elements of secondary structure against each other (see Chapter 2 for helix packing) to derive a very limited number of possible stable globular folds. Consequently, secondary structure prediction lies at the heart of the prediction of tertiary structure from the amino acid sequence.

Figure 17.2 An example of prediction of the conformations of three CDR regions of a monoclonal antibody (*top row*) compared with the unrefined x-ray structure (*bottom row*). L1 and L2 are CDR regions of the light chain, and H1 is from the heavy chain. The amino acid sequences of the loop regions were modeled by comparison with the sequences of loop regions selected from a database of known antibody structures. The three-dimensional structure of two of the loop regions, L1 and L2, were in good agreement with the preliminary x-ray structure, whereas H1 was not. However, during later refinement of the x-ray structure errors were found in the conformations of H1, and in the refined x-ray structure this loop was found to agree with the predicted conformations. In fact, all six loop conformations were correctly predicted in this case. (From C. Chothia et al., *Science* 233: 755–758, 1986.)

Unfortunately for predictive methods, secondary and tertiary structures are closely linked in the sense that global tertiary structure imposes local secondary structure at least in some regions of the polypeptide chain. The ability of a specific short sequence of amino acids to form an α helix, a β strand, or a loop region is dependent not only on the sequence of that region but also on its environment in the three-dimensional structure. For example, by analyzing all the known tertiary structures, it has been shown that peptide regions of up to five residues long with identical amino acid sequences are α-helical in one structure and a β strand or a loop in other structures. While this interdependence of secondary and tertiary structure complicates secondary structure predictions, it can, sometimes, be used to improve such predictions, by an iterative scheme in which a preliminary assignment of secondary structure is used to predict the type of domain structure, for example, a four-helix bundle or an α/β barrel. The structure type of the domain imposes additional constraints on possible secondary structure, which can be used to refine the secondary structure prediction.

Prediction methods for secondary structure benefit from multiple alignment of homologous proteins

Over 20 different methods have been proposed for predictions of secondary structure; they can be categorized in two broad classes. The empirical statistical methods use parameters obtained from analyses of known sequences and tertiary structures. All such methods are based on the assumption that the local sequence in a short region of the polypeptide chain determines local structure; as we have seen, this is not a universally valid assumption. The second group of methods is based on stereochemical criteria, such as compactness of form with a tightly packed hydrophobic core and a polar surface. Three frequently used methods are the empirical approaches of P.Y. Chou and G.D. Fasman and of J. Garnier, D.J. Osguthorpe and B. Robson (the GOR method), and third, the stereochemical method of V.I. Lim.

Although these three methods use quite different approaches to the problem, the accuracy of their secondary structure prediction is about the same. All three methods can be used to assign one of three states to each residue: α helix, β strand, or loop. Random assignment of these three states to residues in a polypeptide chain will give an average score of 33% correctly predicted states. The methods have been assessed in an analysis of single sequences of a large number of known x-ray structures comprising more than 10,000 residues. For the three-state definition of secondary structure, the overall accuracy of prediction was about 55%. Other objective assessments have given similar results.

However, when these predictive methods are used on a set of homologous proteins the predictive power is considerably higher. The underlying assumption is that secondary and tertiary structure has been more conserved during evolution than amino acid sequence; in other words only such changes have been retained during evolution that conserve the structure. Consequently, the pattern of residue changes within homologous proteins contains specific information about the structure. Conserved hydrophobic residues are usually in the interior of the protein with a high probability of belonging to helices or sheet strands. Insertions and deletions almost always occur in loop regions and not in the scaffold built up from helices and strands.

Several programs are now available that use multiple alignment of homologous proteins for prediction of secondary structure. One such program, called PHD, which was developed by Chris Sander and coworkers, EMBL, Heidelberg, has reached a mean accuracy of prediction of 72% for new structures.

A large fraction of the remaining errors occur at the ends of α helices and β strands and, in addition, some errors occur because of occasional difficulties in distinguishing between α helices and β strands. These latter errors can be corrected if the structural class, α, β, or α/β, can be deduced from a combination of physical studies, for example, circular dichroism spectra, and the general features of the secondary structure prediction. For example, if the prediction scheme assigns one or two short α helices among many β strands in a protein of the β class, there is a high probability that the regions of secondary structures are essentially correctly predicted but that they should all be β strands.

These predictive methods are very useful in many contexts; for example, in the design of novel polypeptides for the identification of possible antigenic epitopes, in the analysis of common motifs in sequences that direct proteins into specific organelles (for instance, mitochondria), and to provide starting models for tertiary structure predictions.

Many different amino acid sequences give similar three-dimensional structures

How many completely different amino acid sequences might give a similar three-dimensional structure for an average-sized domain of 150 amino acid residues? Simple combinatorial calculations show that there are a total of 20^{150} or roughly 10^{200} possible amino acid sequences for such a domain, given the 20 different amino acids in natural proteins. This number is much larger than the number of atoms in the known universe. A more laborious calculation shows that out of these 10^{200} possible combinations we can extract about 10^{38} members that have less than 20% amino acid sequence identity with each other and that therefore can be considered to have different sequences. In other words, there are 10^{38} different ways of constructing a domain of 150 amino acids using the 20 standard amino acids as building blocks. We do not know how many of these can form a stable three-dimensional structure but, assuming say that one out of a billion (10^9) can, we are left with 10^{29} folded possible proteins. In the previous chapters we have seen that simple structural motifs arrange themselves into a limited number of topologically different domain structures. It has been estimated on reasonable grounds that there are about 1000 topologically different domain structures. Since there are 10^{29} possible different sequences that might fold into 10^3 different structures, it follows that there are of the order of 10^{26} different side chain arrangements with less than 20% amino acid sequence identity that can give similar polypeptide folds. Only a small fraction of these possible proteins will be found in nature.

For each of the 500 or so different domain structures that have so far been observed, we might at best know about a dozen of these different possible sequences. It is not trivial to recognize the general sequence patterns that are common to specific domain structures from such a limited knowledge base.

Prediction of protein structure from sequence is an unsolved problem

How to predict the three-dimensional structure of a protein from its amino acid sequence is the major unsolved problem in structural molecular biology. We would like to have a computer program that could simulate the action of the processes that operate in a test tube or a living cell when a polypeptide chain with a specific amino acid sequence folds into a precise three-dimensional structure. Why is this prediction of protein folding so difficult? The answer is usually formulated in terms of the complexity of the task of searching through all the possible conformations of a polypeptide chain to find those with low energy. It requires enormous amounts of computing time, in addition to the complication discussed in Chapter 6 that the energy difference between a stable folded molecule and its unfolded state is a small number containing large errors.

With the realization that there are only a limited number of stable folds and many unrelated sequences that have the same fold, biologically oriented computer scientists started to address what is called the **inverse folding problem**; namely, which sequence patterns are compatible with a specific fold? If this question can be answered, such patterns could be used to search through the genome sequence databases and extract those sequences that have a specific fold, such as the α/β barrel or the immunoglobulin fold.

However, given the large number of possible unrelated sequences for each fold and the limited number of known sequences, a variation of this problem has recently been addressed by a large number of groups; namely, which of the known folds, if any, is most compatible with a specific sequence? The methodology used is called **threading** because it involves threading a specific sequence through all known folds and, for each fold, estimating the probability that the sequence can have that fold. Considerable progress has recently been made in threading, and in blind tests several structures have been correctly predicted by different groups.

Threading methods can assign amino acid sequences to known three-dimensional folds

Threading methods, which are also called protein fold assignments or fold recognition, are a promising and rapidly evolving field of computational structural biology. The goal is to assign to each genome-derived protein sequence the protein fold to which it most closely corresponds, or to determine whether there is no known fold to which the sequence belongs. A further goal is to align the new sequence properly to the three-dimensional structure of the fold to which it belongs to provide a low-resolution model. In order to test different methods of threading, blind tests are arranged, called Critical Assessment of Structure Prediction (CASP), in which the participants are given sequences and invited to predict the fold and make an alignment before the structure is determined experimentally. We will briefly describe here the methods used by one of the more successful participants in these tests, the group of David Eisenberg at University of California, Los Angeles.

The first requirement for threading is to have a database of all the known different protein folds. Eisenberg has used his own library of about 800 folds, which represents a minimally redundant set of the more than 6000 structures deposited at the Protein Data Bank. Other groups use databases available on the World Wide Web, where the folds are hierarchically ordered according to structural and functional similarities, such as SCOP, designed by Alexey Murzin and Cyrus Chothia in Cambridge, UK.

For each fold one searches for the best alignment of the target sequence that would be compatible with the fold; the core should comprise hydrophobic residues and polar residues should be on the outside, predicted helical and strand regions should be aligned to corresponding secondary structure elements in the fold, and so on. In order to match a sequence alignment to a fold, Eisenberg developed a rapid method called the 3D profile method. The environment of each residue position in the known 3D structure is characterized on the basis of three properties: (1) the area of the side chain that is buried by other protein atoms, (2) the fraction of side chain area that is covered by polar atoms, and (3) the secondary structure, which is classified in three states: helix, sheet, and coil. The residue positions are rather arbitrarily divided into six classes by properties 1 and 2, which in combination with property 3 yields 18 environmental classes. This classification of environments enables a protein structure to be coded by a sequence in an 18-letter alphabet, in which each letter represents the environmental class of a residue position.

Each of the 20 different amino acids has different preferences for each of the 18 environmental classes; for instance a Leu has a high preference for being in a helical class with a high fraction of buried side chain area, whereas an Asp has a very low preference for that position. Numerical values for these preferences, called 3D-1D scores, were derived from a set of well-refined high-resolution protein structures, together with sets of sequences similar to the sequences of the 3D structures. This produced a scoring table in which for each environmental class a numerical value of preference is associated with each of the 20 amino acids. This table is used to set up a 3D profile table of a protein structure, in which each residue position is assigned an environmental class with corresponding numerical values for preference for each type of amino acid. The essence of this method is that the three-dimensional structure is reduced to a one-dimensional array, which facilitates matching to a one-dimensional sequence.

A target amino acid sequence is aligned against this structure profile in such a way that the best possible match—the highest total score—is obtained, allowing gaps and insertions. Such an alignment is conceptually similar to alignment of two sequences and similar methods have been used. The match of a sequence to a 3D structure profile for a specific fold is expressed quantitatively by a value called the Z-score, which is the number of standard deviations above the mean alignment score for other sequences of similar length. A high Z-score means there is a high probability that the sequence has the corresponding fold.

The methods described here have subsequently been improved and extended by Eisenberg, but the principle remains essentially the same. Other groups use different methods to screen the sequence–structure alignments and different criteria to assess the matches. Manfred Sippl at the University of Salzburg, Austria, has developed a set of potentials to screen and assess the alignments, the essence of which is to maximize the number of hydrophobic interactions and to minimize the number of buried polar atoms that do not participate in hydrogen bonds. These and similar potentials are now used by many groups in their threading programs. Correct folds can be predicted with a reasonably high probability for small and medium-sized proteins. Correct alignment of the sequence to the selected fold is, however, less accurate.

Proteins can be made more stable by engineering

Protein engineering, via site-directed mutagenesis of DNA, can be used to answer very specific questions about protein stability, and the results of these studies are now being used to increase the stability of industrially important enzymes. To illustrate some of the factors of importance for protein stability that have been revealed by protein engineering studies, we have chosen the extensive work on the enzyme lysozyme from bacteriophage T4 that has been done by the group of Brian Mathews, University of Oregon, Eugene.

Lysozyme from bacteriophage T4 is a 164 amino acid polypeptide chain that folds into two domains (Figure 17.3) There are no disulfide bridges; the two cysteine residues in the amino acid sequence, Cys 54 and Cys 97, are far apart in the folded structure. The stability of both the wild-type and mutant proteins is expressed as the melting temperature, Tm, which is the temperature at which 50% of the enzyme is inactivated during reversible heat denaturation. For the wild-type T4 lysozyme the Tm is 41.9 °C.

We will discuss three different approaches to engineer a more thermostable protein than wild-type T4 lysozyme, namely (1) reducing the difference in entropy between folded and unfolded protein, which in practice means reducing the number of conformations in the unfolded state, (2) stabilizing the α helices, and (3) increasing the number of hydrophobic interactions in the interior core.

Disulfide bridges increase protein stability

The greater the number of unfolded conformations of a protein, the higher the entropic cost of folding that protein into its single native state (see Chapter 6). Reducing the number of unfolded conformations therefore increases

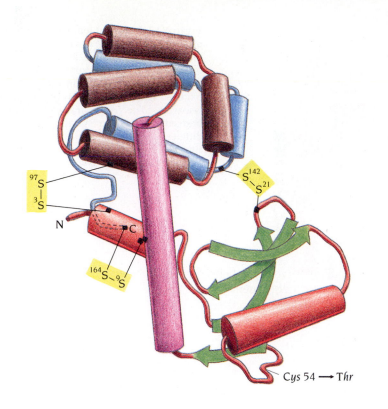

Figure 17.3 The polypeptide chain of lysozyme from bacteriophage T4 folds into two domains. The N-terminal domain is of the α + β type, built up from two α helices (red) and a four-stranded antiparallel β sheet (green). The C-terminal domain comprises seven short α helices (brown and blue) in a rather irregular arrangement. (The last half of this domain is colored blue for clarity.) One long α helix connects the two domains (purple). Thermostable mutants of this protein were constructed by introducing disulfide bridges at three different places (yellow). The position of Cys 54, which was mutated to Thr, is also shown. (Adapted from M. Matsumura et al., *Nature* 342: 291–293, 1989.)

the stability of the native state. The most obvious way to decrease the number of unfolded conformations is to introduce a novel disulfide bond based on knowledge of the tertiary structure of the folded protein. The longer the loop between the cysteine residues, the more restricted is the unfolded polypeptide chain, giving more stabilization of the folded structure. To design such bridges is, however, not a simple task, since the geometry of an unstrained -CH$_2$-S-S-CH$_2$- bridge in proteins is confined to rather narrow conformational limits, and deviations from this geometry will introduce strains into the folded structure and hence reduce rather than increase its stability. It is, therefore, not sufficient to choose at random two residues close together in space to make such a bridge, rather the protein engineer must carefully select pairs of residues with main-chain conformations that fulfill the conditions needed for an unstrained disulfide bridge.

Mathews made a very careful comparison between the geometry of the 295 disulfide bridges in known x-ray structures and all possible pairs of amino acid residues close enough to each other in the refined T4 lysozyme structure to accommodate a disulfide bridge. This was followed by energy minimization of the most likely candidate disulfide bridges and an analysis of stabilizing interactions present in the wild-type structure that would be lost by mutation to a Cys residue. Such losses should be minimized. Three candidate disulfide bridges remained after this filtering, one of which, Cys 3–Cys 97, contained one of the cysteine residues (Cys 97) that is present in the wild type. The five amino acid residues—Ile 3, Ile 9, Thr 21, Thr 142, and Leu 164 (see Figure 17.3)—were mutated to Cys residues in separate experiments so that all single (3–97, 9–164, and 21–142) as well as combinations of double and triple disulfide bonds could be formed. In addition, the second Cys residue of the wild-type enzyme, Cys 54, was mutated to Thr to avoid the formation of incorrect disulfide bonds during folding.

The results of this careful design of novel disulfide bridges were very encouraging (Figure 17.4). All the mutants were more stable in their oxidized forms than wild-type protein. The longer the loop between the cysteine

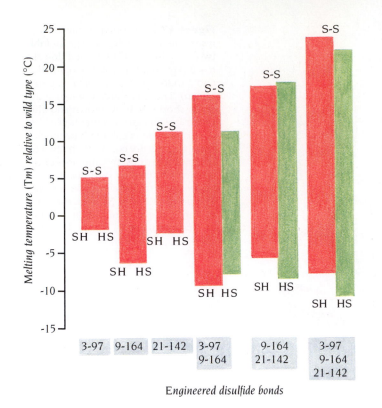

Figure 17.4 Melting temperatures, Tm, of engineered single-, double-, and triple-disulfide-containing mutants of T4 lysozyme relative to wild-type lysozyme. The red bars show the differences in Tm values of the oxidized and reduced forms of the mutant lysozymes. The green bars for the multiple-bridged proteins correspond to the sum of the differences in Tm values for the constituent single-bridged lysozymes. (Adapted from M. Matsumura et al., *Nature* 342: 291–293, 1989.)

residues of the mutants with single disulfide bonds, the larger was the effect on stability. Furthermore, the effects were additive so that the increase in Tm of 23 °C for the mutant with three disulfide bonds was approximately equal to that of the sum of the increases in Tm values for the three mutants with single disulfide bonds (4.8 °C + 6.4 °C + 11.0 °C ≈ 22 °C). The effect on the stability of the protein from reducing the number of possible unfolded structures through introduction of disulfide bridges, the entropic effect, is even larger than these values show because the reduced forms of the mutants had a lower Tm than wild type, which indicates that favorable contacts in the folded structure had been lost by the mutations. These experiments show that engineered disulfide bridges can be combined together to enhance stability dramatically. Needless to say, knowledge of the three-dimensional structure of the protein is a prerequisite to engineer increased stability in this way.

Glycine and proline have opposite effects on stability

Glycine residues have more conformational freedom than any other amino acid, as discussed in Chapter 1. A glycine residue at a specific position in a protein has usually only one conformation in a folded structure but can have many different conformations in different unfolded structures of the same protein and thereby contribute to the diversity of unfolded conformations. Proline residues, on the other hand, have less conformational freedom in unfolded structures than any other residue since the proline side chain is fixed by an extra covalent bond to the main chain. Another way to decrease the number of possible unfolded structures of a protein, and hence stabilize the native structure, is, therefore, to mutate glycine residues to any other residue and to increase the number of proline residues. Such mutations can only be made at positions that neither change the conformation of the main chain in the folded structure nor introduce unfavorable, or cause the loss of favorable, contacts with neighboring side chains.

Both types of mutations have been made in T4 lysozyme. The chosen mutations were Gly 77–Ala, which caused an increase in Tm of 1 °C, and Ala 82–Pro, which increased Tm by 2 °C. The three-dimensional structures of these mutant enzymes were also determined: the Ala 82–Pro mutant had a structure essentially identical to the wild type except for the side chain of residue 82; this strongly indicates that the effect on Tm of Ala 82–Pro is indeed due to entropy changes. Such effects are expected to be additive, so even though each mutation makes only a small contribution to increased stability, the combined effect of a number of such mutations should significantly increase a protein's stability.

Stabilizing the dipoles of α helices increases stability

In Chapter 2 we described the α helix as a dipole with a positive charge at its N-terminus and a negative charge at the C-terminus. Negative ions, such as phosphate groups in coenzymes or substrates, are usually bound to the positive ends of such helical dipoles. The α helices that are not part of a binding site frequently have a negatively charged side chain at the N-terminus or a positively charged residue at the C-terminus that interacts with the dipole of the helix. Such dipole-compensating residues stabilize the helical forms of small synthetic peptides in solution. Do these helix-stabilizing residues also contribute to the overall stability of globular proteins? Of the 11 α helices of T4 lysozyme, 7 helices have negatively charged residues close to their N-termini; two of the remaining four α helices were therefore chosen for engineering studies to answer this question (Figure 17.5).

Two different mutant proteins with single substitutions at the N-terminus of each of these helices, Ser 38–Asp and Asn 144–Asp, were made as well as the corresponding double mutant. The single mutants both showed an increase in Tm of about 2 °C; the effects are additive since the double mutant had a Tm about 4 °C higher than wild type. This corresponds to 1.6 kcal/mol of stabilization energy. From the x-ray structures of these mutants it is apparent that the stabilization is due to electrostatic interactions and not to specific hydrogen bonding between the substituted amino acid and the end of the helix. Alan Fersht in Cambridge, UK has shown, using a different system, the small bacterial ribonuclease, barnase, that a histidine residue at

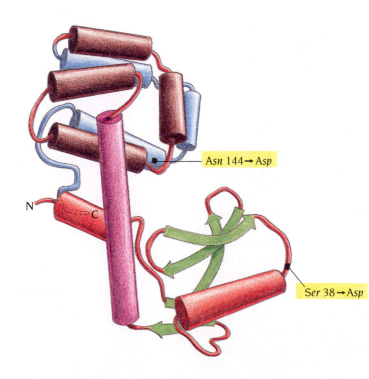

Figure 17.5 Diagram of the T4 lysozyme structure showing the locations of two mutations that stabilize the protein structure by providing electrostatic interactions with the dipoles of α helices. (Adapted from H. Nicholson et al., *Nature* 336: 651–656, 1988.)

the C-terminus of a helix stabilizes the barnase structure by about 2.1 kcal/mol. Significant stabilization of α-helical structures might, therefore, be obtained by combining several such helix-stabilizing mutations.

Mutants that fill cavities in hydrophobic cores do not stabilize T4 lysozyme

We emphasized in Chapter 2 that burying the hydrophobic side chains in the interior of the molecule, thereby shielding them from contact with solvent, is a major determinant in the folding of proteins. The surface that is buried inside a folded protein contributes directly to the stabilization energy of the molecule. Studies of destabilizing mutants in barnase, where **cavities** have been engineered into the hydrophobic core of the wild-type enzyme by mutations such as Ile to Val or Phe to Leu show that the introduction of a cavity the size of one -CH$_2$- group destabilizes the enzyme by about 1 kcal/mol. By analogy it should be possible to stabilize a wild-type protein by making mutations that fill existing cavities in its hydrophobic core. Even though proteins have the atoms of their hydrophobic cores packed approximately as tight as atoms are packed in crystals of simple organic molecules, there are cavities in the cores of almost every protein.

T4 lysozyme has two such cavities in the hydrophobic core of its α helical domain. From a careful analysis of the side chains that form the walls of the cavities and from building models of different possible mutations, it was found that the best mutations to make would be Leu 133–Phe for one cavity and Ala 129–Val for the other. These specific mutants were chosen because the new side chains were hydrophobic and large enough to fill the cavities without making too close contacts with surrounding atoms.

The two single mutants were constructed, purified, analyzed for stability, and crystallized. They were both less stable than wild type by 0.5 to 1.0 kcal/mol. The x-ray structures of the mutants provide a rational explanation for this disappointing result. It turns out that in order to fill the cavities, the new side chains in the mutants adopt energetically unfavorable conformations. This introduces strain in the structure, which obviously costs more energy than is gained by the new hydrophobic interactions. Even careful model building is obviously not sufficient to predict detailed structural and energetic effects of mutations in the hydrophobic core of proteins. Apparently, the observed core structure in T4 lysozyme, and probably in most proteins, reflects a compromise between the hydrophobic effect, which will tend to maximize the core-packing density, and the strain energy that would be incurred in eliminating all packing defects. Therefore, mutations designed to fill existing cavities may be effective in some cases, but they are not likely to provide a general route to substantial improvement in protein stability.

Proteins can be engineered by combinatorial methods

The ultimate goal of protein engineering is to design proteins to carry out predicted functions. However, we do not yet completely understand the rules governing protein folding and molecular recognition, making design of proteins difficult. Protein engineers have therefore invented **combinatorial methods**, in which **libraries** of related proteins are analyzed simultaneously. By sorting these libraries to select for a particular function, the small number of active proteins can be separated from millions of inactive variants. Combinatorial libraries have been used to increase the activity of enzymes, to improve the binding affinity and specificity of proteins, and even to identify novel peptide ligands. Additionally, researchers hope to use the structural and functional data obtained through library selection to improve their ability precisely to engineer molecular interactions.

Combinatorial methods are often referred to as *in vitro* or directed evolution techniques. In nature, the random DNA mutations that lead to changes in protein sequences occur rarely and so evolution is usually a slow

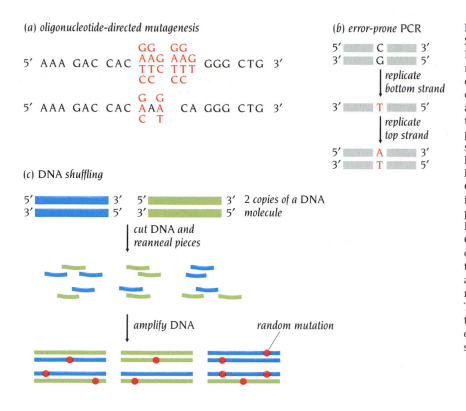

(a) oligonucleotide-directed mutagenesis

5′ AAA GAC CAC GG GG
                AAG AAG GGG CTG 3′
                TTC TTT
                CC CC

5′ AAA GAC CAC G G
                AAA CA GGG CTG 3′
                C T

(c) DNA shuffling

5′ ▬▬▬▬▬ 3′   5′ ▬▬▬▬▬ 3′   2 copies of a DNA
3′ ▬▬▬▬▬ 5′   3′ ▬▬▬▬▬ 5′   molecule

cut DNA and
reanneal pieces

amplify DNA   random mutation

(b) error-prone PCR

5′ ▬▬ C ▬▬ 3′
3′ ▬▬ G ▬▬ 5′

replicate
bottom strand

3′ ▬▬ T ▬▬ 5′

replicate
top strand

5′ ▬▬ A ▬▬ 3′
3′ ▬▬ T ▬▬ 5′

Figure 17.6 Methods of random mutagenesis. Several techniques are available for generating DNA libraries. (a) Oligonucleotides (short molecules of DNA) can be synthesized to contain mixtures of nucleotides at specific codons. An NNS or NNK codon, containing all bases at the first two positions and only two bases at the third position, each allows 20 possible amino acids. Alternatively, a restricted set of bases gives a more focused library; the lower example shown would permit only hydrophilic amino acids (E, K, Q, D, N, H). Oligonucleotide libraries are then incorporated into the gene of interest. (b) Error-prone polymerase chain reaction (PCR) uses a DNA polymerase to replicate the target gene. Conditions are chosen to decrease the fidelity of replication, leading to single base pair errors throughout the gene. (c) In DNA shuffling, a gene is first cut into pieces and then regenerated using a DNA polymerase. The polymerase introduces mutations similar to error-prone PCR; additionally, the pieces of DNA get mixed, so that mutations from separate copies of the gene can be combined.

process. Combinatorial methods accelerate evolution by controlling both the level and location of genetic mutation. Usually, a large number of mutations are concentrated in a single gene through **random mutagenesis**. Because mutagenesis techniques differ in the number and dispersion of mutations introduced to a gene, the appropriate method of mutagenesis depends on what questions the protein engineer seeks to address; the most widely used mutagenesis strategies are outlined in Figure 17.6. The mutated genes are then selected *in vivo*, by conferring a function to cells, or *in vitro*, by binding to an immobilized target. The most common method for *in vitro* selection, **phage display**, is discussed in the following section. The optimal strategies for generating and sorting a library depend on the affinity and specificity of the library for the target. The following examples will therefore illustrate some important combinatorial methods as well as the information that has been gained by using these techniques.

Phage display links the protein library to DNA

In designing a selection strategy, we must consider how to isolate and characterize the functional proteins in a library. Classically, molecular biologists have screened for altered protein function *in vivo*, by measuring effects on whole cells. There are, however, several possible advantages to using *in vitro* selections. For example, we no longer require a selection which modifies the growth of a host organism and can instead focus on an isolated function, such as a ligand binding to its receptor. However, *in vitro* selection does require that we connect each member of the protein library to its gene so that we can readily amplify and identify selectants. Bacteriophage (phage) display provides a simple mechanism to link the protein to its DNA.

Phage display typically utilizes bacteriophage M13. This filamentous phage contains single-stranded DNA encased in a protein coat. In contrast to the spherical viruses discussed in Chapter 16, M13 is long (1–2 μm) and narrow (7 nm) and contains five coat proteins, including approximately 2700 copies of the major coat protein gVIIIp (gene VIII protein) and five copies of the infectivity protein gIIIp. In phage display, the gene encoding the peptide

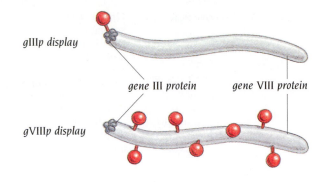

gIIIp display

gene III protein   gene VIII protein

gVIIIp display

Figure 17.7 Proteins displayed on filamentous phage. In phage display, proteins are usually fused to the major coat protein, gVIIIp (2700 copies per phage), or to the infectivity protein, gIIIp (5 copies per phage). During assembly of the virus in bacteria, capsid fusion proteins are incorporated into the virus and are displayed on the surface of the phage. When gVIIIp fusions are produced, there can be many copies of protein per phage particle, leading to multivalent display. By contrast, gIIIp fusions typically give only one copy per phage (monovalent display).

or protein of interest is usually fused to one of these genes (Figure 17.7). When phage particles are produced in bacterial cells, the capsid fusion protein is incorporated into the viral particle and the phage DNA, containing the gene fusion, is packaged into the phage. The protein phenotype on the phage surface is thereby linked to the DNA genotype within the virus. Since progeny phage will not usually assemble from only capsid fusion proteins, wild-type capsid proteins are also expressed in the bacteria using a so-called helper phage that specifies wild-type capsid proteins but which is deficient in phage packaging. Several copies of the gVIIIp fusion protein molecules can be incorporated into each phage, giving multivalent display, but usually only one gIIIp fusion protein molecule is incorporated, giving monovalent display. A phage-display library results from each bacterium producing phage with a different capsid fusion protein. The typical phage display experiment includes three steps, shown in Figure 17.8. First, a library containing around 10^8 phage is screened for binding to an immobilized target. The selected phage are then propagated in bacteria and the phage DNA is characterized to determine the sequence of the gene corresponding to the mutated binding

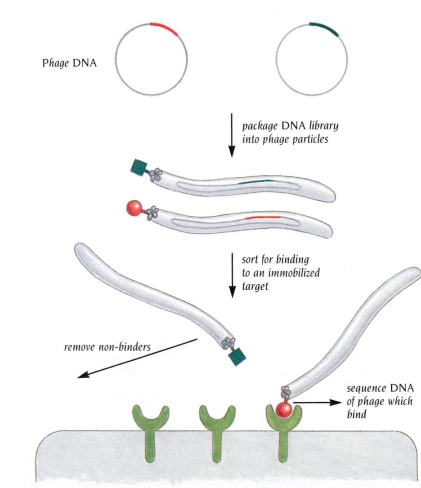

Phage DNA

package DNA library
into phage particles

sort for binding
to an immobilized
target

remove non-binders

sequence DNA
of phage which
bind

Figure 17.8 Sorting phage display libraries. Each phage particle in a library contains one protein–phage fusion and its corresponding DNA code. This phage library is added to an immobilized target protein; if a fusion protein does not bind the target, the phage displaying that protein is washed away. If a fusion protein does bind, its phage is eluted from the target, propagated in bacteria, and resorted under more stringent binding conditions. After several rounds of this sorting procedure, the DNA from phage are sequenced in order to determine the amino acid sequence of the selected protein.

protein. These three steps can then be repeated after mutagenesis of the selected genes to improve further the properties of the selected proteins.

In 1985, George Smith at the University of Missouri first demonstrated that peptide–phage fusions could be selected through the binding of the peptide to an antibody immobilized on a plate. Since that time, phage display has been used to improve the affinity and specificity of both antibodies and antigens, hormones and receptors; researchers have also studied the interactions of proteins with small molecules or nucleic acids. In the sections below, we focus on the use of phage display to characterize the interactions between proteins, including examples in which a protein scaffold is mutated to change the ligand specificity, a truncated protein is mutated to construct a minimized binding domain, and a random peptide library is sorted to identify a novel receptor agonist.

Affinity and specificity of proteinase inhibitors can be optimized by phage display

The blood coagulation cascade involves several trypsin-like serine proteinases (see Chapter 11) including plasmin, kallikrein, factor XIa, and the tissue factor–factor VIIa (TF–FVIIa) complex. While there would be important clinical benefits to engineering specific protease inhibitors, the active sites of these enzymes are highly conserved, showing as high as 81% identity, making the design of specific inhibitors difficult. **Kunitz domains** make up one family of protein inhibitors of the trypsin-like proteinases (Figure 17.9). These protein domains of approximately 60 residues, stabilized by three disulfide bonds, maintain a highly conserved structure with a sequence identity as low as 33%. Each Kunitz domain recognizes one or more proteinases through a set of 10–14 residues, most of which are in the "binding loop" (residues 11–19, green in Figure 17.9).

As described in Chapter 11, the active site cleft of a serine protease forms a row of subsite pockets, named S5 through S4′, which fit the substrate residues, numbered P5 through P4′. Table 17.1 gives the correlation between subsite and residue number for APPI (Alzheimer's amyloid β-protein precursor inhibitor) and LACI-D1 (lipoprotein-associated coagulation inhibitor D1) Kunitz domains. The major binding determinant of inhibitors for all trypsin-like proteinases is the presence of Lys 15 or Arg 15 at the P1 site; the specificity of Kunitz domains for different enzymes is then determined by the other subsite residues. Phage display experiments have identified specific Kunitz domain inhibitors with mutations in the subsite residues of the primary binding loop. From the sequences of these specific inhibitors, we can learn the rules governing the recognition between Kunitz domains and serine proteinases.

Figure 17.9 Structure and protease-binding properties of Kunitz domains. (a) Structure of APPI determined by x-ray crystallography. This 58 residue protein domain is characteristic of Kunitz domains. The three disulfide bonds are colored yellow. Residues 11–19, colored green, constitute the primary loop involved in binding to trypsin-like serine proteinases. Residues 34–39, colored orange, are also thought to be involved in binding to the proteinase or structuring the binding loop. (b) Amino acid sequence of Kunitz domain LACI-D1. Residues in Kunitz domains making the principal contacts with trypsin-like proteinases are shown in dark colors, colored according to (a). While most important proteinase interactions are with the primary binding loop (11–19), the second loop (34–39) contains some residues which contact either the proteinase or the primary binding loop. [(b) Adapted from W. Markland et al., *Biochemistry* 35: 8045–8057, 1996.]

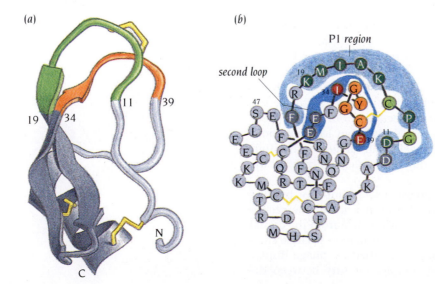

Table 17.1 Phage-optimized sequences of Kunitz domain libraries

wt sequence and target	primary binding loop								secondary loop		K_i for target (M)
	P_5	P_4	P_3	P_1	P_1'	P_2'	P_3'	P_4'			
residue number	11	12	13	15	16	17	18	19	34	39	
LACI-D1	D	G	P	K	A	I	M	K	I	E	
kallikrein	D	G	P	R	A	A	H	P	S	G	40×10^{-12}
APPI	T	G	P	R	A	M	I	S	F	G	
kallikrein	D	G	H	R	A	A	H	P	Y	G	15×10^{-12}
TF-FVIIa + kallikrein	P	G	P	R	A	L	I	L	F	Y	$\sim 2 \times 10^{-9}$
TF-VIIa	P	G	P	K/R	A	L/M	M	K	I	Y/H	$\sim 10 \times 10^{-9}$

The sequences of LACI-D1 and APPI Kunitz domain protease-binding regions are shown, with the sequences of phage-optimized kallikrein and TF-FVIIa-binding variants given below. Variants of both LACI-D1 and APPI were selected for binding to kallikrein, and additionally APPI variants that bound TF-FVIIa and kallikrein were further selected for TF-VIIa preferential binding.

Two groups have selected phage-displayed Kunitz domains for binding to kallikrein. Mark Dennis and Robert Lazarus at Genentech, US, used the human Kunitz domain APPI as a scaffold. They made three libraries, each containing four or five randomized codons, and combined the selected mutations from all libraries to form a consensus sequence. The consensus sequence, given in Table 17.1, showed an inhibition constant (K_i) for kallikrein of 15 pM, which was lower than the K_i of any of the individual libraries. This result is consistent with the additivity principle which states that the effects of noninteracting mutations tend to be independent. Robert Ladner and coworkers at Protein Engineering Corporation, US, used a different human Kunitz domain, LACI-D1 as a scaffold for their kallikrein-binding libraries. These researchers designed DNA libraries using a restricted set of codons (see Figure 17.6) based on the residues commonly found in Kunitz domains at each position. This strategy reduces diversity, thus permitting more residues to be randomized in each library. However, important interactions might be missed when only a subset of amino acids are available. Nevertheless, the phage selectants identified by this method had very similar sequences and K_i values (40 pM) to the proteins selected at Genentech. The fact that these two phage display libraries arrived at very similar sequences, having started from different scaffolds and library designs, can be described as convergent *in vitro* evolution.

Dennis and Lazarus also sorted their APPI libraries against TF–FVIIa. They found that the tightest inhibitors, (K_i approximately 2 nM), also inhibited factor XIa and kallikrein. To identify inhibitors specific for TF–FVIIa, they employed a competitive, or subtractive, sorting strategy in which soluble competitor (factor XIa) was added during sorting. Kunitz domains that had a high affinity for soluble factor XIa thus remained in solution and were not selected for binding to immobilized TF–FVIIa. Selectants from this competitive sorting maintained nanomolar affinity for TF–FVIIa but inhibited both factor XIa and kallikrein ~1000-fold more weakly. Subtractive sorting thus shifted *in vitro* selection towards mutations that were tolerated only by the desired target.

How do the mutations identified by phage display improve binding specificity? There is as yet no direct structural information on the phage-selected inhibitors; however they can be modeled using data from the crystal structures of other Kunitz domains bound to serine proteinases. These studies lead to the conclusion that the mutations identified by phage display improve binding specificity by maximizing complementarity between the primary binding loop and the proteinase active site. Analysis of the models of these new Kunitz domains with altered specificity has given novel insights into the mechanisms of Kunitz domain–protease specificity.

Structural scaffolds can be reduced in size while function is retained

In most proteins, only a small number of residues are directly involved in ligand or receptor binding; the rest of the protein provides the three-dimensional structure to position correctly the functional parts of the molecule. James Wells and colleagues at Genentech have used phage display to ask whether a binding epitope can be transferred to a smaller scaffold. This is an interesting question because protein scaffolds often seem larger than they need to be, and also the production of useful proteins is most efficient if the proteins are small. One such minimization involved the Z domain of bacterial protein A, shown in Figure 17.10. The 59-residue Z domain forms an antiparallel three-helix bundle that binds to the Fc portion of IgG (see Chapter 15) with a dissociation constant (K_d) of 20 nM. The Fc-binding epitope is discontinuous, involving residues from both helix 1 and helix 2. The third helix does not contact Fc, but is required to maintain the structure of the binding domain. Minimizing the Z domain thus poses a difficult design problem: a smaller version must maintain the correct three-dimensional placement of functional groups that are not adjacent in sequence. Furthermore, protein structures smaller than 50 residues are relatively rare, and only recently have 30 residue peptides been designed as stable domains (see below).

Andrew Braisted and J.A. Wells prepared phage containing Z domain helices 1 and 2 and restored Fc binding of this 38 residue minidomain in three iterative stages (see Figure 17.10). The truncated peptide was first randomized at four hydrophobic residues which contact helix 3 in the complete Z domain. The consensus sequence from this library maintained the wild-type residues Ile 17 and Leu 23 while the hydrophobic residues Leu 20 and

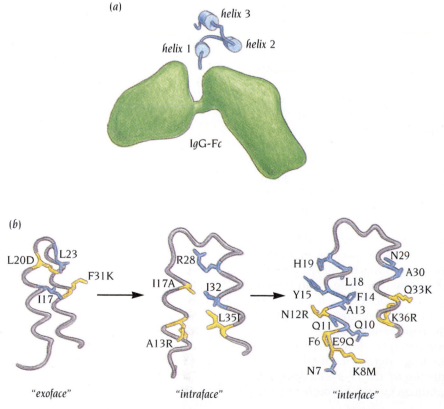

(a)

helix 3

helix 1   helix 2

IgG-Fc

(b)

"exoface"   "intraface"   "interface"

Figure 17.10 Construction of a two helix truncated Z domain. (a) Diagram of the three-helix bundle Z domain of protein A (blue) bound to the Fc fragment of IgG (green). The third helix stabilizes the two Fc-binding helices. (b) Three phage-display libraries of the truncated Z-domain peptide were selected for binding to the Fc. First, four residues at the former helix 3 interface ("exoface") were sorted; the consensus sequence from this library was used as the template for an "intraface" library, in which residues between helices 1 and 2 were randomized. The most active sequence from this library was used as a template for five libraries in which residues on the Fc-binding face ("interface") were randomized. Colored residues were randomized; blue residues were conserved as the wild-type amino acid while yellow residues reached a nonwild-type consensus. [(b) Adapted from A.C. Braisted and J.A. Wells, *Proc. Natl. Acad. Sci. USA* 93: 5688–5692, 1996.]

Phe 31 were mutated to the charged residues Asp and Lys, respectively. This double mutant bound Fc with a K_d of 3.4 μM, a greater than a hundredfold improvement over the unmutated fragment, and had a large increase in α-helical content as shown by circular dichroism spectroscopy. This face of the two-helix peptide was thus converted from a protein core to a protein surface. The second library fixed Asp 20 and Lys 31 and randomized five positions at the "intraface" of helix 1 and helix 2. From this library, three new residues were selected at the open end of the intrahelix interface, and the K_d for this peptide was around 300 nM, a tenfold improvement over the first library. Finally, five libraries were randomized at the Fc-binding face of the two-helix bundle and the consensus mutations were combined. Some libraries yielded mutations that improved binding 2–3 fold, and all together the two-helix bundle contained 12 mutations from the Z domain sequence. A final truncation of the five N-terminal residues yielded a 33 residue peptide with a K_d of 40 nM, very close to the wild-type value of 20 nM. X-ray crystallography and NMR spectroscopy indicated that the two helix bundle had the same three-dimensional structure and the same Fc-binding epitope as the Z domain. Thus, iterative cycles of phage display were used to create a more stable protein surface and to repack the protein core. The resulting scaffold was half the size of the native domain but maintained the three-dimensional arrangement of Fc-binding residues at the interface.

Phage display of random peptide libraries identified agonists of erythropoietin receptor

Many biological processes are activated by hormone–receptor interactions, and a great deal of research has been devoted to identifying peptides or small molecules which either inhibit or simulate hormone function. The idea that a small molecule can mimic the function of a large protein is based on the "hot-spot" principle that a small number of residues at a binding interface contribute most of the binding energy. Nicholas Wrighton and coworkers at Affymax, US, in collaboration with scientists at Johnson & Johnson and at the Scripps Institute, have used phage display methods to isolate a 20 residue peptide, called EMP1, which mimics the activity of erythropoietin (EPO) by promoting dimerization and activation of erythropoietin receptor (EPOR; the extracellular domain of which is called EBP). Erythropoietin is a cytokine hormone which stimulates formation of red blood cells (see Chapter 13).

EMP1 was selected by two cycles of phage display. First, random peptide libraries of the sequence CX_8C, where X is any residue and the cysteines form an intramolecular disulfide bond, were displayed as gVIIIp fusions (see Figure 17.7). Multivalent gVIII protein fusion display permitted selection of weak binders by avidity (multiple) binding to immobilized dimers of EBP. These weakly binding peptides were expanded to the form $X_5CX_8CX_3$ and partially randomized in the X_8 residues. Fusion of the new library to gIIIp yielded a lower valency of display and allowed selection of tight binders. EMP1, isolated from this library, bound to EBP with a K_d of 0.2 μM and stimulated EPOR activity *in vivo*. The dimerization of EMP1, critical for its agonist activity, was probably selected through the interactions of the multivalent peptides with the immobilized EBP dimer.

The crystal structure of EMP1 bound to EBP shows the remarkable structural economy of the EMP1 dimer (Figure 17.11). Each peptide monomer forms a β hairpin, stabilized by an intramolecular disulfide bond. The peptide dimer forms a four-stranded β sheet maintained by four main-chain hydrogen bonds and by the packing of hydrophobic residues. Each monomer makes hydrogen bonds and hydrophobic contacts to both EBP receptors, forming a total of 20 interactions between the peptide dimer and the two EBP proteins; most of the peptide is directly involved in binding (Figure 17.12). Furthermore, EMP1 seems to mimic the binding interactions used by EPO itself. Although there is no structural information on the EPO–EBP complex, the structure of another cytokine hormone, human growth hormone, bound

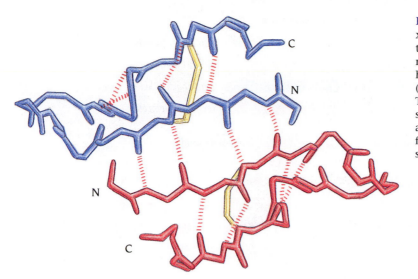

Figure 17.11 Structure of EMP1 dimer from x-ray crystallography. In the presence of EBP, the EMP1 peptide forms a dimer. Each monomer (shown in red and blue) forms a β hairpin structure stabilized by hydrogen bonds (red dashes) and a disulfide bond (yellow). The two peptides form a symmetrical dimer stabilized by four hydrogen bonds (red dashes) and hydrophobic contacts. The two monomers form a four-stranded, anti-parallel pleated sheet.

to its receptor (see Chapter 13) was used as a model. Like EPO, EMP1 contacts four of the six loops on EBP that have sequence similarities to loops of the growth hormone receptor involved in hormone binding. In particular, the three residues of the growth hormone receptor most critical for binding of growth hormone, which correspond to residues Phe 93, Phe 205, and Met 150 in EBP, are well-buried in the EMP1–EBP crystal structure.

EMP1, selected by phage display from random peptide libraries, demonstrates that a dimer of a 20-residue peptide can mimic the function of a monomeric 166-residue protein. In contrast to the minimized Z domain, this selected peptide shares neither the sequence nor the structure of the natural hormone. Thus, there can be a number of ways to solve a molecular recognition problem, and combinatorial methods such as phage display allow us to sort through a multitude of structural scaffolds to discover novel solutions.

DNA shuffling allows accelerated evolution of genes

Natural selection works through the complementary processes of mutation and genetic reassortment by **recombination**. The oligonucleotide-directed mutagenesis methods used in the foregoing examples do not allow for recombination; instead, mutations are combined manually to optimize a protein sequence. Willem Stemmer at Maxygen invented a method of directed evolution that uses both mutation and recombination. This method, called

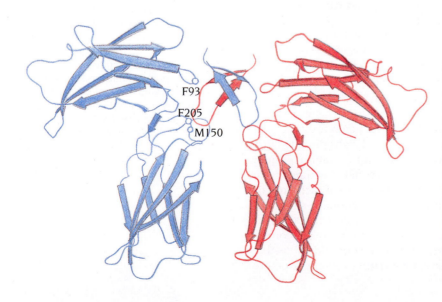

Figure 17.12 Ribbon diagram of EMP1 bound to the extracellular domain of the erythropoietin receptor (EBP). Binding of EMP1 causes dimerization of erythropoietin receptor. The x-ray crystal structure of the EMP1–EBP complex shows a nearly symmetrical dimer complex in which both peptide monomers interact with both copies of EBP. Recognition between the EMP1 peptides and EBP utilizes more than 60% of the EMP1 surface and four of six loops in the erythropoietin-binding pocket of EBP. In particular, three residues thought to be critical for binding erythropoietin (Phe 93, Met 150, and Phe 205) are fully buried in the structure of the peptide–receptor complex. (From J.A. Wells, *Science* 273: 449–450, 1996.)

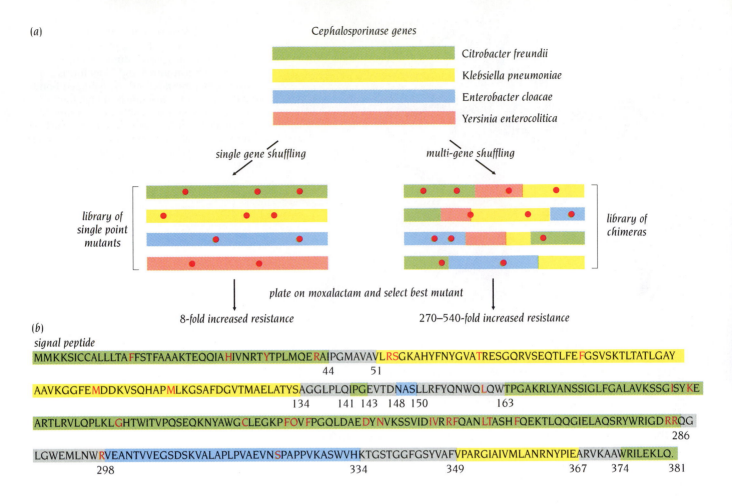

(a)

Cephalosporinase genes

Citrobacter freundii
Klebsiella pneumoniae
Enterobacter cloacae
Yersinia enterocolitica

single gene shuffling   multi-gene shuffling

library of single point mutants   library of chimeras

plate on moxalactam and select best mutant

8-fold increased resistance   270–540-fold increased resistance

(b)

signal peptide

MMKKSICCALLLTAFFSTFAAAKTEQQIAHIVNRTYTPLMQERAIPGMAVAVLRSGKAHYFNYGVATRESGQRVSEQTLFEFGSVSKTLTATLGAY

44   51

AAVKGGFEMDDKVSQHAPMLKGSAFDGVTMAELATYSAGGLPLQIPGEVTDNASLLRFYQNWQLQWTPGAKRLYANSSIGLFGALAVKSSGISYKE

134   141 143 148 150   163

ARTLRVLQPLKLGHTWITVPQSEQKNYAWGCLEGKPFOVFPGQLDAEDYNVKSSVIDIVRRFQANLTASHFQEKTLQQGIELAQSRYWRIGDRRQG

286

LGWEMLNWRVEANTVVEGSDSKVALAPLPVAEVNSPAPPVKASWVHKTGSTGGFGSYVAFVPARGIAIVMLANRNYPIEARVKAAWRILEKLQ.

298   334   349   367 374 381

DNA shuffling (see Figure 17.6), begins by randomly cutting a gene into fragments about 100–300 base pairs in length. This DNA pool is then reassembled with a DNA polymerase, which combines fragments from different copies of the original gene. This process also incorporates point mutations at a defined rate. When point mutations originally in different copies of a gene end up in the same piece of DNA, *in vitro* recombination has occurred. These shuffled DNA libraries are then sorted *in vitro* by phage display or selected *in vivo* through bacterial selection, as described below.

Stemmer and coworkers have extended DNA shuffling to include homologous genes from different organisms. In their first attempt, they mixed the genes encoding class C cephalosporinases from four species (the DNA being 58–82% identical). Their goals were to select for bacterial resistance to the antibiotic moxalactam and to compare the evolution of individual genes with a shuffled gene family. As shown in Figure 17.13, one cycle of evolution using all four genes resulted in recombination of DNA segments as well as incorporation of random point mutations. By contrast, one cycle of DNA shuffling of a single gene only introduced point mutations. The results of recombination of different genes was dramatic: the single genes yielded eightfold increases in antibiotic resistance, while the four genes together gave 270 to 540-fold improvements. The best clone contained eight discrete DNA segments from three of the four genes as well as 33 point mutations. Figure 17.14 shows a three-dimensional model based on the crystal structure of one of the native enzymes. The different colors identify the origin of each protein segment; interestingly, these segments form units of secondary structure. This example demonstrates the utility of recombination and mutation in directed evolution. Additionally, this work underscores the power of combinatorial techniques to construct enzymes whose design would be impossible given our current understanding of the factors governing protein structure and function.

Figure 17.13 DNA shuffling of cephalosporinase genes. (a) Four homologous genes from bacteria were subjected to one cycle of cleavage and recombination, resulting in mutant genes containing point mutations (shown in red circles) and large pieces of each wild-type sequence. These mutant genes were transformed into bacteria and screened for resistance to the antibiotic moxalactam; the most active mutants increased resistance by 540-fold over the wild-type genes. By contrast, DNA shuffling of each individual gene yielded only point mutations; the most active of these mutants only increased resistance to moxalactam eightfold. (b) Amino acid sequence of the most active mutant, colored as in (a) to show the origin of regions of the protein, with point mutations shown in red. The gray regions cannot be unambiguously assigned to one of the original cephalosporinase genes. (From A. Crameri et al., *Nature* 391: 288–291, 1998.)

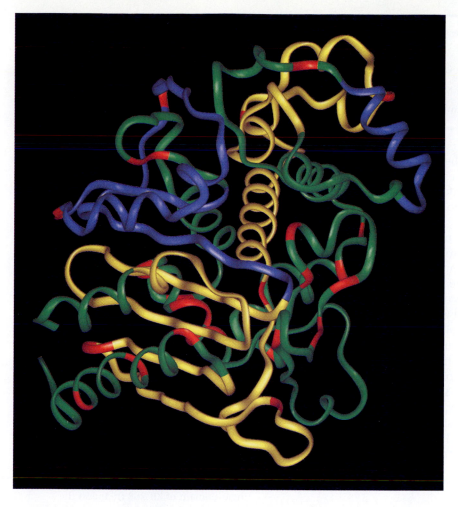

Figure 17.14 Model of evolved mutant from cephalosphorinase shuffling. The sequence of the most active cephalosporinase mutant was modeled using the crystal structure of the class C cephalosporinase from *Enterobacter cloacae*. The mutant and wild-type proteins were 63% identical. This chimeric protein contained portions from three of the starting genes, including *Enterobacter* (blue), *Klebsiella* (yellow), and *Citrobacter* (green), as well as 33 point mutations (red). (Courtesy of A. Crameri.)

Protein structures can be designed from first principles

The ultimate goal of protein engineering is to design an amino acid sequence that will fold into a protein with a predetermined structure and function. Paradoxically, this goal may be easier to achieve than its inverse, the solution of the folding problem. It seems to be simpler to start with a three-dimensional structure and find one of the numerous amino acid sequences that will fold into that structure than to start from an amino acid sequence and predict its three-dimensional structure. We will illustrate this by the design of a stable zinc finger domain that does not require stabilization by zinc.

The classic zinc fingers, the DNA-binding properties of which are discussed in Chapter 10, are small compact domains of about 30 residues that fold into an antiparallel β hairpin followed by an α helix. All known classic zinc fingers have a zinc atom bound to two cysteines in the hairpin and two histidines in the helix, creating a sequence motif common to all zinc finger genes. In the absence of zinc the structure is unfolded.

Stephen Mayo and Bassil Dahiyat at the California Institute of Technology asked the question: Is it possible to design from first principles a sequence whose main chain obtains this zinc finger fold without a zinc atom to stabilize the structure? They chose the second zinc finger of Zif 268 (see Chapter 10) with 28 residues as their target fold and applied their recently developed computer algorithm to the problem. Briefly, this algorithm searches through a very large number of possible sequences using a fast selection procedure for those sequences that stabilize a given fold.

On the basis of the template fold, side chain positions are divided into three categories: core, surface, and boundary positions. Allowed residues at the core positions are Ala, Val, Leu, Ile, Phe, Tyr and Trp, and at the surface positions Ala, Ser, Thr, His, Asp, Asn, Glu, Gln, Lys and Arg. The combined

Table 17.2 Amino acid sequences of the second zinc finger of Zif 268 and the designed peptide FSD-1

	1										11										21							28
FSD-1	Q	Q	**Y**	T	A	K	**I**	K	G	R	T	**F**	R	N	E	K	E	**L**	R	D	**F**	**I**	E	K	**F**	K	G	R
Zif 268	K	P	F	Q	C	R	I	C	M	R	N	F	S	R	S	D	H	L	T	T	H	I	R	T	H	T	G	E

Residues in the hydrophobic core of FSD-1 are green.

core and surface sets (16 residues since Ala belongs to both) are allowed at the boundary positions. They are thus using a restricted set of residues where Pro, Cys, and Met are absent and the only hydrophobic residue allowed at the surface is Ala. Gly is used in special positions to minimize backbone strain due to the conformation of the template.

The selection algorithm used involves energy functions for van der Waals interactions, hydrogen bonding and solvation combined with a secondary structure propensity potential. The total number of amino acid sequences that must be screened is the product of the number of possible amino acids at each residue position. For Zif 268 there is one core position with seven possible amino acids, 20 surface positions each with 10 possible amino acids and seven boundary positions each with 16 possible amino acids, giving a total of about 10^{27} possible sequences. This virtual library size is 15 orders of magnitude larger than an experimental combinatorial library using current methods. A corresponding peptide library consisting of only a single molecule for each 28-residue sequence would have a mass of 11,600 kilograms. The actual virtual search space is, however, even larger since different conformations, rotamers, must be considered for every possible side chain.

The optimal sequence obtained, called FSD-1 for full sequence design, is shown in Table 17.2 and compared with the sequence of the template Zif 268. A search of the FSD-1 sequence against protein databases did not reveal a statistically significant similarity with any other protein, including zinc finger proteins.

In order to examine whether this sequence gave a fold similar to the template, the corresponding peptide was synthesized and its structure experimentally determined by NMR methods. The result is shown in Figure 17.15 and compared to the design target whose main chain conformation is identical to that of the Zif 268 template. The folds are remarkably similar even though there are some differences in the loop region between the two β strands. The core of the molecule, which comprises seven hydrophobic side chains, is well-ordered whereas the termini are disordered. The root mean square deviation of the main chain atoms are 2.0 Å for residues 3 to 26 and 1.0 Å for residues 8 to 26.

In addition to being a remarkable demonstration of the power of computer-based combinatorial design of a protein fold, this designed peptide is the shortest known peptide consisting entirely of naturally occurring amino acids that folds into a well-ordered structure without metal binding, oligomerization or disulfide bond formation.

A β structure has been converted to an α structure by changing only half of the sequence

In 1994 a prize of $1000 called the Paracelsus challenge was offered to any designer of protein structures who could convert one protein fold into another while retaining 50% of the original sequence. The spirit of this challenge was to take one important step towards solving the folding problem by assessing the fraction of a protein's amino acid sequence that is sufficient to specify its structure. On the basis of our knowledge of naturally occurring proteins, this challenge did not look like a feasible project: all proteins of known structure with more than 30% sequence identity had been shown to have identical folds. Yet, it did not take more than three years before the prize

Figure 17.15 Schematic diagrams of the main-chain conformations of the second zinc finger domain of Zif 268 (red) and the designed peptide FSD-1 (blue). The zinc finger domain is stabilized by a zinc atom whereas FSD-1 is stabilized by hydrophobic interactions between the β strands and the α helix. (Adapted from B.I. Dahiyat and S.L. Mayo, *Science* 278: 82–87, 1997.)

(a)

(b)

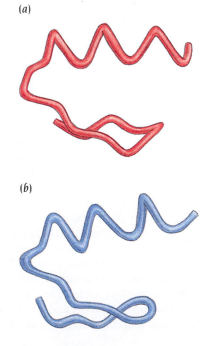

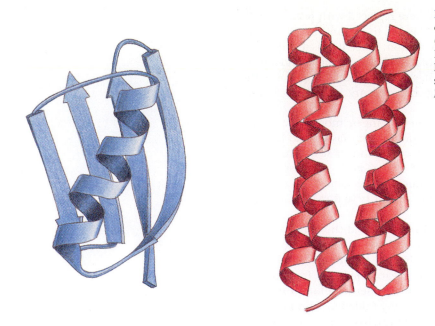

Figure 17.16 Ribbon diagram representations of the structures of domain B1 from protein G (blue) and the dimer of Rop (red). The fold of B1 has been converted to an α-helical protein like Rop by changing 50% of its amino acids sequence. (Adapted from S. Dalal et al., *Nature Struct. Biol.* 4: 548–552, 1997.)

was won. In 1997 the group of Lynne Regan at Yale University converted a protein that folded into a mainly β-sheet structure into an α-helical structure by changing 50% of its sequence.

They started from the sequence of a domain, B1, from an IgG-binding protein called Protein G. This domain of 56 amino acid residues folds into a four-stranded β sheet and one α helix (Figure 17.16). Their aim was to convert this structure into an all α-helical structure similar to that of Rop (see Chapter 3). Each subunit of Rop is 63 amino acids long and folds into two α helices connected by a short loop. The last seven residues are unstructured and were not considered in the design procedure. Two subunits of Rop form a four-helix bundle (Figure 17.16).

Aligning the two sequences from their amino ends produces only three identical residues in the 56 aligned positions (Table 17.3). In order to retain 50% of the original sequence, 28 residues could be changed in the B1 domain for the fold to switch to a Rop-like structure. The rationale for these changes was to use current knowledge about helix formation and stability to identify and change the subset of residues that are key determinants of the fold, considering both local and long-range interactions. Residues in the B1 domain

Table 17.3 Amino acid sequences of domain B1, the designed protein Janus, and Rop

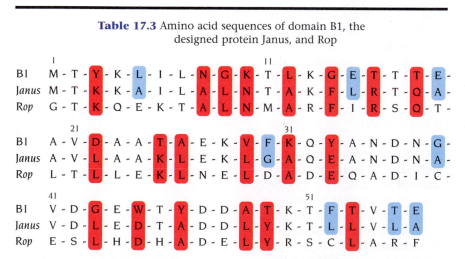

	1										11								
B1	M	T	Y	K	L	I	L	N	G	K	T	L	K	G	E	T	T	T	E
Janus	M	T	K	K	A	I	L	A	L	N	T	A	K	F	L	R	T	Q	A
Rop	G	T	K	Q	E	K	T	A	L	N	M	A	R	F	I	R	S	Q	T

	21										31								
B1	A	V	D	A	A	T	A	E	K	V	F	K	Q	Y	A	N	D	N	G
Janus	A	V	L	A	A	K	L	E	K	L	G	A	Q	E	A	N	D	N	A
Rop	L	T	L	L	E	K	L	N	E	L	D	A	D	E	Q	A	D	I	C

	41										51						
B1	V	D	G	E	W	T	Y	D	D	A	T	K	T	F	T	V	T	E
Janus	V	D	L	E	D	T	A	D	D	L	Y	K	T	L	L	V	L	A
Rop	E	S	L	H	D	H	A	D	E	L	Y	R	S	C	L	A	R	F

Red residues have been changed to the same type as present in Rop, blue residues have been changed to a different type, and black are unchanged.

with high preference for β sheet formation were replaced with residues with high preference for α helix formation, such as Tyr to Lys and Leu to Ala, in regions that were required to switch conformation. Since Rop is a coiled-coil protein, hydrophobic residues were incorporated in appropriate *a* and *d* positions of the heptad repeat (see Chapter 3). Among other changes was the introduction of an intramonomer salt bridge between Arg 16 and Asp 46 that is present between the two helices of Rop.

During this process of designing sequence changes, models were built and assessed to ensure that there were no obvious steric clashes and that the hydrophobic core was well packed. Furthermore, secondary structure prediction was also used to monitor the progress of change and to choose among different possible substitutions. The final sequence (see Table 17.3) contains 28 changes; it had 50% identity to B1 and the similarity to Rop had increased from 5.4% identity to 41%.

A gene encoding this sequence was synthesized and the corresponding protein, called Janus, was expressed, purified, and characterized. The atomic structure of this protein has not been determined at the time of writing but circular dichroic and NMR spectra show very clear differences from B1 and equally clear similarities to Rop. The protein is a dimer in solution like Rop and thermodynamic data indicate that it is a stably folded protein and not a molten globule fold like several other designed proteins.

These results indicate that is it possible to change the fold of a protein by changing a restricted set of residues. They also confirm the validity of the rules for stability of helical folds that have been obtained by analysis of experimentally determined protein structures. One obvious impliction of this work is that it might be possible, by just changing a few residues in Janus, to design a mutant that flip-flops between α helical and β sheet structures. Such a polypeptide would be a very interesting model system for prions and other amyloid proteins.

Conclusion

Homologous proteins have similar three-dimensional structures. They contain a core region, a scaffold of secondary structure elements, where the folds of the polypeptide chains are very similar. Loop regions that connect the building blocks of the scaffolds can vary considerably both in length and in structure. From a database of known immunoglobulin structures it has, nevertheless, been possible to predict successfully the conformation of hypervariable loop regions of antibodies of known amino acid sequence.

Methods for the prediction of the secondary structure of a set of homologous proteins can reach an accuracy of about 75%, most of the errors occur at the ends of α helices or β strands. The central regions of these secondary structure elements are often correctly predicted but the methods do not always correctly distinguish between α helices and β strands.

Prediction of tertiary structure from the amino acid sequence is the major unsolved problem in structural molecular biology. The inverse problem, to predict which amino acid sequences can have a given fold seem to be easier to solve. Significant progress has been made in recent years in threading methods, which assign a known fold to a given sequence by threading the sequence through all known folds.

Protein engineering is now routinely used to modify protein molecules either via site-directed mutagenesis or by combinatorial methods. Factors that are important for the stability of proteins have been studied, such as stabilization of α helices and reducing the number of conformations in the unfolded state. Combinatorial methods produce a large number of random mutants from which those with the desired properties are selected *in vitro* using phage display. Specific enzyme inhibitors, increased enzymatic activity and agonists of receptor molecules are examples of successful use of this method.

Small protein molecules with a predetermined fold can be designed *in silico* by energy calculations of all possible combinations of a restricted set of amino acid residues. A designed zinc finger fold that is stable in the absence of zinc showed no significant sequence similarity to any known protein sequence. Important progress has been made in assessing the fraction of a protein's amino acid sequence that is sufficient to specify its structure. A protein that folds into a mainly β-sheet structure was converted into an α-helical structure by changing only 50% of its sequence.

Selected readings

General

Alber, T. Mutational effects on protein stability. *Annu. Rev. Biochem.* 58: 765–798, 1989.

Barton, G.J. Protein secondary structure prediction. *Curr. Opin. Struct. Biol.* 5: 372–376, 1995.

Blundell, T.L., et al. Knowledge-based prediction of protein structures and the design of novel molecules. *Nature* 326: 347–352, 1987.

Bowie, J.U., Eisenberg, D. Inverted protein structure prediction. *Curr. Opin. Struct. Biol.* 3: 437–444, 1993.

DeGrado, W.F., Wasserman, Z.R., Lear, J.D. Protein design, a minimalist approach. *Science* 243: 622–628, 1989.

Fasman, G.D. Protein conformational prediction. *Trends Biochem. Sci.* 14: 295–299, 1989.

Fersht, A.R. Protein engineering. *Protein Eng.* 1: 7–16, 1986.

Fersht, A.R. The hydrogen bond in molecular recognition. *Trends Biochem. Sci.* 12: 301–304, 1987.

Finkelstein, A.V. Protein structure: what is possible to predict now? *Curr. Opin. Struct. Biol.* **7**: 60-71, 1997.

Jones, D.T., Thornton, J. Potential energy functions for threading. *Curr. Opin. Struct. Biol.* 6: 210–216, 1996.

Kay, B.K, Winter, J., McCafferty, J. eds. *Phage Display of Peptides and Proteins: a Laboratory Manual.* San Diego: Academic Press, 1996.

Moult, J., et al. A large-scale experiment to assess protein–structure prediction methods. *Proteins* 23: ii–iv, 1995.

Murzin, A., et al. Scop: a structural classification of proteins. Database for the investigation of sequences and structures. *J. Mol. Biol.* 247: 536–540, 1995.

Orengo, C.A., et al. CATH—a hierarchic classification of protein domain structures. *Structure* 5: 1093–1108, 1997.

Richardson, J.S., Richardson, D.C. The *de novo* design of protein structures. *Trends Biochem. Sci.* 14: 304–309, 1989.

Sippl, M.J. Knowledge-based potentials for proteins. *Curr. Opin. Struct. Biol.* 5: 229–235, 1995.

Sonnhammer, E.L., Eddy, S.R., Durbin, R. Pfam: a comprehensive database of protein domain families based on seed alignments. *Proteins* 28: 405–420, 1997.

Thornton, J.M., Gardner, S.P. Protein motifs and data-base searching. *Trends Biochem. Sci.* 14: 300–304, 1989.

von Heijne, G. *Sequence Analysis in Molecular Biology: Treasure Trove or Trivial Pursuit.* San Diego: Academic Press, 1987.

Wells, J.A. Additivity of mutational effects in proteins. *Biochemistry* 29: 8509–8517, 1990.

Specific structures

Aszodi, A., Taylor, W.R. Homology modelling by distance geometry. *Fold. Des.* 1: 325–334, 1996.

Bowie, J.U. Deciphering the message in protein sequences: tolerance to amino acid substitutions. *Science* 247: 1306–1310, 1990.

Bowie, J.U., Luthy, R., Eisenberg, D. A method to identify protein sequences that fold into a known three-dimensional structure. *Science* 253: 164–170, 1991.

Braisted, A.C., Wells, J.A. Minimizing a binding domain from protein A. *Proc. Natl. Acad. Sci. USA* 93: 5688–5692, 1996.

Bryant, S.H. Evaluation of threading specificity and accuracy. *Proteins* 26: 172–185, 1996.

Chothia, C., et al. Conformations of immunoglobulin hypervariable regions. *Nature* 342: 877–883, 1989.

Chothia, C., et al. The predicted structure of immunoglobulin D 1.3 and its comparison with the crystal structure. *Science* 233: 755–758, 1986.

Chothia, C., Lesk, A. The relation between the divergence of sequence and structure in proteins. *EMBO J.* 5: 823–826, 1986.

Chou, P.Y., Fasman, G.D. Prediction of protein conformation. *Biochemistry* 13: 222–245, 1974.

Cohen, C., Parry, D.A.D. α-helical coiled coils and bundles: how to design an α-helical protein. *Proteins: Struct. Funct. Gen.* 7: 1–15, 1990.

Crameri, A., Raillard, S.-A., Bermudez, E., Stemmer, W.P.C. DNA shuffling of a family of genes from diverse species accelerates directed evolution. *Nature* 391: 288–291, 1998.

Dahiyat, B.I., Mayo, S.L. De novo protein design: fully automated sequence selection. *Science* 278: 82–87, 1997.

Dalal, S., Balasubramanian, S., Regan, L. Protein alchemy: changing β sheet into α helix. *Nature Struct. Biol.* 4: 548–552, 1997.

DeGrado, W.F., Regan, L., Ho, S.P. The design of a four-helix bundle protein. *Cold Spring Harbor Symp. Quant. Biol.* 52: 521–526, 1987.

Dennis, M.S., Herzka, A., Lazarus, R.A. Potent and selective Kunitz domain inhibitors of plasma kallikrein designed by phage display. *J. Biol. Chem.* 270: 25411–25417, 1995.

Dennis, M.S., Lazarus, R.A. Kunitz domain inhibitors of tissue factor-factor VIIa. I. Potent inhibitors selected from libraries by phage display. *J. Biol. Chem.* 269: 22129–22136, 1994.

Dennis, M.S., Lazarus, R.A. Kunitz domain inhibitors of tissue factor-factor VIIa. II. Potent and specific inhibitors by competitive phage selection. *J. Biol. Chem.* 269: 22137–22144, 1994.

Faber, H.R., Matthews, B.W. A mutant T4 lysozyme displays five different crystal conformations. *Nature* 348: 263–266, 1990.

Fersht, A.R., et al. Hydrogen bonding and biological specificity analyzed by protein engineering. *Nature* 314: 235–238, 1985.

Garnier, J., Osguthorpe, D.J., Robson, B. Analysis of the accuracy and implications of simple methods for predicting the secondary structure of globular proteins. *J. Mol. Biol.* 120: 97–120, 1978.

Gribskov, M., McLaschlan, A.D., Eisenberg, D.E. Profile analysis: detection of distantly related proteins. *Proc. Natl. Acad. Sci. USA* 84: 4355–4358, 1987.

Jones, D.T., Taylor, W.R., Thornton, J.M. A new approach to protein fold recognition. *Nature* 358: 86–89, 1992.

Karpusas, M., et al. Hydrophobic packing in T4 lysozyme probed by cavity-filling mutants. *Proc. Natl. Acad. Sci. USA* 86: 8237–8241, 1989.

Kellis, J.T., et al. Contribution of hydrophobic interactions to protein stability. *Nature* 333: 784–786, 1988.

Lim, V.I. Algorithms for prediction of α-helical and β-structural regions in globular proteins. *J. Mol. Biol.* 88: 873–894, 1974.

Livnah, O., Stura, E.A., Johnson, D.L., Middleton, S.A., Mulcahy, L.S., Wrighton, N.C., Dower, W.J., Jolliffe, L.K., Wilson, I.A. Functional mimicry of a protein hormone by a peptide agonist: the EPO receptor complex at 2.8 Å. *Science* 273: 464–471, 1996.

Matsumura, M., Signor, G., Matthews, B.W. Substantial increase of protein stability by multiple disulfide bonds. *Nature* 342: 291–293, 1989.

Matthews, B.W., Nicholson, H., Becktel, W.J. Enhanced protein thermostability from site-directed mutations that decrease the entropy of unfolding. *Proc. Natl. Acad. Sci. USA* 84: 6663–6667, 1987.

Nicholson, H., Becktel, W.J., Matthews, B.W. Enhanced protein thermostability from designed mutations that interact with α-helix dipoles. *Nature* 336: 651–656, 1988.

Regan, L., DeGrado, W.F. Characterization of a helical protein designed from first principles. *Science* 241: 976–978, 1988.

Rice, D.W., et al. A 3D-1D substitution matrix for protein fold recognition that includes predicted secondary structure of the sequence. *J. Mol. Biol.* 267: 1026–1038, 1997.

Rice, D.W., et al. Fold assignments for amino acid sequences of the CASP2 experiment. *Proteins* (Suppl. 1) 113–122, 1997.

Richardson, J.S., Richardson, D.C. Amino acid preferences for specific locations at the ends of α helices. *Science* 240: 1648–1652, 1988.

Rost, B., Sander, C. Combining evolutionary information and neural networks to predict secondary structure. *Proteins* 19: 55–72, 1994.

Rost, B., Sander, C., Schneider, R. Redefining the goals of protein secondary structure prediction. *J. Mol. Biol.* 235: 13–26, 1994.

Smith, G.P. Filamentous fusion phage: novel expression vectors that display cloned antigens on the virion surface. *Science* 228: 1315–1317, 1985.

Starovasnik, M.A., Braisted, A.C., Wells, J.A. Structural mimicry of a native protein by a minimized binding domain. J. A. *Proc. Natl. Acad. Sci. USA* 94: 10080–10085, 1997.

Stemmer, W.P.C. DNA shuffling by random fragmentation and reassembly: *In vitro* recombination for molecular evolution. *Proc. Natl. Acad. Sci. USA* 91: 10747–10751, 1994.

Weaver, L.H., Matthews, B.W. Structure of bacteriophage T4 lysozyme refined at 1.7 Å resolution. *J. Mol. Biol.* 193: 189–199, 1987.

Wells, T.N.C., Fersht, A.R. Hydrogen bonding in enzymatic catalysis analyzed by protein engineering. *Nature* 316: 656–657, 1985.

Wetzel, R. Harnessing disulfide bonds using protein engineering. *Trends Biochem. Sci.* 12: 478–482, 1987.

Wrighton, N.C., et al. Small peptides as potent mimetics of the protein hormone erythropoietin. *Science* 273: 458–463, 1996.

Zvelebil, M.J., et al. Prediction of protein secondary structure and active sites using the alignment of homologous sequences. *J. Mol. Biol.* 195: 957–961, 1987.

Determination of Protein Structures

18

The structures described in this book have been determined by physical methods: most of them by x-ray crystallography, some of the smaller ones by nuclear magnetic resonance (NMR). We conclude the book with a short description of these techniques. It is not our aim to convert biologists into x-ray crystallographers and NMR spectroscopists; a complete explanation of the physical basis of these techniques and of the methods as currently practiced would fill more than one textbook. Our purpose is rather to convey the essence of the principles and procedures involved, so as to provide a general understanding of what is entailed in solving protein structures by these means. We will see how deriving a three-dimensional protein structure from x-ray or NMR data depends not only on the quality of the data themselves, but also on biochemical and sometimes genetic information that are essential to their interpretation.

Several different techniques are used to study the structure of protein molecules

Different techniques give different and complementary information about protein structure. The primary structure is obtained by biochemical methods, either by direct determination of the amino acid sequence from the protein or indirectly, but more rapidly, from the nucleotide sequence of the

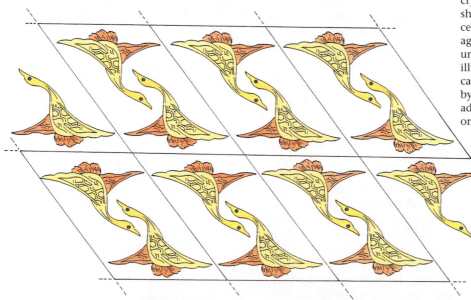

Figure 18.1 A crystal is built up from many billions of small identical units, or unit cells. These unit cells are packed against each other in three dimensions much as identical boxes are packed and stored in a warehouse. The unit cell may contain one or more than one molecule. Although the number of molecules per unit cell is always the same for all the unit cells of a single crystal, it may vary between different crystal forms of the same protein. The diagram shows in two dimensions several identical unit cells, each containing two objects packed against each other. The two objects within each unit cell are related by twofold symmetry to illustrate that each unit cell in a protein crystal can contain several molecules that are related by symmetry to each other. (The pattern is adapted from a Japanese stencil of unknown origin from the nineteenth century.)

(a)

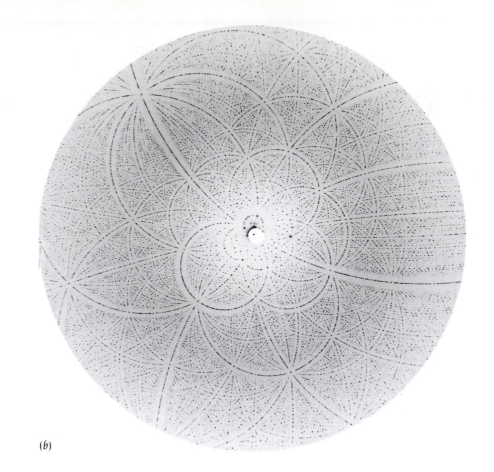

(b)

corresponding gene or cDNA. The quaternary structure of large proteins or aggregates such as virus particles, ribosomes, or gap junctions can be determined by **electron microscopy**. In general, this method gives structural information at very low resolution, with no atomic details, although, if one can obtain ordered two-dimensional arrays of the object, the noise in the electron microscopic image can be reduced enough to reveal the shape of individual subunits or, in rare cases, even determine the path of the polypeptide chain within a protein molecule.

To obtain the secondary and tertiary structure, which requires detailed information about the arrangement of atoms within a protein, the main method so far has been x-ray crystallography. In recent years NMR methods have been developed to obtain three-dimensional models of small protein molecules, and NMR is becoming increasingly useful as it is further developed.

Protein crystals are difficult to grow

The first prerequisite for solving the three-dimensional structure of a protein by **x-ray crystallography** is a well-ordered crystal that will diffract x-rays strongly. The crystallographic method depends, as we will see, upon directing a beam of x-rays onto a regular, repeating array of many identical molecules (Figure 18.1) so that the x-rays are diffracted from it in a pattern, a **diffraction pattern**, from which the structure of an individual molecule can be retrieved. The repeating unit forming the crystal is called the **unit cell**, and each unit cell may contain one or more molecules. Well-ordered crystals (Figure 18.2) are difficult to grow because globular protein molecules are large, spherical, or ellipsoidal objects with irregular surfaces, and it is impossible to pack them into a crystal without forming large holes or channels between the individual molecules. These channels, which usually occupy

Figure 18.2 Well-ordered protein crystals diffract x-rays and produce diffraction patterns that can be recorded on film. The crystal shown in (a) is of the enzyme RuBisCo from spinach and the photograph in (b) is a recording (Laue photograph) of the diffraction pattern of a similar crystal of the same enzyme. The diffraction pattern was obtained using polychromatic radiation from a synchrotron source in the wavelength region 0.5 to 2.0 Å. More than 100,000 diffracted beams have been recorded on this film during an exposure of the crystal to x-rays for less than one second. (The Laue photograph was recorded by Janos Hajdu, Oxford, and Inger Andersson, Uppsala, at the synchrotron radiation source in Daresbury, England.)

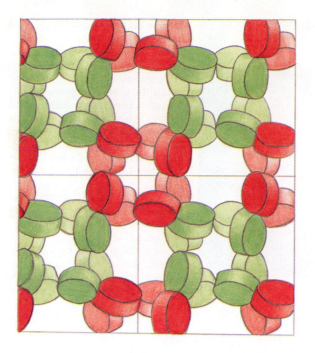

Figure 18.3 Protein crystals contain large channels and holes filled with solvent molecules, as shown in this diagram of the molecular packing in crystals of the enzyme glycolate oxidase. The subunits (colored disks) form octamers of molecular weight around 300 kDa, with a hole in the middle of each of about 15 Å diameter. Between the molecules there are channels (white) of around 70 Å diameter through the crystal. (Courtesy of Ylva Lindqvist, who determined the structure of this enzyme to 2.0 Å resolution in the laboratory of Carl Branden, Uppsala.)

more than half the volume of the crystal, are filled with disordered solvent molecules (Figure 18.3). The protein molecules are in contact with each other at only a few small regions, and even in these regions many interactions are indirect, through one or several layers of solvent molecules, usually water. This is one reason why structures of proteins determined by x-ray crystallography are the same as those for the proteins in solution.

Crystallization is usually quite difficult to achieve, and crystal growth can be slow; in some cases it may require months for sufficiently large crystals (~0.5 mm) to grow from microcrystals. The formation of crystals is also critically dependent on a number of different parameters, including pH, temperature, protein concentration, the nature of the solvent and precipitant as well as the presence of added ions or ligands to the protein. Many crystallization experiments are therefore required to screen all these parameters for the few combinations that might give crystals suitable for x-ray diffraction analysis. Crystallization robots and commercially available crystallization kits automate and speed up the tedious work of reproducibly setting up large numbers of crystallization experiments.

A pure and homogeneous protein sample is crucial for successful crystallization, and recombinant DNA techniques have been a major breakthrough in this regard. Proteins obtained from cloned genes in efficient expression vectors can be purified quickly to homogeneity in large quantities in a few purification steps. As a rule of thumb, a protein to be crystallized should ideally be more than 97% pure according to standard criteria of homogeneity. Crystals form when molecules are precipitated very slowly from **supersaturated solutions**. The most frequently used procedure for making protein crystals is the **hanging-drop** method (Figure 18.4), in which a drop of protein solution is brought very gradually to supersaturation by loss of water from the droplet to the larger reservoir that contains salt or polyethylene glycol solution.

Since there are so few direct packing interactions between protein molecules in a crystal, small changes in, for example, the pH of the solution can cause the molecules to pack in different ways to produce different crystal forms. The structures of some protein molecules such as lysozyme and myoglobin have been determined in different crystal forms and found to be essentially similar, except for a few side chains involved in packing interactions. Because they are so few, these interactions between protein molecules in a crystal do not change the overall structure of the protein. However,

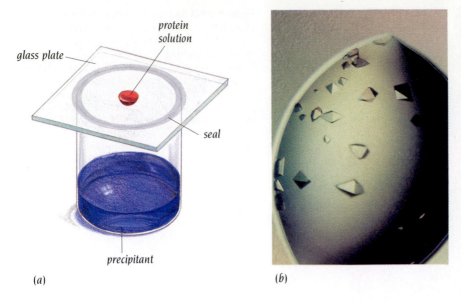

Figure 18.4 The hanging-drop method of protein crystallization. (a) About 10 μl of a 10 mg/ml protein solution in a buffer with added precipitant—such as ammonium sulfate, at a concentration below that at which it causes the protein to precipitate—is put on a thin glass plate that is sealed upside down on the top of a small container. In the container there is about 1 ml of concentrated precipitant solution. Equilibrium between the drop and the container is slowly reached through vapor diffusion, the precipitant concentration in the drop is increased by loss of water to the reservoir, and once the saturation point is reached the protein slowly comes out of solution. If other conditions such as pH and temperature are right, protein crystals will occur in the drop. (b) Crystals of recombinant enzyme RuBisCo from *Anacystis nidulans* formed by the hanging-drop method. (Courtesy of Janet Newman, Uppsala, who produced these crystals.)

different crystal forms can be more or less well ordered and hence give diffraction patterns of different quality. As a general rule, the more closely the protein molecules pack, and consequently the less water the crystals contain, the better is the diffraction pattern because the molecules are better ordered in the crystal.

X-ray sources are either monochromatic or polychromatic

X-rays are electromagnetic radiation at short wavelengths, emitted when electrons jump from a higher to a lower energy state. X-rays can be produced by high-voltage tubes in which a metal plate, the anode, is bombarded with accelerating electrons and thereby caused to emit x-rays of a specific wavelength, so-called **monochromatic x-rays**. The high voltage rapidly heats up the metal plate, which therefore has to be cooled. Efficient cooling is achieved by so-called rotating anode x-ray generators, where the metal plate revolves during the experiment so that different parts are heated up. Rotating anode x-ray generators are the conventional equipment used in most protein crystallography laboratories.

More powerful x-ray beams can be produced in synchrotron storage rings where electrons (or positrons) travel close to the speed of light. These particles emit very strong radiation at all wavelengths from short gamma rays to visible light. When used as an x-ray source, only radiation within a window of suitable wavelengths is channeled from the storage ring. **Polychromatic x-ray beams** are produced by having a broad window that allows through x-ray radiation with wavelengths of 0.2–2.0 Å. Such beams were used to record the Laue diffraction picture shown in Figure 18.2b. These very intense beams allow extremely short exposure times in diffraction experiments and can be used to collect data in experiments designed to observe changes in protein structure over very short periods of time; for example, electron transfer occurring in nanoseconds. Such studies are called **time-resolved crystallography.**

A very narrow window produces monochromatic radiation that is still several orders of magnitude more intense than the beam from conventional rotating anode x-ray sources. Such beams allow crystallographers to record diffraction patterns from very small crystals of the order of 50 micrometers or smaller. In addition, the diffraction pattern extends to higher resolution and consequently more accurate structural details are obtained as described later in this chapter. The availability and use of such beams have increased enormously in recent years and have greatly facilitated the x-ray determination of protein structures.

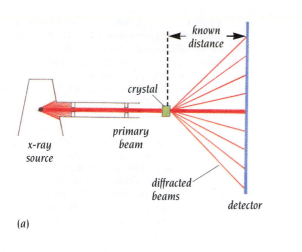

(a)

(b)

In diffraction experiments a narrow and parallel beam of x-rays is taken out from the x-ray source and directed onto the crystal to produce diffracted beams (Figure 18.5a). The primary beam must strike the crystal from many different directions to produce all possible diffraction spots; and so the crystal is rotated in the beam during the experiment. Rotating the crystal is much easier than rotating the x-ray source, especially when it is a synchrotron.

The incident primary beam causes damage to both protein and solvent molecules. This produces free radicals that in turn damage other molecules in the crystal. In addition, heat is generated, especially from synchrotron radiation, and eventually the primary beam burns through the crystal. To minimize this damage the crystal in the x-ray beam can be cooled to about –150 °C Such cooling does not prevent the formation of free radicals but greatly reduces the rate at which they can diffuse in the crystal, and greatly prolongs the life of the crystal in the x-ray beam. To prevent ice forming and destroying the crystal, it is essential to replace some of the water in the crystal with cryoprotectants. The crystal is suspended in the cryoprotectant, then chilled in liquid nitrogen and transferred to the x-ray beam where it is kept in a fine jet of nitrogen gas from boiling liquid nitrogen. Cryocooling has proved to be one very important technical innovation and during recent years has been adopted by most crystallographic laboratories.

Figure 18.5 Schematic view of a diffraction experiment. (a) A narrow beam of x-rays (red) is taken out from the x-ray source through a collimating device. When the primary beam hits the crystal, most of it passes straight through, but some is diffracted by the crystal. These diffracted beams, which leave the crystal in many different directions, are recorded on a detector, either a piece of x-ray film or an area detector. (b) A diffraction pattern from a crystal of the enzyme RuBisCo using monochromatic radiation (compare with Figure 18.2b, the pattern using polychromatic radiation). The crystal was rotated one degree while this pattern was recorded.

X-ray data are recorded either on image plates or by electronic detectors

Today the diffracted spots are usually recorded on an image plate rather than on x-ray film, the classical method (see Figure 18.5b), or by an electronic detector. The **image plate** is in effect a reusable film. The diffraction pattern recorded on the plate is scanned and stored in a computer. The image plate is then erased and ready for reuse. Electronic **area detectors** feed the signals they detect directly in a digitized form into a computer, and can therefore be regarded as an electronic film. They significantly reduce the time required to collect and measure diffraction data. To determine the structure of a protein, as we will see, it is necessary to compare x-ray data from native crystals of the protein with those from crystals in which different atoms of the protein are complexed with heavy metals. Moreover, to elucidate a protein's function x-ray data must also be collected from complexes with different types of bound ligands. In total, therefore, several hundred thousand diffraction spots are usually collected and measured for each protein.

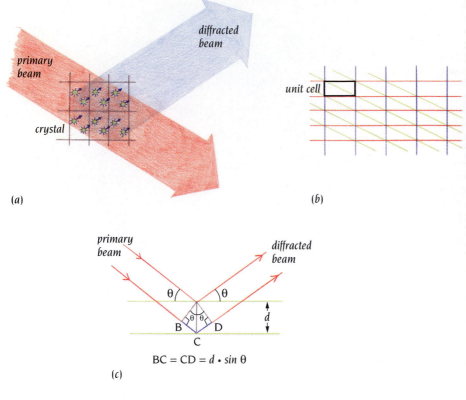

(a)

(b)

primary
beam

diffracted
beam

θ   θ

θ θ

B   D

C

d

$BC = CD = d \cdot sin\ \theta$

(c)

Figure 18.6 Diffraction of x-rays by a crystal. (a) When a beam of x-rays (red) shines on a crystal all atoms (green) in the crystal scatter x-rays in all directions. Most of these scattered x-rays cancel out, but in certain directions (blue arrow) they reinforce each other and add up to a diffracted beam. (b) Different sets of parallel planes can be arranged through the crystal so that each corner of all unit cells is on one of the planes of the set. The diagram shows in two dimensions three simple sets of parallel lines: red, blue, and green. A similar effect is seen when driving past a plantation of regularly spaced trees. One sees the trees arranged in different sets of parallel rows. (c) X-ray diffraction can be regarded as reflection of the primary beam from sets of parallel planes in the crystal. Two such planes are shown (green), separated by a distance d. The primary beam strikes the planes at an angle θ and the reflected beam leaves at the same angle, the reflection angle. X-rays (red) that are reflected from the lower plane have traveled farther than those from the upper plane by a distance BC + CD, which is equal to $2d \cdot sin\theta$. Reflection can only occur when this distance is equal to the wavelength λ of the x-ray beam and Bragg's law—$2d \cdot sin\theta = \lambda$—gives the conditions for diffraction. To determine the size of the unit cell, the crystal is oriented in the beam so that reflection is obtained from the specific set of planes in which any two adjacent planes are separated by the length of one of the unit cell axes. This distance, d, is then equal to $\lambda/(2 \cdot sin\theta)$. The wavelength, λ, of the beam is known since we use monochromatic radiation. The reflection angle, θ, can be calculated from the position of the diffracted spot on the film, using the relation derived in Figure 18.7, where the crystal to film distance can be easily measured. The crystal is then reoriented, and the procedure is repeated for the other two axes of the unit cell.

The rules for diffraction are given by Bragg's law

When the primary beam from an x-ray source strikes the crystal, most of the x-rays travel straight through it. Some, however, interact with the electrons on each atom and cause them to oscillate. The oscillating electrons serve as a new source of x-rays, which are emitted in almost all directions. We refer to this rather loosely as **scattering**. When atoms and hence their electrons are arranged in a regular three-dimensional array, as in a crystal, the x-rays emitted from the oscillating electrons interfere with one another. In most cases, these x-rays, colliding from different directions, cancel each other out; those from certain directions, however, will add together to produce diffracted beams of radiation that can be recorded as a pattern on a photographic plate or detector (Figure 18.6a).

How is the diffraction pattern obtained in an x-ray experiment such as that shown in Figure 18.5b related to the crystal that caused the diffraction? This question was addressed in the early days of x-ray crystallography by Sir Lawrence Bragg of Cambridge University, who showed that diffraction by a crystal can be regarded as the reflection of the primary beam by sets of parallel planes, rather like a set of mirrors, through the unit cells of the crystal (see Figure 18.6b and c).

X-rays that are reflected from adjacent planes travel different distances (see Figure 18.6c), and Bragg showed that diffraction only occurs when the difference in distance is equal to the wavelength of the x-ray beam. This distance is dependent on the reflection angle, which is equal to the angle between the primary beam and the planes (see Figure 18.6c).

The relationship between the reflection angle, θ, the distance between the planes, d, and the wavelength, λ, is given by **Bragg's law**: $2d \cdot sin\theta = \lambda$. This relation can be used to determine the size of the unit cell (see legend to Figure 18.6c and Figure 18.7). Briefly, the position on the film of the diffraction data relates each spot to a specific set of planes through the crystal. By using Bragg's law, these positions can be used to determine the size of the unit cell.

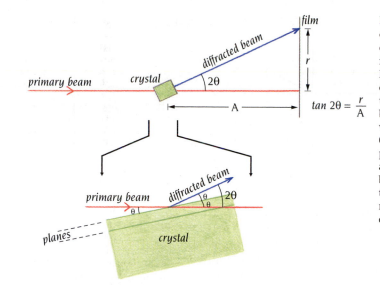

Figure 18.7 The reflection angle, θ, for a diffracted beam can be calculated from the distance (*r*) between the diffracted spot on a film and the position where the primary beam hits the film. From the geometry shown in the diagram the tangent of the angle $2\theta = r/A$. *A* is the distance between crystal and film that can be measured on the experimental equipment, while *r* can be measured on the film. Hence θ can be calculated. The angle between the primary beam and the diffracted beam is 2θ as can be seen on the enlarged insert at the bottom. It shows that this angle is equal to the angle between the primary beam and the reflecting plane plus the reflection angle, both of which are equal to θ.

Phase determination is the major crystallographic problem

Each atom in a crystal scatters x-rays in all directions, and only those that positively interfere with one another, according to Bragg's law, give rise to diffracted beams (see Figure 18.6a) that can be recorded as a distinct **diffraction spot** above background. Each diffraction spot is the result of interference of all x-rays with the same diffraction angle emerging from all atoms. For a typical protein crystal, myoglobin, each of the about 20,000 diffracted beams that have been measured contains scattered x-rays from each of the around 1500 atoms in the molecule. To extract information about individual atoms from such a system requires considerable computation. The mathematical tool that is used to handle such problems is called the **Fourier transform**, invented by the French mathematician Jean Baptiste Joseph Fourier while he served as a bureaucrat in the government of Napoleon Bonaparte.

Each diffracted beam, which is recorded as a spot on the film, is defined by three properties: the **amplitude**, which we can measure from the intensity of the spot; the **wavelength**, which is set by the x-ray source; and the **phase**, which is lost in x-ray experiments (Figure 18.8). We need to know all three properties for all of the diffracted beams to determine the position of the atoms giving rise to the diffracted beams. How do we find the phases of the diffracted beams? This is the so-called phase problem in x-ray crystallography.

In small-molecule crystallography the phase problem was solved by so-called direct methods (recognized by the award of a Nobel Prize in chemistry to Jerome Karle, US Naval Research Laboratory, Washington, DC, and Herbert Hauptman, the Medical Foundation, Buffalo). For larger molecules, protein crystallographers have stayed at the laboratory bench using a method pioneered by Max Perutz and John Kendrew and their co-workers to circumvent the phase problem. This method, called **multiple isomorphous replacement**

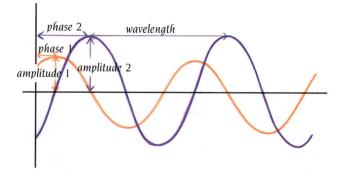

Figure 18.8 Two diffracted beams (purple and orange), each of which is defined by three properties: amplitude, which is a measure of the strength of the beam and which is proportional to the intensity of the recorded spot; phase, which is related to its interference, positive or negative, with other beams; and wavelength, which is set by the x-ray source for monochromatic radiation.

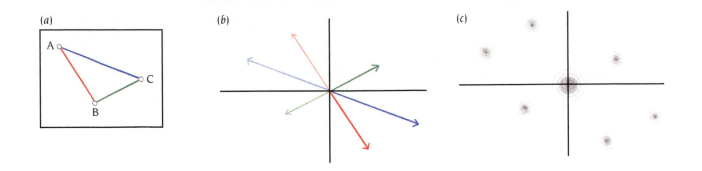

(a)   (b)   (c)

(MIR), requires the introduction of new x-ray scatterers into the unit cell of the crystal. These additions should be heavy atoms (so that they make a significant contribution to the diffraction pattern); there should not be too many of them (so that their positions can be located); and they should not change the structure of the molecule or of the crystal cell—in other words, the crystals should be isomorphous. In practice, isomorphous replacement is usually done by diffusing different heavy-metal complexes into the channels of preformed protein crystals. With luck the protein molecules expose side chains in these solvent channels, such as SH groups, that are able to bind heavy metals. It is also possible to replace endogenous light metals in metalloproteins with heavier ones, e.g., zinc by mercury or calcium by samarium.

Since such heavy metals contain many more electrons than the light atoms, H, N, C, O, and S, of the protein, they scatter x-rays more strongly. All diffracted beams would therefore increase in intensity after heavy-metal substitution if all interference were positive. In fact, however, some interference is negative; consequently, following heavy-metal substitution, some spots measurably increase in intensity, others decrease, and many show no detectable difference.

How do we find phase differences between diffracted spots from intensity changes following heavy-metal substitution? We first use the intensity differences to deduce the positions of the heavy atoms in the crystal unit cell. Fourier summations of these intensity differences give maps of the vectors between the heavy atoms, the so-called **Patterson maps** (Figure 18.9). From these vector maps it is relatively easy to deduce the atomic arrangement of the heavy atoms, so long as there are not too many of them. From the positions of the heavy metals in the unit cell, one can calculate the amplitudes and phases of their contribution to the diffracted beams of the protein crystals containing heavy metals.

How is that knowledge used to find the phase of the contribution from the protein in the absence of the heavy-metal atoms? We know the phase and amplitude of the heavy metals and the amplitude of the protein alone. In addition, we know the amplitude of protein plus heavy metals (i.e., protein heavy-metal complex); thus we know one phase and three amplitudes. From this we can calculate whether the interference of the x-rays scattered by the heavy metals and protein is constructive or destructive (Figure 18.10). The extent of positive or negative interference plus knowledge of the phase of the heavy metal together give an estimate of the phase of the protein.

Figure 18.9 Fourier summations of the intensity differences between diffracted spots from crystals of the protein alone and protein plus heavy metals give vector maps between the heavy atoms. Three atoms—A, B, and C—are at specific positions in the unit cell in (a). They give vectors A–B, A–C, and B–C, which are drawn from a common origin in (b) in dark colors. They also give the same vectors in the opposite directions as shown in light colors. The experimentally observed vector map is shown in (c) with a large peak at the origin corresponding to zero vectors between an atom and itself. It is straightforward to deduce the map in (c) from the atomic arrangement in (a). It is more difficult to do the reverse, to deduce·the atomic arrangement in (a) from the vector map in (c), especially if there are many atoms in the unit cell that give rise to a large number of peaks in the vector map. For example, with 10 atoms in the unit cell there are 90 different vectors between the atoms.

Figure 18.10 The diffracted waves from the protein part (red) and from the heavy metals (green) interfere with each other in crystals of a heavy-atom derivative. If this interference is positive as illustrated in (a), the intensity of the spot from the heavy-atom derivative (blue) crystal will be stronger than that of the protein (red) alone (larger amplitude). If the interference is negative as in (b), the reverse is true (smaller amplitude).

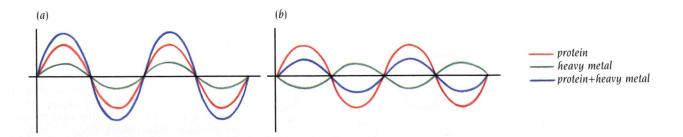

(a)   (b)

— protein
— heavy metal
— protein+heavy metal

380

Unfortunately, the problem is underdetermined so that two different phase angles are equally good solutions. To distinguish between these two possible solutions, a second heavy-metal complex must be used, which also gives two possible phase angles. Only one of these will have the same value as one of the two previous phase angles; it therefore represents the correct phase angle. In practice, more than two different heavy-metal complexes are needed to give a reasonably good phase determination for all reflections. Each individual phase estimate contains experimental errors arising from errors in the measured amplitudes; furthermore, for many reflections, the intensity differences are too small to measure after one particular isomorphous replacement, and others must be tried.

Phase information can also be obtained by Multiwavelength Anomalous Diffraction experiments

For certain x-ray wavelengths, the interaction between the x-rays and the electrons of an atom causes the electrons to absorb the energy of the x-ray. This causes a change in the x-ray scattering of the atom, called **anomalous scattering**, that depends both on the type of atom and on the wavelength of the x-rays. The size of this change is small and negligible for light atoms such as hydrogen, carbon, nitrogen and oxygen but is measurable for heavier atoms such as selenium, iron, zinc, and mercury. Due to this effect, small changes of the wavelength of the incident x-ray beam around the absorption edge of the heavy atom produce measurable intensity differences in the diffraction pattern. It is sufficient to have only one such heavy atom bound to each protein molecule of medium size (about 200 amino acid residues) for the effect to be measurable.

The intensity differences obtained in the diffraction pattern by illuminating such a crystal by x-rays of different wavelengths can be used in a way similar to the method of multiple isomorphous replacement to obtain the phases of the diffracted beams. This method of phase determination which is called **Multiwavelength Anomalous Diffraction, MAD**, and which was pioneered by Wayne Hendrickson at Columbia University, US, is now increasingly used by protein cystallographers.

The MAD method requires access to synchrotron radiation since different wavelengths are used, and it also requires that the crystal contains heavy atoms. Some protein molecules, such as metalloenzymes, contain intrinsic metal atoms but most proteins do not. However, using recombinant DNA technology it is possible to incorporate selenomethionine instead of methionine into recombinant proteins, thereby fulfilling the requirements for using the MAD method. Proteins with selenomethionine have very similar structures to the methionine-containing proteins. The structure of a number of selenomethionine-containing proteins, including several described in earlier chapters, has been determined using data collected at experimental stations specifically designed for MAD experiments at several synchrotron sources. Alternatively, proteins can be soaked in heavy metal solutions as for conventional x-ray structure determination (as discussed above).

Building a model involves subjective interpretation of the data

The amplitudes and the phases of the diffraction data from the protein crystals are used to calculate an **electron-density map** of the repeating unit of the crystal. This map then has to be interpreted as a polypeptide chain with a particular amino acid sequence. The interpretation of the electron-density map is complicated by several limitations of the data. First of all, the map itself contains errors, mainly due to errors in the phase angles. In addition, the quality of the map depends on the **resolution** of the diffraction data, which in turn depends on how well-ordered the crystals are. This directly influences the image that can be produced. The resolution is measured in Å

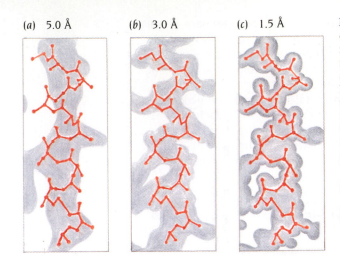

Figure 18.11 Electron-density maps at different resolution show more detail at higher resolution. (a) At low resolution (5.0 Å) individual groups of atoms are not resolved, and only the rodlike feature of an α helix can be deduced. (b) At medium resolution (3.0 Å) the path of the polypeptide chain can be traced, and (c) at high resolution (1.5 Å) individual atoms start to become resolved. Relevant parts of the protein chain (red) are superimposed on the electron densities (gray). The diagrams show one α helix from a small protein, myohemerythrin. [Adapted from W.A. Hendrickson in *Protein Engineering* (eds. D.L. Oxender and C.F. Fox.), p. 11. New York: Liss, 1987.]

units; the smaller this number is, the higher the resolution and therefore the greater the amount of detail that can be seen (Figure 18.11).

From a map at low resolution (5 Å or higher) one can obtain the shape of the molecule and sometimes identify α-helical regions as rods of electron density. At medium resolution (around 3 Å) it is usually possible to trace the path of the polypeptide chain and to fit a known amino acid sequence into the map. At this resolution it should be possible to distinguish the density of an alanine side chain from that of a leucine, whereas at 4 Å resolution there is little side chain detail. Gross features of functionally important aspects of a structure usually can be deduced at 3 Å resolution, including the identification of active-site residues. At 2 Å resolution details are sufficiently well resolved in the map to decide between a leucine and an isoleucine side chain, and at 1 Å resolution one sees atoms as discrete balls of density. However, the structures of only a few small proteins have been determined to such high resolution.

Building the initial **model** is a trial-and-error process. First, one has to decide how the polypeptide chain weaves its way through the electron-density map. The resulting chain trace constitutes a hypothesis, by which one tries to match the density of the side chains to the known sequence of the polypeptide. This sounds easy, but it is not; a map showing continuous density from N-terminus to C-terminus is rare. More usually one produces a number of matches between the electron density and discontinuous regions of the sequence that may initially account for only a small fraction of the molecule and may be internally inconsistent. When a reasonable chain trace has finally been obtained, an initial model is built to give the best fit of the atoms to the electron density (Figure 18.12). Today, computer graphics are exploited both for chain tracing and for model building to present the data and manipulate the models.

Figure 18.12 The electron-density map is interpreted by fitting into it pieces of a polypeptide chain with known stereochemistry such as peptide groups and phenyl rings. The electron density (blue) is displayed on a graphics screen in combination with a part of the polypeptide chain (red) in an arbitrary orientation (a). The units of the polypeptide chain can then be rotated and translated relative to the electron density until a good fit is obtained (b). Notice that individual atoms are not resolved in such electron densities, there are instead lumps of density corresponding to groups of atoms. [Adapted from A. Jones *Methods Enzym.* (eds. H.W. Wyckoff, C.H. Hirs, and S.N. Timasheff) 115B: 162, New York: Academic Press, 1985.]

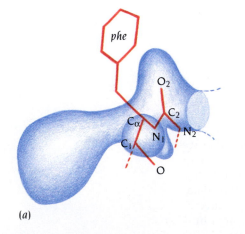

(a)

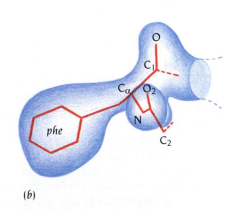

(b)

Errors in the initial model are removed by refinement

The initial model will contain errors. Provided the protein crystals diffract to high enough resolution (better than ~2.5 Å), most of the errors can be removed by crystallographic refinement of the model. In this process the model is changed to minimize the difference between the experimentally observed diffraction amplitudes and those calculated for a hypothetical crystal containing the model instead of the real molecule. This difference is expressed as an **R factor**, residual disagreement, which is 0.0 for exact agreement and around 0.59 for total disagreement.

In general, the R factor is between 0.15 and 0.20 for a well-determined protein structure. The residual difference rarely is due to large errors in the model of the protein molecule, but rather it is an inevitable consequence of errors and imperfections in the data. These derive from various sources, including slight variations in conformation of the protein molecules and inaccurate corrections both for the presence of solvent and for differences in the orientation of the microcrystals from which the crystal is built. This means that the final model represents an average of molecules that are slightly different both in conformation and orientation, and not surprisingly the model never corresponds precisely to the actual crystal.

The atoms of a protein's structure are usually defined by four parameters, three coordinates that give their position in space and one quantity, B, which is called the temperature factor. For well refined, correct structures these B-values are of the order of 20 or less. High B-values, 40 or above, in a local region can be due to flexibility or slight disorder, but also serve as a warning that the model of this region may be incorrect.

In refined structures at high resolution (around 2 Å) there are usually no major errors in the orientation of individual residues, and the estimated errors in atomic positions are around 0.1–0.2 Å provided the amino acid sequence is known. Hydrogen bonds both within the protein and to bound ligands can be identified with a high degree of confidence.

At medium resolution (around 3 Å) it is possible to make serious errors in the interpretation of the electron-density map, and there are, unfortunately, a number of them in the literature. Errors usually arise because elements of secondary structure are wrongly connected by the loop regions. Alpha helices and β strands in the interior of the protein are rigid in structure and well defined in the electron-density map. The loop regions, however, are usually more flexible, and therefore the corresponding electron density is less well defined. It is easy to make errors in such regions in the preliminary interpretations of electron-density maps at medium resolution. These errors are usually caught and corrected before publication since such models will not refine properly and are likely to be incompatible with existing biochemical data.

However, some models containing serious errors have been published. They all have been based on data to only medium resolution together with insufficient phase information, which gives large errors in the electron density. It should, therefore, be kept in mind that unrefined structures with R values higher than 0.30 at medium resolution may contain errors, although the overwhelming majority of such published structures have survived subsequent refinement at high resolution.

Recent technological advances have greatly influenced protein crystallography

In the early days of protein crystallography the determination of a protein structure was laborious and time consuming. The diffracted beams were obtained from weak x-ray sources and recorded on films that had to be manually scanned and measured. The available computers were far from adequate for the problem, with a computing power roughly equal to present-day pocket calculators. Computer graphics were not available, and models of the protein had to be built manually from pieces of steel rod. To determine the structure of even a small protein molecule, therefore, required many years of work and entailed time-consuming bottlenecks at almost every stage.

The situation is radically different today. The diffraction pattern can now be recorded on electronic area detectors coupled to powerful microcomputers that immediately interpret and process the recorded signals. Data collection that only a few years ago required many months of work is now done in a few days. If the in-house x-ray source is too weak for the problem, there are synchrotron sources available in several centers around the world that provide x-ray beams that are brighter by several orders of magnitude. Powerful computers in the laboratory provide the crystallographer with immediate access to almost all the computing power he or she needs. The electron-density maps are interpreted, and models of the protein molecules are built by the crystallographer sitting in front of a computer graphics screen. He or she is greatly aided by sophisticated software that involves semiautomatic methods for the model building using knowledge from databases of previously determined and refined protein structures.

These technical advances have greatly facilitated the use of crystallography for protein structure determination. One significant problem, however, still remains: obtaining crystals that diffract to high resolution. Some protein molecules give excellent crystals after the first few trials, others may require several months of screening for the proper crystallization conditions, and many have so far resisted all attempts to crystallize them. Fortunately, it is now possible to determine the structure of small protein molecules in solution by nuclear magnetic resonance (NMR) methods, and of large complexes by a combination of x-ray or NMR studies of individual or smaller pieces and fiber diffraction or electron microscopy studies of the complete complex.

X-ray diffraction can be used to study the structure of fibers as well as crystals

As described in Chapter 2, the first complete protein structure to be determined was the globular protein myoglobin. However, the α helix that was recognized in this structure, and which has emerged as a persistent structural motif in the many hundreds of globular proteins determined subsequently, was first observed in x-ray diffraction studies of fibrous proteins.

Polymer molecules that have a high degree of regularity in their monomer sequence tend to assume helical rather than globular conformations, and fibers are formed when these helices become aligned with each other. This is well illustrated by the proteins keratin and collagen and by the nucleic acid, DNA. The regularity in the DNA double-helix is so high that fibers drawn from a concentrated DNA gel are highly ordered; the long, thread-like DNA molecules extend parallel to the length of the fiber with a high degree of regularity in their side-by-side packing extending over many molecules. These regularly packed molecules form structures know as **crystallites**, and a typical fiber of 100 microns diameter contains a large number of such crystallites separated by less ordered regions where the molecules, while still largely parallel to the fiber length, are much less regularly packed. Because the crystallites are in random orientation about the direction of the fiber length, a diffraction pattern recorded from the fiber is similar to the pattern obtained when a single crystal is rotated 360° about the vertical axis while the data is being recorded. Therefore the diffraction pattern from a crystalline fiber can be analyzed using standard crystallographic techniques. The power of crystalline fiber diffraction analysis is illustrated by the detailed stereochemical information on the A and B conformations of DNA, reviewed in Chapter 7.

The polymer molecules in fibers typically assume helical structures with one pitch of the helix forming the repeating unit. This symmetry is the origin of the characteristic "cross-like" variation in overall intensity across the diffraction pattern from helical molecules. As the degree of regularity in the arrangement of repeating units in an array decreases, the diffraction spots become broader. For a completely irregular array, such as a crystalline powder,

384

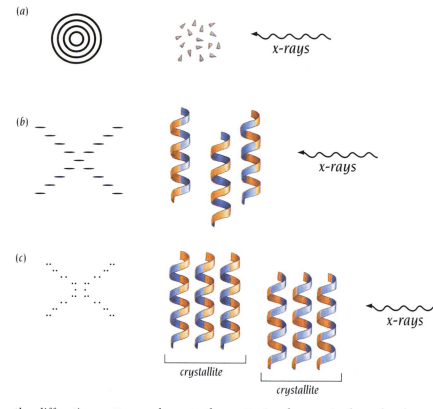

(a)

(b)

x-rays

(c)

x-rays

x-rays

crystallite

crystallite

Figure 18.13 Typical fiber diffraction patterns (left) for differently ordered arrays of molecules (right). (a) A randomly oriented array, such as a powder of small crystals. The molecules within each crystal are highly ordered, but the crystals are in random orientation with respect to each other. (b) A fiber formed from an array of poorly ordered molecules. Three molecules within the array are shown (in a typical x-ray specimen there would be many more); they are parallel but have little regularity in their side-by-side packing. (c) A fiber consisting of crystallites of well ordered molecules. Two crystallites are shown, each with only three of the many molecules they would typically contain. The molecules within each crystallite are aligned with a high degree of order in their side-by-side packing and relative orientation, but the crystallites are rotated with respect to each other. A similar pattern is obtained for a single crystal that is rotated during the data collection.

the diffraction pattern reduces to the scattering from a single molecule averaged over all orientations and is observed as a continuous distribution without any characteristic diffraction spots (Figure 18.13a). However, the chains of extended helical molecules in a fiber tend to remain parallel even when there is no regularity in the orientation about the helix axis. This parallelism is reflected in the overall intensity distribution within the diffraction pattern and in particular the restriction of diffraction to layer lines (see Figure 18.13b,c). The main features in the cross-like diffraction pattern of a helical structure can be illustrated by considering diffraction by a single slit that has the shape of a sine curve and that therefore corresponds to a projection of a helix. The diffraction patterns for projections of two helices differing in pitch length are illustrated in Figure 18.14. The separation between the layer lines is determined by the helix pitch: as the helix pitch increases the layer lines move closer together (compare parts a and b of Figure 18.14).

In addition to the reciprocal relationship between the helix pitch and layer line spacing, Figure 18.14 illustrates the reciprocal relationship between the orientation of the arms of the cross and the angle of climb of the helix: as the helix becomes steeper the arms of the cross become more horizontal.

Figure 18.14 The diffraction pattern of helices in fiber crystallites can be simulated by the diffraction pattern of a single slit with the shape of a sine curve (representing the projection of a helix). Two such simulations are given in (a) and (b), with the helix shown to the left of its diffraction pattern. The spacing between the layer lines is inversely related to the helix pitch, P and the angle of the cross arms in the diffraction pattern is related to the angle of climb of the helix, θ. The helix in (b) has a smaller pitch and angle of climb than the helix in (a). (Courtesy of W. Fuller.)

(a)

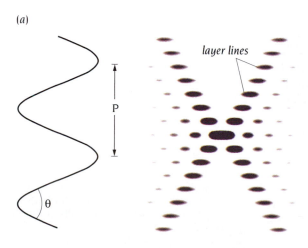

layer lines

P

θ

(b)

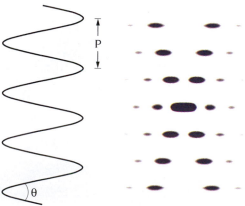

P

θ

The structure of biopolymers can be studied using fiber diffraction

A number of important protein molecules occur *in vivo* as fibrous structures. In keratin and collagen, groups of extended polymer molecules are wound around each other like strands in a rope, while muscle fibers contain a mixture of extended and globular components in a highly ordered array (see Chapter 14). The periodicities in these complex arrays of polymer molecules may be as large as hundreds or thousands of angstroms, whereas the periodicities of structural features in individual polymer molecules such as the pitch of the DNA double helix are typically a few tens of angstroms. The fibrous state offers a less constrained environment than a single crystal so that it is possible to follow changes in the x-ray fiber diffraction pattern as a consequence of structural transitions. In the case of DNA, changes of helix pitch and the number of nucleotide-pairs per pitch have been followed, while in muscle such time-resolved techniques have allowed changes in the diffraction pattern to be recorded during the contractile cycle.

The diffraction from helical molecules can be illustrated by the diffraction patterns from DNA (Figure 18.15). The pattern from the A form on the left has sharp diffraction spots across the whole pattern, indicating the regular packing of molecules in crystallites. In contrast, the pattern from the B form consists almost entirely of continuous diffraction along layer lines and the structure is said to be semi-crystalline. The pitch of the A form is 28 Å compared with 34 Å for the B form and this is reflected in Figure 18.15 by the smaller spacing between layer lines in the B form. The strong meridional diffraction near the top of the pattern from the B form is due to scattering from the stack of base pairs that form the core of the DNA double helix. Its position indicates that the distance between successive base pairs is 3.4 Å and therefore that there are 10 base pairs per helix pitch.

Knowing the helix pitch, it is possible to determine an approximate value for the radius of the helix in the B form from the inclination of the

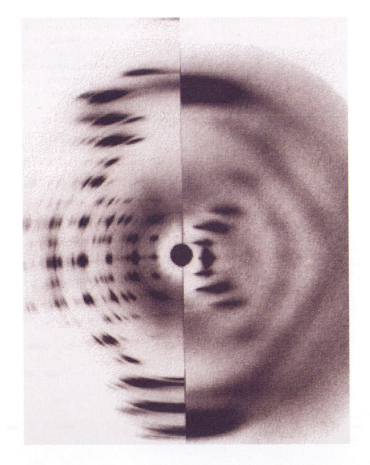

Figure 18.15 Diffraction patterns of DNA, showing the patterns obtained for both A-DNA (left half) and B-DNA (right half). (Courtesy of W. Fuller.)

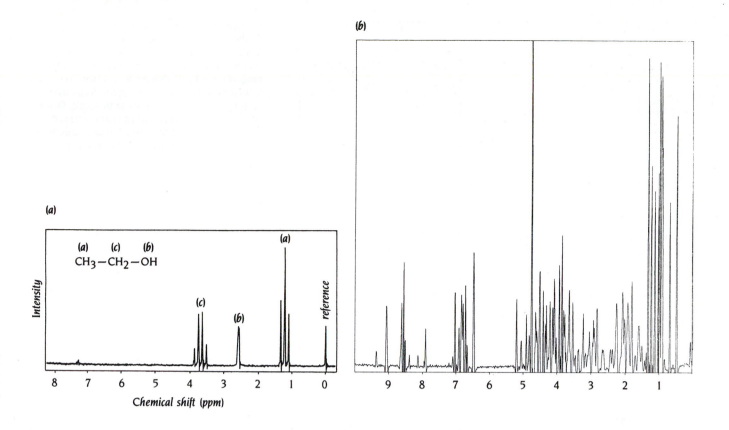

(a)

CH₃ — CH₂ — OH

(a)   (c)   (b)

Intensity

Chemical shift (ppm)

(b)

Figure 18.16 One-dimensional NMR spectra. (a) ¹H-NMR spectrum of ethanol. The NMR signals (chemical shifts) for all the hydrogen atoms in this small molecule are clearly separated from each other. In this spectrum the signal from the CH₃ protons is split into three peaks and that from the CH₂ protons into four peaks close to each other, due to the experimental conditions. (b) ¹H-NMR spectrum of a small protein, the C-terminal domain of a cellulase, comprising 36 amino acid residues. The NMR signals from many individual hydrogen atoms overlap and peaks are obtained that comprise signals from many hydrogen atoms. (Courtesy of Per Kraulis, Uppsala, from data published in Kraulis et al., *Biochemistry* 28: 7241–7257, 1989.)

arms of the cross in the B diffraction pattern. Arguments such as those outlined above, based on x-ray fiber diffraction patterns obtained by Maurice Wilkins and Rosalind Franklin, were used by Jim Watson and Francis Crick in the construction of their double-helical model for DNA.

NMR methods use the magnetic properties of atomic nuclei

Certain atomic nuclei, such as ¹H, ¹³C, ¹⁵N, and ³¹P have a magnetic moment or **spin**. The chemical environment of such nuclei can be probed by **nuclear magnetic resonance** (**NMR**) and this technique can be exploited to give information on the distances between atoms in a molecule. These distances can then be used to derive a three-dimensional model of the molecule. Most structure determinations of protein molecules by NMR have used the spin of ¹H, since hydrogen atoms are abundant in proteins. Small proteins can be analyzed by ¹H (proton) NMR but to study larger proteins and to obtain sufficient data to determine side chain conformations it is necessary to introduce ¹³C and ¹⁵N into the protein. This is usually done by producing the protein in microorganisms grown in media enriched with these isotopes. NMR studies of proteins containing one of the isotopes are called 3-D NMR, and when both ¹³C and ¹⁵N are present they are called 4-D NMR.

When protein molecules are placed in a strong magnetic field, the spin of their hydrogen atoms aligns along the field. This equilibrium alignment can be changed to an excited state by applying **radio frequency** (**RF**) pulses to the sample. When the nuclei of the protein molecule revert to their equilibrium state, they emit RF radiation that can be measured. The exact frequency of the emitted radiation from each nucleus depends on the molecular environment of the nucleus and is different for each atom, unless they are chemically equivalent and have the same molecular environment (Figure 18.16a). These different frequencies are obtained relative to a reference signal and are called **chemical shifts**. The nature, duration, and combination of applied RF pulses can be varied enormously, and different molecular properties of the sample can be probed by selecting the appropriate combination of pulses.

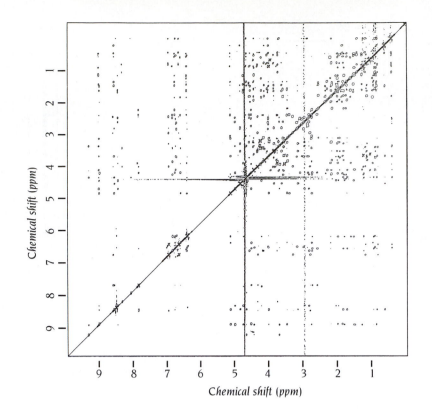

Figure 18.17 Two-dimensional NMR spectrum of the C-terminal domain of a cellulase. The peaks along the diagonal correspond to the spectrum shown in Figure 18.16b. The off-diagonal peaks in this NOE spectrum represent interactions between hydrogen atoms that are closer than 5 Å to each other in space. From such a spectrum one can obtain information on both the secondary and tertiary structures of the protein. (Courtesy of Per Kraulis, Uppsala.)

In principle, it is possible to obtain a unique signal (chemical shift) for each hydrogen atom in a protein molecule, except those that are chemically equivalent, for example, the protons on the CH_3 side chain of an alanine residue. In practice, however, such one-dimensional NMR spectra of protein molecules (see Figure 18.16b) contain overlapping signals from many hydrogen atoms because the differences in chemical shifts are often smaller than the resolving power of the experiment. In recent years this problem has been bypassed by designing experimental conditions that yield a two-dimensional NMR spectrum, the results of which are usually plotted in a diagram as shown in Figure 18.17.

The diagonal in such a diagram corresponds to a normal one-dimensional NMR spectrum. The peaks off the diagonal result from interactions between hydrogen atoms that are close to each other in space. By varying the nature of the applied RF pulses these off-diagonal peaks can reveal different types of interactions. A **COSY** (*correlation spectroscopy*) experiment gives peaks between hydrogen atoms that are covalently connected through one or two other atoms, for example, the hydrogen atoms attached to the nitrogen and C_α atoms within the same amino acid residue (Figure 18.18a). An **NOE** (*nuclear Overhauser effect*) spectrum, on the other hand, gives peaks between pairs of hydrogen atoms that are close together in space even if they are from amino acid residues that are quite distant in the primary sequence (see Figure 18.18b).

Figure 18.18 (a) COSY NMR experiments give signals that correspond to hydrogen atoms that are covalently connected through one or two other atoms. Since hydrogen atoms in two adjacent residues are covalently connected through at least three other atoms (for instance, HCα-C'-NH), all COSY signals reveal interactions within the same amino acid residue. These interactions are different for different types of side chains. The NMR signals therefore give a "fingerprint" of each amino acid. The diagram illustrates fingerprints (red) of residues Ala and Ser. (b) NOE NMR experiments give signals that correspond to hydrogen atoms that are close together in space (less than 5 Å), even though they may be far apart in the amino acid sequence. Both secondary and tertiary structures of small protein molecules can be derived from a collection of such signals, which define distance constraints between a number of hydrogen atoms along the polypeptide chain.

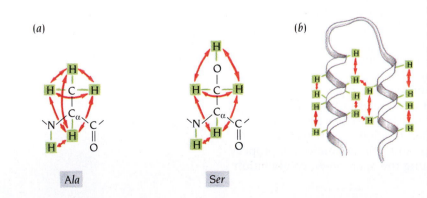

(a)   Ala   Ser   (b)

Two-dimensional NMR spectra of proteins are interpreted by the method of sequential assignment

Two-dimensional NOE spectra, by specifying which groups are close together in space, contain three-dimensional information about the protein molecule. It is far from trivial, however, to assign the observed peaks in the spectra to hydrogen atoms in specific residues along the polypeptide chain because the order of peaks along the diagonal has no simple relation to the order of amino acids along the polypeptide chain. This problem has in principle been solved in the laboratory of Kurt Wüthrich in the ETH, Zürich, where the method of **sequential assignment** was developed.

Sequential assignment is based on the differences in the number of hydrogen atoms and their covalent connectivity in the different amino acid residues. Each type of amino acid has a specific set of covalently connected hydrogen atoms that will give a specific combination of **cross-peaks**, a "fingerprint," in a COSY spectrum (see Figure 18.18a). From the COSY spectrum it is therefore possible to identify the H atoms that belong to each amino acid residue and, in addition, determine the nature of the side chain of that residue. However, the order of these fingerprints along the diagonal has no relation to the amino acid sequence of the protein. For example, when the fingerprint in one specific region of the COSY spectrum of the *lac*-repressor segment was assigned to a Ser residue, it was not known whether this fingerprint corresponded to Ser 16, Ser 28, or Ser 31 in the amino acid sequence.

The sequence-specific assignment, however, can be made from NOE spectra (see Figures 18.17 and 18.18b) that record signals from H atoms that are close together in space. In addition to the interactions between H atoms that are far apart in the sequence, these spectra also record interactions between H atoms from sequentially adjacent residues, specifically, interactions from the H atom attached to the main chain N of residue number $i + 1$ to H atoms bonded to N, C_α, and C_β of residue number i (Figure 18.19a).

Figure 18.19 (a) Adjacent residues in the amino acid sequence of a protein can be identified from NOE spectra. The H atom attached to residue $i + 1$ (orange) is close to and interacts with (purple arrows) the H atoms attached to N, $C_{\alpha'}$ and C_β of residue i (light green). These interactions give cross-peaks in the NOE spectrum that identify adjacent residues and are used for sequence-specific assignment of the amino acid fingerprints derived from a COSY spectrum. (b) Regions of secondary structure in a protein have specific interactions between hydrogen atoms in sequentially nonadjacent residues that give a characteristic pattern of cross-peaks in an NOE spectrum. In antiparallel β-sheet regions there are interactions between C_α-H atoms of adjacent strands (pink arrows), between N-H and C_α-H atoms (dark purple arrows), and between N-H atoms of adjacent strands (light purple arrows). The corresponding pattern of cross-peaks in an NOE spectrum identifies the residues that form the antiparallel β sheet. Parallel β sheets and α helices are identified in a similar way.

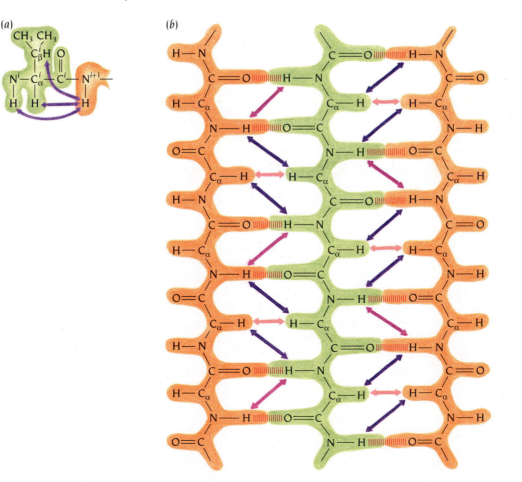

These signals in the NOE spectra therefore in principle make it possible to determine which fingerprint in the COSY spectrum comes from a residue adjacent to the one previously identified. For example, in the case of the *lac*-repressor fragment the specific Ser residue that was identified from the COSY spectrum was shown in the NOE spectrum to interact with a His residue, which in turn interacted with a Val residue. Comparison with the known amino acid sequence revealed that the tripeptide Ser-His-Val occurred only once, for residues 28–30.

In practice, it is difficult to make unique assignments for longer pieces than di- or tri-peptides, since NOE signals also occur between residues close together in space but far apart in the sequence. Therefore, the peptide segments that have been uniquely identified by NMR are usually matched with corresponding segments in the independently determined amino acid sequence of the protein. Thus knowledge of the amino acid sequence is just as essential for the correct interpretation of NMR spectra as it is for the interpretation of electron-density maps in x-ray crystallography. Whereas x-ray crystallography directly gives an image of a three-dimensional model of the protein molecule, NMR spectroscopy identifies H atoms in the protein that are close together in space, and this information is then used to derive, indirectly, a three-dimensional model of the protein.

Distance constraints are used to derive possible structures of a protein molecule

The final result of the sequence-specific assignment of NMR signals, preferably done using interactive computer graphics, is a list of **distance constraints** from specific hydrogen atoms in one residue to hydrogen atoms in a second residue. The list contains a large number of such distances, which are usually divided into three intervals within the region 1.8 Å to 5 Å, depending on the intensity of the NOE peak. This list immediately identifies the secondary structure elements of the protein molecule because both α helices and β sheets have very specific sets of interactions of less than 5 Å between their hydrogen atoms (see Figure 18.19b). It is also possible to derive models of the three-dimensional structure of the protein molecule. However, usually a set of possible structures rather than a unique structure (Figure 18.20) is obtained, each of the possible structures obeying the distance constraints equally well. The sets of possible structures, which are frequently seen in NMR articles, do not, therefore, represent different actual conformations of a protein molecule present in solution. Rather they are simply the different structures that are compatible with data obtained by current methods. The primary source of this ambiguity is an insufficient number of measured distance constraints. Because of this ambiguity, the accuracy of an NMR structure is not constant over the whole molecule and is also difficult to quantify.

In addition to the problem of ambiguity, there are other limitations to the use of NMR methods for the determination of protein structures. The most severe concerns the size of the protein molecules whose structures can be determined. Currently, the upper limit is molecules with molecular weights of around 25 kDa, but this limit will be increased in the future by using improved methods and equipment. Furthermore, the method requires highly concentrated protein solutions, on the order of 1–2 mM, with the additional requirement that the protein molecules must not aggregate at these concentrations. In addition, the pH of the solution should be lower than about 6 for proton NMR experiments. The exchanges of the NH protons in the main chain become so fast at higher pH that it is very difficult to observe them with NMR, and the signals from these hydrogen atoms are essential for the sequential assignment procedure.

How well do NMR-derived structures agree with those determined by x-ray methods? The structures of some different globular proteins that have been independently obtained by the two methods—such as bovine pancreatic

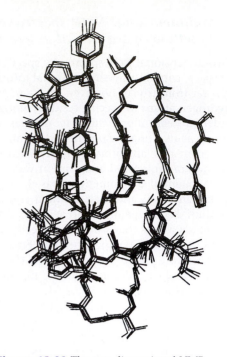

Figure 18.20 The two-dimensional NMR spectrum shown in Figure 18.17 was used to derive a number of distance constraints for different hydrogen atoms along the polypeptide chain of the C-terminal domain of a cellulase. The diagram shows 10 superimposed structures that all satisfy the distance constraints equally well. These structures are all quite similar since a large number of constraints were experimentally obtained. (Courtesy of P. Kraulis, Uppsala, from data published in P. Kraulis et al., *Biochemistry* 28: 7241–7257, 1989, by copyright permission of the American Chemical Society.)

trypsin inhibitor (see Figure 2.14a), plastocyanin (see Figure 2.11c) and thioredoxin from *E. coli* (see Figure 2.7)—show that NMR and x-ray crystallography give nearly identical results. The minor differences that exist are of the same order of magnitude as usually seen between x-ray structures of unrelated crystal forms of the same protein or determinations made under different experimental conditions. In other words, they are mostly small differences in loop regions of the main chain and different conformations of exposed side chains.

The situation is different for other examples—for example, the peptide hormone glucagon and a small peptide, metallothionein, which binds seven cadmium or zinc atoms. Here large discrepancies were found between the structures determined by x-ray diffraction and NMR methods. The differences in the case of glucagon can be attributed to genuine conformational variability under different experimental conditions, whereas the disagreement in the metallothionein case was later shown to be due to an incorrectly determined x-ray structure. A re-examination of the x-ray data of metallothionein gave a structure very similar to that determined by NMR.

NMR and x-ray crystallography are in many respects complementary. X-ray crystallography deals with the structure of proteins in the crystalline state, while NMR determines the structure in solution. The time scales of the measurements are different: NMR is more suitable for investigation of various dynamic processes such as those during folding, while x-ray crystallography is more suitable for characterization of protein surfaces and the water structure around the protein. X-ray crystallography remains the only method available to determine the structure of large protein molecules, whereas NMR is the method of choice for small protein molecules that might be difficult to crystallize.

Biochemical studies and molecular structure give complementary functional information

Our current knowledge of the relation between structure and function of protein molecules is insufficient to deduce the function of a protein from its structure alone, although, as we have seen, structural homology with proteins of known function can sometimes allow this. It is necessary to combine biochemical studies with structural information. Biochemical and cell biological studies can tell us if a protein is a receptor, a transport molecule, or an enzyme and, in addition, which ligands can bind to it, as well as the functional effects of such ligand binding. Studies of the three-dimensional structure of complexes between specific ligands and the protein will then give detailed information on how the active site is constructed and which amino acid residues are involved in ligand binding. Examples that we have described include protein–DNA interaction in Chapters 8, 9, and 10, sugar binding to a sugar transport protein in Chapter 4, and binding of inhibitors to enzymes that cleave peptide bonds in Chapter 11.

The specific role of each amino acid residue for the function of the protein can be tested by making specific mutations of the residue in question and examining the properties of the mutant protein. By combining in this way functional studies in solution, site-directed mutagenesis by recombinant DNA techniques, and three-dimensional structure determination, we are now in a position to gain fresh insights into the way protein molecules work.

Conclusion

The three-dimensional structure of protein molecules can be experimentally determined by two different methods, x-ray crystallography and NMR. The interaction of x-rays with electrons in molecules arranged in a crystal is used to obtain an electron-density map of the molecule, which can be interpreted in terms of an atomic model. Recent technical advances, such as powerful computers including graphics work stations, electronic area detectors, and very strong x-ray sources from synchrotron radiation, have greatly facilitated the use of x-ray crystallography.

Crystallization of proteins can be difficult to achieve and usually requires many different experiments varying a number of parameters, such as pH, temperature, protein concentration, and the nature of solvent and precipitant. Protein crystals contain large channels and holes filled with solvents, which can be used for diffusion of heavy metals into the crystals. The addition of heavy metals is necessary for the phase determination of the diffracted beams.

X-ray structures are determined at different levels of resolution. At low resolution only the shape of the molecule is obtained, whereas at high resolution most atomic positions can be determined to a high degree of accuracy. At medium resolution the fold of the polypeptide chain is usually correctly revealed as well as the approximate positions of the side chains, including those at the active site. The quality of the final three-dimensional model of the protein depends on the resolution of the x-ray data and on the degree of refinement. In a highly refined structure, with an R value less than 0.20 at a resolution around 2.0 Å, the estimated errors in atomic positions are around 0.1 Å to 0.2 Å, provided the amino acid sequence is known.

Biological fibers, such as can be formed by DNA and fibrous proteins, may contain crystallites of highly ordered molecules whose structure can in principle be solved to atomic resolution by x-ray crystallography. In practice, however, these crystallites are rarely as ordered as true crystals, and in order to locate individual atoms it is necessary to introduce stereochemical constraints in the x-ray analysis so that the structure can be refined by molecular modeling.

In NMR the magnetic-spin properties of atomic nuclei within a molecule are used to obtain a list of distance constraints between those atoms in the molecule, from which a three-dimensional structure of the protein molecule can be obtained. The method does not require protein crystals and can be used on protein molecules in concentrated solutions. It is, however, restricted in its use to small protein molecules.

Selected readings

Bernstein, F.C., et al. The protein data bank: a computer-based archival file for macromolecular structures. *J. Mol. Biol.* 112: 535–542, 1977.

Blundell, T.L., Johnson, L.N. *Protein Crystallography.* London: Academic Press, 1976.

Branden, C.-I., Jones, A. Between objectivity and subjectivity. *Nature* 343: 687–689, 1990.

Clore, G.M., Gronenborn, A.M. Determination of three-dimensional structures of proteins and nucleic acids in solution by nuclear magetic resonance spectroscopy. *CRC Crit. Rev. Biochem.* 24: 479–564, 1989.

Drenth, J. *Principles of protein x-ray crystallogrphy.* Berlin: Springer-Verlag, 1994.

Eisenberg, D., Hill, C.P. Protein crystallography: more surprises ahead. *Trends Biochem. Sci.* 14: 260–264, 1989.

Ferre-D'Amare, A.R., Burley, S.K. Use of dynamic light scattering to assess crystallizability of macromolecules and macromolecular assemblies. *Structure* 2: 357–359, 1994.

Hendrickson, W. Determination of macromolecular structures from anomalous diffraction of synchrotron radiation. *Science* 254: 51–58, 1991.

Kleywegt, G.J., Jones, T.A. Where freedom is given, liberties are taken. *Structure* 3: 535–540, 1995.

McPherson, A. *The Preparation and Analysis of Protein Crystals.* New York: Wiley, 1982.

McPherson, A., et al. The science of macromolecular crystallization. *Structure* 3: 759–768, 1995.

Rodgers, D.W. Cryocrystallography. *Structure* 2: 1135–1140, 1994.

Walter, R.L., et al. High resolution macromolecular structure determination using CCD detectors and synchrotron radiation. *Structure* 3: 835–844, 1995.

Wilson, H.R. *Diffraction of X-rays by Proteins, Nucleic Acids and Viruses.* London: Edward Arnold, 1966.

Wright, P. What can two-dimensional NMR tell us about proteins? *Trends Biochem. Sci.* 14: 255–260, 1989.

Wüthrich, K. *NMR of Proteins and Nucleic Acids.* New York: Wiley, 1986.

Wüthrich, K. Protein structure determination in solution by nuclear magnetic resonance spectroscopy. *Science* 243: 45–50, 1989.

Wyckoff, H.W., Hirs, C.H.W., Timasheff, S.N. Diffraction methods for biological macromolecules. *Methods Enzymol.* 114: 330–386, 1985.

Protein Structure on the World Wide Web

The World Wide Web has transformed the way in which we obtain and analyze published information on proteins. What only a few years ago would take days or weeks and require the use of expensive computer workstations can now be achieved in a few minutes or hours using personal computers, both PCs and Macintosh, connected to the internet. The Web contains hundreds of sites of interest to molecular biologists, many of which are listed in **Pedro's BioMolecular Research Tools (http:// www.fmi.ch/biology/research_tools.html)**. Many sites provide free access to databases that make it very easy to obtain information on structurally related proteins, the amino acid sequences of homologous proteins, relevant literature references, medical information and metabolic pathways. This development has opened up new opportunities for even non-specialists to view and manipulate a structure of interest or to carry out amino-acid sequence comparisons, and one can now rapidly obtain an overview of a particular area of molecular biology. We shall here describe some Web sites that are of interest from a structural point of view. Updated links to these sites can be found in the *Introduction to Protein Structure* **Web site (http:// www.ProteinStructure.com/)**.

Many Web sites offer the opportunity to view the structures of proteins interactively, using a protein's atomic coordinates to produce images of different types that can be rotated and zoomed. Both the atomic coordinates and the computer software required for this can be freely either accessed from the Web or downloaded to desktop computers for off-line use. Three commonly used and free programs to view proteins structures on personal computers are RasMol, Chime and Mage. **RasMol (http://www.umass.edu/mi crobio/rasmol/)** runs on personal computers and produces interactive molecular images from a molecule's atomic coordinates. **Chime (http//www.mdli.com/tech/chemscape.html)** is a plug-in for Web browsers that allows interactive RasMol-like images to be embedded within Web pages. **Kinemages (http://www.faseb.org/protein/kinemages/kin page.html)** are interactive molecular images produced by the program Mage. Many on-line journals use the kinemage format to display protein structures, and kinemages illustrating many of the protein structures discussed in this book can be found in the *Introduction to Protein Structure* Kinemage supplement disks (for further information, see http://www.Protein Structure.com).

The **Brookhaven Protein Data Bank, PDB (http://www.pdb.bnl.gov)**, is the primary store of experimentally determined atomic coordinates of proteins. Each coordinate set has a unique identification code that can be retrieved together with the coordinates and other information about the structure, including interactive images, by searching for the protein's name or publication details. The ENTREZ search engine at the **National Center for Biotechnology Information (NCBI; http://www3.ncbi.nlm.nih.gov/Entrez/)** allows the searching of integrated databases for, amongst other things, protein sequence data, protein structures and bibliographic data. This site also provides the Vector Alignment Tool (VAST), allowing one to find and view similar structures, and the Basic Alignment Search Tool (BLAST), allowing one to find similar sequences.

There are also several databases that have arranged all known protein structures into some classification scheme. **SCOP (http://scop.mrc lmb.cam.ac.uk/scop/)** is a database of structural domains arranged in a hierarchical manner according to structural and evolutionary relatedness. The SCOP site allows structures to be viewed interactively and enables the search for proteins with similar amino acid sequences using BLAST. **CATH (http://www.biochem.ucl.ac.uk/bsm/cath/)** arranges domains according to class (secondary structure composition), architecture (orientations of the secondary structures) and topology (shape and connectivity of secondary structure). The CATH site also includes a useful glossary of terms used in the description of protein structures. Protein topologies are extensively discussed in **TOPS (http://www3.ebi.ac.uk/tops/)**, a site that searches for specific topologies and which can determine the topology of a new protein. **FSSP (http://www2.ebi.ac.uk/dali/fssp/)** is based on an all-against-all comparison of known structures. FSSP allows one to view similar structures superimposed and to obtain sequences of homologous proteins. A network service that allows the comparison of three-dimensional structures is **DALI (http://www2.ebi.ac.uk/dali/)**. The coordinates of a new structure can be submitted to the site and DALI will check these coordinates against known structures to reveal biologically interesting similarities.

The ExPASy molecular biology Web site of the **Swiss Institute of Bioinformatics (http://www.expasy.ch)** covers many aspects of protein sequence and structure. It includes SWISS-PROT, which is an annotated protein sequence database, and SWISS-MODEL, an automated knowledge-based protein modeling server that allows one to model the three-dimensional structure of a protein whose sequence is known, based on the known structure of a homologous protein. Two very useful databases concerning the compilation and multiple sequence alignment of homologous domains are **Pfram (http://www.sanger.ac.uk/Pfam/)** and **ProDom (http://protein.tou louse.inra.fr/prodom.html)**.

There are now a large number of Web sites devoted to particular fields of research. The **Nucleic Acid Database atlas (http://ndbserver. rutgers.edu/NDB/ndb.html)** provides a resource for viewing many DNA, RNA, protein–DNA and protein–RNA structures that have been solved by x-ray crystallography. In addition, many researchers produce Web pages that are concerned with their own particular research interests. As examples, an interesting site to learn about viruses is http://www.bocklabs.wisc.edu/ Welcome.html, and an introduction to the disulfide bond-coupled folding pathway of bovine pancreatic trypsin inhibitor can be found at http://www. biology.utah.edu/People/regfaculty/~goldenberg/GoldenbergLab/research/bp ti.html.

Index

Page numbers in **bold** refer to a major text discussion; page numbers with an F refer to a figure; page numbers with a T refer to a table.

β-loop-β units 31F, 68
b/HLH transcription factors 175, 196F, 197, 202
    amino acid sequences 201F
    consensus sequence recognized 197, 199, 201
    homodimer and heterodimers 196–197, 196F
    structure 197, 200
b/HLH/zip transcription factors 196F, 199–200
    amino acid sequences 201F
    motif structure 200
    Myc 199
Biliverdin 70
Biochemical studies 391
Biopolymers, fiber diffraction 386–387
Björkman, Pamela 312
Blake, Colin 288
Blow, David 59, 210
Bluetongue virus 326
Blundell, Tom 74, 76
Bovine pancreatic trypsin inhibitor (BPTI) 26, 26F, 96–97, 96F
    folding pathway 96–97, 96F
    NMR and x-ray crystallography comparison 390–391
Bragg, Lawrence 378
Bragg, W.L. 13
Bragg's law 378, 379F
Braisted, Andrew 363
Branden, Carl 20F, 49F, 53F, 59, 97
Breathing, proteins 105
Brennan, Richard 143
Bugg, Charles 109
Burley, Stephen 154, 159, 199–200
b/zip family, Fos and Jun 199
b/zip transcription factor 196F, 197

C_6-zinc cluster family 190–191, 190F, 202
Calcium 25F
Calcium-binding domain 29F
Calcium-binding motif 24, 25F
    amino acid sequences 26F
Calcium-binding proteins 25
Calmodulin 24, 26, **109–110**
    calcium binding 26
    domains and structure 110, 110F
    peptide binding **109–110**, 110F
Campbell, Ian 274
Canonical loop structures 311
CAP see Catabolite gene activating protein (CAP)
Capsid 325, 326
    bacteriophage MS2 339–340
    see also specific viruses
Carbon atoms 4
Carbonyl groups, GAL4 binding to DNA 188, 189F
Carboxyl groups 4, 4F
Carboxypeptidase
    active site 62F
    α/β protein with mixed β sheet 60–62, 61F
    zinc environment 62F
Cardioviruses 333
Carrel, Robin 111
Caspar, Don 330, 342
Catabolite gene activating protein (CAP) 132
    cAMP–DNA complex 146, 146F
    DNA bending 146–147, 156
    helix-turn-helix motif 146, 146F
Catalysis 205
    reactions, a/β barrels in 51
    substrate-assisted 218–219, 218F
    without catalytic triad in subtilisin 217–218
    see also Serine proteinases

Catalytic triad 209
    alphavirus coat protein protease 341
    catalysis without 217–218
    chymotrypsin 211, 211F, 212F
    subtilisin 216, 217
CCG triplets 191
    GAL4 binding 188, 189
    zinc-containing motifs binding to 190, 191
CD4 168, 319, 319F
CDK see Cyclin-dependent protein kinases (CDKs)
CDR see Complementarity determining regions
Cell cycle 105–106, 106F
    G_0 phase 106
    Gap 1 (G_1 phase) 105
    Gap 2 (G_2 phase) 105–106
    M phase 105
    protein kinase conformational changes **105–109**
    regulation by p21 166
    S phase 105
Cell growth/differentiation 271, 272F
Cellulase, NMR 387F, 390F
Cephalosporinase genes, DNA shuffling 366, 366F, 367F
C_H see Immunoglobins, constant domains
Chaperones 89
    hsc70 293
Chaperonin
    definition 100
    GroEL see GroEL
    protein folding/unfolding in 99–100
Chemical shifts, in NMR methods 387, 387F, 388
Chimeric protein 190
Chiral forms, amino acids 5
Chlorophyll
    accessory molecules 238, 238F, 239
    arrangement 238F
    circular rings in light-harvesting complex LH2 241, 241F, 242F, 243
    photon absorption 239
    'special pair' 236, 238, 238F, 239, 244
Cholera toxin 254
Chothia, Cyrus 31, 32, 42, 311, 317
cH-ras p21 254F
Chymotrypsin 29F
    active site structure 211–212, 211F, 212F
    domains 211, 211F
    evolution 210
        gene duplication 212
    preferential cleavage mechanism 212–213
    specificity mechanism 209
    specificity pockets 212–213, 213F
    structure 210–211, 210F
    subtilisin similarity 216–217
    superfamily 210, 212
    see also Serine proteinases
cis-peptide 98, 98F
cis-retinoid acid receptor (RXR) 185
    heterodimer formation 186, 186F
Citrate synthase 17T
Classification of protein structure
    classes **31–32**
    topology diagrams 23
Clonal selection theory 299F, 300
Coagulation cascade 361
Cogdell, Richard 241
Coiled-coil α helix see α helix
Collagen **284–286**
    alanine mutation 285, 285F
    fibers 283
    polypeptide chains 284, 284F, 285
    superhelix of left-handed helices 284–286,

284F
    hydrogen bonding 286, 286F
    synthesis 284
Collectins 36
Colman, Peter 71
Combinatorial control, leucine zipper dimerization 193
Combinatorial design, FSD-1 peptide 368
Combinatorial joining 302, 302F, 303
Combinatorial libraries 358
Combinatorial methods
    definition 358
    protein engineering **358–359**
    in vitro selection see Bacteriophage display
Combinatorial screening, sequence recognition by SH3 274
Common cold, drugs 337–338
Complementarity determining regions (CDR) 301, 302F
    CDR1 305
    CDR2 305, 311
    CDR3 302–303, 305, 310, 311
        loop conformations and sequences 350
    conformation prediction 350, 350F
    conformational changes 311–312
        limited range 311–312, 350
    lysozyme and Fab binding 309–310, 310F
    T-cell receptor 317F
    see also Hypervariable regions, immunoglobulins
Computer-generated diagrams
    γ-crystallin structure 74, 74F
    myoglobin 22F
Computer-generated models 23
    building from x-ray diffraction data 382, 382F, 384
Concanavalin A 77
Concerted model 113–114
Conformational changes 105
    calmodulin and peptide binding 109–110, 110F
    complementarity determining regions 311–312
    ligand-induced 142–143
    protein kinase **105–109**
    R and T states of allosteric proteins 113–114
        phosphofructokinase 114–117, 117F
    serpins 111–113, 112F
    switch regions in G_α 257–259
    trp repressor 142–143
Consensus motif, sequence recognition by SH3 274
Consensus sequence
    b/HLH transcription factors binding to 197, 199, 201
    α/β-horseshoe fold 55
    TATA box 154F
Continuous lipidic cubic phase 225
Control module 151
Cooperative binding 113
Coreceptors, CD4 319
Corepressor 142–143, 143F
COSY (correlation spectroscopy) NMR experiments 388, 388F, 389
    cross-peaks ('fingerprints') 388F, 389
Covalent bonds
    native/denatured state of proteins 90
    peptide units 8
Craik, Craig 213
Creighton, Thomas 96
Creutzfeldt-Jacob disease 113
Crick, Francis 13, 35, 36, 121, 285, 387
Critical Assessment of Structure Prediction (CASP) 353
Cro gene, repression by repressor protein 130
Cro protein **129**

Page numbers in **bold** refer to a major text discussion; page numbers with an F refer to a figure; page numbers with a T refer to a table.

400

Page numbers in **bold** refer to a major text discussion; page numbers with an F refer to a figure; page numbers with a T refer to a table.

prediction 21
Ras protein 255–256
serpins 111
spherical viruses 335–336
three-dimensional structure 21
variable in homologous proteins 349–350
Low, Barbara 27
Lysine
GAL4 binding to DNA 188, 189F
recognition helix of glucocorticoid
receptor 184–185
specificity of serine proteinases 213
structure 6F
trypsin mutation 215
Lysozyme
antilysozyme complex, structure
310–311, 310F
Fab binding 309–310, 310F
folding pathways 95–96, 95F
structure 310–311
T4 bacteriophage see Bacteriophage T4
α-Lytic protease 92
folding with prosegment 92
Lytic–lysogenic cycle switch 130–131,
130F, 133

M subunit, photosynthetic reaction center
235, 236–237, 246F
conservation between species 246–247
pigments bound to 237–239
MacKinnon, Roderick 232, 234
'Mad cow disease' 113
Magnesium
G$_α$ activation 258
GTP linking to Ras protein by 255, 255F
LH2 light-harvesting complex 241
Main-chain
formation 4, 4F
modeling of protein structures 349
polarity 14
Major histocompatibility complex 300
see also MHC molecules
Malaria, resistance and sickle-cell hemoglobin
43–45, 44F
Maltoporin 230
Mandelate, conversion to benzoate 54–55,
54F
Mandelate racemase 54–55, 54F
Mariuzza, Roy 317
Mat α1 162
Mat α2 gene, homeodomain 162, 162F
Mat α2 repressor 160
Mat α2–Mat a1 complex 163, 163F
Matthews, Brian 132, 134, 354, 355
Max 175, 192F
binding to DNA 200F
heterodimer with Myc 199
homodimers 199–200
monomer structure 200F
sequence-specific interactions with DNA
201, 201F
Mayo, Stephen 367
Mcm1 162
McPherson, Alexander 312
Melting temperature (Tm) 354, 356F
Membrane fusogen, hemagglutinin as 80
Membrane lipids 223, 246–247, 253
Membrane proteins **223–250**
crystallization
difficulties 224
novel methods 224–225
functions 224
signal transduction 251
solubilization by detergents 224, 225F
two-dimensional crystals and EM
225–226, 226F

types 223, 223F
see also specific proteins
Membrane-bound proteins, α helices 35
Membranes 223
functions 224
Mengo virus 333, 336
Menten, Maud 206
Met repressor 175
Metal atoms, in proteins 11, 11F
Metallo proteinases 205
Metallo proteins 11
Methallothionein, NMR and x-ray
crystallography comparison 391
Methionine, structure 7F
3-Methylisoxazole groups 338
Methylmalonyl-coenzyme A mutase, α/β
barrel domain 50–51, 50T
MHC genes 314–315
polymorphism 315
MHC molecules 300, **312–313**
antigen recognition 314–315, 316
class I 300
antigen-binding site 314
peptide binding 315F, 316
peptide complexes 318, 318F
structure 312, 313, 313F
class II 300
domains 315
peptide binding 315–316, 315F
domains 313–314, 313F
peptide complex as ligand for T-cell
receptor 318, 318F
structures 312–313
synthesis 316
Michaelis, Leonor 206
Michaelis–Menten equation 206F
Michaelis–Menten scheme 206, 206F
Michel, Hartmut 234, 241
β$_2$ Microglobulin 313, 314–315
Microtubules 284
Milligan, Ronald 295
Mineralocorticoid receptor 181F
'Miniglobular protein' 177
Model building
antigen-binding sites of immunoglobulins
349–350
Cro–DNA interactions 134–135
hypervariable regions, immunoglobulins
349
x-ray diffraction data 381–382, 382F
see also specific models
Modeling of protein structures 349
Molecular chaperones see Chaperones
Molecular disease see Sickle-cell anemia
Molecular dynamics simulations 105
Molten globular proteins 89, 92, 92F
barnase folding intermediate 94
Monoclonal antibodies, CDR conformation
prediction 350, 350F
Monod, Jacques 113, 117, 142, 143
Monomeric proteins 29
Motifs **13–34**, 29
in barrel and sheet structures 47–48, 49F
β-α-β motif see β-α-β motif
combined into domains 29, 30
Greek key see Greek key motif
hairpin β see Hairpin β motif
jelly roll see Jelly roll motifs
simple **24–26**
combination into complex motifs 30–31
see also α helices; β sheets; Loop regions;
other specific motifs
Muconate lactonizing enzyme 54, 54F
Muirhead, Hilary 51
Multimeric proteins 29
Multiple isomorphous replacement (MIR)

379–380
Multiwavelength Anomalous Diffraction
(MAD) 381
μ-oxo bridge 11, 11F
Muramidase, bacterial 39, 39F
Muscle contraction 292
ATP role 296–297
Muscle fibers **290–291**
thick and thin filaments 290, 291
see also Actin; Myosin
Mutagenesis
oligonucleotide-directed 359F
random 359, 359F
site-directed 163–164
Mutations
DNA shuffling method 365–366
enzyme evolution 55
point 366
protein folding studies 93–95
Myc 191
heterodimer with Max 199
myc gene 199
Myeloma proteins 309
MyoD 197
binding to DNA 198F
dimerization region structure 197F
α helix region 197, 198–199
sequence-specific interactions with DNA
201
Myofibrils 291F
Myogenic proteins 197
Myoglobin
breathing of molecule 105
as α domain structure 35
globin fold in 40
oxygen binding 105
structural irregularity 13
structure 384
computer-generated schematic diagram
22F
early results 13, 13F
schematic diagram 23F
two-dimensional 22F
x-ray diffraction 379
Myohemerythrin 37, 381F
Myosin 36, 197, 256, **290–291**, 291F
actin complex, structure 295, 295F
conformational change 294–295, 296
cross-bridge movement 291–292, 295
confirmation 292–293, 295–296
nucleotide-binding cleft 295, 296
S1 fragment 294, 294F, 295
sliding filament model 291, 291F
structure 292, 294–295, 294F
swinging cross-bridge hypothesis 292, 292F,
295–296, 296F

Nef protein 275, 275F, 276F
Neuraminidase 70–71
active site 71F, 72
amino acids 71
folding motifs in propeller-like structure
71–72, 71F, 73F
function 70–71
subunit structure 71, 71F, 72F
Neurofilament proteins 287F
Neurospora crassa, PRA isomerase and IGP
synthase 53
Neutrofil elastase 110
NF-κB 168–169
NMR 374, **387–388**
advantages 391
COSY 388, 388F, 389
distance constraints 390–391
folded protein flexibility 105
homeodomain binding to DNA 162

Page numbers in **bold** refer to a major text discussion; page numbers with an F refer to a figure; page numbers with a T refer to a table.

Page numbers in **bold** refer to a major text discussion; page numbers with an F refer to a figure; page numbers with a T refer to a table.